CW00672989

FPGA-Based System Design

Prentice Hall Modern Semiconductor Design Series

James R. Armstrong and F. Gail Gray
VHDL Design Representation and Synthesis

Jayaram Bhasker
A VHDL Primer, Third Edition

Mark D. Birnbaum
Essential Electronic Design Automation (EDA)

Eric Bogatin
Signal Integrity: Simplified

Douglas Brooks
Signal Integrity Issues and Printed Circuit Board Design

Alfred Crouch
Design-for-Test for Digital IC's and Embedded Core Systems

Tom Granberg
Handbook of Digital Techniques for High-Speed Design

Howard Johnson and Martin Graham
High-Speed Digital Design: A Handbook of Black Magic

Howard Johnson and Martin Graham
High-Speed Signal Propagation: Advanced Black Magic

Farzad Nekoogar and Faranak Nekoogar
From ASICs to SOCs: A Practical Approach

Samir Palnitkar
Design Verification with **e**

Christopher T. Robertson
Printed Circuit Board Designer's Reference: Basics

Chris Rowen and Steve Leibson
Engineering the Complex SOC

Wayne Wolf
FPGA-Based System Design

Wayne Wolf
Modern VLSI Design: System-on-Chip Design, Third Edition

Brian Young
Digital Signal Integrity: Modeling and Simulation with Interconnects and Packages

FPGA-Based System Design

Wayne Wolf

Prentice Hall PTR
Upper Saddle River, New Jersey 07458

Library of Congress Cataloging-in-Publication Data

A catalog record for this book can be obtained from the Library of Congress

Editorial/production supervision: *Nicholas Radhuber*
Publisher: *Bernard Goodwin*
Cover design director: *Jerry Votta*
Cover design: *Talar Boorujy*
Manufacturing manager: *Maura Zaldivar*
Editorial assistant: *Michelle Vincenti*
Marketing manager: *Dan DePasquale*

Published by Prentice Hall Professional Technical Reference
PRENTICE HALL PTR Pearson Education, Inc.
Upper Saddle River, New Jersey 07458

Prentice Hall books are widely used by corporations and government agencies for training, marketing, and resale.

Prentice Hall offers excellent discounts on this book when ordered in quantity for bulk purchases or special sales. For more information, please contact:
U.S. Corporate and Government Sales
1-800-382-3419
corpsales@pearsontechgroup.com

For sales outside of the U.S., please contact:
International Sales
1-317-581-3793
international@pearsontechgroup.com

Illustrated and typeset by the author. This book was typeset using FrameMaker. Illustrations were drawn using Adobe Illustrator.

Printed in the United States of America

4th Printing

ISBN 0-13-142461-0

Pearson Education LTD.
Pearson Education Australia PTY, Limited
Pearson Education Singapore, Pte. Ltd.
Pearson Education North Asia Ltd.
Pearson Education Canada, Ltd.
Pearson Educación de Mexico, S.A. de C.V.
Pearson Education — Japan
Pearson Education Malaysia, Pte. Ltd.

for Nancy and Alec

Table of Contents

Preface

This book is an experiment. Shortly after completing the third edition of *Modern VLSI Design*, I came to realize that an increasing number of digital designs that used to be built in custom silicon are now implemented in field programmable gate arrays (FPGAs). While traditional VLSI system design won't go away any time soon, an increasing number of designers will work with FPGAs and many of them will never design a custom chip.

However, designers of large FPGA-based systems really do need to understand the fundamentals of VLSI in order to make best use of their FPGAs. While it is true that many system designers simply treat the FPGA as a black box, that approach makes the system designer miss many opportunities to optimize the design to fit within the FPGA. The architecture of FPGAs is largely determined by VLSI constraints: logic element structures, programmable interconnect structures, interconnection networks, configuration, pinout, etc. Understanding how the characteristics of VLSI devices influence the design of FPGA fabrics helps the designer better understand how to take advantage of the FPGA's best characteristics and to minimize the influence of its limitations.

Consider, for example, the interconnection networks in FPGAs. Most modern FPGA architectures provide designers with several different types of wires: local, general-purpose, and global. Why do all these different types of connections exist? Because wires become harder to drive as they grow in length; the extra circuitry required to drive long wires makes them more expensive. Understanding how these different types of interconnect work helps a designer decide whether a particular logic connection requires one of the more expensive types of wires.

Today's FPGAs are truly amazing. High-end FPGAs can hold several million gates. Several FPGAs incorporate one or more CPUs on-chip to provide a complete embedded computing platform. Many of the techniques for designing such large systems are the same whether they are built using FPGAs or custom silicon. This is particularly true when we want to make best use of the silicon characteristics of VLSI structures.

As a result of these advances in VLSI systems, I decided to use *Modern VLSI Design* as a starting point for a new book on FPGA-based system design. Readers of *Modern VLSI Design* will recognize material from most of the chapters in that book. I have extracted material on VLSI characteristics, circuits, combinational and sequential logic design, and system architectures. However, I have also added quite a bit of new material, some of which is specific to FPGAs and some of which is simply a new (and I hope better) look at digital system design.

One of my major goals in writing this book was to provide a useful text for both designers interested in VLSI and those who simply want to use FPGAs to build big systems. Chapter 2 of this book is devoted to a review of VLSI: fabrication, circuits, interconnect characteristics, etc. Throughout the rest of the book, I have tried to break out most details of VLSI design into separate sections that can be skipped by the reader who is not interested in VLSI. However, those who want to understand more about the design of FPGAs as VLSI devices can read this material at their leisure.

Chapter 3 is devoted to a survey of FPGA fabrics—the basic programmable structures of FPGAs. The commercial offerings of companies change all the time, so this chapter is not meant to be a replacement for a data book. Its goal is to introduce some basic concepts in FPGAs and to compare different approaches to solving the basic problems in programmable logic. What to do with these FPGA structures is the subject of the rest of the book.

Chapters 4 and 5 go into detail about combinational and sequential logic design, respectively. They describe methods for the specification and optimization of digital logic to meet the major goals in most design efforts: size, speed, and power consumption. We introduce both Verilog and VHDL in this book. While this book is not intended as a definitive reference on either language, hardware description languages are the medium of choice today for designing digital systems. A basic understanding of these languages, as well as of the fundamentals of hardware description languages, is essential to success in digital system design. We also study the tools for optimizing logic and sequential machine designs in order to understand how to best make use of logic and physical synthesis.

Chapter 6 looks at the structure of large digital systems. This chapter introduces register-transfer design as a methodology for structuring digital designs. It uses a simple DSP as a design example. This DSP is not intended as a state-of-the-art CPU design, but it does allow us to consider a large number of different design problems in a single example.

Chapter 7 caps off the book by studying large-scale systems built with FPGAs. Platform FPGAs that include CPUs and FPGA fabrics allow designers to mix hardware and software on a single chip to solve difficult design problems. Multi-FPGA systems can be used to implement very large designs; a single multi-FPGA system can be programmed to implement many different logic designs.

So what will happen to ASIC design? I don't think it will go away—people will still need the high density and high performance that only custom silicon provides. But I think that FPGAs will become one of the major modes of implementation for digital systems.

Xilinx has graciously allowed us to include CDs that contain the Xilinx Student Edition (XSE) tools. The examples in this book were prepared with these tools and you can follow along with the examples using the tools. You can also use them to create your own examples. Having a working set of tools makes it much easier to practice concepts and I greatly appreciate Xilinx's help with including these books.

You can find additional materials for the book at the Web site:

http://www.princeton.edu/~wolf/fpga-book

The Web site includes overheads for the chapters, pointer to additional Web materials, some sample labs, and errata. Properly accredited instructors can obtain a copy of the instructor's manual by contacting the publisher. I hope you enjoy this book; please feel free to email me at wolf@princeton.edu.

I'd like to thank the students of ELE 462 in the spring of 2003, who were patient with my experimentation on the traditional VLSI course. I'd also like to thank Jiang Xu and Li Shang, my teaching assistants that semester, who improved our infrastructure for FPGA design and helped me debug the DSP design. Mike Butts and Mohammed Khalid gave valuable advice on partitioning algorithms. Steven Brown, Jonathan Rose, Zvonko Vranesic, and William Yu provided figures from their papers to be put directly into the book, providing data straight from the source for the reader as well as simplifying my life as a typesetter. I greatly appreciate the thorough review of a draft of this manuscript given by Andre De Hon, Carl Ebeling, Yankin Tanurhan, and Steve Trimberger; they all made many excellent suggestions on how to improve this book. I greatly appreciate the efforts of Ivo Bolsens, Anna Acevedo, and Jeff Weintraub for access to the knowledge of Xilinx in general and permission to include the Xilinx ISE disks with this book in particular. And, of course, I'd like to thank my editor Bernard Goodwin for his tireless efforts on behalf of this book. All the problems remaining in the book, both small and large, are of course my responsibility.

Wayne Wolf

Princeton, New Jersey

1 FPGA-Based Systems

FPGAs in system design.

FPGAs *vs*. ASICs.

Design methodologies.

1.1 Introduction

This chapter will set the stage for the rest of the book. The next section talks about some basic concepts in Boolean algebra and schematics. Section 1.3 introduces FPGAs and describes their importance. Section 1.4 describes how we use FPGAs to design complex digital systems.

1.2 Basic Concepts

This section introduces some basic concepts in logic design. If this material is review for you it will help to establish some terminology that we will use throughout the rest of the book. If the material is not review then it should be of use for the rest of the book.

1.2.1 Boolean Algebra

combinational logic functions

We use Boolean algebra to represent the logical functions of digital circuits. Shannon [Sha38] showed that networks of switches (such as light switches) could be modeled by Boolean functions. Today's logic gates are usually not built from switches but we still consider Boolean algebra to be a fundamental representation. We use Boolean algebra to describe **combinational** (not combinatorial) **logic** functions. Our Boolean functions describe combinations of inputs; we do not use functions with existential ($\exists x\, f(x)$) or universal ($\forall x\, g(x)$) quantification.

Figure 1-1 Symbols for functions in logical expressions.

name	symbol
NOT	', ~
AND	$\cdot$, $\wedge$, &
NAND	|
OR	+ , $\vee$
NOR	NOR
XOR	$\oplus$
XNOR	XNOR

notation

We will use fairly standard notation for logic expressions: if a and b are variables, then a' (or $\bar{a}$) is the complement of a, $a \cdot b$ (or ab) is the AND of the variables, and $a + b$ is the OR of the variables. In addition, for the NAND function $(ab)'$ we will use the | symbol [1], for the NOR function $(a + b)'$ we will use a *NOR* b, and for exclusive-or ($a\ XOR\ b = ab' + a'b$) we will use the $\oplus$ symbol. (Students of algebra know that XOR and AND form a ring.) Figure 1-1 summarizes the names and symbols for common logic functions.

We use the term **literal** for either the true form (a) or complemented form (a') of a variable. Understanding the relationship between logical expressions and gates lets us study problems in the model that is simplest for that problem, then transfer the results.

1. The Scheffer stroke is a dot with a negation line through it. C programmers should note that this character is used as OR in the C language.

algebraic rules

Let's review some basic rules of algebra that we can use to transform expressions. Some are similar to the rules of arithmetic while others are particular to Boolean algebra:

- **idempotency**: $a \cdot a = a$, $a + a = a$.
- **inversion**: $a + a' = 1$, $a \cdot a' = 0$.
- **identity elements**: $a + 0 = a$, $a \cdot 1 = a$.
- **commutativity**: $a + b = b + a$, $a \cdot b = b \cdot a$.
- **null elements**: $a \cdot 0 = 0$, $a + 1 = 1$.
- **involution** : $(a')' = a$.
- **absorption:** $a + ab = a$.
- **associativity**: $a + (b + c) = (a + b) + c$, $a \cdot (b \cdot c) = (a \cdot b) \cdot c$.
- **distributivity**: $a \cdot (b + c) = ab + ac$, $a + bc = (a + b)(a + c)$.
- **De Morgan's laws**: $(a + b)' = a' \cdot b'$, $(a \cdot b)' = a' + b'$.

completeness and irredundancy

Although Boolean algebra may seem abstract, some of the mathematical results it provides directly relate to the physical properties of logic circuits. Two problems in Boolean algebra that are of importance to logic design are **completeness** and **irredundancy**.

completeness

A set of logical functions is **complete** if we can generate every possible Boolean expression using that set of functions—that is, if for every possible function built from arbitrary combinations of +, ·, and ', an equivalent formula exists written in terms of the functions we are trying to test. We generally test whether a set of functions is complete by inductively testing whether those functions can be used to generate all logic formulas. It is easy to show that the NAND function is complete, starting with the most basic formulas:

- 1: $a|(a|a) = a|a' = 1$.
- 0: $\{a|(a|a)\}|\{a|(a|a)\} = 1|1 = 0$.
- a': $a|a = a'$.
- ab: $(a|b)|(a|b) = ab$.
- $a + b$:$(a|a)|(b|b) = a'|b' = a + b$.

From these basic formulas we can generate all the formulas. So the set of functions {|} can be used to generate any logic function. Similarly, any formula can be written solely in terms of NORs.

The combination of AND and OR functions, however, is not complete. That is fairly easy to show: there is no way to generate either 1 or 0 directly from any combination of AND and OR. If NOT is added to the set, then we can once again generate all the formulas: $a + a' = 1$, etc. In fact, both $\{', \cdot\}$ and $\{',+\}$ are complete sets.

Any circuit technology we choose to implement our logic functions must be able to implement a complete set of functions. Static, complementary circuits naturally implement NAND or NOR functions, but some other circuit families do not implement a complete set of functions. Incomplete logic families place extra burdens on the logic designer to ensure that the logic function is specified in the correct form.

irredundancy and minimality

A logic expression is **irredundant** if no literal can be removed from the expression without changing its truth value. For example, $ab + ab'$ is redundant, because it can be reduced to a. An irredundant formula and its associated logic network have some important properties: the formula is smaller than a logically equivalent redundant formula; and the logic network is guaranteed to be testable for certain kinds of manufacturing defects. However, irredundancy is not a panacea. Irredundancy is not the same as **minimality**—there are many irredundant forms of an expression, some of which may be smaller than others, so finding one irredundant expression may not guarantee you will get the smallest design. Irredundancy often introduces added delay, which may be difficult to remove without making the logic network redundant. However, simplifying logic expressions before designing the gate network is important for both area and delay. Some obvious simplifications can be done by hand; CAD tools can perform more difficult simplifications on larger expressions.

on-sets and off-sets

We often speak of the **on-set** and **off-set** of a function. The on-set is simply the set of input values for which the function is true; the off-set is the input values for which the function is false. We typically specify the function with the on-set but using the off-set is equally valid (so long as it is clear which is being used).

don't-cares

We often make use of **don't-cares** in combinational logic. There are two types of don't-cares: **input don't-cares** and **output don't-cares**. Despite their similar names they serve very different functions.

An input don't-care is a shorthand notation. Consider the function shown in Figure 1-2. In this function, the inputs a=0, b=0 and a=1, b=1 give the same result. When we write the truth table we can collapse those two rows into a single function using the don't-care symbol -. We can simplify the terms $ab + ab'$ to simply a' with b being an input don't-

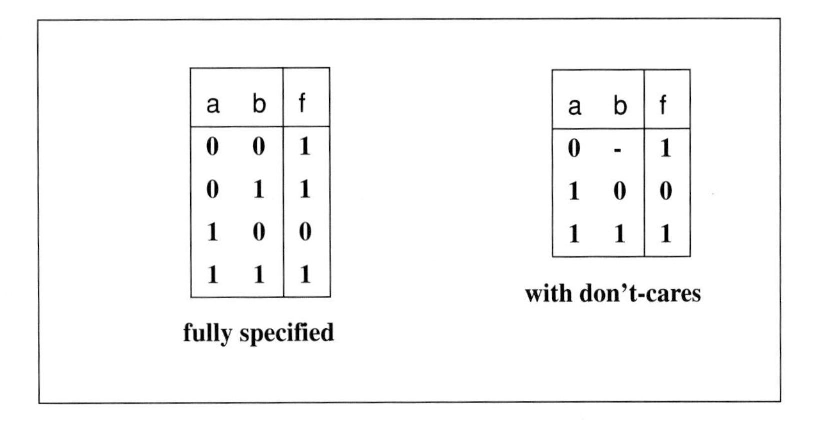

a	b	f
0	0	1
0	1	1
1	0	0
1	1	1

fully specified

a	b	f
0	-	1
1	0	0
1	1	1

with don't-cares

Figure 1-2 An example of input don't-cares.

care to this term. (The complete logic function is $f = a' + ab$. You can verify this by drawing a Karnaugh map.)

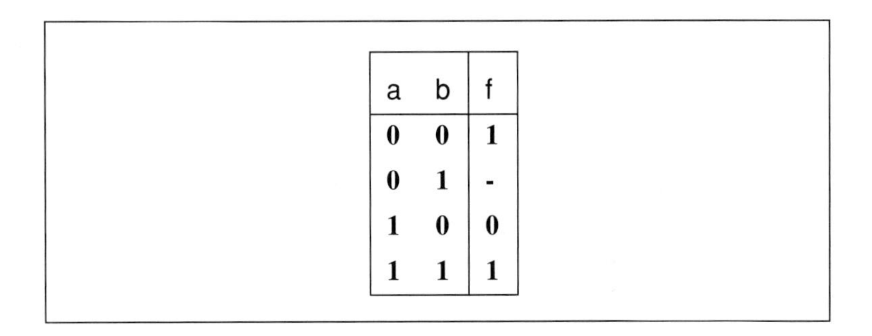

a	b	f
0	0	1
0	1	-
1	0	0
1	1	1

Figure 1-3 An example of output don't-cares.

An output don't-care is a means of incompletely specifying a logic function. An output don't-care is a combination of inputs that is not used. Consider the function shown in Figure 1-3. In this case the function's value for the input combination a=0, b=1 is the don't-care symbol -. (We can use the same symbol for input don't-cares.) This means that we do not care whether the output's value is 0 or 1 for this combination of inputs. When implementing this function in gates, we can choose the output value that gives us the best implementation. We saw in the last paragraph that choosing the function's output to be 1 in this case lets us simplify the expression.

We often write Boolean expressions in **factored** form. For example, we can write our function $f = a' + ab'$ as a single function. If we factor out a new function $g = ab'$ we can rewrite f as $f = a' + g$. This transformation in itself isn't very interesting, but if the ab' term is used elsewhere then the g function becomes a common factor. The opposite of factoring is collapsing; we collapse g into f when we rewrite f as $f = a' + ab'$.

1.2.2 Schematics and Logic Symbols

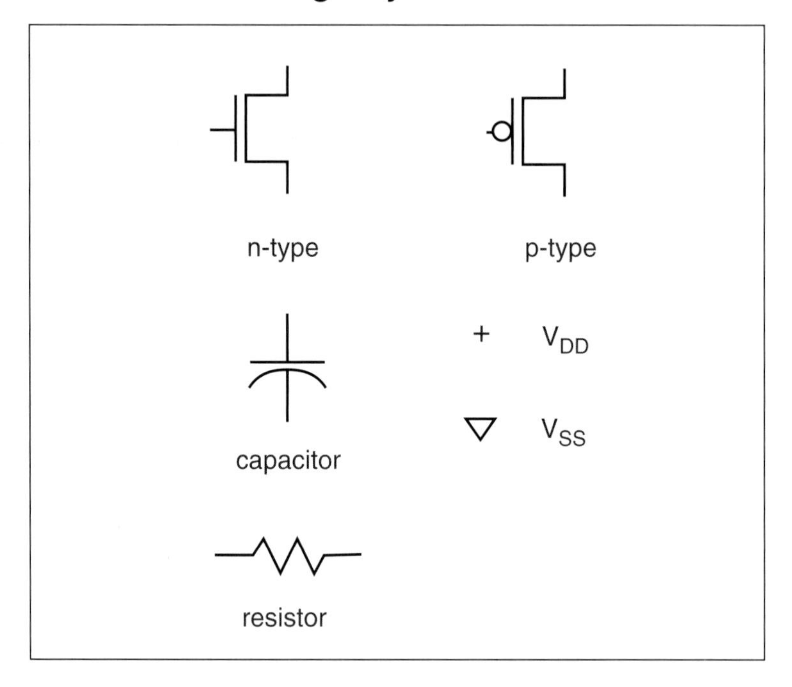

Figure 1-4 Schematic symbols for electrical components.

Since we will use a variety of schematic symbols let's briefly review them here. Figure 1-4 shows the symbols for some electrical components: n-type and p-type transistors, a capacitor, a resistor, and power supply connections (V_{DD}, the positive terminal and V_{SS}, the negative terminal).

Figure 1-5 shows some symbols for logic gates: NAND, NOR, etc. Figure 1-6 shows the symbols for two common register-transfer components: the multiplexer (mux) and arithmetic logic unit (ALU).

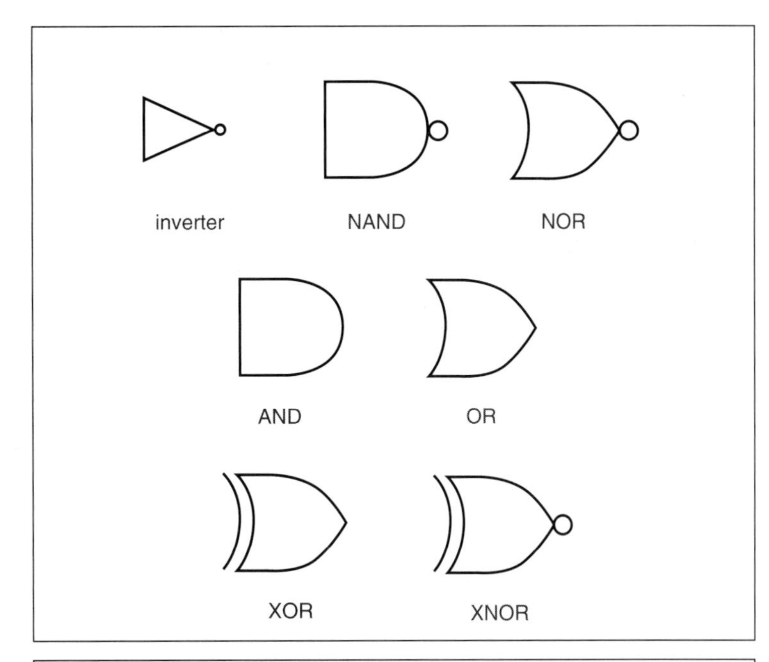

Figure 1-5 Schematic symbols for logic gates.

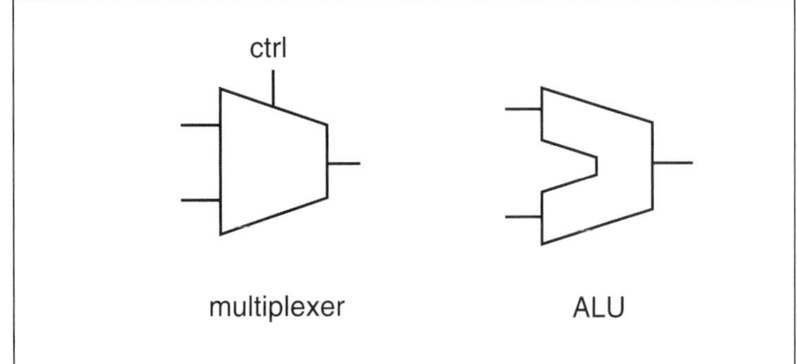

Figure 1-6 Schematic symbols for register-transfer components.

1.3 Digital Design and FPGAs

1.3.1 The Role of FPGAs

Field-programmable gate arrays (**FPGAs**) fill a need in the design space of digital systems, complementary to the role played by microprocessors. Microprocessors can be used in a variety of environments, but because they rely on software to implement functions, they are generally slower and more power-hungry than custom chips. Similarly, FPGAs are not custom parts, so they aren't as good at any particular function as a dedicated chip designed for that application. FPGAs are generally slower and burn more power than custom logic. FPGAs are also relatively expensive; it is often tempting to think that a custom-designed chip would be cheaper.

advantages of FPGAs

However, they have compensating advantages, largely due to the fact that they are standard parts.

- There is no wait from completing the design to obtaining a working chip. The design can be programmed into the FPGA and tested immediately.
- FPGAs are excellent prototyping vehicles. When the FPGA is used in the final design, the jump from prototype to product is much smaller and easier to negotiate.
- The same FPGA can be used in several different designs, reducing inventory costs.

The area filled by FPGAs has grown enormously in the past twenty years since their introduction. **Programmable logic devices (PLDs)** had been on the market since the early 1970s. These devices used two-level logic structures to implement programmed logic. The first level of logic, the AND plane, was generally fixed, while the second level, known as the OR plane, was programmable. PLDs are generally programmed by antifuses, which are programmed through large voltages to make connections.

They were most often used as **glue logic**—logic that was needed to connect together the major components of the system. They were often used in prototypes because they could be programmed and inserted into a board in a few minutes, but they did not always make it into the final product. Programmable logic devices were usually not seen as the principal components of the systems in which they were used. As digital systems became more complex, more dense programmable logic was needed, and the limitations of PLDs' two-level logic became clear.

Two-level logic is useful for relatively small logic functions, but as levels of integration grew, two-level structures became too inefficient. FPGAs provided programmable logic using multi-level logic of arbitrary depth. They used both programmable logic elements and programmable interconnect to build the multi-level logic functions.

history of the FPGA

Ross Freeman is generally credited with the invention of the FPGA. His FPGA [Fre89] included both programmable logic elements and a programmable interconnect structure. His FPGA was also programmed using SRAM, not antifuses. This allowed the FPGA to be manufactured on standard VLSI fabrication processes, saving money and providing more manufacturing options. It also allowed the FPGA to be reprogrammed while it was in-circuit; this was a particularly interesting feature since flash memory was not yet in common use.

Xilinx and Altera both sold early SRAM-based FPGAs. An alternative architecture was introduced by Actel, which used an antifuse architecture. This architecture was not reprogrammable in the field, which arguably was an advantage in situations that did not require reconfiguration. The Actel FPGAs used a mux-oriented logic structure organized around wiring channels.

For many years FPGAs were seen primarily as glue logic and prototyping devices. Today, they are used in all sorts of digital systems:

- as part of high-speed telecommunications equipment;
- as video accelerators in home personal video recorders (PVRs).

FPGAs have become mainstream devices for implementing digital systems.

1.3.2 FPGA Types

We have so far avoided defining the FPGA. A good definition should distinguish the FPGA on the one hand from smaller programmable devices like PLDs and on the other hand from custom chips. Here are some defining characteristics of FPGAs:

- **They are standard parts**. They are not designed for any particular function but are programmed by the customer for a particular purpose.
- **They implement multi-level logic**. The logic blocks inside FPGAs can be connected in networks of arbitrary depth. PLDs, in contrast, use two levels of NAND/NOR functions to implement all their logic.

Because FPGAs implement multi-level logic, they generally need both programmable logic blocks and programmable interconnect. PLDs use fixed interconnect and simply change the logic functions attached to the wires. FPGAs, in contrast, require programming logic blocks and con-

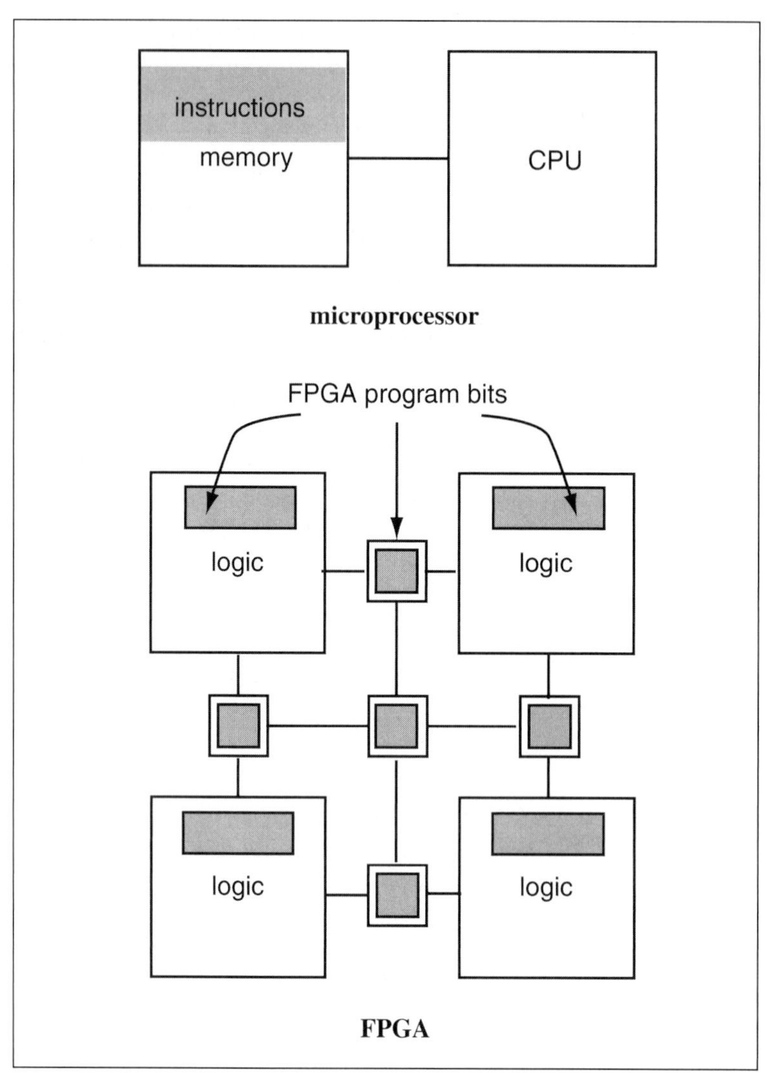

Figure 1-7 CPUs *vs*. FPGAs.

necting them together in order to implement functions. The combination of logic and interconnect is known as a **fabric** because it possesses a regular structure that can be efficiently utilized by design tools that map the desired logic onto the FPGA.

FPGA programming

One of the major defining characteristics of the FPGA is that it can be programmed. As illustrated in Figure 1-7, programming an FPGA is very different from programming a microprocessor. A microprocessor is a stored-program computer. A computer system includes both a CPU and a separate memory that stores instructions and data. The FPGA's program (also called a **personality**) is interwoven into the logic structure of the FPGA. An FPGA does not fetch instructions—the FPGA's programming directly implements logic functions and interconnections.

A variety of technologies are used to program FPGAs. Some FPGAs are permanently programmed; others can be reprogrammed. Reprogrammable FPGAs are also known as **reconfigurable** devices. Reconfigurable FPGAs are generally favored in prototype building because the device doesn't need to be thrown away every time a change is made. Reconfigurable systems can also be reprogrammed on-the-fly during system operation. This allows one piece of hardware to perform several different functions. Of course, those functions cannot be performed at the same time, but reconfigurability can be very useful when a system operates in different modes. For example, the Radius computer display operated in both horizontal (landscape) and vertical (portrait) modes. When the user rotated the display, a mercury switch caused the FPGA that ran the display to be reprogrammed for the new mode.

FPGAs have traditionally used fine-grained logic. A combinational logic element in a traditional FPGA implements the function of a handful of logic gates plus a register. As chips become larger, coarser-grained FPGAs are coming into use. A single logic element in these chips may implement a multi-bit ALU and register. Coarser-grained FPGAs may make more efficient use of chip area for some types of functions.

platform FPGAs

A newer category of FPGAs includes more than just the FPGA fabric itself. **Platform FPGAs** include several different types of structures so that each part of a large system can be efficiently implemented on the type of structure best suited to it. A platform FPGA typically includes a CPU so that some functions can be run in software. It may also include specialized bus logic so that, for example, a PCI bus interface can easily be included into the system.

1.3.3 FPGAs *vs.* Custom VLSI

The main alternative to an FPGA is an **application-specific IC** (**ASIC**). Unlike an FPGA, an ASIC is designed to implement a particular logical function. The design of an ASIC goes down to the masks used to fabricate an IC. The ASIC must be fabricated on a manufacturing line, a process that takes several months, before it can be used or even tested. ASICs are typically distinguished from full custom designs: a full-custom design has a custom layout, while an ASIC uses pre-designed layouts for logic gates. Today, few large digital chips other than microprocessors include a significant amount of custom layout.

As we noted above, ASICs have some significant advantages because they are designed for a particular purpose: they are generally faster and lower power than FPGA equivalents; when manufactured in large volumes, they are also cheaper. However, two trends are leading many system designers to use FPGAs rather than ASICs for custom logic.

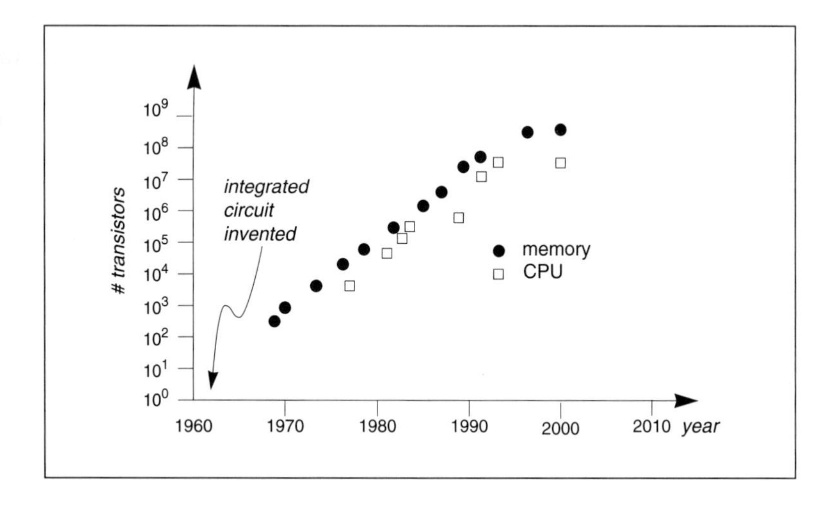

Figure 1-8 Moore's Law.

Moore's Law

On the one hand, **Moore's Law** has provided substantial increases in the capacity of integrated circuits. Moore's Law states that the number of transistors that can be manufactured on a chip will double every 18 months. As shown in Figure 1-8, the semiconductor industry has managed to keep up this pace for several decades. Even though Moore's Law will not continue forever, chips with tens of millions of transistors can already be fabricated in large quantities. With so many transistors on a single chip, it is often tempting—and increasingly necessary—to throw away transistors in order to simplify the task of designing the logic on the chip. FPGAs use more transistors for a given function than do ASICs, but an FPGA can be designed in days compared to the year-long design cycle required for garden-variety FPGAs.

On the other hand, the cost of the masks used to manufacture an ASIC is skyrocketing. The basic figure of merit of an IC manufacturing line is the width of the smallest transistor it can build. As line widths shrink, the cost of manufacturing goes up—a modern semiconductor manufacturing plant costs several billion dollars. However, those manufacturing costs can be spread over all the parts built on the line. In contrast, the

Table 1-1 Mask cost as a function of line width.

technology	mask cost
0.09 μm	$1,000,000
0.18 μm	$250,000
0.25 μm	$120,000
0.35 μm	$60,000

masks that define the transistors and wires for a particular chip are useful only for that chip. The cost of the mask must be absorbed entirely in the cost of that chip. As Table 1-1 shows, the costs of masks are growing exponentially. As VLSI line widths shrink, mask costs will start to overshadow the costs of the designers who create the chip. As mask costs soar, standard parts become more attractive. Because FPGAs are standard parts that can be programmed for many different functions, we can expect them to take a larger share of the IC market for high-density chips.

1.4 FPGA-Based System Design

1.4.1 Goals and Techniques

The logical function to be performed is only one of the goals that must be met by an FPGA or any digital system design. Many other attributes must be satisfied for the project to be successful:

- **Performance.** The logic must run at a required rate. Performance can be measured in several ways, such as throughput and latency. Clock rate is often used as a measure of performance.
- **Power/energy.** The chip must often run within an energy or power budget. Energy consumption is clearly critical in battery-powered systems. Even if the system is to run off the power grid, heat dissipation costs money and must be controlled.
- **Design time.** You can't take forever to design the system. FPGAs, because they are standard parts, have several advantages in design time. They can be used as prototypes, they can be programmed quickly, and they can be used as parts in the

final design

- **Design cost.** Design time is one important component of design cost, but other factors, such as the required support tools, may be a consideration. FPGA tools are often less expensive than custom VLSI tools.
- **Manufacturing cost.** The manufacturing cost is the cost of replicating the system many times. FPGAs are generally more expensive than ASICs thanks to the overhead of programming. However, the fact that they are standard parts helps to reduce their cost.

design challenges

Design is particularly hard because we must solve several problems:

- **Multiple levels of abstraction**. FPGA design requires refining an idea through many levels of detail. Starting from a specification of what the chip must do, the designer must create an architecture which performs the required function, and then expand the architecture into a logic design.
- **Multiple and conflicting costs**. Costs may be in dollars, such as the expense of a particular piece of software needed to design some piece. Costs may also be in performance or power consumption of the final FPGA.
- **Short design time**. Electronics markets change extremely quickly. Getting a chip out faster means reducing your costs and increasing your revenue. Getting it out late may mean not making any money at all.

requirements and specifications

A design project may start out with varying amounts of information. Some projects are revisions of earlier designs; some are implementations of published standards. In these cases, the function is well-specified and some aspects of the implementation may be clear. Other projects are less well-founded and may start only with some basic notions of what the chip should do. We use the term **requirements** for a written description of what the system is to do—"ADSL modem baseband functions," for example. We use the term **specification** for a more formal description of the function, such as a simulator program of the ADSL baseband functionality. Attributes like performance, power consumption, and cost are known as **non-functional requirements**.

1.4.2 Hierarchical Design

Hierarchical design is a standard method for dealing with complex digital designs. It is commonly used in programming: a procedure is written not as a huge list of primitive statements but as calls to simpler procedures. Each procedure breaks down the task into smaller operations until each step is refined into a procedure simple enough to be written directly. This technique is commonly known as **divide-and-conquer**—the procedure's complexity is conquered by recursively breaking it down into manageable pieces.

Figure 1-9 Pins on a component.

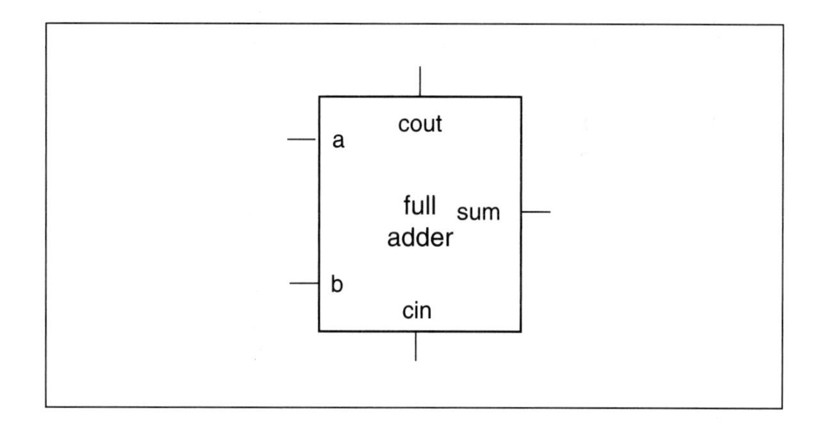

component types

Chip designers divide and conquer by breaking the chip into a hierarchy of components. As shown in Figure 1-9, a component consists of a **body** and a number of **pins**—this full adder has pins *a*, *b*, *cin*, *cout*, and *sum*. If we consider this full adder the definition of a **type**, we can make many **instances** of this type. Repeating commonly used components is very useful, for example, in building an n-bit adder from n full adders. We typically give each component instance a **name**. Since all components of the same type have the same pins, we refer to the pins on a particular component by giving the component instance name and pin name together; separating the instance and pin names by a dot is common practice. If we have two full adders, *add1* and *add2*, we can refer to *add1.sum* and *add2.sum* as distinct **terminals** (where a terminal is a component-pin pair).

nets and components

Figure 1-10 shows a slightly more complex component that contains two other components. We show both the instance name and type name of each component, such as *large(bx)* where *large* is the instance name and *bx* is the type name. We also show the name of each pin on the components and every net that connects the pins. We can list the electrical

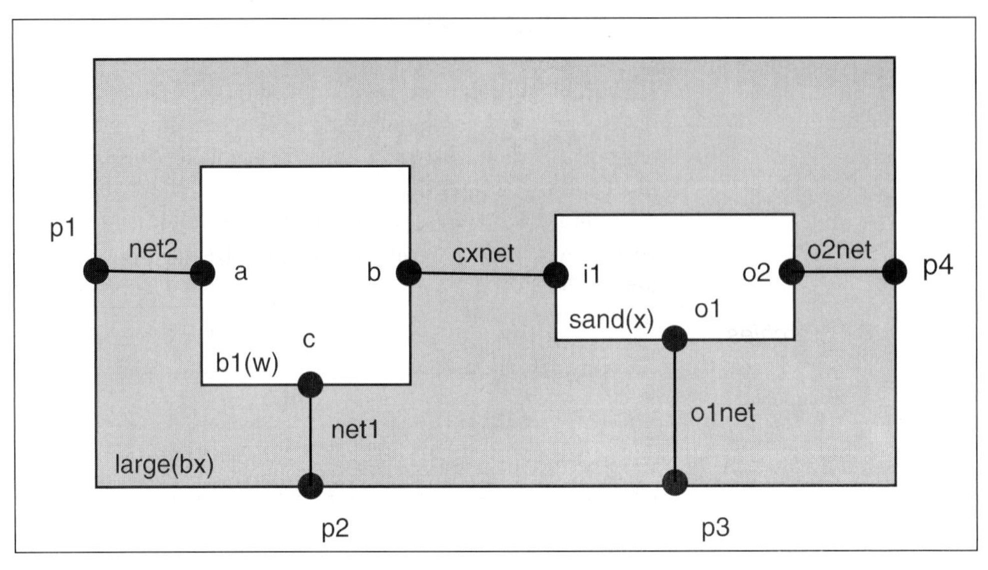

Figure 1-10 A hierarchical logic design.

Figure 1-11 A component hierarchy.

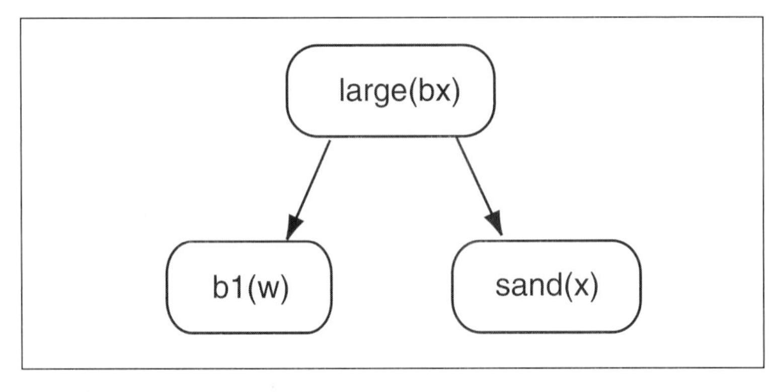

connections which make up a circuit in either of two equivalent ways: a **net list** or a **component list**. A net list gives, for each net, the terminals connected to that net. Here is a net list for the top component of Figure 1-10:

```
net2: large.p1, b1.a;
net1: large.p2, b1.c;
cxnet: b1.b, sand.i1;
o1net: large.p3, sand.o1;
o2net: sand.o2, large.p4.
```

A component list gives, for each component, the net attached to each pin. Here is a component list version of the same circuit:

```
large: p1: net2, p2: net1, p3: o1net, p4: o2net;
b1: a: net2, b: cxnet, c: net1;
sand: i1: cxnet, o1: o1net, o2: o2net;
```

Given one form of connectivity description, we can always transform it into the other form. Which format is used depends on the application—some searches are best performed net-by-net and others component-by-component. As an abuse of terminology, any file which describes electrical connectivity is usually called a **netlist file**, even if it is in component list format.

component hierarchies

Component ownership forms a hierarchy. The component hierarchy of Figure 1-10's example is shown in Figure 1-11. Each rounded box represents a component; an arrow from one box to another shows that the component pointed to is an element in the component which points to it. We may need to refer to several instance names to differentiate components. In this case, we may refer to either *large/bw* or *large/sand*, where we trace the component ownership from the most highest-level component and separate component names by slashes (/). (The resemblance of this naming scheme to UNIX file names is intentional—some design tools use files and directories to model component hierarchies.)

Each component is used as a black box—to understand how the system works, we only have to know each component's input-output behavior, not how that behavior is implemented inside the box. To design each black box, we build it out of smaller, simpler black boxes. The internals of each type define its behavior in terms of the components used to build it. If we know the behavior of our primitive components, such as transistors, we can infer the behavior of any hierarchically described component.

People can much more easily understand a 10,000,000-gate hierarchical design than the same design expressed directly as a million gates wired together. The hierarchical design helps you organize your thinking—the hierarchy organizes the function of a large number of transistors into a particular, easy-to-summarize function. Hierarchical design also makes it easier to reuse pieces of chips, either by modifying an old design to perform added functions or by using one component for a new purpose.

1.4.3 Design Abstraction

uses of abstractions

Design abstraction is critical to hardware system design. Hardware designers use multiple levels of design abstraction to manage the design process and ensure that they meet major design goals, such as speed and power consumption. The simplest example of a design abstraction is the logic gate. A logic gate is a simplification of the nonlinear circuit used to build the gate: the logic gate accepts binary Boolean values. Some design tasks, such as accurate delay calculation, are hard or impossible when cast in terms of logic gates. However, other design tasks, such as logic optimization, are too cumbersome to be done on the circuit. We choose the design abstraction that is best suited to the design task.

We may also use higher abstractions to make first-cut decisions that are later refined using more detailed models: we often, for example, optimize logic using simple delay calculations, then refine the logic design using detailed circuit information. Design abstraction and hierarchical design aren't the same thing. A design hierarchy uses components at the same level of abstraction—an architecture built from Boolean logic functions, for example—and each level of the hierarchy adds complexity by adding components. The number of components may not change as it is recast to a lower level of abstraction—the added complexity comes from the more sophisticated behavior of those components.

The next example shows how a slightly more complex hardware design is built up from circuit to complex logic.

Example 1-1 Digital logic abstractions

The basic element of the FPGA is the logic block. We can build function units like adders out of logic blocks:

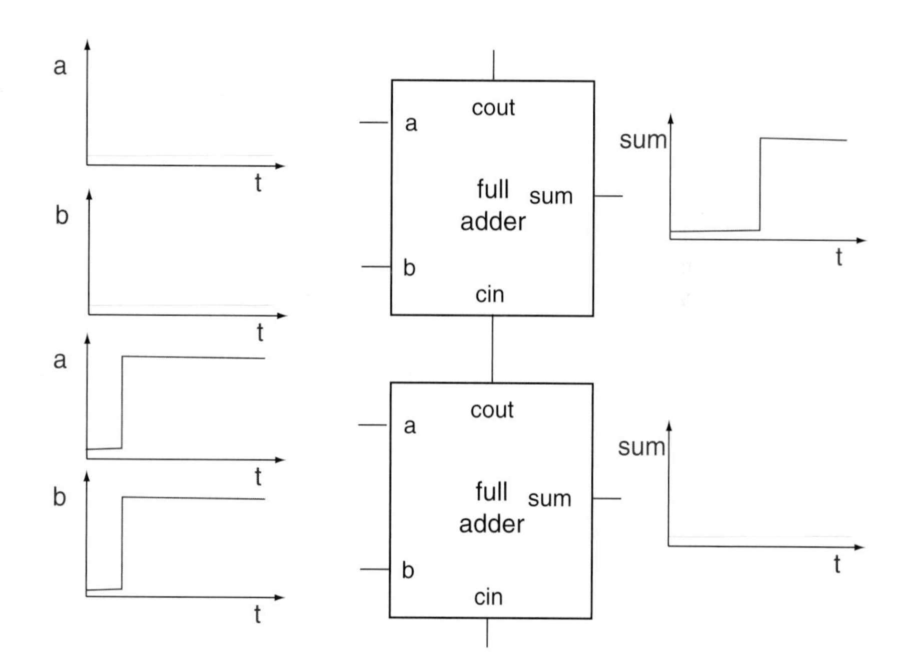

Determining waveforms for all the signals in an FPGA isn't practical. However, by making reasonable assumptions we can determine approximate delays through logic networks in FPGAs.

When designing large register-transfer systems, such as data paths, we may abstract one more level to generic adders:

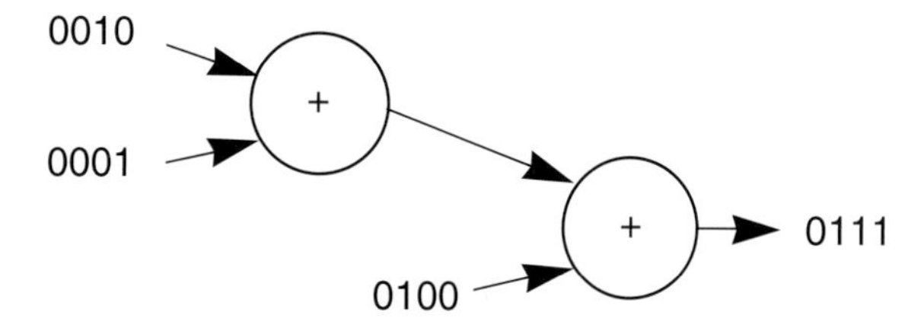

At this point, since we don't know how the adders are built, we don't have any delay information. These components are pure combinational elements—they produce an output value given an input value. The adder abstraction helps us concentrate on the proper function before we worry about the details of performance.

FPGA abstractions

Figure 1-12 shows a design abstraction ladder for FPGAs:

- **Behavior.** A detailed, executable description of what the chip should do, but not how it should do it. A C program, for example, may be used as a behavioral description. The C program will not mimic the clock cycle-by-clock cycle behavior of the chip, but it will allow us to describe in detail what needs to be computed, error and boundary conditions, etc.
- **Register-transfer**. The system's time behavior is fully-specified—we know the allowed input and output values on every clock cycle—but the logic isn't specified as gates. The system is specified as Boolean functions stored in abstract memory elements. Only the vaguest delay and area estimates can be made from the Boolean logic functions.
- **Logic**. The system is designed in terms of Boolean logic gates, latches, and flip-flops. We know a lot about the structure of the system but still cannot make extremely accurate delay calculations.
- **Configuration**. The logic must be placed into logic elements around the FPGA and the proper connections must be made between those logic elements. Placement and routing perform these important steps.

top-down and bottom-up design

Design always requires working down from the top of the abstraction hierarchy and up from the least abstract description. Obviously, work must begin by adding detail to the abstraction—**top-down** design adds functional detail. But top-down design decisions are made with limited information: there may be several alternative designs at each level of abstraction; we want to choose the candidate which best fits our speed, area, and power requirements. We often cannot accurately judge those costs until we have an initial design. **Bottom-up** analysis and design percolates cost information back to higher levels of abstraction; for instance, we may use more accurate delay information from the circuit design to redesign the logic. Experience will help you judge costs before you complete the implementation, but most designs require cycles of top-down design followed by bottom-up redesign.

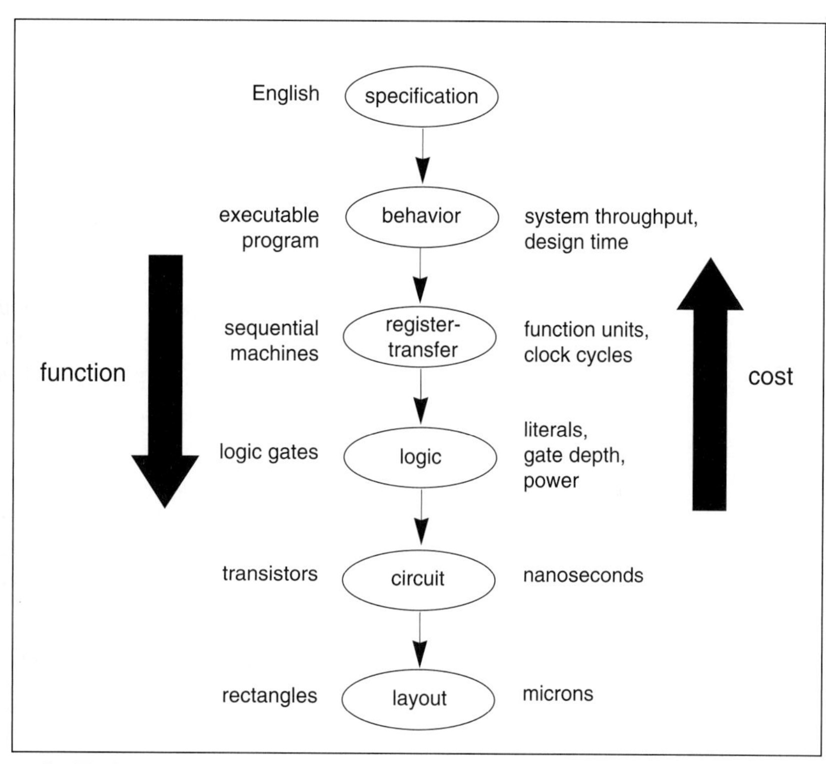

Figure 1-12 Abstractions for FPGA design.

1.4.4 Methodologies

Complex problems can be tackled using **methodologies**. We can use the lessons we learn on one system to help us design the next system. Digital system design methodologies have been built up over several decades of experience. A methodology provides us with a set of guidelines for what to do, when to do it, and how to know when we are done. Some aspects of FPGA design methodologies are unique to FPGAs, but many aspects are generic to all digital systems.

hardware description languages

Modern digital designers rely on **hardware description languages (HDLs)** to describe digital systems. Schematics are rarely used to describe logic; block diagrams are often drawn at higher levels of the system hierarchy but usually as documentation, not as design input. HDLs are tied to **simulators** that understand the semantics of hardware.

They are also tied to **synthesis tools** that generate logic implementations.

Methodologies have several goals:

- **Correct functionality**. Simply speaking, the chip must work.
- **Satisfaction of performance and power goals**. Another aspect of making the chip work.
- **Catching bugs early**. The later a bug is caught, the more expensive it is to fix. Good methodologies check pieces of the design as they are finished to be sure they are correct.
- **Good documentation**. The methodology should produce a paper trail that allows the designers to check when mistakes were made and to ensure that things were done properly.

A functional model of some sort is often a useful aid. Such a model may be written in a programming language such as C. Functional models are useful because they can be written without regard to the added detail needed to create a functional design. The output of the functional model can be used to check the register-transfer design. When a design is based on an existing standard, the standard typically includes an executable specification that can be used to verify the implementation.

design optimization

Turning the functional description into a register-transfer design takes many steps. The system must be architected into its major components. The components must be successively refined into blocks that can be designed individually. The design tasks must be parceled out to people on the team. Each piece must be functionally verified. As pieces are put together, the new modules must also be functionally verified.

Some timing and power analysis can be performed on the raw register-transfer design. However, most of that work requires information about the placement and interconnect of the logic on the chip. Place-and-route tools must be run repeatedly to get information on the decomposition of logic into the logic elements and the interconnections on the chip.

1.5 Summary

The VLSI manufacturing industry has delivered on the promise of Moore's Law for decades, giving us chips that can hold hundreds of millions of transistors. One use of these transistors is to simplify chip design. FPGAs take advantage of complex chips to improve the chip design in several ways: FPGAs can be quickly programmed, their performance characteristics are predictable, and the same part can be used for several different logic designs. Throughout the rest of this book we will study the characteristics of FPGAs and design techniques for building digital systems in FPGAs.

1.6 Problems

Q1-1. Describe each of these functions in truth table format:

a. a & b.

b. a | ~b.

c. (a & b) | ~c.

d. w | (x & ~y).

e. w | (~x & ~y) | z.

Q1-2. Hierarchy helps you reuse pieces of a design. A large CPU today has roughly 10 million transistors.

a. How many days would it take you to design a CPU if you typed in each transistor by hand into a hardware description language, taking only one second per transistor?

b. How long would it take you if half the transistors consisted of cache memory, which was designed from a six-transistor cell which could be replicated to complete the cache?

c. How long would it take if half of the non-cache transistors were implemented in 32 identical bit slices, requiring one copy of the bit slice to be drawn, with the rest created by replication?

Q1-3. Draw a logic diagram for a full adder (consult Chapter 4 if you don't know how a full adder works). Name each logic gate. Draw a four-bit adder from four full adders. Name each component in the four-bit adder and define the four-bit adder as a type. Draw the component hierarchy, showing the four-bit adder, the full adder, and the logic gates; when a component is repeated, you can draw its sub-components once and refer to them elsewhere in the diagram.

Q1-4. Plot the cost of FPGAs as a function of capacity. Choose an FPGA family, go to the catalog (usually on a Web site) of an electronics component supplier, and look up the prices and capacities of several members of the FPGA family. Plot your data. Is cost a linear function of capacity?

2 VLSI Technology

Integrated circuit fabrication.

Transistors.

Logic gates.

Memory elements.

I/O pins and packaging.

2.1 Introduction

In this chapter we review some basic concepts in integrated circuit technology. Since FPGAs are standard parts we have no control over the detailed design of the chips. However, many aspects of VLSI design do determine the characteristics of the FPGA and how we must use it to implement our logic. In many cases, a qualitative knowledge of VLSI principles is sufficient to understand why certain problems occur and what we can do about them.

You can make use of most of the material on FPGAs without reading this chapter. However, a basic knowledge of VLSI technology will help you appreciate some of the finer points of FPGA design. FPGAs also make a good vehicle for illustrating some aspects of VLSI design.

We will start with a brief introduction to semiconductor fabrication methods. We will next look at the basic characteristics of transistors. Based on that understanding, we will look at how to design and analyze logic gates. We will then study wires, which are a major source of delay in modern CMOS integrated systems. Based on our understanding of these fundamental components, we will go on to study the design of registers and memories. We will close with a discussion of I/O circuitry and the packages used to house integrated circuits.

2.2 Manufacturing Processes

wafers and chips

Integrated circuits are built on a silicon **substrate** provided by the wafer. Wafer sizes have steadily increased over the years. Larger wafers mean more chips per wafer and higher productivity. The key figure of merit for a fabrication process is the size of the smallest transistor it can manufacture. Transistor size helps determine both circuit speed and the amount of logic that can be put on a single chip. Fabrication technologies are usually identified by their minimum transistor length, so a process which can produce a transistor with a 0.13 μm minimum channel length is called a 0.13 μm process. (Below 0.10 μm we switch to nanometer units, such as a 90 nm process.)

electrons and holes

A pure silicon substrate contains equal numbers of two types of electrical carriers: electrons and holes. While we will not go into the details of device physics here, it is important to realize that the interplay between electrons and holes is what makes transistors work. The goal of doping is to create two types of regions in the substrate: an **n-type** region which contains primarily electrons and a **p-type** region which is dominated by holes. (Heavily doped regions are referred to as n+ and p+.) Transistor action occurs at properly formed boundaries between n-type and p-type regions.

chip cross-section

A cross-section of an integrated circuit is shown in Figure 2-1. Components are formed by a combination of processes:

- **doping** the substrate with impurities to create areas such as the n+ and p+ regions;
- adding or cutting away insulating glass (**silicon dioxide**, or SiO_2) on top of the substrate;
- adding wires made of polycrystalline silicon (**polysilicon**, also known as **poly**) or metal, insulated from the substrate by SiO_2.

The n-type and p-type regions and the polysilicon can be used to make wires as well as transistors, but metal (either copper or aluminum) is the primary material for wiring together transistors because of its superior electrical properties. There may be several levels of metal wiring to ensure that enough wires can be made to create all the necessary connections. Glass insulation lets the wires be fabricated on top of the substrate using processes like those used to form transistors. The integration of wires with components, which eliminates the need to manually wire together components on the substrate, was one of the key inventions that made the integrated circuit feasible.

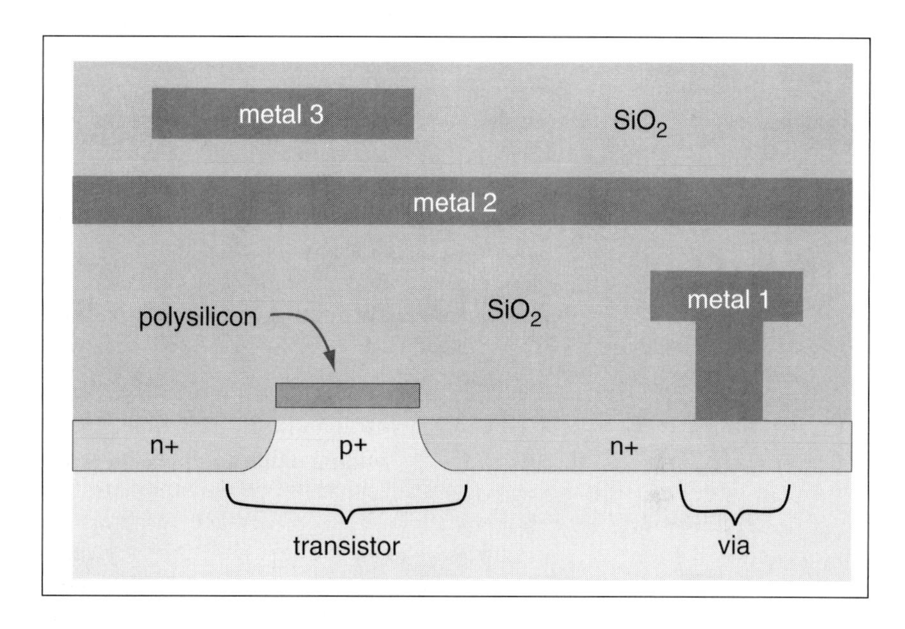

Figure 2-1 Cross-section of an integrated circuit.

manufacturing processes

Features are patterned on the wafer by a photolithographic process; the wafer is covered with light-sensitive material called **photoresist**, which is then exposed to light with the proper pattern. The patterns left by the photoresist after development can be used to control where SiO_2 is grown or where materials are placed on the surface of the wafer.

A layout contains summary information about the patterns to be made on the wafer. Photolithographic processing steps are performed using **masks** which are created from the layout information supplied by the designer. In simple processes there is roughly one mask per layer in a layout, though in more complex processes some masks may be built from several layers while one layer in the layout may contribute to several masks.

Transistors may be fabricated on the substrate by doping the substrate; transistors may also be fabricated within regions called **tubs** or **wells**. An n-type transistor is built in a p-doped, and a p-type transistor is built in an n-doped region. The wells prevent undesired conduction from the drain to the substrate. (Remember that the transistor type refers to the minority carrier which forms the inversion layer, so an n-type transistor pulls electrons out of a p-tub.) The **twin-tub process,** which starts from an undoped wafer and creates both types of tubs, has become the most commonly used CMOS process because it produces tubs with better electrical characteristics.

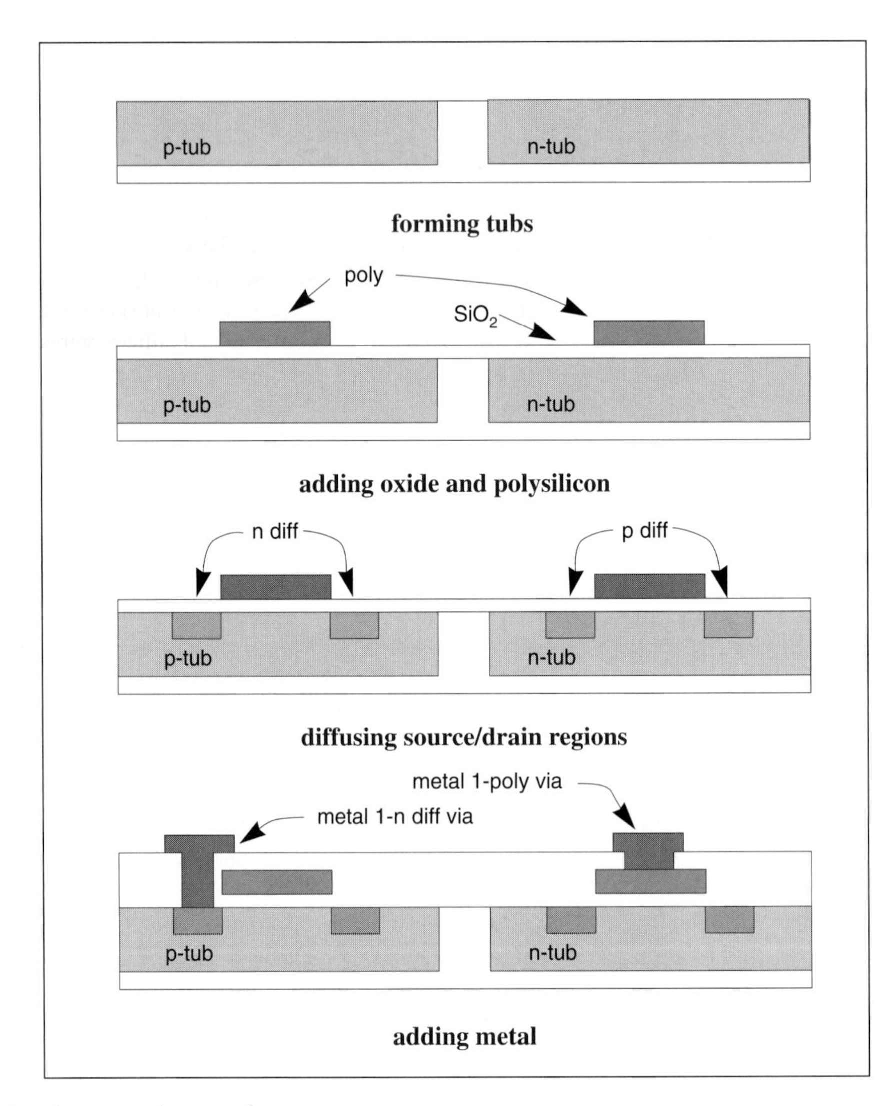

Figure 2-2 Some steps in processing a wafer.

Figure 2-2 illustrates important steps in a twin-tub process. Details can vary from process to process, but these steps are representative. The first step is to put tubs into the wafer at the appropriate places for the n-type and p-type wafers. Regions on the wafer are selectively doped by implanting ionized dopant atoms into the material, then heating the wafer to heal damage caused by ion implantation and further move the dopants by diffusion. The tub structure means that n-type and p-type wires cannot directly connect. Since the two diffusion wire types must exist in different type tubs, there is no way to build a via which can directly connect them. Connections must be made by a separate wire, usually metal, which runs over the tubs.

The next steps form an oxide covering of the wafer and the polysilicon wires. The oxide is formed in two steps: first, a thick field oxide is grown over the entire wafer. The field oxide is etched away in areas directly over transistors; a separate step grows a much thinner oxide which will form the insulator of the transistor gates. After the field and thin oxides have been grown, the polysilicon wires are formed by depositing polysilicon crystalline directly on the oxide.

Note that the polysilicon wires have been laid down before the diffusion wires were made—that order is critical to the success of MOS processing. Diffusion wires are laid down immediately after polysilicon deposition to create **self-aligned** transistors—the polysilicon masks the formation of diffusion wires in the transistor channel. For the transistor to work properly, there must be no gap between the ends of the source and drain diffusion regions and the start of the transistor gate. If the diffusion were laid down first with a hole left for the polysilicon to cover, it would be very difficult to hit the gap with a polysilicon wire unless the transistor were made very large. Self-aligned processing allows much smaller transistors to be built.

After the diffusions are complete, another layer of oxide is deposited to insulate the polysilicon and metal wires. Aluminum has long been the dominant interconnect material, but copper has now moved into mass production. Copper is a much better conductor than aluminum, but even trace amounts of it will destroy the properties of semiconductors. Chips with copper interconnect include a special protection layer between the substrate and the first layer of copper. That layer prevents the copper from entering the substrate during processing.

Multiple layers of metal interconnect are separated by silicon dioxide. Each layer of SiO_2 must be very smooth to allow the next layer of metal to be deposited without breaks. The deposition process may be somewhat uneven; in addition, the existing layers of metal form hills and valleys underneath the silicon dioxide. After an insulating layer is deposited, it is polished to a smooth surface using processes similar to those used to grind optical glass. This ensures that the next layer of interconnect will not have to form itself over an uneven surface that may cause breaks in the metal.

Holes are cut in the field oxide where vias to the substrate are desired. The metal 1 layer is then deposited where desired. The metal fills the cuts to make connections between layers. The metal 2 layer requires an additional oxidation/cut/deposition sequence. Another layer of silicon dioxide is deposited and then polished to form the base for the next layer of interconnect. Most modern processes offer at least four layers of metal.

After all the important circuit features have been formed, the chip is covered with a final **passivation layer** of SiO_2 to protect the chip from chemical contamination.

Modern VLSI fabrication processes in fact take hundreds of steps and dozens of masks. Generating very deep submicron structures with adequate electrical properties is much more challenging than was the task of building early MOS ICs whose transistors were multiple microns across. The large number of fabrication steps causes manufacturing times for 90 nm processes to take six to eight months; those lead times are likely to grow as line widths shrink.

2.3 Transistor Characteristics

During transistor operation we use the gate voltage to modulate the current through the channel. An ideal transistor would act as a switch, but realistic transistors have more complex characteristics that we can best understand by looking at the structure of the transistor. We will build only simple models of transistors here. They will be good enough for basic analysis of digital behavior but are far from complete models of transistors.

transistor structure

Figure 2-3 shows the cross-section of an n-type MOS transistor. (The name MOS is an anachronism. The first such transistors used a metal wire for a gate, making the transistor a sandwich of metal, silicon dioxide, and the semiconductor substrate. Even though transistor gates are now made of polysilicon, the name MOS has stuck.) An n-type transistor is embedded in a p-type substrate; it is formed at the intersection of an n-type diffusion region and a polysilicon wire. The region at the intersection, called the **channel**, is where the transistor action takes place. The channel connects to the two n-type wires which form the source and drain, but is itself doped to be p-type. The insulating silicon dioxide at the channel (called the **gate oxide**) is much thinner than it is away from the channel (called the **field oxide**); having a thin oxide at the channel is critical to the successful operation of the transistor.

transistor operation

The transistor works as a switch because the gate-to-source voltage modulates the amount of current that can flow between the source and

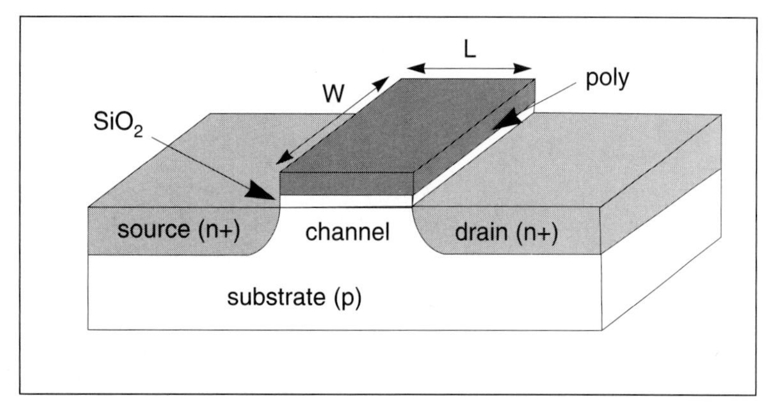

Figure 2-3 Cross-section of an n-type transistor.

drain. When the gate voltage (V_{gs}) is zero, the p-type channel is full of holes, while the n-type source and drain contain electrons. The p-n junction at the source terminal forms a diode, while the junction at the drain forms a second diode that conducts in the opposite direction. As a result, no current can flow from the source to the drain.

As V_{gs} rises above zero, the situation starts to change. While the channel region contains predominantly p-type carriers, it also has some n-type **minority carriers**. The positive voltage on the polysilicon which forms the gate attracts the electrons. Since they are stopped by the gate oxide, they collect at the top of the channel along the oxide boundary. At a critical voltage called the **threshold voltage** (V_t), enough electrons have collected at the channel boundary to form an **inversion layer**—a layer of electrons dense enough to conduct current between the source and the drain.

The size of the channel region is labeled relative to the direction of current flow: the channel **length** (L) is along the direction of current flow between source and drain, while the **width** (W) is perpendicular to current flow. The amount of current flow is a function of the W/L ratio, for the same reasons that bulk resistance changes with the object's width and length: widening the channel gives a larger cross-section for conduction, while lengthening the channel increases the distance current must flow through the channel.

P-type transistors have identical structures but complementary materials: trade p's and n's in Figure 2-3 and you have a picture of a p-type transistor. The p-type transistor conducts by forming an inversion region of holes in the n-type channel; therefore, the gate-to-source voltage must be negative for the transistor to conduct current.

transistor equations

We will do most of our design work with a simple model of the transistor's operation. This model ignores many effects but is good enough for basic digital design and simple enough for hand calculation. The behavior of both n-type and p-type transistors is described by two equations and two physical constants; the sign of one of the constants distinguishes the two types of transistors. The variables that describe a transistor's behavior are:

- V_{gs}—the gate-to-source voltage;
- V_{ds}—the drain-to-source voltage (remember that $V_{ds} = -V_{sd}$);
- I_d—the current flowing between the drain and source.

The constants that determine the magnitude of source-to-drain current in the transistor are:

- V_t—the transistor threshold voltage, which is positive for an n-type transistor and negative for a p-type transistor;
- k'—the transistor transconductance, which is positive for both types of transistors;
- W/L—the width-to-length ratio of the transistor.

Both V_t and k' are measured, either directly or indirectly, for a fabrication process. W/L is determined by the layout of the transistor, but since it does not change during operation, it is a constant of the device equations.

linear and saturated regions

The equations that govern the transistor's behavior are traditionally written to show the drain current as a function of the other parameters. A reasonably accurate model for the transistor's behavior, written in terms of the drain current I_d, divides operation into **linear** and **saturated**. For an n-type transistor, we have:

- *Linear region* $V_{ds} < V_{gs} - V_t$:

$$I_d = k'\frac{W}{L}\left[(V_{gs}-V_t)V_{ds}-\frac{1}{2}V_{ds}^2\right] \qquad \text{(EQ 2-1)}$$

- *Saturated region* $V_{ds} \geq V_{gs} - V_t$:

$$I_d = \frac{1}{2}k'\frac{W}{L}(V_{gs}-V_t)^2 \qquad \text{(EQ 2-2)}$$

For a p-type transistor, the drain current is negative and the device is on when V_{gs} is below the device's negative threshold voltage. Figure 2-4 plots these equations over some typical values for an n-type device.

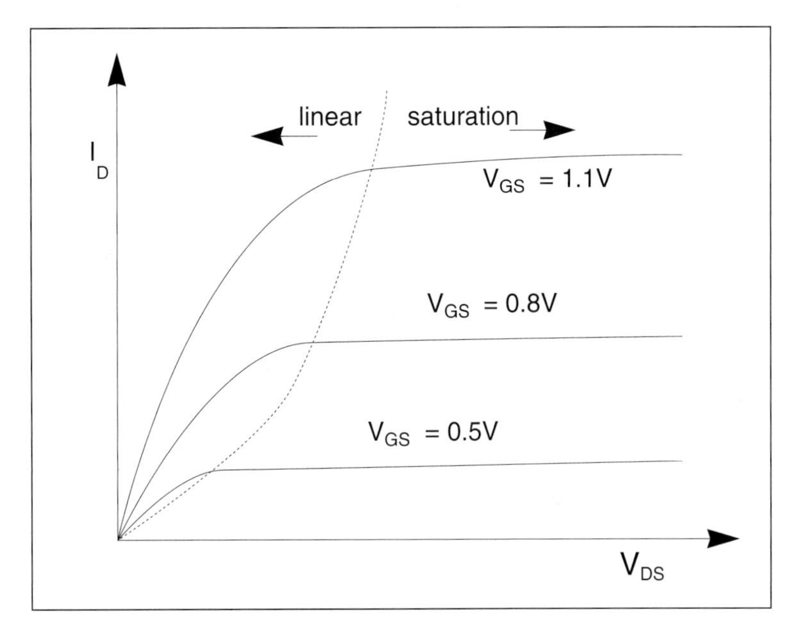

Figure 2-4 The I_d curves of an n-type transistor.

Each curve shows the transistor current as V_{gs} is held constant and V_{ds} is swept from 0 V to a large voltage.

The transistor's switch action occurs because the density of carriers in the channel depends strongly on the gate-to-substrate voltage. For $|V_{gs}| < |V_t|$, there are not enough carriers in the inversion layer to conduct an appreciable current. Beyond that point and until saturation, the number of carriers is directly related to V_{gs}: the greater the gate voltage applied, the more carriers are drawn to the inversion layer and the greater the transistor's conductivity.

The relationship between W/L and source-drain current is equally simple. As the channel width increases, more carriers are available to conduct current. As channel length increases, however, the drain-to-source electric field diminishes. V_{ds} is the potential energy available to push carriers from drain to source; as the distance from drain to source increases, it takes longer to push carriers across the transistor for a fixed V_{ds}, reducing current flow.

transistor parameters

Table 2-1 shows some sample values of k' and V_t for a 90 nm process. The next example calculates the current through a transistor.

Table 2-1 Sample transistor parameters for a 90 nm process.

	k'	V_t
n-type	$k'_n = 13 \mu A / V^2$	$0.14 V$
p-type	$k'_p = 7 \mu A / V^2$	$-0.21 V$

Example 2-1 Current through a transistor

A minimum-size transistor has a channel with aspect ratio of $L = 2$ and $W = 3$. Given this size of transistor and the 90 nm transistor characteristics, the current through a minimum-sized n-type transistor at the boundary between the linear and saturation regions when the gate is at the low voltage $V_{gs} = 0.25 V$ would be

$$I_d = \frac{1}{2}\left(13\frac{\mu A}{V^2}\right)\left(\frac{3\lambda}{2\lambda}\right)(0.25V\text{-}0.14V)^2 = 0.12\mu A .$$

The saturation current when the n-type transistor's gate is connected to a 1.0 V power supply would be

$$I_d = \frac{1}{2}\left(13\frac{\mu A}{V^2}\right)\left(\frac{3\lambda}{2\lambda}\right)(1.0V\text{-}0.14V)^2 = 7.2\mu A .$$

transistor capacitances

Real devices have parasitic elements that are necessary artifacts of the device structure. The transistor itself introduces significant **gate capacitance**, C_g. This capacitance, which comes from the parallel plates formed by the poly gate and the substrate, forms the majority of the capacitive load in small logic circuits. The total gate capacitance for a transistor is computed by measuring the area of the active region (or $W \times L$) and multiplying the area by the unit capacitance C_g.

Smaller but important parasitics are the **source/drain overlap capacitances**. During fabrication, the dopants in the source/drain regions diffuse in all directions, including under the gate as shown in Figure 2-5. The source/drain overlap region tends to be a larger fraction of the channel area in deep submicron devices.

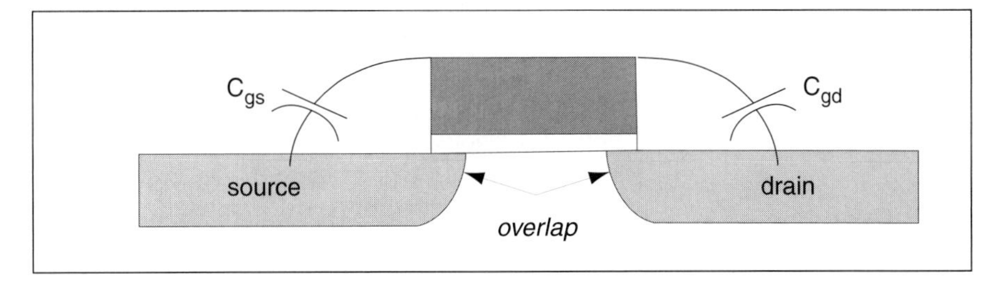

Figure 2-5 Parasitic capacitances from the gate to the source/drain overlap regions.

The overlap region is independent of the transistor length, so it is usually given in units of Farads per unit gate width. Then the total source overlap capacitance for a transistor would be

$$Cgs = C_{ol}W. \quad \text{(EQ 2-3)}$$

There is also a **gate/bulk overlap capacitance** due to the overhang of the gate past the channel and onto the bulk. The source and drain regions also have a non-trivial capacitance to the substrate and a very large resistance.

tub ties

An MOS transistor is actually a four-terminal device, but we have up to now ignored the electrical connection to the substrate. The substrates underneath the transistors must be connected to power supply terminals: the p-tub (which contains n-type transistors) to V_{SS} and the n-tub to V_{DD}. These connections are made by special vias called **tub ties**.

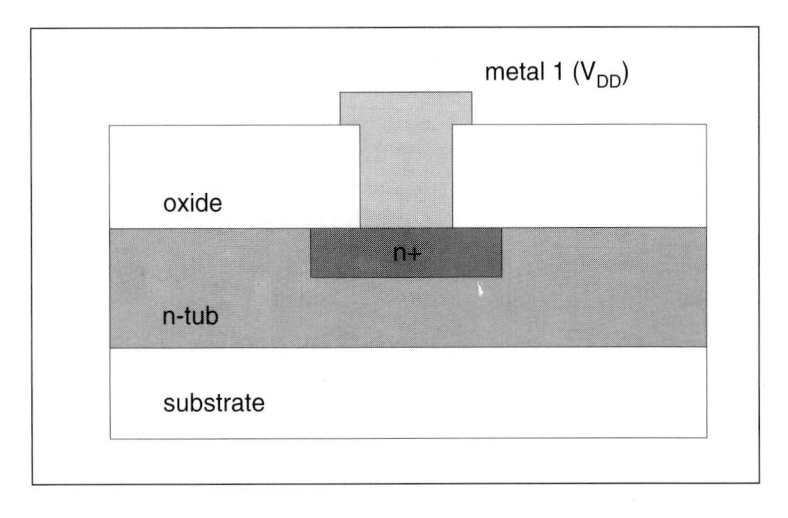

Figure 2-6 Cross-section of an n-tub tie.

Figure 2-6 shows the cross-section of a tub tie connecting to an n-tub. The tie connects a metal wire connected to the V_{DD} power supply directly to the substrate. The connection is made through a standard via cut. The substrate underneath the tub tie is heavily doped with n-type dopants (denoted as n+) to make a low-resistance connection to the tub. Using many tub ties in each tub makes a low-resistance connection between the tub and the power supply.

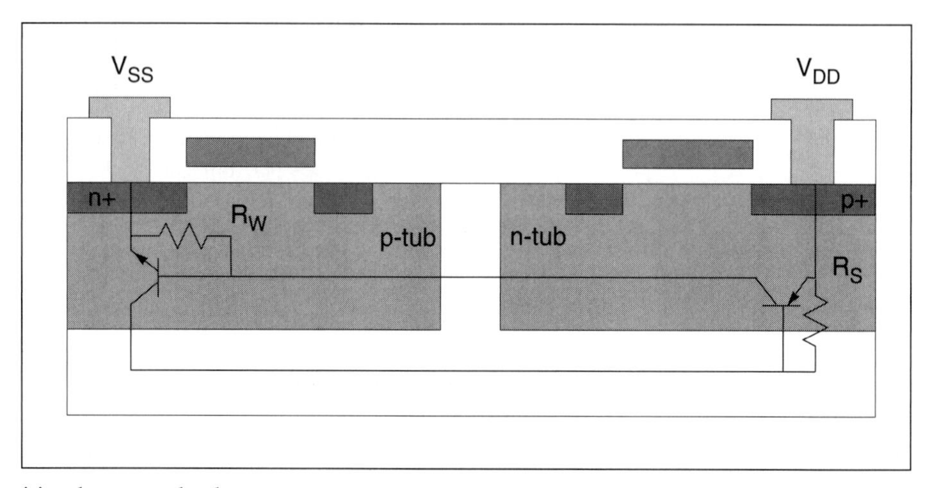

Figure 2-7 Parasitics that cause latch-up.

latch-up

If the connection has higher resistance, parasitic bipolar transistors can cause the chip to **latch-up**, inhibiting normal chip operation. Figure 2-7 shows a chip cross-section which might be found in an inverter or other logic gate. The MOS transistor and tub structures form parasitic bipolar transistors: npn transistors are formed in the p-tub and pnp transistors in the n-tub. Since the tub regions are not physically isolated, current can flow between these parasitic transistors along the paths shown as wires. Since the tubs are not perfect conductors, some of these paths include parasitic resistors; the key resistances are those between the power supply terminals and the bases of the two bipolar transistors.

parasitic SCRs and latch-up

The parasitic bipolar transistors and resistors create a parasitic **silicon-controlled rectifier**, or SCR. The schematic for the SCR and its behavior are shown in Figure 2-8. The SCR has two modes of operation. When both bipolar transistors are off, the SCR conducts essentially no current between its two terminals. As the voltage across the SCR is raised, it may eventually turn on and conducts a great deal of current with very little voltage drop. The SCR formed by the n- and p-tubs, when turned on, forms a high-current, low-voltage connection between

Figure 2-8 Characteristics of a silicon-controlled rectifier.

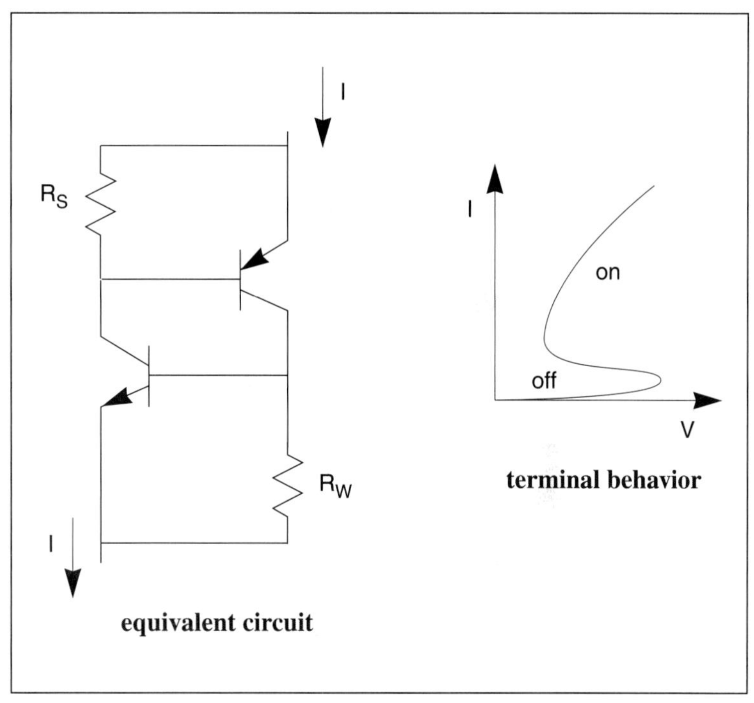

V_{DD} and V_{SS}. Its effect is to short together the power supply terminals. When the SCR is on, the current flowing through it floods the tubs and prevents the transistors from operating properly. In some cases, the chip can be restored to normal operation by disconnecting and then reconnecting the power supply; in other cases the high currents cause permanent damage to the chip.

leakage current

The drain current through the transistor does not drop to zero once the gate voltage goes below the threshold voltage. A variety of **leakage currents** continue to flow through various parts of the transistor, including a **subthreshold current** through the channel. Those currents are small, but they are becoming increasingly important in low-power applications. Not only do many circuits need to operate under very low current drains, but subthreshold currents are becoming relatively larger as transistor sizes shrink.

sources of leakage currents

Leakage currents come from a variety of effects within the transistor [Roy00]:

- The most important source of leakage is the **weak inversion current** (also known as the subthreshold current). This current is carried through the channel when the gate is below threshold.

- Reverse-biased pn junctions in the transistor, such as the one between the drain and its well, carry small **reverse bias currents**.
- **Drain-induced barrier lowering** is an interaction between the drain's depletion region and the source that causes the source's potential barrier to be lowered.
- **Gate-induced drain leakage** current happens around the high electric field under the gate/drain overlap.
- **Punchthrough currents** flow when the source and drain depletion regions connect within the channel.
- **Gate oxide tunneling currents** are caused by high electric fields in the gate.
- **Hot carriers** can be injected into the channel.

Different mechanisms dominate at different drain voltages, with weak inversion dominating at low drain voltages.

subthreshold current

The subthreshold current can be written as [Roy00]:

$$I_{sub} = ke^{\left(\frac{V_{gs}-V_t}{S/\ln 10}\right)}\left[1-e^{-qV_{ds}/kT}\right]. \quad \text{(EQ 2-4)}$$

The **subthreshold slope** S characterizes the magnitude of the weak inversion current in the transistor. The subthreshold slope is determined by a plot of log I_d *vs.* V_{gs}. An S value of 100 mV/decade indicates a very leaky transistor, with lower values indicating lower leakage currents.

The subthreshold current is a function of the threshold voltage V_t. The threshold voltage is primarily determined by the process. However, since the threshold voltage is measured relative to the substrate, we can adjust V_t by changing the substrate bias.

2.4 CMOS Logic Gates

In this section we will learn about CMOS logic, the basic building blocks of digital design. In order to understand logic gates, we also need to understand the basic characteristics of transistors. First, we will introduce the basic design of gates, since simple gates can be understood by thinking of the transistors as switches. We will then use simple models to analyze the performance and power consumption of CMOS logic. We will briefly consider how to size the transistors in a logic gate to drive large capacitive loads. We will also look at low-power gates and switch logic.

2.4.1 Static Complementary Gates

We can understand the basic structure of CMOS logic gates without a detailed understanding of transistor characteristics. For the moment, we will consider a transistor to be an ideal switch—when its gate is on the switch conducts, when the gate is on the switch does not conduct. The on condition is opposite for n-type and p-type transistors: an n-type is on when its gate voltage is positive relative to the substrate while a p-type is on when its gate voltage is negative relative to the substrate.

pullup and pulldown networks

The basic CMOS logic gate is known as the **static complementary gate**. It is divided into a **pullup network** made of p-type transistors and a **pulldown network** made of n-type transistors. The gate's output can be connected to V_{DD} by the pullup network or V_{SS} by the pulldown network. The two networks are complementary to ensure that the output is always connected to exactly one of the two power supply terminals at any time: connecting the output to neither would cause an indeterminate logic value at the output, while connecting it to both would cause not only an indeterminate output value, but also a low-resistance path from V_{DD} to V_{SS}.

The simplest type of gate is an inverter which computes the logical complement of its input—given an input value a the inverter computes a'. The structure of an inverter is shown in Figure 2-9; + stands for V_{DD} and the triangle stands for V_{SS} in the schematic. Its pulldown network is one n-type transistor that is on when a high voltage is presented to the inverter's input. Its pulldown transistor is a single p-type transistor that is on when the gate is presented a low input voltage.

The schematic for a two-input NAND gate is shown in Figure 2-10. The NAND function is (ab)'. Because the two n-type transistors in the pulldown network are in series, the output will be pulled down when both a and b are high. When either a or b is low, the pulldown network will be off and at least one of the p-type pullup transistors will be on. If both pullups are on, the gate still works and nothing is harmed. (We will see later that parallel and series transistors in the pullup/pulldown networks do affect delay.)

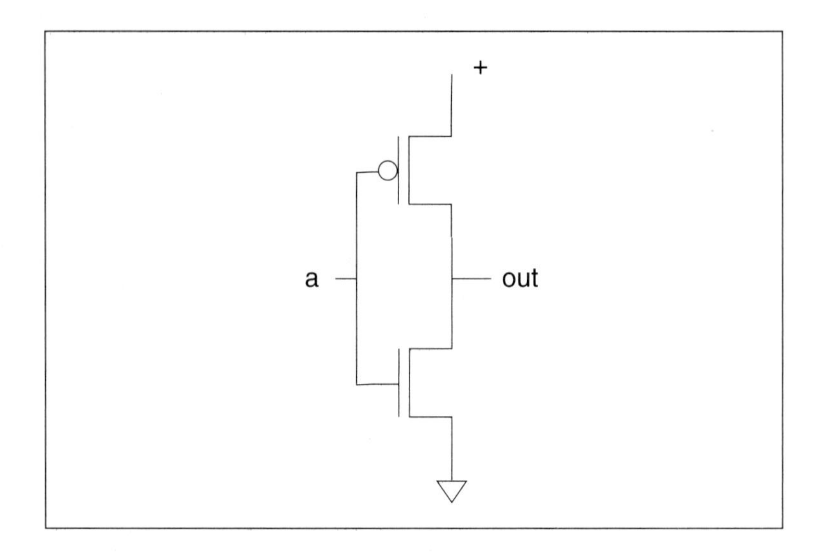

Figure 2-9 Transistor schematic of a static complementary inverter.

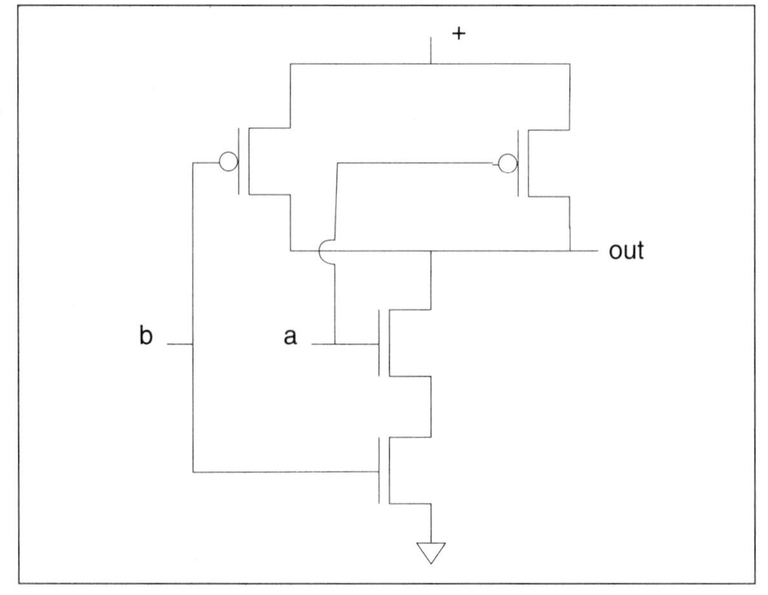

Figure 2-10 A static complementary NAND gate.

The schematic for a two-input NOR gate, which computes the function (a+b)', is shown in Figure 2-11. When either input is high, at least one of the pulldowns is on, providing a low output voltage. When both inputs are low the pullup network provides a path from V_{DD} to the output.

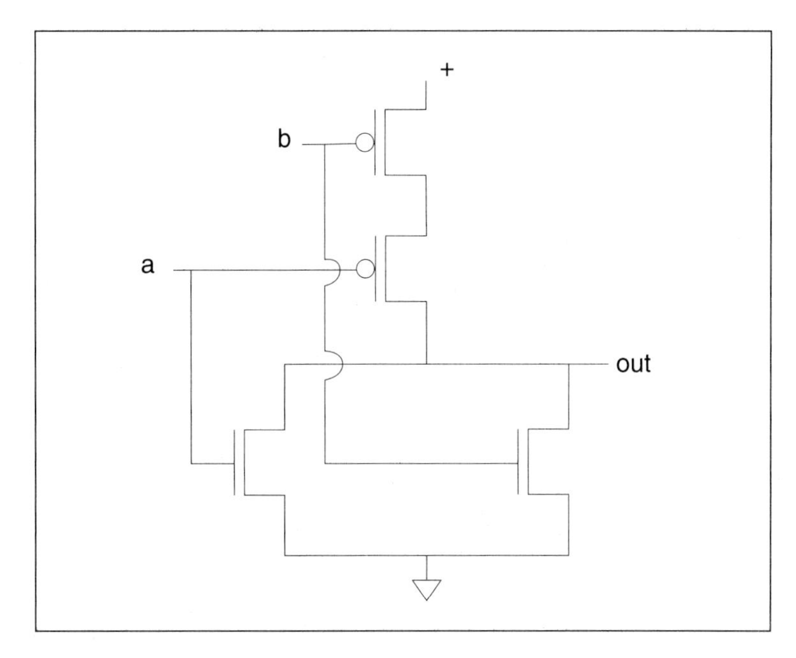

Figure 2-11 A static complementary NOR gate.

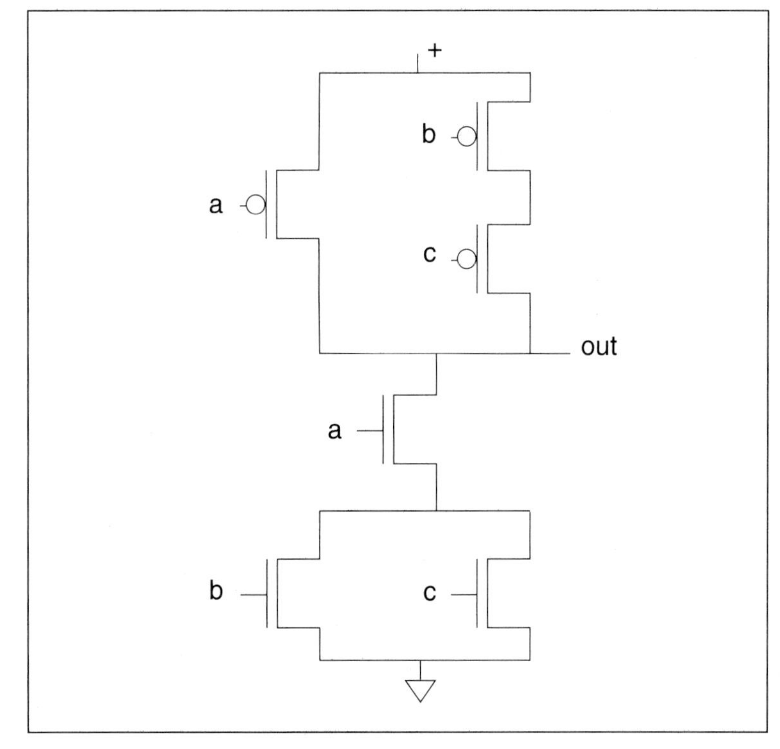

Figure 2-12 A static complementary gate that computes [a(b+c)]’.

Gates can be designed for functions other than NAND and NOR by designing the proper pullup and pulldown networks. Networks that are series-parallel combinations of transistors can be designed directly from the logic expression the gate is to implement. In the pulldown network, series-connected transistors or subnetworks implement AND functions in the expression and parallel transistors or subnetworks implement OR functions. The converse is true in the pullup network because p-type transistors are off when their gates are high. Consider the design of a two-input NAND gate as an example. To design the pulldown network, write the gate's logic expression to have negation at the outermost level: *(ab)*' in the case of the NAND. This expression specifies a series-connected pair of n-type transistors. To design the pullup network, rewrite the expression to have the inversion pushed down to the innermost literals: *a*' + *b*' for the NAND. This expression specifies a parallel pair of p-type transistors, completing the NAND gate design of Figure 2-10. Figure 2-12 shows the topology of a gate which computes [*a*(*b*+*c*)]': the pulldown network is given by the expression, while the rewritten expression *a*' + (*b*'*c*') determines the pullup network.

Figure 2-13 Constructing the pullup network from the pulldown network.

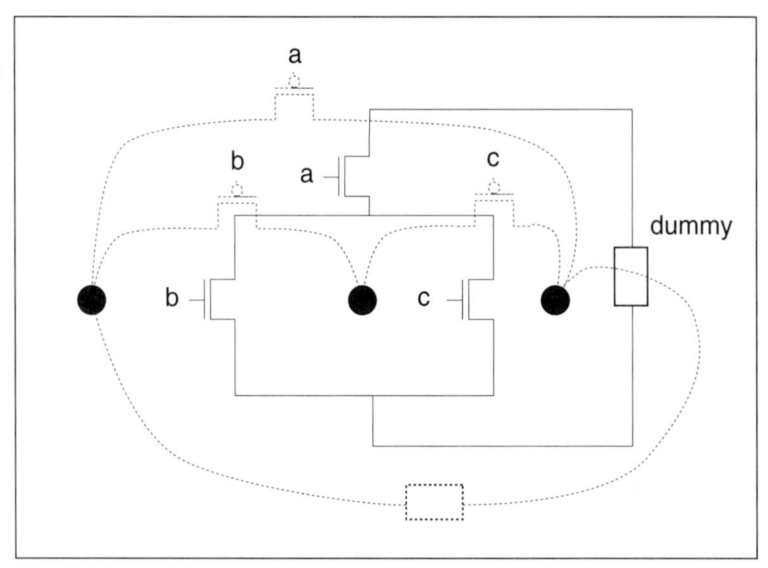

You can also construct the pullup network of an arbitrary logic gate from its pulldown network, or vice versa, because they are **duals**. Figure 2-13 illustrates the dual construction process using the pulldown network of Figure 2-12. First, add a dummy component between the output and the V_{SS} (or V_{DD}) terminals. Assign a node in the dual network for each region, including the area not enclosed by wires, in the non-dual graph. Finally, for each component in the non-dual network, draw a dual component which is connected to the nodes in the regions separated by the non-dual component. The dual component of an n-type transistor is a p-type, and the dual of the dummy is the dummy. You can check your work by noting that the dual of the dual of a network is the original network.

Common forms of complex logic gates are **and-or-invert** (AOI) and **or-and-invert** (OAI) gates, both of which implement sum-of-products/product-of-sums expressions. The function computed by an AOI gate is best illustrated by its logic symbol, shown in Figure 2-14: groups of inputs are ANDed together, then all products are ORed together and inverted for output. An AOI-21 gate, like that shown in the figure, has two inputs to its first product and one input (effectively eliminating the AND gate) to its second product; an AOI-121 gate would have two one-input products and one two-input product. An OAI gate, in comparison, computes an expression in product-of-sums form: it generates sums in the first stage which are then ANDed together and inverted.

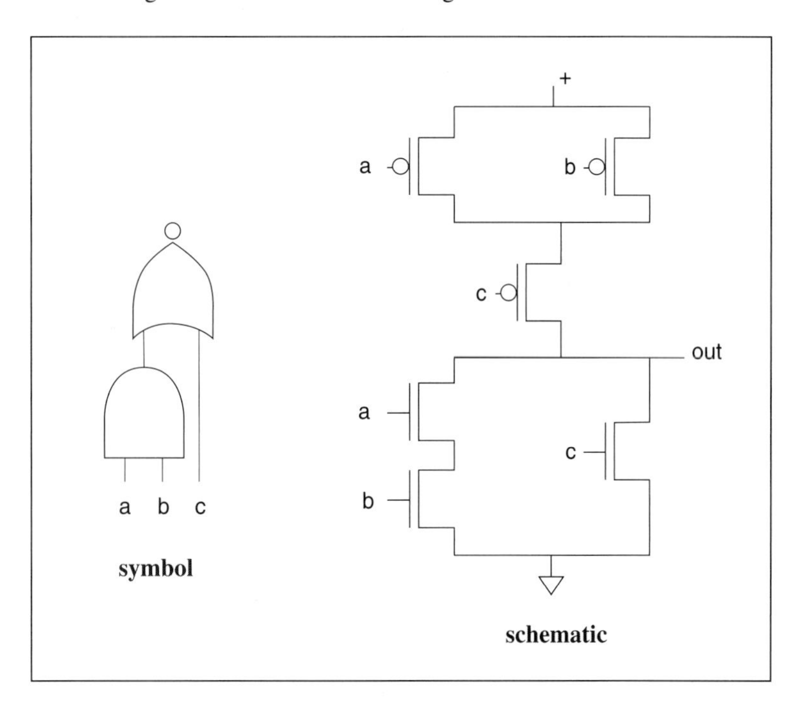

Figure 2-14 An and-or-invert-21 (AOI-21) gate.

2.4.2 Gate Delay

Delay is one of the most important properties of a logic gate. Gate delay is measured as the time required for the gate to change its output voltage from a logic 0 to a logic 1 or *vice versa.* So before analyzing gate delay in detail, let's consider the mapping between logic values and voltages.

logic 0 and 1 voltages

As Figure 2-15 shows, a range of voltages near V_{DD} corresponds to logic 1 and a band around V_{SS} corresponds to logic 0. The range in between is X, the unknown value. Although signals must swing through the X region while the chip is operating, no node should ever achieve X as its final value.

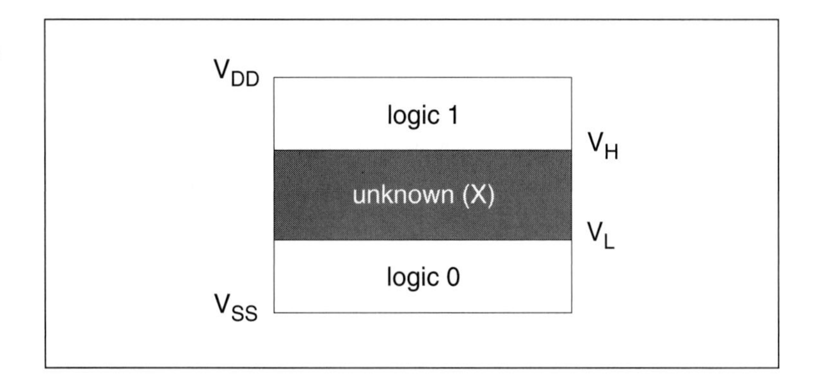

Figure 2-15 How voltages correspond to logic levels.

We want to calculate the upper boundary of the logic 0 region and the lower boundary of the logic 1 region. In fact, the situation is slightly more complex, as shown in Figure 2-16, because we must consider the logic levels produced at outputs and required at inputs. Given our logic gate design and process parameters, we can guarantee that the maximum voltage produced for a logic 0 will be some value V_{OL} and that the minimum voltage produced for a logic 0 will be V_{OH}. These same constraints place limitations on the input voltages which will be interpreted as a logic 0 (V_{IL}) and logic 1 (V_{IH}). If the gates are to work together, we must ensure that $V_{OL} < V_{IL}$ and $V_{OH} > V_{IH}$.

The output voltages produced by a static, complementary gate are V_{DD} and V_{SS}, so we know that the output voltages will be acceptable as input voltages to the next stage of logic. (That isn't true of all gate circuits; ratioed logic circuits like pseudo-nMOS produce a logic 0 level well above V_{SS}.) We need to compute the values of V_{IL} and V_{IH}, and to do the computation, we need to define those values. A standard definition is based on the transfer characteristic of the inverter—its output voltage as

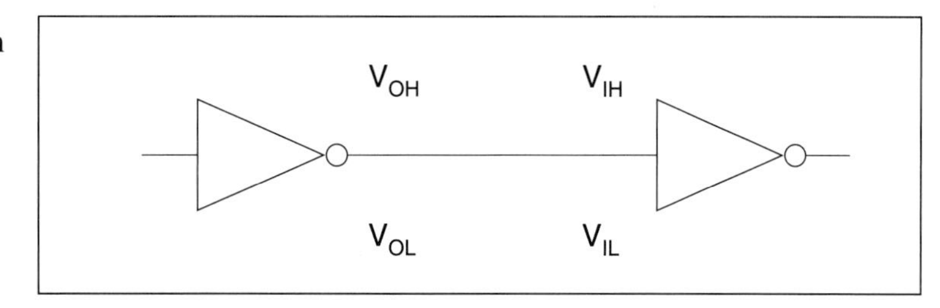

Figure 2-16 Logic levels on cascaded gates.

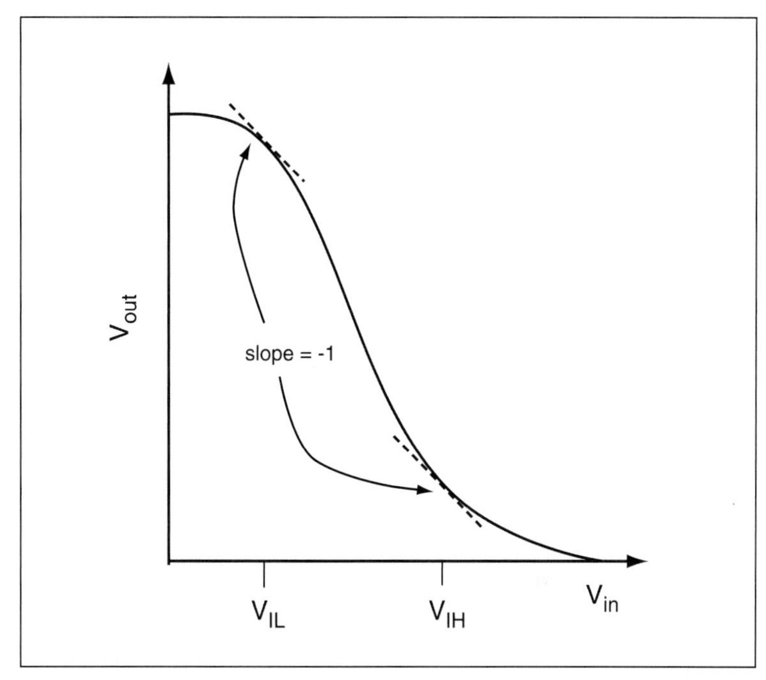

Figure 2-17 Voltage transfer curve of an inverter.

a function of its input voltage, assuming that the input voltage and all internal voltages and currents are at equilibrium. Figure 2-18 shows the circuit we will use to measure an inverter's transfer characteristic. We apply a sequence of voltages to the input and measure the voltage at the output. Alternatively, we can solve the circuit's voltage and current equations to find V_{out} as a function of V_{in}: we equate the drain currents of the two transistors and set their gate voltages to be complements of each other (since the n-type's gate voltage is measured relative to V_{SS} and the p-type's to V_{DD}).

transfer curve and threshold voltages

Figure 2-17 shows a typical **transfer characteristic** of an inverter with minimum-size transistors for both pullup and pulldown. We choose values for V_{IL} and V_{IH} as the points at which the curve's tangent has a slope

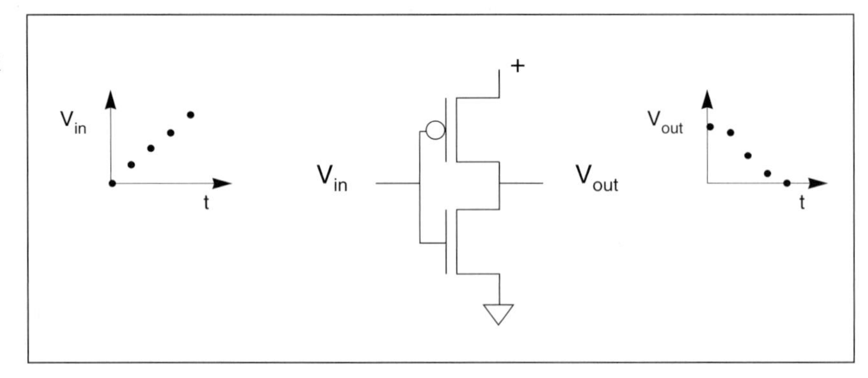

Figure 2-18 The inverter circuit used to measure transfer characteristics.

of -1. Between these two points, the inverter has high gain—a small change in the input voltage causes a large change in the output voltage. Outside that range, the inverter has a gain less than 1, so that even a large change at the input causes only a small change at the output, attenuating the noise at the gate's input.

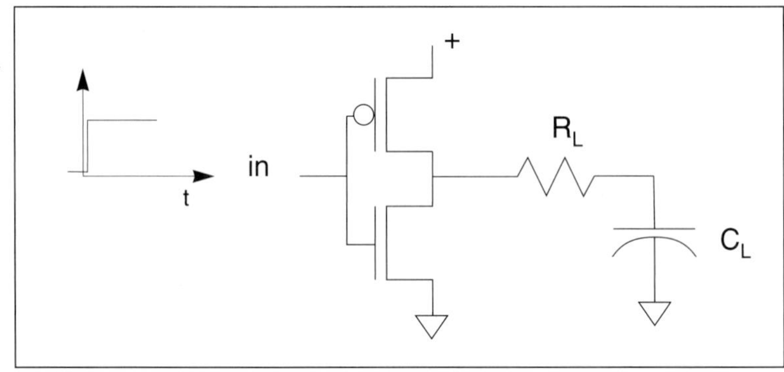

Figure 2-19 An idealized input and load used for delay analysis.

delay measurement

When analyzing delay, we will assume an idealized input. As shown in Figure 2-19, we will assume that the inverter's input is a step such as might be generated by an ideal voltage source. In practice, the input will be generated by another logic gate and the input waveform will have a more complex shape. We will also assume that the load presented by the gate can be modeled by a single resistor and capacitor. Once again, the load on a real gate is created by another logic gate and is of somewhat more complex form. But the RC load model is good enough to make a variety of basic design decisions.

tau delay model

A simple but effective model of delay in logic gates is the τ **(tau) model** of Mead and Conway [Mea80]. This model reduces the delay of the gate to an RC time constant which is given the name τ. As the sizes of the transistors in the gate are increased, the delay scales as well.

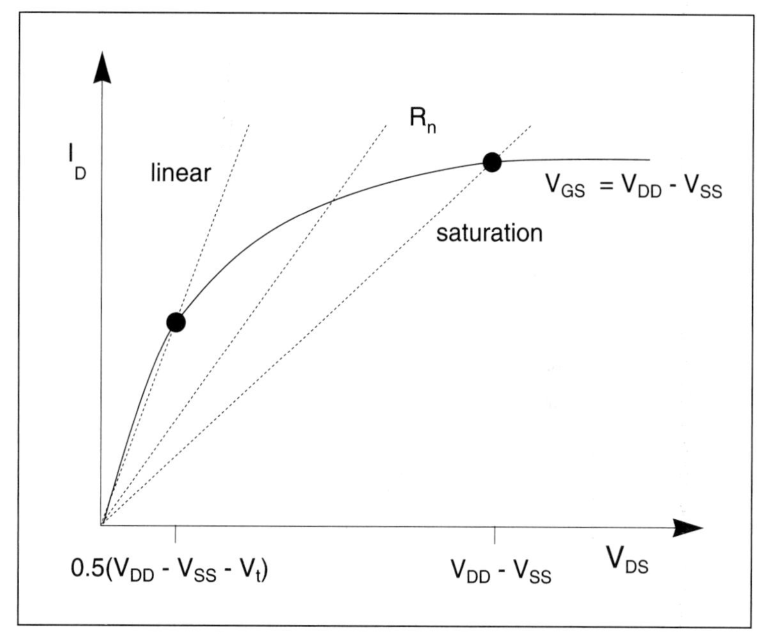

Figure 2-20 How to approximate a transistor with a resistor.

Table 2-2 Effective resistance values for minimum-size transistors in our 90 nm process.

type	V_{DD}-V_{SS} = 1V
R_n	**11.1 kΩ**
R_p	**24.0 kΩ**

At the heart of the τ model is the assumption that the pullup or pulldown transistor can be modeled as a resistor. The transistor does not obey Ohm's law as it drives the gate's output, of course. As Figure 2-21 shows, the pulldown spends the first part of the 1→ 0 transition in the saturation region, then moves into the linear region. But the resistive model will give sufficiently accurate results to both estimate gate delay and to understand the sources of delay in a logic circuit.

equivalent resistance of transistor

How do we choose a resistor value to represent the transistor over its entire operating range? A standard resistive approximation for a transistor is to measure the transistor's resistance at two points in its operation and take the average of the two values [Hod83]. We find the resistance by choosing a point along the transistor's I_d *vs.* V_{ds} curve and computing

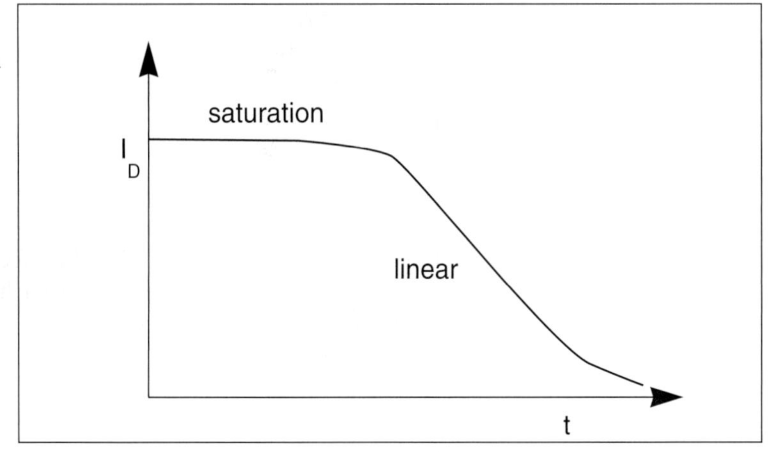

Figure 2-21 Current through the pulldown during a $1 \rightarrow 0$ transition.

the ratio V/I, which is equivalent to measuring the slope of a line between that point and the origin. Figure 2-20 shows the approximation points for an n-type transistor: the inverter's maximum output voltage, $V_{DS} = V_{DD} - V_{SS}$, where the transistor is in the saturation region; and the middle of the linear region, $V_{DS} = (V_{DD}-V_{SS}-V_t)/2$. We will call the first value $R_{sat} = V_{sat}/I_{sat}$ and the second value $R_{lin} = V_{lin}/I_{lin}$. This gives the basic formula

$$R_n = \left(\frac{V_{sat}}{I_{sat}} + \frac{V_{lin}}{I_{lin}}\right)/2 \quad \text{(EQ 2-5)}$$

for which we must find the Vs and Is.

The current through the transistor at the saturation-region measurement point is

$$I_{sat} = \frac{1}{2}k'\frac{W}{L}(V_{DD}-V_{SS}-V_t)^2 . \quad \text{(EQ 2-6)}$$

The voltage across the transistor at that point is

$$V_{sat} = V_{DD} - V_{SS}. \quad \text{(EQ 2-7)}$$

At the linear region point,

$$V_{lin} = (V_{DD}-V_{SS}-V_t)/2, \quad \text{(EQ 2-8)}$$

so the drain current is

$$I_{lin} = k'\frac{W}{L}\left[\frac{1}{2}(V_{DD}-V_{SS}-V_t)^2-\frac{1}{2}\left(\frac{V_{DD}-V_{SS}-V_t}{2}\right)^2\right]$$
$$= \frac{3}{8}k'\frac{W}{L}(V_{DD}-V_{SS}-V_t)^2 \qquad \text{(EQ 2-9)}$$

We can compute the effective resistances of transistors in the 90 nm process by plugging in the technology values of Table 2-1. The resistance values for minimum-size n-type and p-type transistors are shown in Table 2-2 for a 1V power supply. The effective resistance of a transistor is scaled by *L*/*W*. The p-type transistor has about twice the effective resistance of an n-type transistor for this set of process parameters.

Given these resistance values, we can then analyze the delay and transition time of the gate.

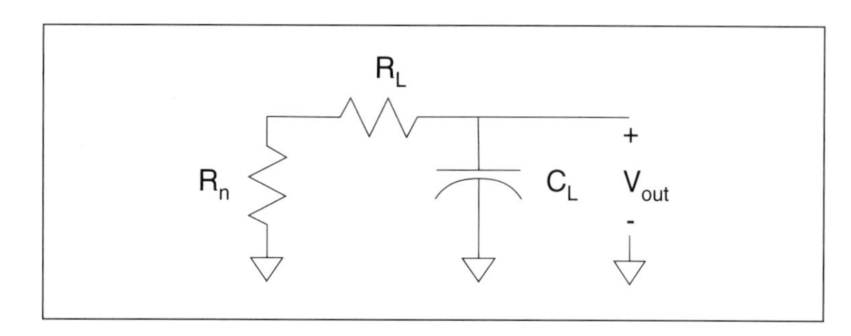

Figure 2-22 The circuit model for τ model delay.

rise/fall time calculation

We can now develop the τ model that helps us compute transition time. Figure 2-22 shows the circuit model we use: R_n is the transistor's effective resistance while R_L and C_L are the load. The capacitor has an initial voltage of V_{DD}. The transistor discharges the load capacitor from V_{DD} to V_{SS}; the output voltage as a function of time is

$$V_{out}(t) = V_{DD}e^{-t/[(R_n + R_L)C_L]} . \qquad \text{(EQ 2-10)}$$

We typically use R_L to represent the resistance of the wire which connects the inverter to the next gate; in this case, we'll assume that $R_L = 0$, simplifying the total resistance to $R = R_n$.

We generally measure transition time as the interval between the time at which $V_{out} = 0.9V_{DD}$ and $V_{out} = 0.1V_{DD}$; let's call these times t_1 and t_2.

Then

$$t_f = t_2 - t_1 = -(R_n + R_L)C_L ln\frac{0.1}{0.9} = 2.2(R_n + R_L)C_L. \quad \text{(EQ 2-11)}$$

The next example illustrates how to compute the transition time using the τ model.

Example 2-2 Inverter delay and transition time using the τ model

Once the effective resistance of a transistor is known, delay calculation is easy. What is a minimum inverter delay and fall time with 90 nm process parameters? Assume a minimum-size pulldown, no wire resistance, and a capacitive load equal to two minimum-size transistors' gate capacitance. First, the τ model parameters:

$$R_n = 11.1k\Omega$$
$$C_L = 0.12\text{fF}$$

Then the fall time is

$$t_f = 2.2 \cdot 11.1k\Omega \cdot 0.12\times10^{-15} = 2.9ps$$

If the transistors are not minimum size, their effective resistance is scaled by *L*/*W*. To compute the delay through a more complex gate, such as a NAND gate, compute the effective resistance of the pullup/pulldown network using the standard Ohm's law simplifications, then plug the effective *R* into the delay formula.

---

gate delay and transistor sizes

This simple RC analysis tells us two important facts about gate delay. First, if the pullup and pulldown transistor sizes are equal, the $0 \rightarrow 1$ transition will be about one-half to one-third the speed of the $1 \rightarrow 0$ transition. That observation follows directly from the ratio of the n-type and p-type effective resistances. Put another way, to make the high-going and low-going transition times equal, the pullup transistor must be twice to three times as wide as the pulldown. Second, complex gates like NANDs and NORs require wider transistors where those transistors are connected in series. A NAND's pulldowns are in series, giving an effective pulldown resistance of $2R_n$. To give the same delay as an inverter, the NAND's pulldowns must be twice as wide as the inverter's pulldown. The NOR gate has two p-type transistors in series for the pullup network. Since a p-type transistor must be two to three times wider than an n-type transistor to provide equivalent resistance, the pullup network of a NOR can take up quite a bit of area.

50% point model

Another way to measure delay is to calculate the time required to reach the 50% point. Then

$$0.5 = e^{-t_d/[(R_n + R_L)C_L]}, \quad \text{(EQ 2-12)}$$

$$t_d = -(R_n + R_L)C_L \ln 0.5 = 0.69(R_n + R_L)C_L. \quad \text{(EQ 2-13)}$$

Notice that this formula has the same basic form as the fall time calculation but uses a different constant.

power modeling using current source

Yet another model is the **current source model**, which is sometimes used in power/delay studies because of its tractability. If we assume that the transistor acts as a current source whose V_{gs} is always at the maximum value, then the delay can be approximated as

$$t_f = \frac{C_L(V_{DD}-V_{SS})}{I_d} = \frac{C_L(V_{DD}-V_{SS})}{0.5k'(W/L)(V_{DD}-V_{SS}-V_t)^2}. \quad \text{(EQ 2-14)}$$

fitted model

A third type of model is the **fitted model**. This approach measures circuit characteristics and fits the observed characteristics to the parameters in a delay formula. This technique is not well-suited to hand analysis but it is easily used by programs that analyze large numbers of gates.

The fundamental reason for developing an RC model of delay is that we often can't afford to use anything more complex. Full circuit simulation of even a modest-size chip is infeasible: we can't afford to simulate even one waveform, and even if we could, we would have to simulate all possible inputs to be sure we found the worst-case delay. The RC model lets us identify sections of the circuit which probably limit circuit performance; we can then, if necessary, use more accurate tools to more closely analyze the delay problems of that section. However, if you are interested in the detailed design of a particular circuit, tools such as Spice provide much more accurate results on the behavior of circuits.

body effect and delay

Body effect is the modulation of threshold voltage by a difference between the voltage of the transistor's source and the substrate—as the source's voltage rises, the threshold voltage also rises. This effect can be modeled by a capacitor from the source to the substrate's ground as shown in Figure 2-23. To minimize body effect, we want to drive that capacitor to 0 voltage as soon as possible. If there is one transistor between the gate's output and the power supply, body effect is not a problem, but series transistors in a gate pose a challenge. Not all of the gate's input signals may reach their values at the same time—some signals may arrive earlier than others. If we connect early-arriving signals to the transistors nearest the power supply and late-arriving signals to transistors nearest the gate output, the early-arriving signals will discharge the body effect capacitance of the signals closer to the output. This simple optimization can significantly improve gate delay [Hil89].

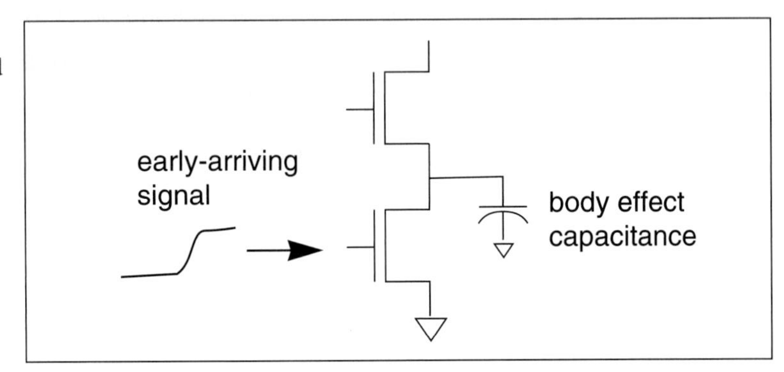

Figure 2-23 Body effect and signal ordering.

2.4.3 Power Consumption

Analyzing the power consumption of an inverter provides an alternate window into the cost and performance of a logic gate. Static, complementary CMOS gates are remarkably efficient in their use of power to perform computation. However, leakage increasingly threatens to drive up chip power consumption.

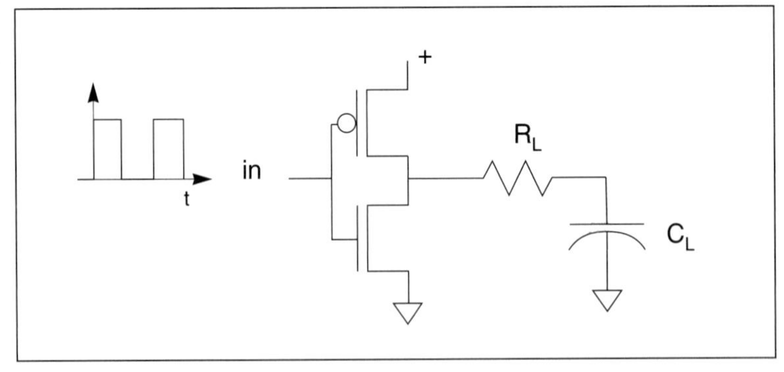

Figure 2-24 Circuit used for power consumption analysis.

power consumption of inverter

We will analyze an inverter with a capacitor connected to its output. However, to analyze power consumption we must consider both the pullup and pulldown phases of operation. The model circuit is shown in Figure 2-24. The first thing to note about the circuit is that it has almost no steady-state power consumption. After the output capacitance has been fully charged or discharged, only one of the pullup and pulldown transistors is on. The following analysis ignores the leakage current.

Power is consumed when gates drive their outputs to new values. Surprisingly, the power consumed by the inverter is independent of the sizes/resistances of its pullup and pulldown transistors—power consumption depends only on the size of the capacitive load at the output and the rate at which the inverter's output switches.

The current through the capacitor and the voltage across it are:

$$i_{CL}(t) = \frac{V_{DD}-V_{SS}}{R_p} e^{-(t/R_pC_L)}, \quad \text{(EQ 2-15)}$$

$$v_{CL}(t) = (V_{DD}-V_{SS})[1-e^{-(t/R_pC_L)}]. \quad \text{(EQ 2-16)}$$

energy consumption

So, the energy required to charge the capacitor is:

$$E_C = \int_0^{\infty} i_{C_L}(t) v_{C_L}(t) dt \quad \text{(EQ 2-17)}$$

$$= C_L(V_{DD}-V_{SS})^2 \left(e^{-t/R_pC_L} - \frac{1}{2} e^{-2t/R_pC_L} \right) \Bigg|_0^{\infty}$$

$$= \frac{1}{2} C_L (V_{DD}-V_{SS})^2$$

The energy consumed in discharging the capacitor can be calculated the same way. The discharging energy consumption is equal to the charging power consumption: $1/2\, C_L(V_{DD}-V_{SS})^2$. A single cycle requires the capacitor to both charge and discharge, so the total energy consumption is $C_L(V_{DD}-V_{SS})^2$.

power consumption

Power is energy per unit time, so the power consumed by the circuit depends on how frequently the inverter's output changes. The worst case is that the inverter alternately charges and discharges its output capacitance. This sequence takes two clock cycles. The clock frequency is $f = 1/t$. The total power consumption is

$$fC_L(V_{DD}-V_{SS})^2 . \quad \text{(EQ 2-18)}$$

power and operating frequency

Power consumption in CMOS circuits depends on the frequency at which they operate—this source of power consumption is known as **dynamic power**. Power consumption depends on the sizes of the transistors in the circuit only in that the transistors largely determine C_L. The current through the transistors, which is determined by the transistor *W/Ls*, doesn't determine power consumption, though the available transistor current does determine the maximum speed at which the circuit can run, which indirectly determines power consumption.

This analysis assumes that the CMOS transistor is a perfect switch. While traditional CMOS technologies provided transistors that leaked very little current when they were off, very deep submicron technologies are not quite so good. These transistors can leak substantial amounts of current. In a 90 nm process a transistor with a standard threshold leaks about 10 nA per micron of width. (A high-threshold voltage will leak only about 1 nA per micron of width but it also increases the delay of the gate.) While a traditional CMOS gate's power consumption can be cut to negligible levels simply by not changing its inputs, a gate built from leaky transistors will consume current even when its inputs are stable. As a result, the power supply voltages must be removed in order to stop leakage.

operating voltage and power

Static complementary gates can operate over a wide range of voltages, allowing us to trade delay for power consumption. To see how performance and power consumption are related, let's consider changing the power supply voltage from its original value V to a new V'. It follows directly from Equation 2-18 that the ratio of power consumptions P'/P is proportional to V'^2/V^2. When we compute the ratio of rise times t'_r/t_r the only factor to change with voltage is the transistor's equivalent resistance R, so the change in delay depends only on R'/R. If we compute the new effective resistance, we find that $t'_r/t_r \propto V/V'$. So as we reduce power supply voltage, dynamic power consumption goes down faster than does delay.

speed-power product

The **speed-power product**, also known as the **power-delay product**, is an important measure of the quality of a logic circuit family. Since delay can in general be reduced by increasing power consumption, looking at either power or delay in isolation gives an incomplete picture.

The speed-power product for static CMOS is easy to calculate. If we ignore leakage current and consider the speed and power for a single inverter transition, then we find that the speed-power product SP is

$$SP = \frac{1}{f}P = CV^2. \quad \text{(EQ 2-19)}$$

voltage scaling

The speed-power product for static CMOS is independent of the operating frequency of the circuit. It is, however, a quadratic function of the power supply voltage. This result suggests an important method for power consumption reduction known as **voltage scaling**: we can often reduce power consumption by reducing the power supply voltage and adding parallel logic gates to make up for the lower performance. Since the power consumption shrinks more quickly than the circuit delay when the voltage is scaled, voltage scaling is a powerful technique.

2.4.4 Driving Large Loads

Logic delay increases as the capacitance attached to the logic's output becomes larger. In many cases, one small logic gate is driving an equally small logic gate, roughly matching drive capability to load. However, there are several situations in which the capacitive load can be much larger than that presented by a typical gate:

- driving a signal connected off-chip;
- driving a long signal wire;
- driving a clock wire which goes to many points on the chip.

The obvious answer to driving large capacitive loads is to increase current by making wider transistors. However, this solution begs the question—those large transistors simply present a large capacitive load to the gate which drives them, pushing the problem back one level of logic. It is inevitable that we must eventually use large transistors to drive the load, but we can minimize delay along the path by using a sequence of successively larger drivers.

exponentially-tapered drivers

The driver chain with the smallest delay to drive a given load is exponentially tapered—each stage supplies e times more current than the last [Jae75]. In the chain of inverters of Figure 2-25, each inverter can produce α times more current than the previous stage (implying that its pullup and pulldown are each α times larger). If C_g is the minimum-size load capacitance, the number of stages n is related to α by the formula

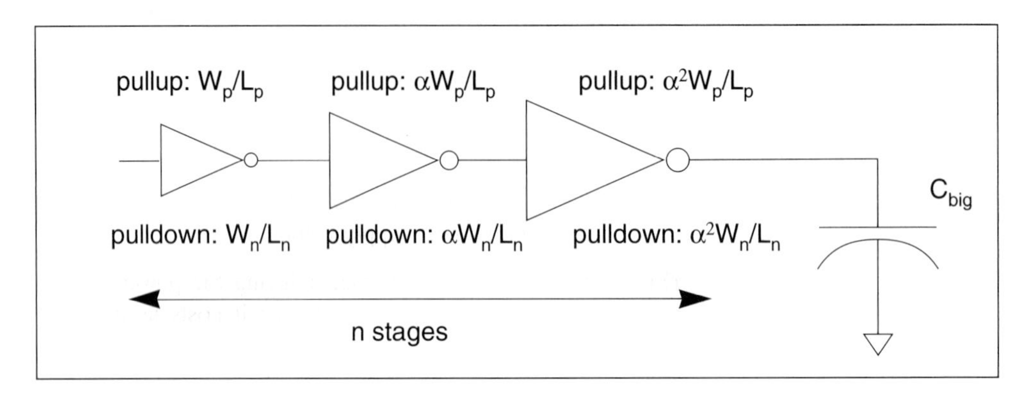

Figure 2-25 Cascaded inverters driving a large capacitive load.

$\alpha = (C_{big}/C_g)^{1/n}$. The time to drive a minimum-size load is t_{min}. We want to minimize the total delay through the driver chain:

$$t_{tot} = n\left(\frac{C_{big}}{C_g}\right)^{1/n} t_{min}. \quad \text{(EQ 2-20)}$$

To find the minimum, we set $\frac{dt_{tot}}{dn} = 0$, which gives

$$n_{opt} = ln\left(\frac{C_{big}}{C_g}\right). \quad \text{(EQ 2-21)}$$

When we substitute the optimal number of stages back into the definition of α, we find that the optimum value is at $\alpha = e$. Of course, n must be an integer, so we will not in practice be able to implement the exact optimal circuit. However, delay changes slowly with n near the optimal value, so rounding n to the floor of n_{opt} gives reasonable results.

2.4.5 Low-Power Gates

There are several different strategies for building low-power gates. Which one is appropriate for a given design depends on the required performance and power as well as the fabrication technology. In very deep submicron technologies leakage current has become a major consumer of power.

power supply voltage and power consumption

Of course, the simplest way to reduce the operating voltage of a gate is to connect it to a lower power supply. We saw the relationship between power supply voltage and power consumption in Section 2.4.3:

- For large V_t, Equation 2-14 tells us that delay changes linearly with power supply voltage.
- Equation 2-18 tells us that power consumption varies quadratically with power supply voltage.

This simple analysis tells us that reducing the power supply saves us much more in power consumption than it costs us in gate delay. Of course, the performance penalty incurred by reducing the power supply voltage must be taken care of somewhere in the system. One possible solution is architecture-driven voltage scaling, which replicates logic to make up for slower operating speeds.

multiple power supply voltages

It is also possible to operate different gates in the circuit at different voltages: gates on the critical delay path can be run at higher voltages while gates that are not delay-critical can be run at lower voltages. However, such circuits must be designed very carefully since passing logic values between gates running at different voltages may run into noise limits.

low-power gate topologies

After changing power supply voltages, the next step is to use different logic gate topologies. An example of this strategy is the differential current switch logic (DCSL) gate [Roy00] shown in Figure 2-26. It uses nMOS pulldown networks for both logic 0 and logic 1. The DCSL gate disconnects the n-type networks to reduce their power consumption. This gate is precharged with Q and Q' low. When the clock goes high, one of Q or Q' will be pulled low by the n-type evaluation tree, and that value will be latched by the cross-coupled inverters.

leakage and turning gates off

After these techniques have been tried, two techniques can be used: reducing leakage current and turning off gates when they are not in use. Leakage current is becoming increasingly important in very deep submicron technologies. One simple approach to reducing leakage currents in gates is to choose, whenever possible, don't-care conditions on the inputs to reduce leakage currents. Series chains of transistors pass much lower leakage currents when both are off than when one is off and the other is on. If don't-care conditions can be used to turn off series combinations of transistors in a gate, the gate's leakage current can be greatly reduced.

The key to low leakage current is low threshold voltage. Unfortunately, there is an essential tension between low leakage and high performance. Remember from Equation 2-4 that leakage current is an exponential

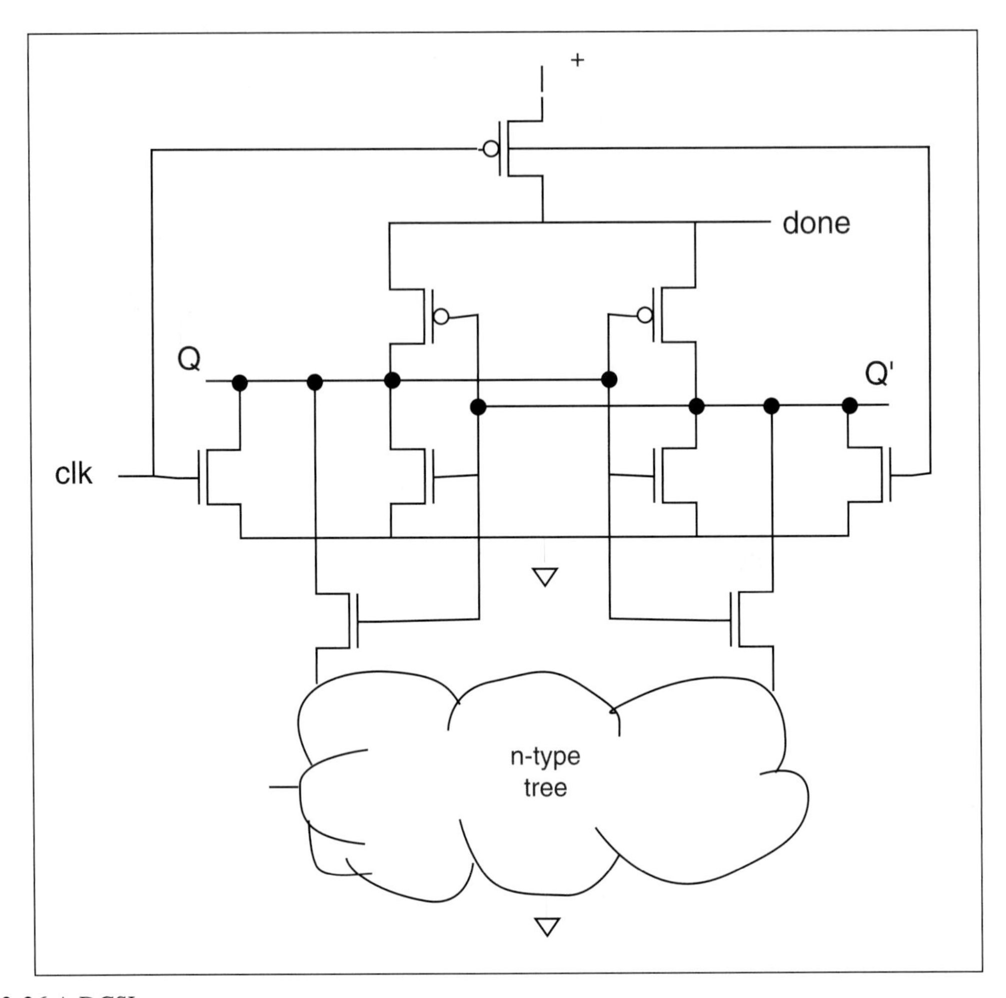

Figure 2-26 A DCSL gate.

function of V_{gs} - V_t. As a result, increasing V_t decreases the subthreshold current when the transistor is off. However, a high threshold voltage increases the gate's delay since the transistor turns on later in the input signal's transition. One solution to this dilemma is to use transistors with different thresholds at different points in the circuit.

Turning off gates when they are not used saves even more power, particularly in technologies that exhibit significant leakage currents. Care must be used in choosing which gates to turn off, since it often takes 100 μs for the power supply to stabilize after it is turned on. However, turning off gates is a very useful technique that becomes increasingly important in very deep submicron technologies with high leakage currents.

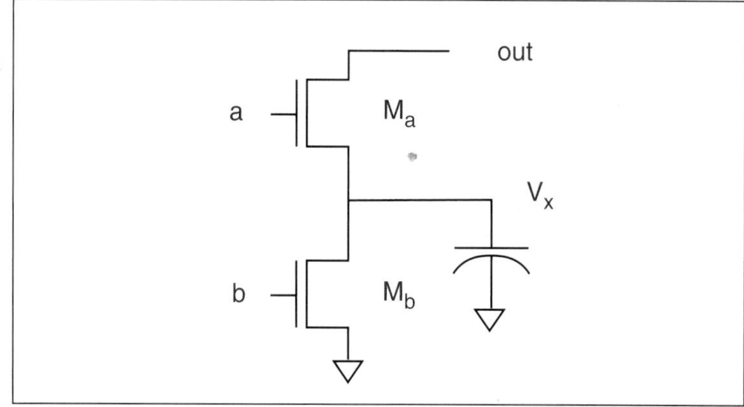

Figure 2-27 Leakage through transistor stacks.

leakage through a chain of transistors

The leakage current through a chain of transistors in a pulldown or pullup network is lower than the leakage current through a single transistor [De01]. It also depends on whether some transistors in the stack are also on. Consider the pulldown network of a NAND gate shown in Figure 2-27. If both the a and b inputs are 0, then both transistors are off. Because a small leakage current flows through transistor M_a, the parasitic capacitance between the two transistors is charged, which in turns holds the voltage at that node above ground. This means that V_{gs} for M_a is negative, thus reducing the total leakage current. The leakage current is found by simultaneously solving for the currents through the two transistors. The leakage current through the chain can be an order of magnitude lower than the leakage current through a single transistor. But the total leakage current clearly depends on the gate voltages of the transistors in the chain; if some of the gate's inputs are logic 1, then there may not be chains of transistors that are turned off and thus have reduced input voltages. Algorithms can be used to find the lowest-leakage input values for a set of gates; latches can be used to hold the gates' inputs at those values in standby mode to reduce leakage.

MTCMOS gate

Figure 2-28 shows a **multiple-threshold logic** (**MTCMOS**) [Mut98] gate that can be powered down. This circuit family uses low-leakage transistors to turn off gates when they are not in use. A **sleep transistor** is used to control the gate's access to the power supply; the gated power supply is known as a **virtual** $\boldsymbol{V_{DD}}$. The gate uses low-threshold transistors to increase the gate's delay time. However, lowering the threshold voltage also increases the transistors' leakage current, which causes us to introduce the sleep transistor. The sleep transistor has a high thresh-

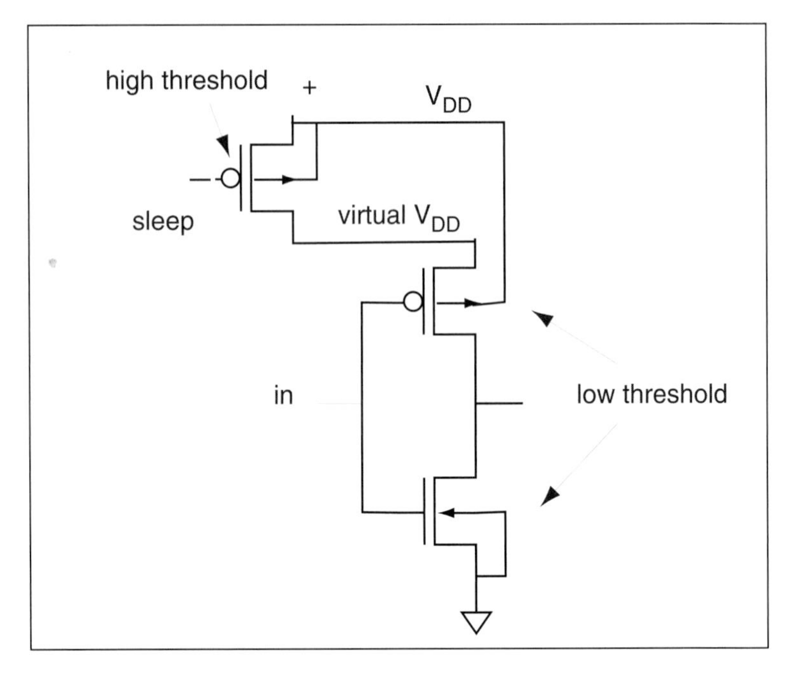

Figure 2-28 A multiple-threshold (MTCMOS) inverter.

old to minimize its leakage. The fabrication process must be able to build transistors with low and high threshold voltages.

The layout of this gate must include both V_{DD} and virtual V_{DD}: virtual V_{DD} is used to power the gate but V_{DD} connects to the pullup's substrate. The sleep transistor must be properly sized. If the sleep transistor is too small, its impedance would cause virtual V_{DD} to bounce. If the sleep transistor is too large, the sleep transistor would occupy too much area and it would use more energy when switched.

It is important to remember that some other logic must be used to determine when a gate is not used and to control the gate's power supply. This logic must be used to watch the state of the chip's inputs and memory elements to know when logic can safely be turned off. It may also take more than one cycle to safely turn on a block of logic.

MTCMOS flip-flop

Figure 2-29 shows an MTCMOS flip-flop. The storage path is made of high V_t transistors and is always on. The signal is propagated from input to output through low V_t transistors. The sleep-control transistors on the second inverter in the forward path prevent a short-circuit path between V_{DD} and virtual V_{DD} that could flow through the storage inverter's pullup and the forward chain inverter's pullup.

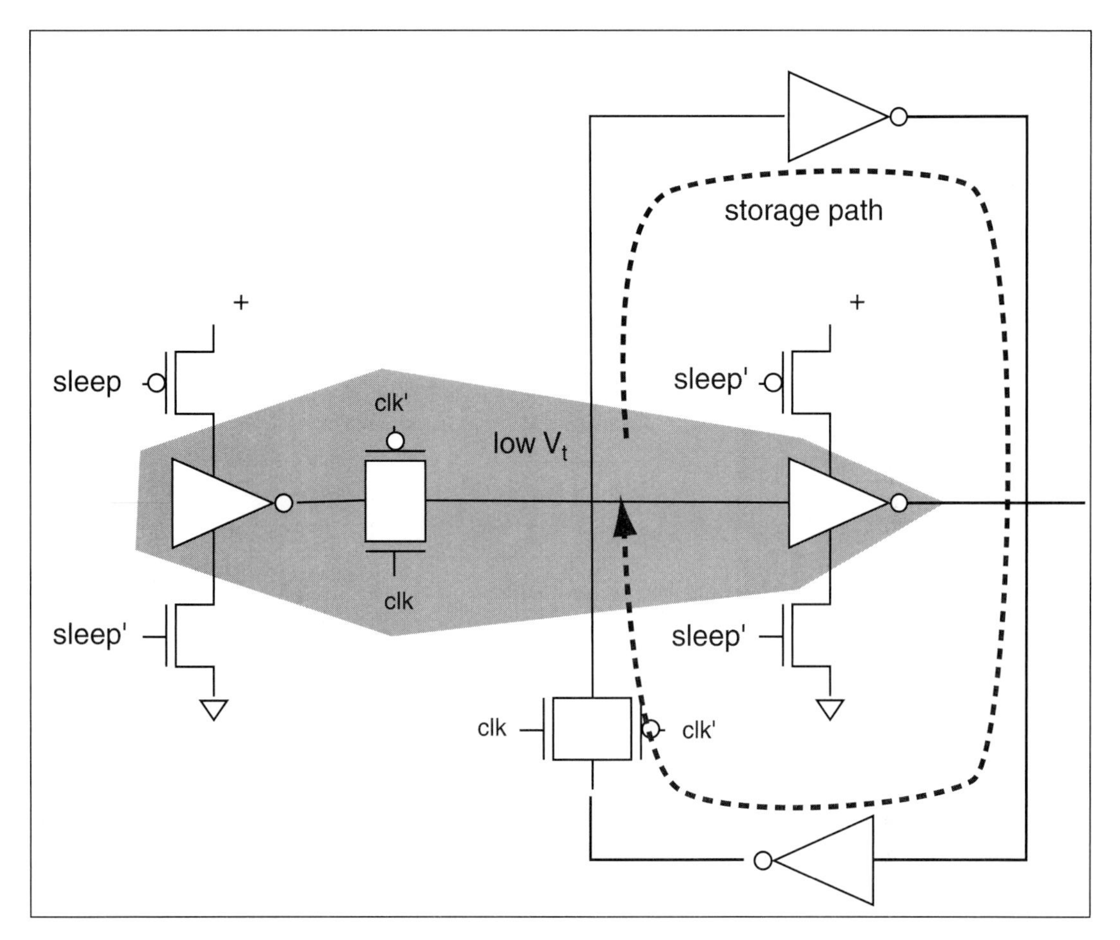

Figure 2-29 An MTCMOS flip-flop.

VTCMOS gate

A more aggressive method is **variable threshold CMOS (VTCMOS)** [Kur96], which actually can be implemented in several ways. Rather than fabricating fixed-threshold voltage transistors, the threshold voltages of the transistors in the gate are controlled by changing the voltages on the substrates. Figure 2-30 shows the structure of a VTCMOS gate. The substrates for the p- and n-type transistors are each connected to their own threshold supply voltages, $V_{BB,p}$ and $V_{BB,n}$. V_{BB} is raised to put the transistor in standby mode and lowered to put it into active mode. Rather sophisticated circuitry is used to control the substrate voltages.

VTCMOS logic comes alive faster than it falls asleep. The transition time to sleep mode depends on how quickly current can be pulled out of

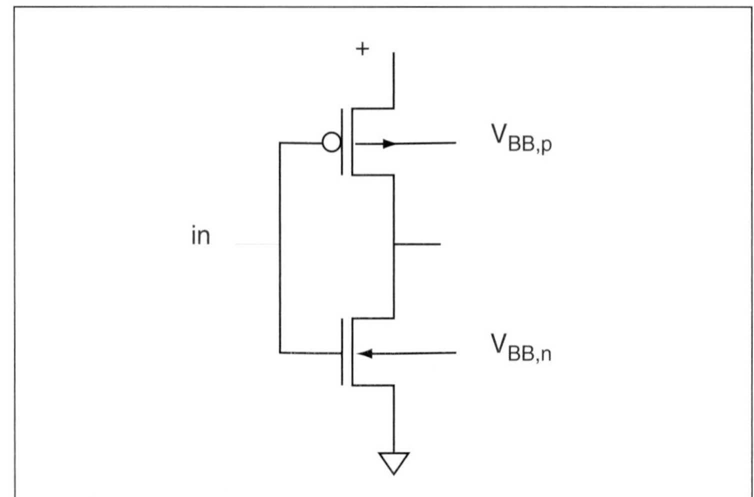

Figure 2-30 A variable-threshold CMOS (VTCMOS) gate.

the substrate, which is typically tens to hundreds of microseconds. Returning the gate to active mode requires injecting current back into the substrate, which can be done 100 to 1000 times faster than pulling that current out of the substrate. In most applications, a short wake-up time is important—the user generally gives little warning that the system is needed.

2.4.6 Switch Logic

switches and Boolean algebra

We have used MOS transistors to build complementary logic gates, which we use to construct combinational logic functions. But MOS transistors are good switches—a switch being a device which makes or breaks an electrical connection—and switches themselves can be used to directly implement Boolean functions [Sha38]. Switch logic isn't universally useful: large switch circuits are slow and switches introduce hard-to-trace electrical problems; and the lack of drive current presents particular problems when faced with the relatively high parasitics of deep-submicron processes. But building logic directly from switches can help save area and parasitics in some cases and switch logic is particularly useful in FPGA design.

Figure 2-31 shows how to build AND and OR functions from switches. The control inputs control the switches—a switch is closed when its control input is 1. The switch drains are connected to constants (V_{DD} or V_{SS}). A pseudo-AND is computed by series switches: the output is a

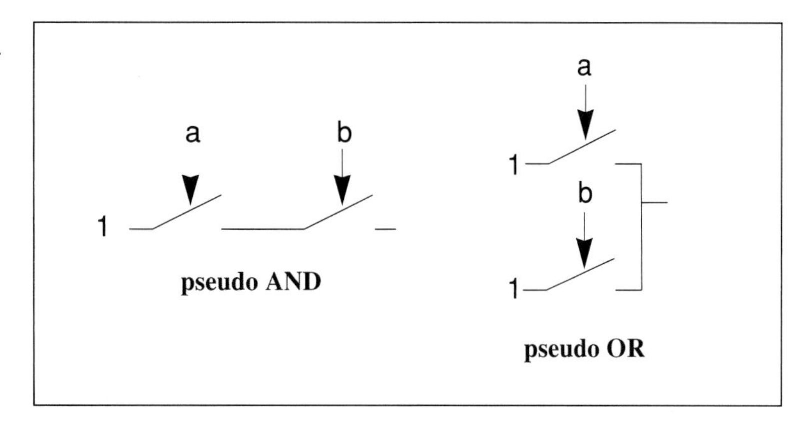

Figure 2-31 Boolean functions built from switches.

logic 1 if and only if both inputs are 1. Similarly, a pseudo-OR is computed by parallel switches: the output is logic 1 if either input is 1. We call these functions *pseudo* because when none of the switches are turned on by the input variables, the output is not connected to any constant source and its value is not defined. As we will see shortly, this property causes havoc in real circuits with parasitic capacitance. Switch logic is not complete—we can compute AND and OR but we cannot invert an input signal. If, however, we supply both the true and complement forms of the input variables, we can compute any function of the variables by combining true and complement forms with AND and OR switch networks.

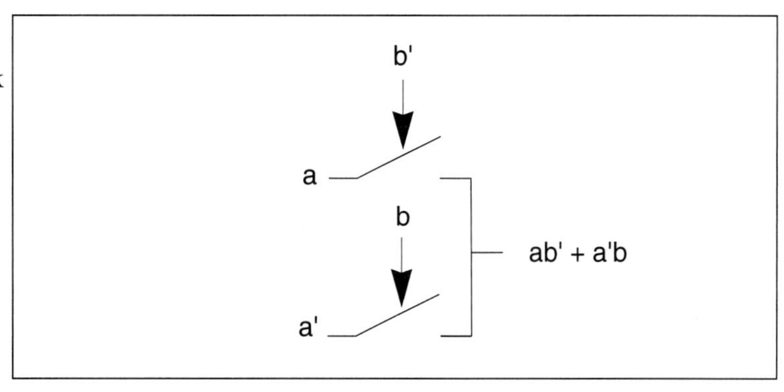

Figure 2-32 A switch network with non-constant source inputs.

non-constant inputs

We can reduce the size of a switch network by applying some of the input variables to the switches' gate inputs. The network of Figure 2-32, for example, computes the function $ab' + a'b$ using two switches by using one variable to select another. This network's output is also defined for all input combinations. Switch networks which apply the inputs to both the switch gate and drain are especially useful because some functions can be computed with a very small number of switches.

The next example shows how to build a multiplexer from switches.

Example 2-3 Switch implementation of a multiplexer

We want to design a multiplexer (commonly called a *mux*) with four data inputs and four select inputs—the two select bits s_1, s_0 and their complements are all fed into the switch network. The network's structure is simple:

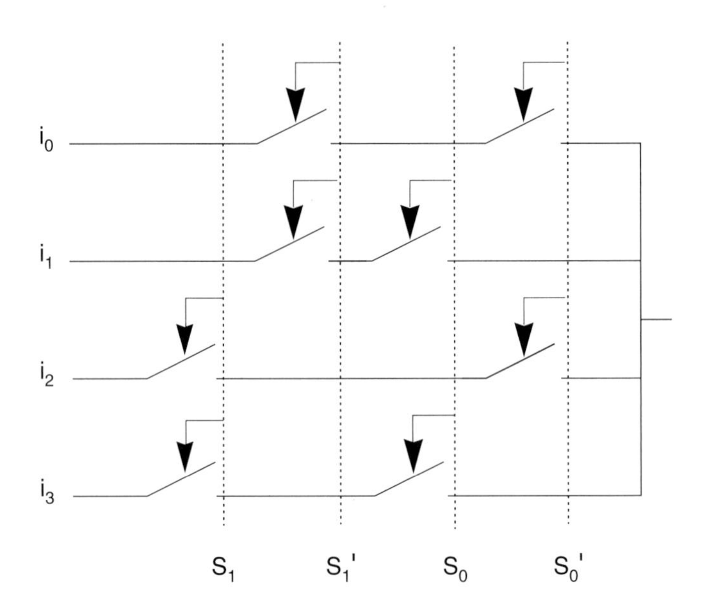

In practice, the number of select lines limits the useful size of the multiplexer.

building switches from transistors

How do we build switches from MOS transistors? One way is the **transmission gate** shown in Figure 2-33, built from parallel n-type and p-type transistors. This switch is built from both types of transistors so that it transmits logic 0 and 1 from drain to source equally well: when you put a V_{DD} or V_{SS} at the drain, you get V_{DD} or V_{SS} at the source. But it requires two transistors and their associated tubs; equally damning, it requires both true and complement forms of the gate signal.

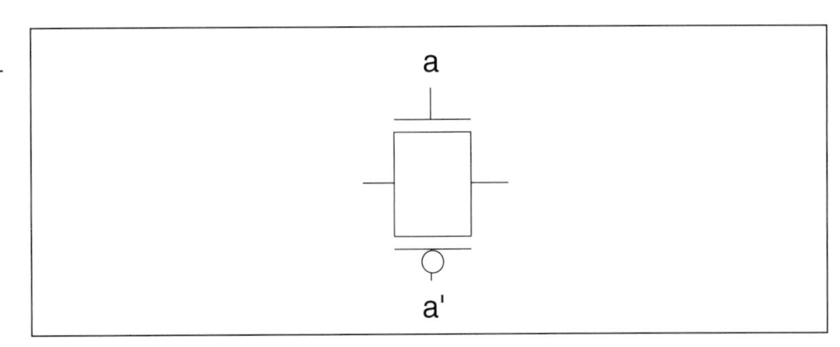

Figure 2-33 A complementary transmission gate.

An alternative is the **n-type switch**—a solitary n-type transistor. It requires only one transistor and one gate signal, but it is not as forgiving electrically. The n-type switch can transmit a logic 0 well as the transistor continues to conduct as the source voltage falls to V_{SS}. But when V_{DD} is applied to the drain, the voltage at the source will rise only to $V_{DD} - V_{tn}$. When switch logic drives gate logic, n-type switches can cause electrical problems. An n-type switch driving a complementary gate causes the complementary gate to run slower when the switch input is 1: since the n-type pulldown current is weaker when a lower gate voltage is applied, the complementary gate's pulldown will not suck current off the output capacitance as fast.

series switches

When several switches are connected in series, whether they are implemented as pass transistors or transmission gates, each switch can be approximately modeled by a resistor. As we will see in the next section, the delay through a series of resistors grows with the square of the resistance. This implies that no more than three switches should be connected in series before a restoring static gate is placed in the path. This resistor model is only approximate, so for accurate delay analysis, you should perform a more accurate circuit or timing simulation.

charge sharing

The most insidious electrical problem in switch networks is **charge sharing**. Switches built from MOS transistors have parasitic capacitances at their sources and drains thanks to the source/drain diffusion; capacitance can be added by wires between switches. While this capacitance is too small to be of much use (such as building a memory element), it is enough to cause trouble.

When we look at the network's behavior over several cycles, we see that much worse things can happen. As shown in Figure 2-34, when a switch connects two capacitors not driven by the power supply, current flows to place the same voltage across the two capacitors. The final amounts of charge depend on the ratio of the capacitances. Charge division can produce arbitrary voltages on intermediate nodes. These bad logic values

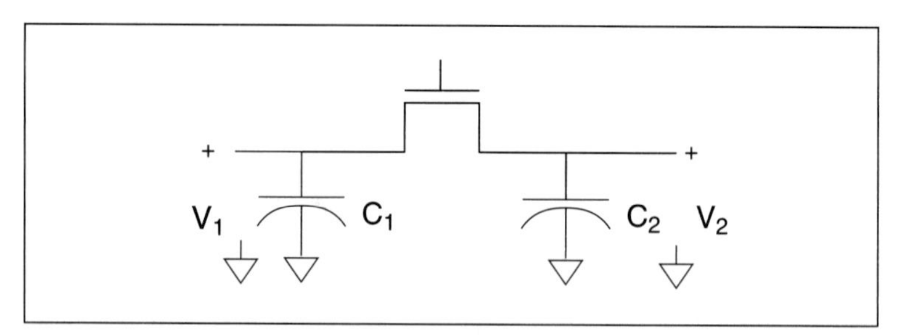

Figure 2-34 Charge division across a switch.

can be propagated to the output of the switch network and wreak havoc on the logic connected there. Consider the value of each input and of the parasitic capacitance between each pair of switches/terminals over time:

time	i	C_{ia}	a	C_{ab}	b	C_{bc}	c	C_{co}
0	1	1	1	1	1	1	1	1
1	0	0	1	0	0	1	0	1
2	0	0	0	1/2	1	1/2	0	1
3	0	0	0	1/2	0	3/4	1	3/4
4	0	0	1	0	0	3/4	0	3/4
5	0	0	0	3/8	1	3/8	0	3/4

The switches can shuttle charge back and forth through the network, creating arbitrary voltages, before presenting the corrupted value to the network's output.

Charge sharing can be easily avoided—design the switch network so that its output is always driven by a power supply. There must be a path from V_{DD} or V_{SS} through some set of switches to the output for every possible combination of inputs. Since charge can be divided only between undriven capacitors, always driving the output capacitance ensures that it receives a valid logic value.

The severity of charge sharing suggests that strong measures be used to ensure the correct behavior of switch logic networks. One way to improve the reliability of transmission gates is to insert buffers before and after them.

2.5 Wires

In this section we study the characteristics of wires. In many modern chips, the delay through long wires is larger than the delay through gates, so studying the electrical characteristics of wires is critical to high performance and low power chip design. We will then consider analytical models for wire delay. We will also consider the problem of where to insert buffers along wires to minimize delay.

2.5.1 Wire Structures

wires and vias

Figure 2-35 illustrates the cross-section of a nest of wires and vias. N-diffusion and p-diffusion wires are created by doping regions of the substrate. Polysilicon and metal wires are laid over the substrate, with silicon dioxide to insulate them from the substrate and each other. Wires are added in layers to the chip, alternating with SiO_2: a layer of wires is added on top of the existing silicon dioxide, then the assembly is covered with an additional layer of SiO_2 to insulate the new wires from the next layer. Vias are simply cuts in the insulating SiO_2; the metal flows through the cut to make the connection on the desired layer below.

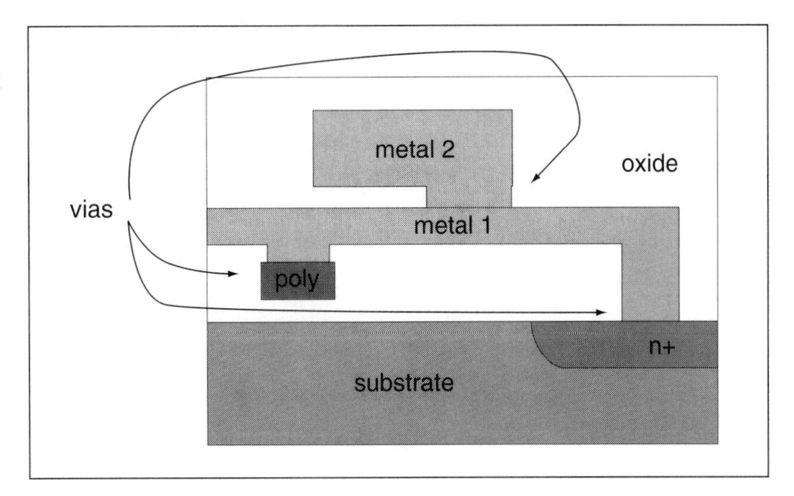

Figure 2-35 A cross-section of a chip showing wires and vias.

Copper interconnect can now be produced in volume thanks to a special protection layer that keeps the copper from poisoning the semiconductors in the substrate. The fabrication methods, and therefore the design rules, for copper interconnect are similar to those used for aluminum wires. However, the circuit characteristics of copper differ radically from those of aluminum.

power distribution

In addition to carrying signals, metal lines are used to supply power throughout the chip. On-chip metal wires have limited current-carrying capacity, as does any other wire. (Poly and diffusion wires also have current limitations, but since they are not used for power distribution those limitations do not affect design.) Electrons drifting through the voltage gradient on a metal line collide with the metal grains which form the wire. A sufficiently high-energy collision can appreciably move the metal grain. Under high currents, electron collisions with metal grains cause the metal to move; this process is called **metal migration** (also known as **electromigration**) [Mur93].

metal current capacity

Metal wires can handle about 2-3 mA of current per micron of wire width. (Width is measured perpendicular to current flow.) In larger designs, sizing power supply lines is critical to ensuring that the chip does not fail once it is installed in the field.

2.5.2 Wire Parasitics

Wires and vias introduce parasitic elements into our circuits. It is important to understand the structural properties of our components that introduce parasitic elements, and how to measure parasitic element values from layouts.

poly and metal wire capacitance

The capacitance mechanism for poly and metal wires is, in contrast, the parallel plate capacitor from freshman physics. We must also measure area and perimeter on these layers to estimate capacitance, but for different reasons. The **plate capacitance** per unit area assumes infinite parallel plates. We take into account the changes in the electrical fields at the edges of the plate by adding in a **fringe capacitance** per unit perimeter. These two capacitances are illustrated in Figure 2-39. Capacitances can form between signal wires. In conservative technologies, the dominant parasitic capacitance is between the wire and the substrate, with the silicon dioxide layer forming the insulator between the two parallel plates.

wire-to-wire capacitance

However, as the number of metal levels increases and the substrate capacitance decreases, wire-to-wire parasitics are becoming more important. Both capacitance between two different layers and between two wires on the same layer are basic parallel plate capacitances. The parasitic capacitance between two wires on different layers, such as C_{m1m2} in Figure 2-36, depends on the area of overlap between the two wires. When two wires run together for a long distance, with one staying over the other, the layer-to-layer capacitance can be very large. The capacitance between two wires on the same layer, C_{w1w2} in the figure, is formed by the vertical sides of the metal wires. Metal wires can be very tall in relation to their width, so the vertical wall coupling is non-negligible. However, this capacitance depends on the distance between two wires. The values given in process specifications are for minimum-separation wires, and the capacitance decreases by a factor of $1/x$ as distance increases. When two wires on the same layer run in parallel for a long distance, the coupling capacitance can become very large.

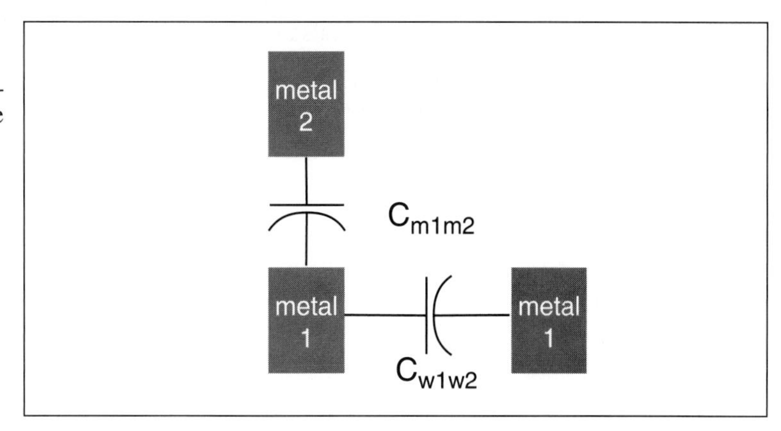

Figure 2-36 Capacitive coupling between signals on the same and different layers.

wire resistance

Wire resistance is also computed by measuring the size of the wire in the layout, but the unit of resistivity is **ohms per square** ($\Omega/\square$), not ohms per square micron. The resistance of a square unit of material is the same for a square of any size; to understand, consider Figure 2-37. Assume that a unit square of material has a resistance of 1Ω. Two squares of material connected in parallel have a total resistance of $1/2\Omega$. Connecting two such rectangles in series creates a 2×2 square with a resistance of 1Ω. We can therefore measure the resistance of a wire by measuring its aspect ratio.

In general, diffusion wires have higher resistivity than polysilicon wires and metal wires have low resistivities.

diffusion wire capacitance

Diffusion wire capacitance is introduced by the p-n junctions at the boundaries between the diffusion and underlying tub or substrate. While these capacitances change with the voltage across the junction, which varies during circuit operation, we generally assume worst-case values. An accurate measurement of diffusion wire capacitance requires separate calculations for the bottom and sides of the wire—the doping density, and therefore the junction properties, vary with depth. To measure total capacitance, we measure the diffusion area, called **bottomwall** capacitance, and perimeter, called **sidewall** capacitance, as shown in Figure 2-38, and sum the contributions of each.

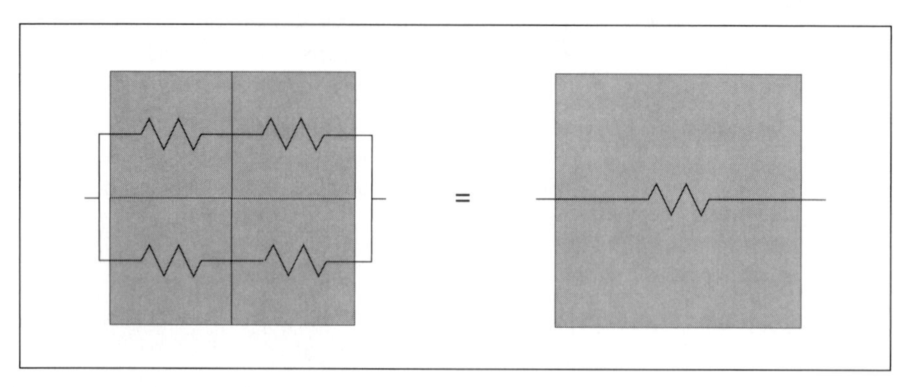

Figure 2-37 Resistance per unit square is constant.

Figure 2-38 Sidewall and bottomwall capacitances of a diffusion region.

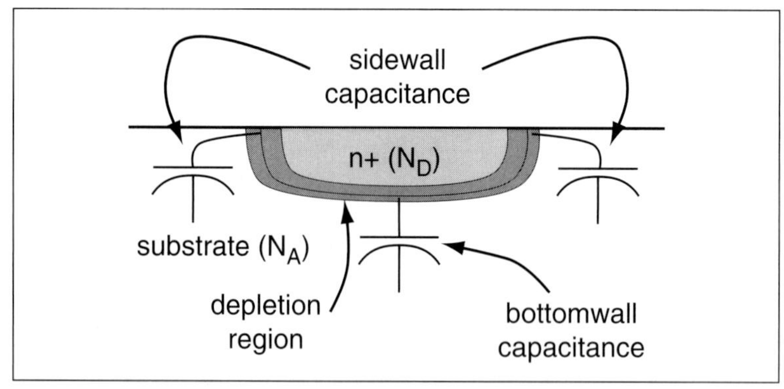

depletion region capacitance

The **depletion region capacitance** value is given by

$$C_{j0} = \frac{\varepsilon_{si}}{x_d} . \qquad \text{(EQ 2-22)}$$

This is the **zero-bias depletion capacitance**, assuming zero voltage and an abrupt change in doping density from N_a to N_d. The depletion region width x_{d0} is shown in Figure 2-38 as the dark region; the depletion region is split between the n+ and p+ sides of the junction. It can be shown that the junction capacitance decreases as the reverse bias voltage increases.

source/drain parasitics

The source and drain regions of a transistor have significant capacitance and resistance. These parasitics are, for example, entered into a Spice circuit simulation as device characteristics rather than as separate wire models. However, we measure the parasitics in the same way we would

Figure 2-39 Plate and fringe capacitances of a parallel-plate capacitor.

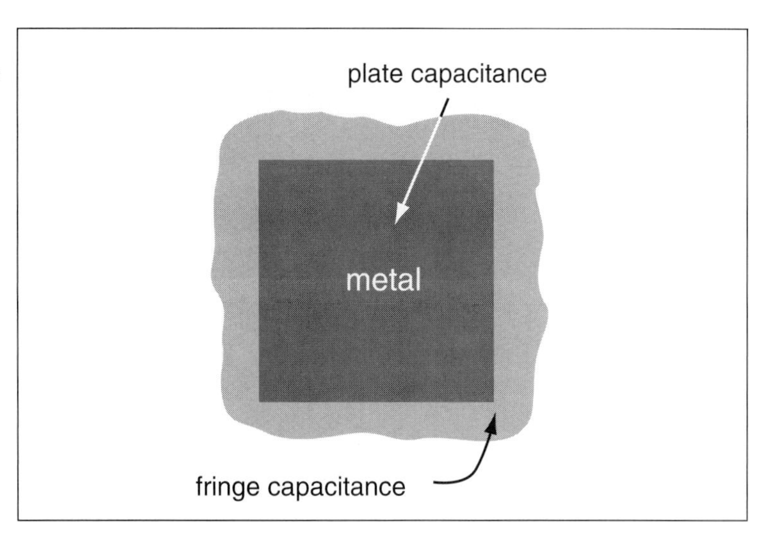

measure the parasitics on an isolated wire, measuring area and perimeter up to the gate-source/drain boundary.

via resistance

Vias have added resistance because the cut between the layers is smaller than the wires it connects and because the materials interface introduces resistance. The resistance of the via is usually determined by the resistance of the materials: a metal 1-metal 2 via has a typical resistance of less than 0.5 Ω while a metal 1-poly contact has a resistance of 2.5 Ω. We may use several vias in parallel to reduce the total resistance of a connection between two wires. We generally try to avoid introducing unnecessary vias in current paths for which low resistance is critical.

skin effect in copper

Low-resistance conductors like copper not only exhibit inductance, they also display a more complex resistance relationship due to a phenomenon called **skin effect** [Ram65]. The skin effect causes a copper conductor's resistance to increase (and its inductance to decrease) at high frequencies.

An ideal conductor would conduct current only on its surface. The current at the surface is a boundary effect—any current within the conductor would set up an electromagnetic force that would induce an opposing and cancelling current. The copper wiring used on ICs is a non-ideal conductor; at low frequencies, the electromagnetic force is low enough and resistance is high enough that current is conducted throughout the wire's cross-section. However, as the signal's frequency increases, the electromagnetic forces increase. As illustrated in Figure 2-40, the current through an isolated conductor migrates toward the

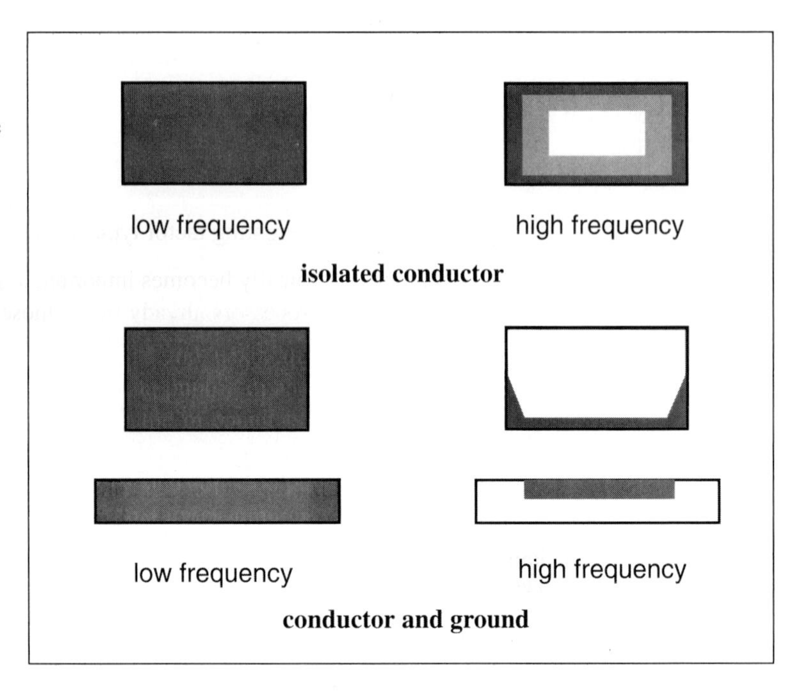

Figure 2-40 How current changes with frequency due to skin effect.

edges as frequency increases; when the conductor is close to a ground, the currents in both move toward each other.

Skin effect causes the conductor's resistance to increase with frequency. The **skin depth** δ is the depth at which the conductor's current is reduced to $1/e = 37\%$ of its surface value [Che00]:

$$\delta = \frac{1}{\sqrt{\pi f \mu \sigma}}, \quad \text{(EQ 2-23)}$$

where f is the signal frequency, μ is the magnetic permeability, and σ is the wire's conductivity. The skin depth goes down as the square root of frequency.

delay caused by skin effect

Cheng et al. [Che00] provide an estimation of the delay per unit length of a wire suffering from skin effect. Two values, R_{dc} and R_{hf}, estimate the resistance at low and high frequencies:

$$\left(R_{dc} = \frac{1}{\sigma w t}\right), R_{hf} = \frac{1}{2\sigma\delta(w+t)}, \quad \text{(EQ 2-24)}$$

where w and t are the width and height of the conductor, respectively. The skin depth δ ensures that R_{hf} depends on frequency. The resistance per unit length can be estimated as

$$R_{ac} = \sqrt{R_{dc}^2 + (\kappa R_{hf})^2} \,, \qquad \text{(EQ 2-25)}$$

where κ is a weighting factor typically valued at 1.2.

Skin effect typically becomes important at gigahertz frequencies in ICs. Some microprocessors already run at those frequencies and more chips will do so in the near future.

2.5.3 Models for Wires

Table 2-3 Wire capacitance values for a 90 nm process.

layer	capacitance to ground (aF/μm)	coupling capacitance (aF/μm)	resistance/ length (Ω/μm)
metal 3	18	9	0.02
metal 2	47	24	0.03
metal 1	76	36	0.03

wire capacitance and resistance

The most important electrical characteristics of wires (assuming they don't exhibit significant inductance) are capacitance and resistance. Table 2-3 gives some sample capacitance and resistance per unit length values for minimum-width wires in a 90 nm process. Note that capacitance to ground goes down as we go to higher layers—metal 3 has about a quarter the capacitance of metal 1. Also note that coupling capacitance, the capacitance to adjacent wires on the same layer, is smaller than capacitance to ground but very significant. Resistance is measured per unit length given our assumption of minimum-width wires; the height of wiring varies from layer to layer.

One of the most basic delay questions we can ask is the length of a wire for which the delay through the wire equals the delay through the gate. The next example examines this question.

Example 2-4
Gate delay *vs.* wire delay

Let us first compute the delay through the smallest, simplest logic gate circuit that we can build. That circuit consists of a minimum-size inverter that is driving another minimum-size inverter. Luckily, we already computed that value in Example 2-2. We obtained a fall time value of 2.9 ps.

We now want to find the length of a minimum-width wire that doubles the total delay from the inverter's input to the far end of the wire. We assume that a minimum-size inverter is connected to the far end of the wire. If we ignore the wire's resistance, then the delay formula tells us that we double the total delay by doubling the load capacitance. (We will see that the wire will be short enough that ignoring its resistance is reasonable.) This means that a wire whose capacitance equals the total gate capacitance of the minimum-size inverter will provide the delay value we seek. We calculated the inverter's capacitance as 0.12 fF. We can look up the wire capacitance for a metal 3 wire in Table 2-3. If we assume that all the capacitance is to ground and there is no coupling capacitance, then a wire of 6.7 μm is long enough to give a delay equal to the delay of the inverter itself. This is only about 75 times the width of a minimum-size transistor, which is not very long compared to the size of the chip, which can range well over a half a centimeter.

2.5.4 Delay Through an RC Transmission Line

transmission line model

An **RC transmission line** models a wire as infinitesimal RC sections, each representing a differential resistance and capacitance. Since we are primarily concerned with RC transmission lines, we can use the transmission line model to compute the delay through very long wires. We can model the transmission line as having unit resistance r and unit capacitance c. The standard schematic for the RC transmission line is shown in Figure 2-41. The transmission line's voltage response is modeled by a differential equation:

$$\frac{1}{r}\frac{d^2V}{dx^2} = c\frac{dV}{dt}. \qquad \text{(EQ 2-26)}$$

This model gives the voltage as a function of both x position along the wire and time.

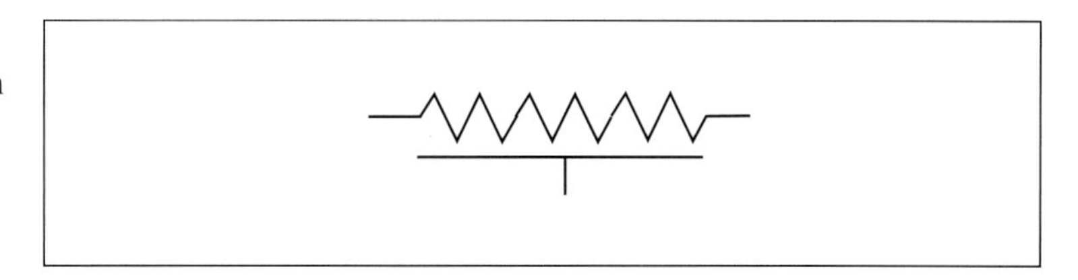

Figure 2-41 Symbol for a distributed RC transmission line.

Elmore delay

The raw differential equation, however, is unwieldy for many circuit design tasks. **Elmore delay** [Elm48] is the most widely used metric for RC wire delay and has been shown to model the results of simulating RC wires on integrated circuits with sufficient accuracy [Boe93]. Elmore defined the delay through a linear network as the first moment of the impulse response of the network:

$$\delta_E = \int_0^\infty t V_{out}(t) dt \quad . \qquad \text{(EQ 2-27)}$$

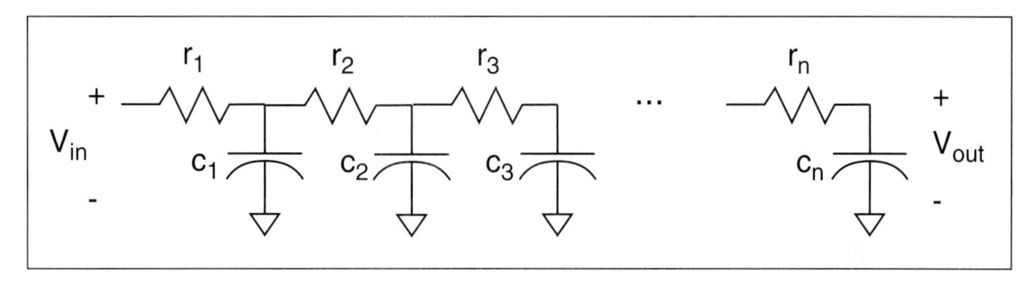

Figure 2-42 An RC transmission line for Elmore delay calculations.

Elmore modeled the transmission line as a sequence of n sections of RC, as shown in Figure 2-42. In the case of a general RC network, the Elmore delay can be computed by taking the sum of RC products, where each resistance R is multiplied by the sum of all the downstream capacitors. Since all the transmission line section resistances and capacitances in an n-section are identical, this reduces to

Elmore delay for RC network

$$\delta_E = \sum_{i=1}^{n} r(\mathrm{n\text{-}i})c = \frac{1}{2} rc \times n(\mathrm{n\text{-}1}) \, . \qquad \text{(EQ 2-28)}$$

One consequence of this formula is that wire delay grows as the square of wire length, since n is proportional to wire length. Since the wire's delay also depends on its unit resistance and capacitance, it is imperative to use the material with the lowest RC product (which will almost always be metal) to minimize the constant factor attached to the n^2 growth rate.

continuous transmission line model

Although the Elmore delay formula is widely used, we will need some results from the analysis of continuous transmission lines for our later discussion of crosstalk. The normalized voltage step response of the transmission line can be written as

$$V(t) = 1 + \sum_{k=1}^{\infty} K_k e^{-\sigma_k t/RC} \approx 1 + K_1 e^{-\sigma_1 t/RC}, \tag{EQ 2-29}$$

where R and C are the total resistance and capacitance of the line. We will define R_t as the internal resistance of the driving gate and C_t as the load capacitance at the opposite end of the transmission line.

Sakurai [Sak93] estimated the required values for the first-order estimate of the step response as:

$$K_1 = \frac{-1.01(R_T + C_T + 1)}{R_T + C_T + \pi/4}, \tag{EQ 2-30}$$

$$\sigma_1 = \frac{1.04}{R_T C_T + R_T + C_T + (2/\pi)^2}, \tag{EQ 2-31}$$

where R_T and C_T are R_t/R and C_t/C, respectively.

tapered wires

So far, we have assumed that the wire has constant width. In fact, tapered wires provide smaller delays. Consider the first resistance element in the transmission line—the current required to charge all the capacitance of the wire must flow through this resistance. In contrast, the resistance at the end of the wire handles only the capacitance at the end. Therefore, if we can decrease the resistance at the head of the wire, we can decrease the delay through the wire. Unfortunately, increasing the resistance by widening the wire also increases its capacitance, making this a non-trivial problem to solve.

Fishburn and Schevon [Fis95] proved that the optimum-shaped wire has an exponential taper. If the source resistance is R_0, the sink capacitance is C_0, and the unit resistance and capacitance are R_s and C_s, the width of the wire as a function of distance is

$$w(x) = \frac{2C_0}{C_s L} W\left(\frac{L}{2}\sqrt{\frac{R_s C_s}{R_0 C_0}}\right) e^{2W\left(\frac{L}{2}\sqrt{\frac{R_s C_s}{R_0 C_0}}\right)\frac{x}{L}}, \quad \text{(EQ 2-32)}$$

where W is the function that satisfies the equality $W(x)e^{W(x)} = x$. The advantage of optimal tapering is noticeable. Fishburn and Schevon calculate that, for one example, the optimally tapered wire has a delay of 3.72 ns while the constant-width wire with minimum delay has a delay of 4.04 ns. In this example, the optimally tapered wire shrinks from 30.7 μm at the source to 7.8 μm at the sink.

Of course, exponentially tapered wires are impossible to fabricate exactly, but it turns out that we can do nearly as well by dividing the wire into a few constant width sections. Figure 2-43 shows that a few segments of wire can be used to approximate the exponential taper reasonably well. This result also suggests that long wires which can be run on several layers should run on the lowest-resistance layer near the driver and can move to the higher-resistance layers as they move toward the signal sink.

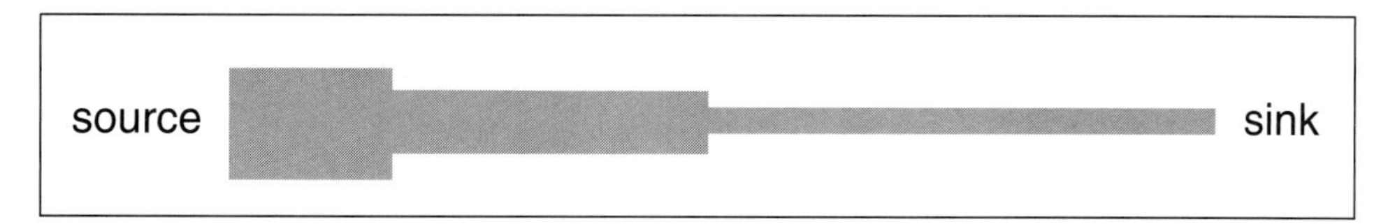

Figure 2-43 A step-tapered wire.

2.5.5 Buffer Insertion in RC Transmission Lines

buffer insertion for minimum delay

We do not obtain the minimum delay through an RC transmission line by putting a single large driver at the transmission line's source. Rather, we must put a series of buffers equally spaced through the line to restore the signal. Bakoglu [Bak90] derived the optimal number of repeaters and repeater size for an RC transmission line. As shown in Figure 2-44, we want to divide the line into k sections, each of length l. Each buffer will be h times larger than a minimum-sized buffer.

Let's first consider the case in which $h = 1$ and the line is broken into k sections. R_{int} and C_{int} are the total resistance and capacitance of the transmission line. R_0 is the driver's equivalent resistance and C_0 its input capacitance. Then the 50% delay formula is

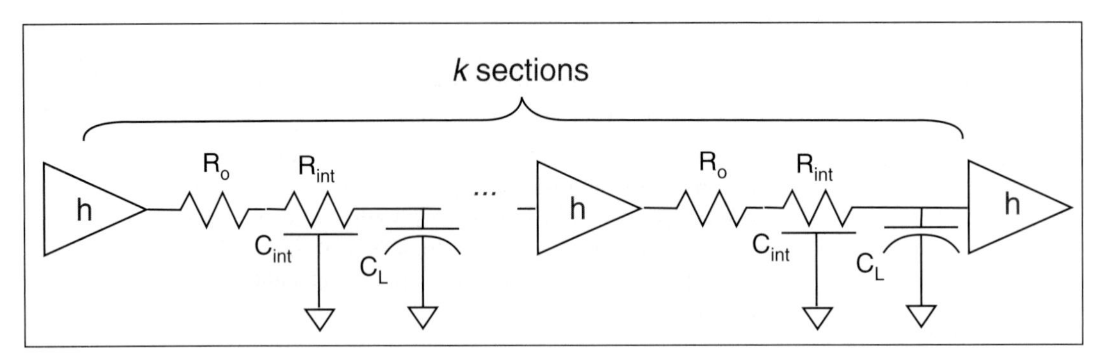

Figure 2-44 An RC transmission line with repeaters.

$$T_{50\%} = k\left[0.7R_0\left(\frac{C_{\text{int}}}{k} + C_0\right) + \frac{R_{\text{int}}}{k}\left(0.4\frac{C_{\text{int}}}{k} + 0.7C_0\right)\right] \quad \text{(EQ 2-33)}$$

The various coefficients are due to the distributed nature of the transmission line. We find the minimum delay by setting $dT/dk = 0$. This gives the number of repeaters as

$$k = \sqrt{\frac{0.4R_{\text{int}}C_{\text{int}}}{0.7R_0C_0}}. \quad \text{(EQ 2-34)}$$

When we free the size of the repeater to be an arbitrary value h, the delay equation becomes

$$T_{50\%} = k\left[0.7\frac{R_0}{h}\left(\frac{C_{\text{int}}}{k} + hC_0\right) + \frac{R_{\text{int}}}{k}\left(0.4\frac{C_{\text{int}}}{k} + 0.7hC_0\right)\right]. \quad \text{(EQ 2-35)}$$

We solve for minimum delay by setting $\frac{dT}{dk} = 0$ and $\frac{dT}{dh} = 0$. This gives the optimal values for k and h as

optimum number of buffers

$$k = \sqrt{\frac{0.4R_{\text{int}}C_{\text{int}}}{0.7R_0C_0}}, \quad \text{(EQ 2-36)}$$

optimum buffer size

$$h = \sqrt{\frac{R_0C_{\text{int}}}{R_{\text{int}}C_0}}. \quad \text{(EQ 2-37)}$$

The total delay at these values is

$$T_{50\%} = 2.5\sqrt{R_0 C_0 R_{int} C_{int}}\,. \quad \text{(EQ 2-38)}$$

Example 2-5 Buffer insertion in an RC line

Let's calculate the buffers required when an inverter that is 10 times minimum width drives a metal 3 wire that is 5000 μm long. In this case, R_0 = 1.11 kΩ and C_0 = 1.2 fF (starting from Example 2-2, resistance goes down 10X and capacitance goes up 10X). We can estimate the wire resistance and capacitance from Table 2-3: R_{int} = 100 Ω; C_{int} = 36 fF + 18 fF = 135 fF, taking into account the capacitance to ground and coupling capacitance.

The optimal number of buffers is

$$k = \sqrt{\frac{0.4 \times 100 \times 135\times10^{-15}}{0.7 \times 1.11\times10^{3} \times 1.2\times10^{-15}}} = 2.4 \approx 2\,.$$

The optimal buffer size is

$$h = \sqrt{\frac{1.11\times10^{3} \times 135\times10^{-15}}{100 \times 1.2\times10^{-15}}} = 35.1\,.$$

The 50% delay is

$$T_{50\%} = 2.5\sqrt{1.11\times10^{3} \times 1.2\times10^{-15} \times 100 \times 135\times10^{-15}} = 11\times10^{-12}s\,.$$

If we increase the size of the driver by a factor of 4, reducing its resistance by 4X and increasing its capacitance by 4X, what happens? k and $T_{50\%}$ remain unchanged, but the buffer size drops by a factor of 4.

2.5.6 Crosstalk Between RC Wires

Crosstalk is important to analyze because it slows down signals—the crosstalk noise increases the signal's settling time. Crosstalk can become a major component of delay if wiring is not carefully designed.

aggressor and victim nets

Figure 2-45 shows the basic situation in which crosstalk occurs. Two nets are coupled by parasitic capacitance. One net is the **aggressor net** that interferes with a **victim net** through that coupling capacitance. A transition in the aggressor net is transmitted to the victim net causing the victim to glitch. The glitch causes the victim net to take longer to settle

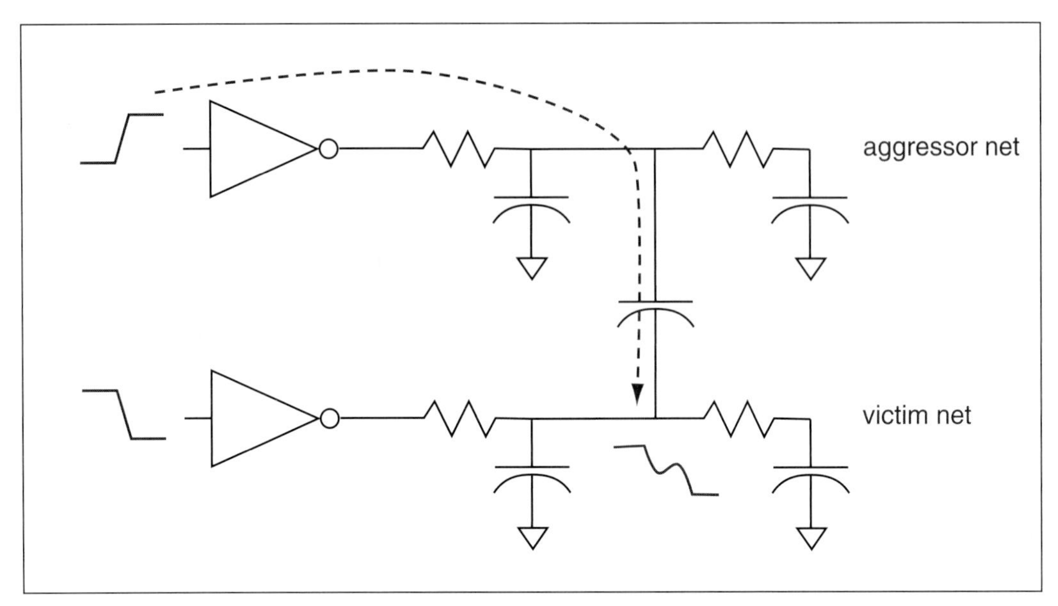

Figure 2-45 Aggressor and victim nets.

to its final value. In static combinational logic, crosstalk increases the delay across a net.

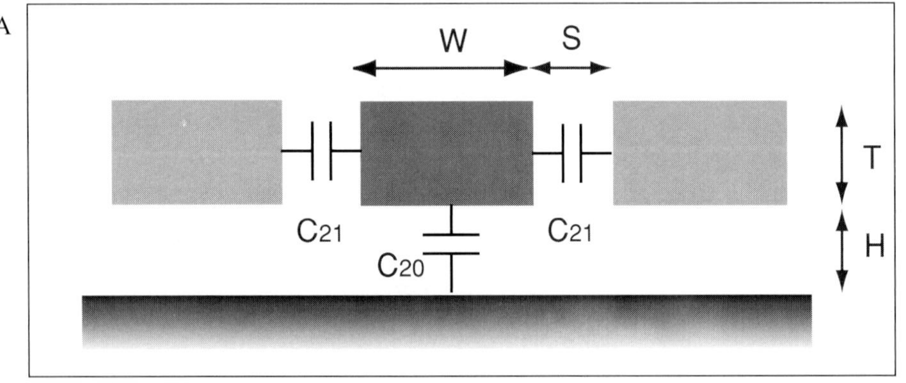

Figure 2-46 A simple crosstalk model (after Sakurai [Sak93], © 1993 IEEE).

The simplest case to consider is a set of three wires [Sak93], as shown in Figure 2-46. The middle wire carries the signal of interest, while the other two capacitively inject crosstalk noise. Each wire is of height T and width W, giving an aspect ratio of W/T. Each wire is height H above the substrate and the wires are spaced a distance S apart. The wire pitch P is the sum of width W and spacing S. We must consider three capaci-

Figure 2-47 Delay vs. wire aspect ratio and spacing (after Sakurai [Sak93] © 1993 IEEE).

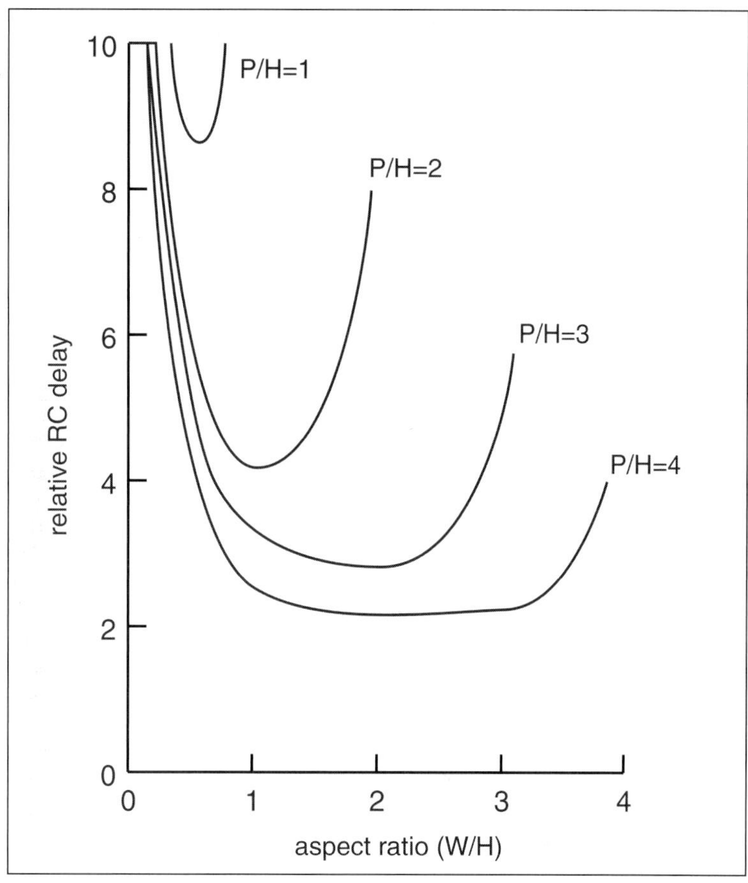

tances: C_{20} between the signal wire and the substrate, and two capacitances of equal value, C_{21}, to the two interfering wires. We denote the sum of these three capacitances as C_3. Sakurai estimates the RC delay through the signal wire in arbitrary time units as

$$t_r = \frac{(C_{20} + 4C_{21})}{W/H} . \qquad \text{(EQ 2-39)}$$

Using this simple model, Figure 2-47 shows Sakurai's calculation of relative RC delay in arbitrary units for a 0.5 μm technology for the signal wire. This plot assumes that $T/H = 1$ and that the aspect ratio varies from near 0 through 4; the delay is shown for four different spacings between the wires, as given by the P/H ratio. This plot clearly shows two important results. First, there is an optimum wire width for any given wire spacing, as shown by the U shape of each curve. Second, the optimum width increases as the spacing between wires increases.

2.6 Registers and RAM

Memory of various forms is very important in digital design and is particularly important in FPGAs. In this section we will look at registers—memory elements designed for clocked operation—and random-access memory.

2.6.1 Register Structures

Building a sequential machine requires **registers** that read a value, save it for some time, and then can write that stored value somewhere else, even if the register's input value has subsequently changed.

register characteristics

In CMOS circuits the memory for a register is formed by some kind of capacitance or by positive feedback of energy from the power supply. Access to the internal memory is controlled by the *clock* input—the memory element reads its *data* input value when instructed by the clock and stores that value in its memory. The output reflects the stored value, probably after some delay. Registers differ in many key respects:

- exactly what form of clock signal causes the input data value to be read;
- how the behavior of *data* around the read signal from *clock* affects the stored value;
- when the stored value is presented to the output;
- whether there is ever a combinational path from the input to the output.

latches and flip-flops

We will describe the latches and flip-flops in more detail in Section 5.4.1. For the moment, all we need to know is that latches are transparent while flip-flops are not. A latch's output follows its input while the latch's clock input is active; a flip-flop, in contrast, allows the input to affect the output only in a narrow window around a clock edge event.

dynamic latch

The simplest register in CMOS technology is the **dynamic latch** shown in Figure 2-48. It is called dynamic because the memory value is not refreshed by the power supply and a latch because its output follows its

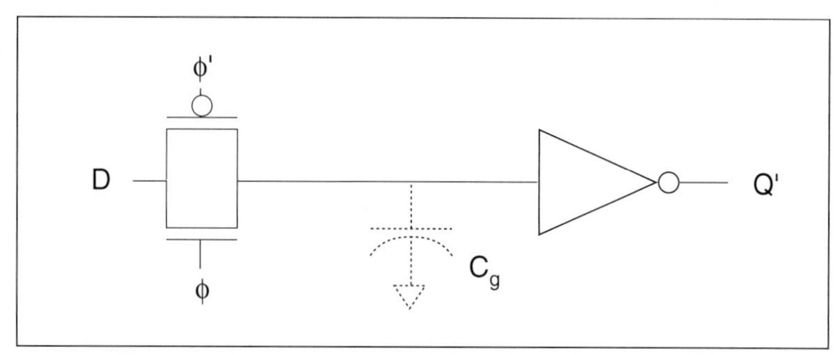

Figure 2-48 A dynamic latch circuit.

input under some conditions. The latch is a D-type, so its input is D and its output is Q'. The inverter connected to the output should be familiar. The **storage capacitance** has been shown in dotted lines since it is a parasitic component; this capacitance has been named C_g since most of the capacitance comes from the gates of the transistors in the inverter.

The latch's operation is straightforward. When the transmission gate is on, whatever logic gate is connected to the D input is allowed to charge or discharge C_g. As the voltage on C_g changes, Q' follows in complement—as C_g goes to low voltages, Q' follows to high voltages, and *vice versa*. When the transmission gate opens, C_g is disconnected from any logic gate that could change its value. Therefore, the value of latch's output Q' depends on the voltage of the storage capacitor: if the capacitor has been discharged, the latch's output will be a logic 1; if the storage capacitor has been charged, the latch's output will be a 0. Note that the value of Q' is the logical complement of the value presented to the latch at D; we must take this inversion into account when using the latch. To change the value stored in the latch, we can close the transmission gate by setting $\phi = 1$ and $\phi' = 0$ and change the voltage on C_g.

flip-flops

The structure of an **edge-triggered** flip-flop is shown in Figure 2-49. It is built from two back-to-back latches. The first latch reads the data input when the clock is high. Meanwhile, the internal inverter assures that the second latch's clock input is low, insulating the second latch from changes in the first latch's output and leaving the flip-flop's output value stable. After the clock has gone low, the second latch's clock input is high, making it transparent, but a stable value is presented to the second latch by the first one. When the clock moves back from 0 to 1, the second latch will save its value before the first's output has a chance to change.

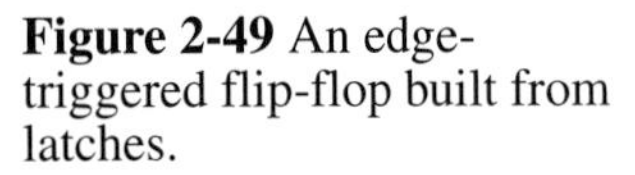

Figure 2-49 An edge-triggered flip-flop built from latches.

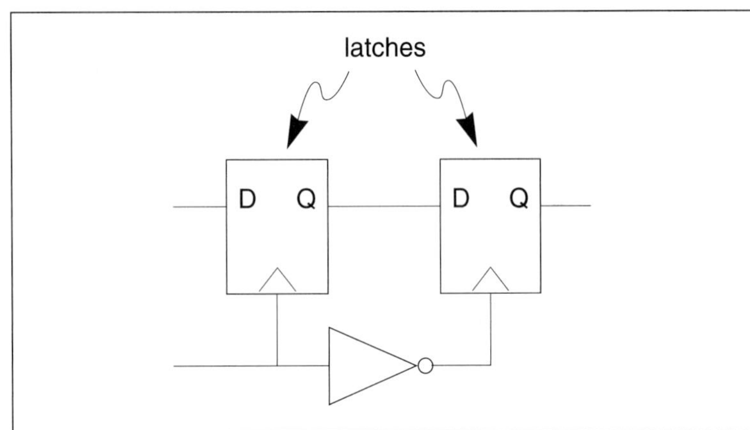

2.6.2 Random-Access Memory

Random-access memory (**RAM**) is often used in FPGAs because they efficiently implement large blocks of memory. FPGAs use **static RAM** (**SRAM**) because the dynamic RAM commonly used for bulk memory requires specialized capacitor structures that can't be built on the same chip along with high-performance logic transistors.

So far, we have built registers out of circuits that exhibit mostly digital behavior. By taking advantage of analog design methods, we can build memories that are both smaller and faster. Memory design is usually best left to expert circuit designers; however, understanding how these memories work will help you learn how to use them in system design.

read-only vs. random access memory

Read-only memory (**ROM**), as the name implies, can be read but not written. It is used to store data or program values that will not change, because it is the densest form of memory available. An increasing number of digital logic processes support **flash memory**, which is the dominant form of electrically erasable PROM memory.

There are two types of read-write **random access memories**: **static** (**SRAM**) and **dynamic** (**DRAM**). SRAM and DRAM use different circuits, each of which has its own advantages: SRAM is faster but uses more power and is larger; DRAM has a smaller layout and uses less power. DRAM cells are also somewhat slower and require the dynamically stored values to be periodically refreshed, just as in a dynamic latch.

Some types of memory are available for integration on a chip, while others that require special processes are generally used as separate chips. Commodity DRAMs are based on a one-transistor memory cell. That cell requires specialized structures, such as poly-poly capacitors, which are built using special processing steps not usually included in ASIC processes.

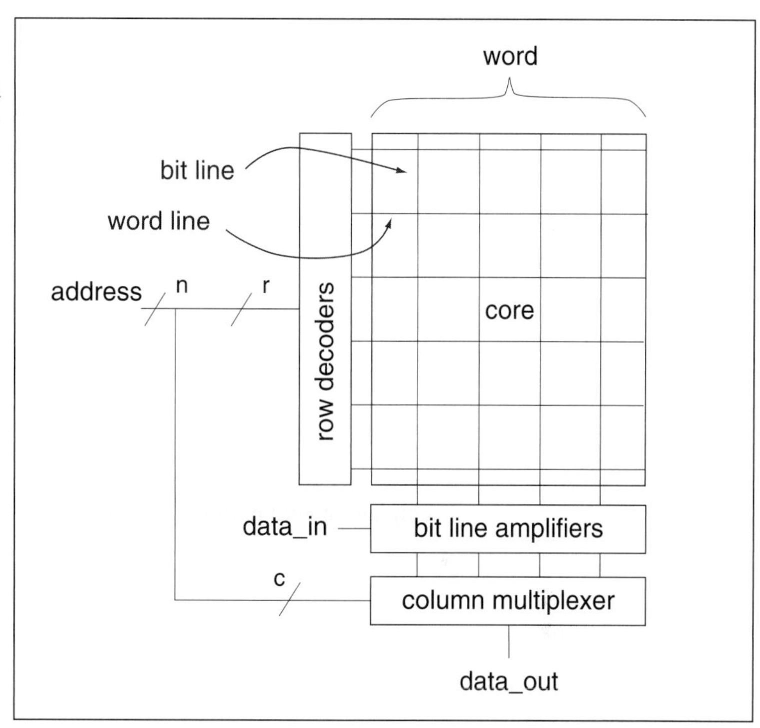

Figure 2-50 Architecture of a high-density memory system.

memory operation

A RAM or ROM is used by presenting it with an address and receiving the value stored at that address some time later. Details differ, of course: large memories often divide the address into row and column sections, which must be sent to the memory separately, for example.

memory architecture

The architecture of a generic RAM/ROM system is shown in Figure 2-50. Think of the data stored in the memory core as being organized into n bit-wide words. Vertical wires (known as **bit lines**) and horizontal wires (known as **word lines**) intersect at each memory cell. The address decoders (also known as row decoders) translate the binary address into a unary address—exactly one word in the core is selected. A read is performed by having each cell in the selected word set the bit lines to precharged values; the bit line amplifiers then determine if the selected cell pulled down the bit lines. These circuits also amplify and restore the signal to the proper voltage levels. A write is performed by setting the bit lines to the desired values and driving that value from the bit lines into the cell. If the core word width is narrower than the final word width (for example, a one-bit wide RAM typically has a core much wider than one bit to make a more compact layout), a multiplexer uses the bottom few bits of the address to select the desired bits out of the word.

row decoders

The row decoders are not very complex: they typically use NOR gates to decode the address, followed by a chain of buffers to allow the circuit to drive the large capacitance of the word line. There are two major choices for circuits to implement the NOR function: pseudo-nMOS and precharged. Pseudo-nMOS gates do not have a fully complementary pullup network; they instead use a biased transistor as a substitute for a pullup resistor. They are adequate for small memories, but precharged circuits offer better performance (at the expense of control circuitry) for larger memory arrays.

Figure 2-51 A precharged row decoder.

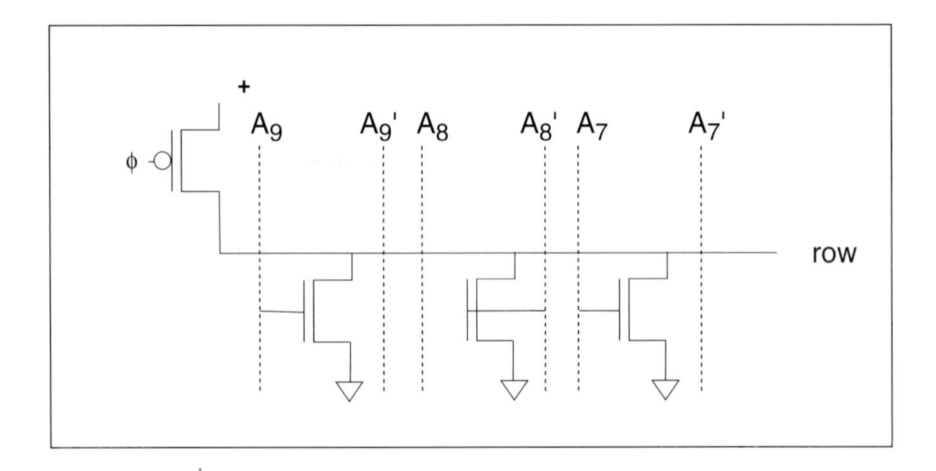

Figure 2-51 shows a precharged row decoder in which the row output is precharged high on one phase and then selectively discharged on the other phase, depending on whether the desired row output is 0 or 1. The true and complement forms of the address lines can be distributed vertically through the decoders and connected to the NOR pulldowns as appropriate for the address to be decoded at that row.

column multiplexers

The column multiplexers are typically implemented as pass transistors on the bit lines. As shown in Figure 2-52, each output bit will be selected from several columns. The multiplexer control signals can be

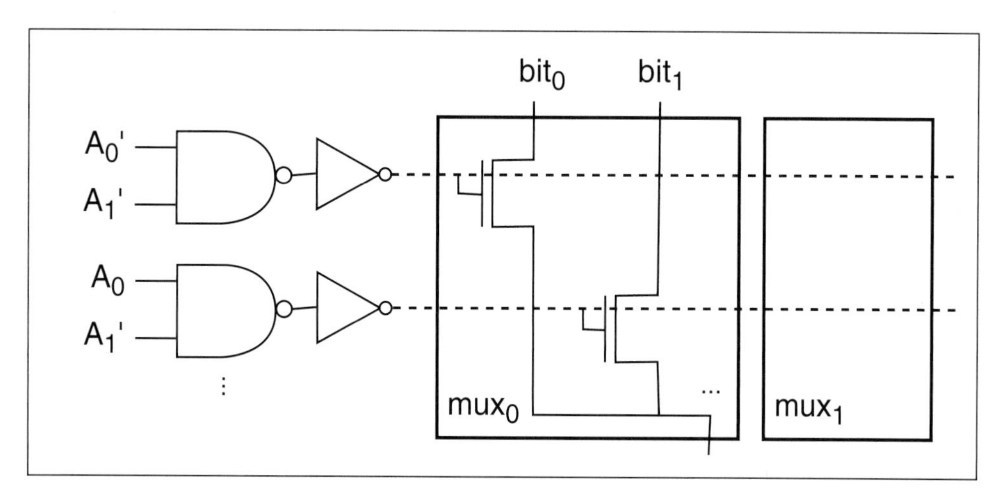

Figure 2-52 A column decoding scheme.

generated at one end of the string of multiplexers and distributed to all the mux cells.

ROM memory core

A read-only memory is programmed with transistors to supply the desired values. A common circuit is the NOR array shown in Figure 2-53. It uses a pseudo-nMOS NOR gate: a transistor is placed at the word-bit line intersection for which *bit'* = 0. Programmable ROMs (PROMs) are also available using flash storage which uses a double capacitor to remember the bit value.

SRAM memory core

The SRAM core circuit is shown in Figure 2-54. The value is stored in the middle four transistors, which form a pair of inverters connected in a loop (try drawing a gate-level version of this schematic). The other two transistors control access to the memory cell by the bit lines. When *select* = 0, the inverters reinforce each other to store the value. A read or write is performed when the cell is selected:

- To read, *bit* and *bit'* are precharged to V_{DD} before the *select* line is allowed to go high. One of the cell's inverters will have its output at 1, and the other at 0; which inverter is 1 depends on the value stored. If, for example, the right-hand inverter's output is 0, the *bit'* line will be drained to V_{SS} through that inverter's pulldown and the *bit* line will remain high. If the opposite value is stored in the cell, the *bit* line will be pulled low while *bit'* remains high.
- To write, the *bit* and *bit'* lines are set to the desired values, then *select* is set to 1. Charge sharing forces the inverters to switch

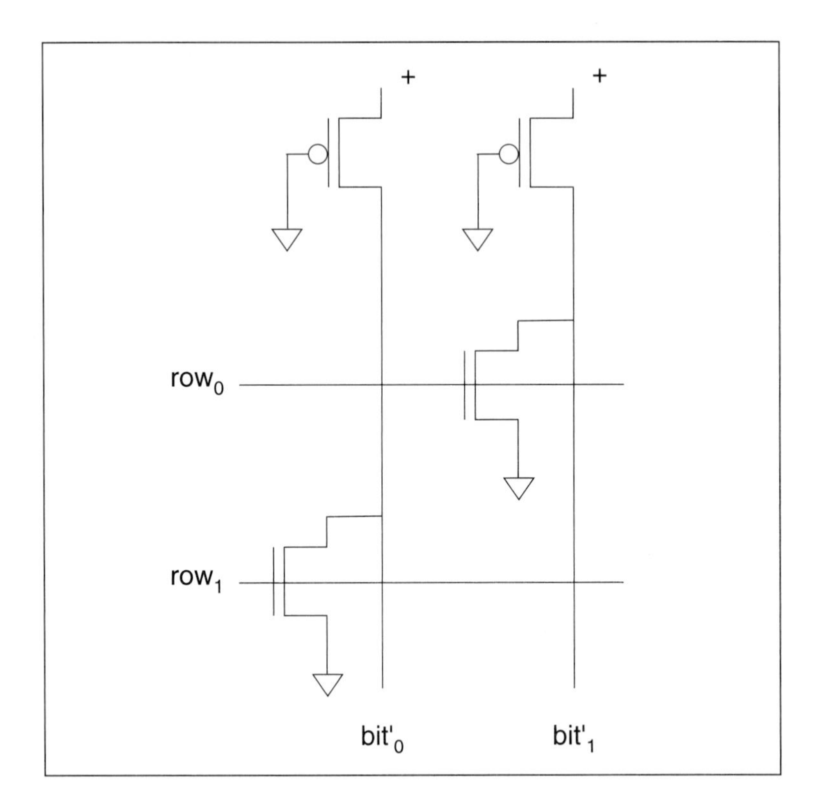

Figure 2-53 Design of a ROM core.

values, if necessary, to store the desired value. The bit lines have much higher capacitance than the inverters, so the charge on the bit lines is enough to overwhelm the inverter pair and cause it to flip state.

multi-port SRAM

Many designs require multi-ported RAMs. For example, a register file is often implemented as a multi-port SRAM. A multi-ported SRAM can be built using several access transistors. Each address in the SRAM can be accessed from only one port, but different addresses can be accessed simultaneously by using the different ports. Each port consists of address input, data outputs, select and read/write lines. When select is asserted on the ith port, the ith address is used to read or write the addressed cells using the ith set of data lines. Reads and writes on separate ports are independent, although the effect of simultaneous writes to a port are undefined.

The circuit schematic for a two-port SRAM core cell is shown in Figure 2-55. Each port has its own pair of access transistors. The transistors in the cross-coupled inverters must be resized moderately to ensure that

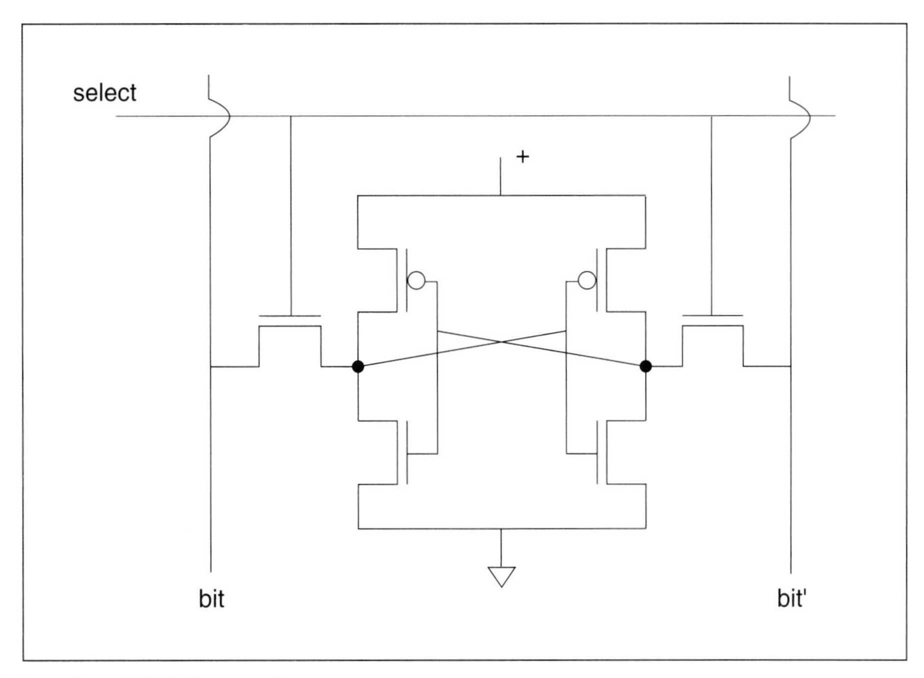

Figure 2-54 Design of an SRAM core cell.

multiple port activations do not affect the stored value, but the circuit and layout designs do not differ radically from the single-ported cell.

A simplified SRAM core circuit is shown in Figure 2-56. The value is stored in the cross-coupled inverters just as in the six-transistor cell. An access transistor controls whether the bit can be read or written. A bit is read by sensing the value at the data pin when the access transistor is addressed. A bit is written by driving the data pin while the access transistor is on, possibly forcing the inverter pair to flip its state. Separate logic decodes addresses to determine when each bit is accessed.

sense amplifier

A **sense amplifier**, shown in Figure 2-57, makes a reasonable bit line receiver for modest-sized SRAMs. The n-type transistor at the bottom acts as a switchable current source—when turned on by the *sense* input, the transistor pulls a fixed current *I* through the sense amp's two arms. Kirchoff's current law tells us that the currents through the two branches must sum to *I*. When one of the bit lines goes low, the current through that leg of the amplifier goes low, increasing the current in the other leg. P-type transistors are used as loads. For an output of the opposite polarity, both the output and the pullup bias connection must be switched to

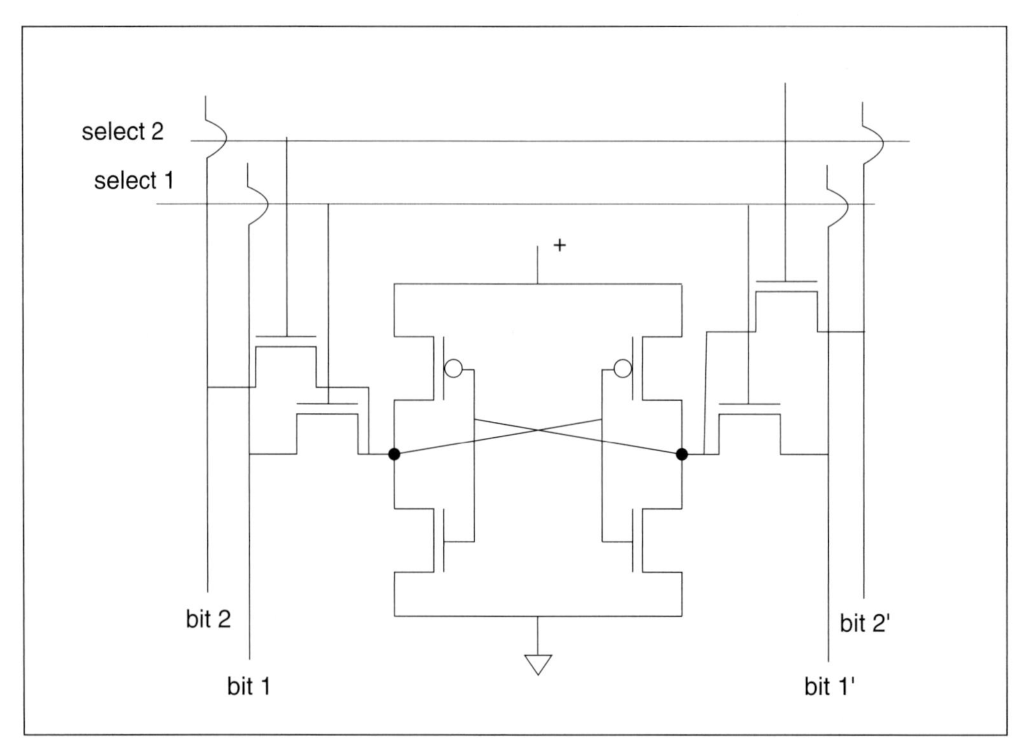

Figure 2-55 A dual-ported SRAM core cell.

Figure 2-56 A simplified SRAM cell.

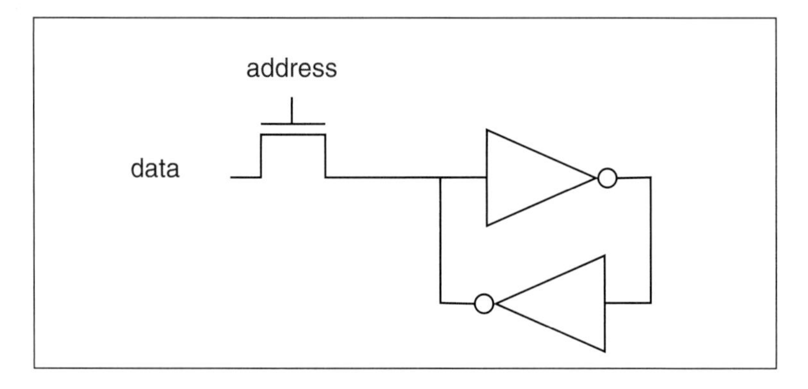

the opposite sides of the circuit. More complex circuits can determine the bit line value more quickly [Gla85].

A precharging circuit for the bit lines is shown in Figure 2-58. Precharging is controlled by a single line. The major novelty of this circuit is the

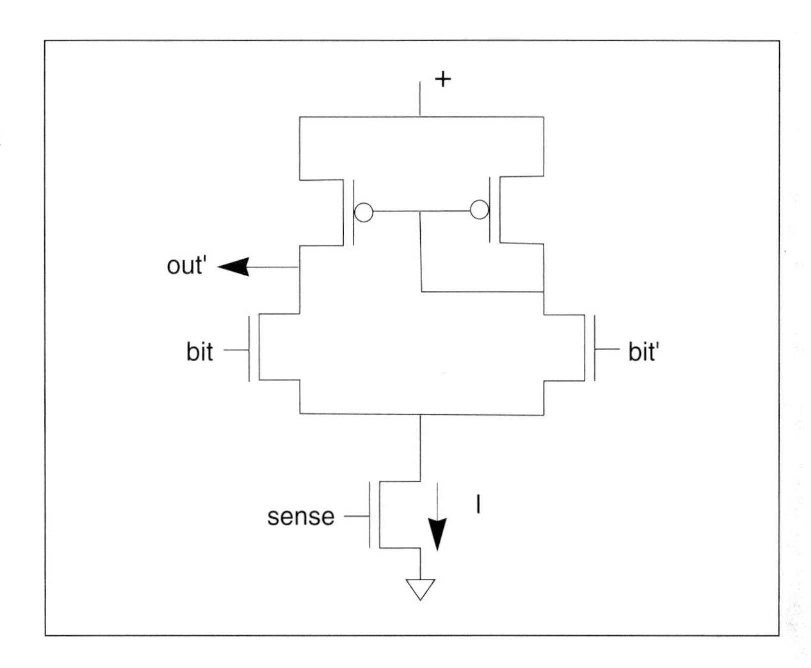

Figure 2-57 A differential pair sense amplifier for an SRAM.

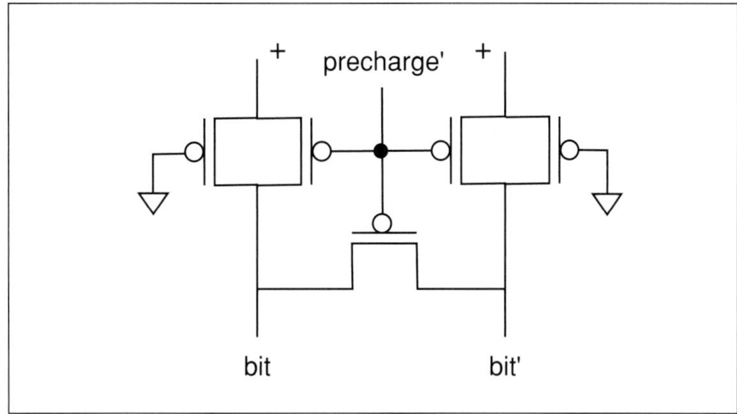

Figure 2-58 An SRAM precharge circuit.

transistor between the bit and bit' lines, which is used to equalize the charge on the two lines.

dynamic RAM core cells

The simplest dynamic RAM cell uses a three-transistor circuit [Reg70]. This circuit is fairly large and slow. It is sometimes used in ASICs because it is denser than SRAM and, unlike one-transistor DRAM, does not require special processing steps. The three-transistor DRAM circuit is shown in Figure 2-59. The value is stored on the gate capacitance of t_1; the other two transistors are used to control access to that value:

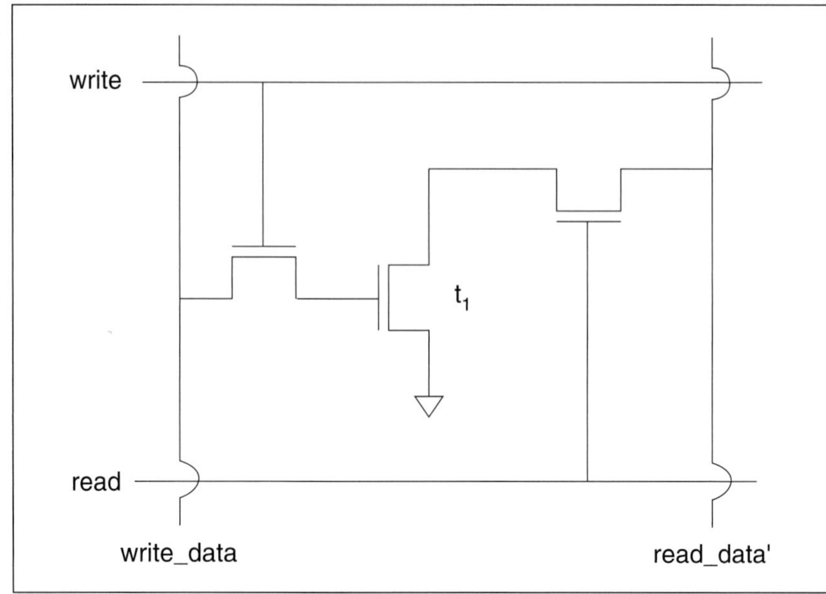

Figure 2-59 Design of a three-transistor DRAM core cell.

- To read, *read_data'* is precharged to V_{DD}. We then set *read* to 1 and write to 0. If t_1's gate has a stored charge, then t_1 will pull down the *read_data'* signal else *read_data'* will remain charged. *Read_data'*, therefore, carries the complement of the value stored on t_1.
- To write, the value to be written is set on *write_data, write* is set to 1, and *read* to 0. Charge sharing between *write_data* and t_1's gate capacitance forces t_1 to the desired value.

DRAM refresh

Substrate leakage will cause the value in this cell to decay. The value must be **refreshed** periodically—a refresh interval of 10 ms is consistent with the approximate leakage rate of typical processes. The value is refreshed by rewriting it into the cell, being careful of course to rewrite the original value.

one-transistor DRAM

The one-transistor DRAM circuit quickly supplanted the three-transistor circuit because it could be packed more densely, particularly when advanced processing techniques are used. The term one-transistor is somewhat of a misnomer—a more accurate description would be one-transistor/one-capacitor DRAM, since the charge is stored on a pure capacitor rather than on the gate capacitance of a transistor. The design of one-transistor DRAMs is an art beyond the scope of this book. But since embedded DRAM is becoming more popular, it is increasingly likely that designers will build chips with one-transistor DRAM subsystems, so it is useful to understand the basics of this memory circuit.

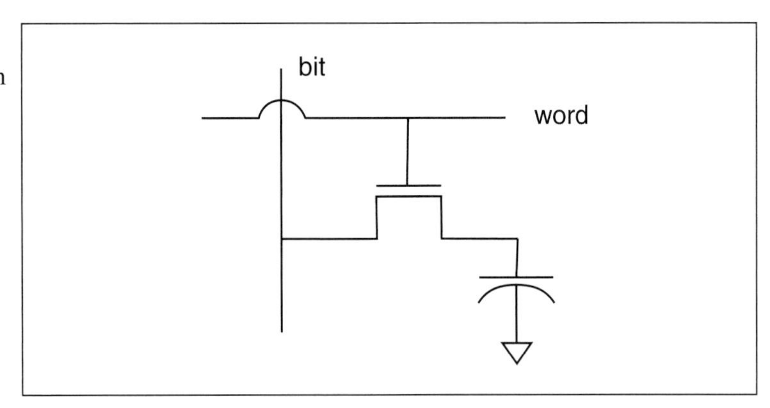

Figure 2-60 Circuit diagram for a one-transistor DRAM core cell.

Figure 2-60 shows the circuit diagram of a one-transistor DRAM core cell. The cell has two external connections: a bit line and a word line. The value is stored on a capacitor guarded by a single transistor. Setting the word line high connects the capacitor to the bit line. To write a new value, the bit line is set accordingly and the capacitor is forced to the proper value. When reading the value, the bit line is first precharged before the word line is activated. If the storage capacitor is discharged, then charge will flow from the bit line to the capacitor, lowering the voltage on the bit line. A sense amp can be used to detect the dip in voltage; since the bit line provides only a single-ended input to the bit line, a reference voltage may be used as the sense amp's other input. One common way to generate the reference voltage is to introduce dummy cells which are precharged but not read or written. This read is destructive—the zero on the capacitor has been replaced by a one during reading. As a result, additional circuitry must be placed on the bit lines to pull the bit line and storage capacitor to zero when a low voltage is detected on the bit line. This cell's value must also be refreshed periodically, but it can be refreshed by reading the cell.

Modern DRAMs are designed with three-dimensional structures to minimize the size of the storage cell. The two major techniques for DRAM fabrication are the **stacked capacitor** and the **trench capacitor**. The cross-section of a pair of stacked capacitor cells is shown in Figure 2-61 [Tak85]. The cell uses three layers of polysilicon and one level of metal: the word line is fabricated in poly 1, the bottom of the capacitor in poly 2, and the top plate of the capacitor in poly 3. The bit line is run in metal above the capacitor structures. The capacitor actually wraps around the access transistor, packing a larger parallel plate area in a smaller surface

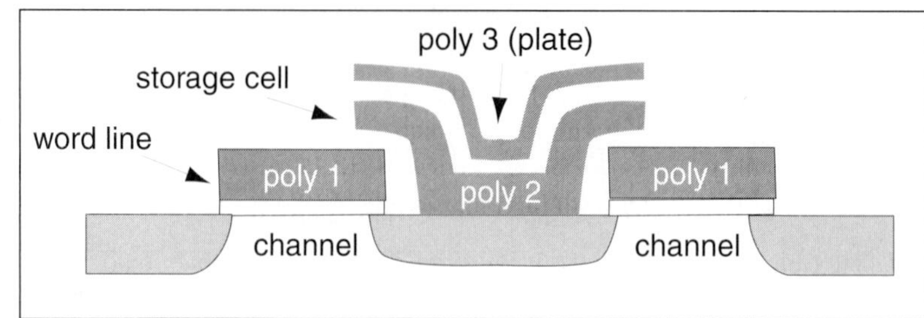

Figure 2-61 Cross-section of a pair of stacked-capacitor DRAM cells.

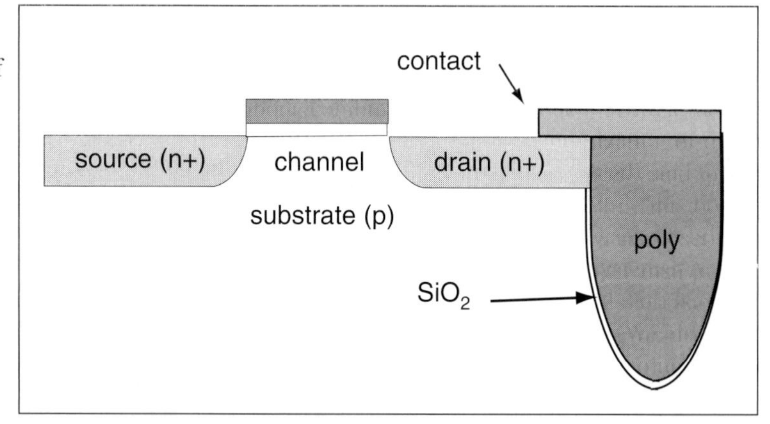

Figure 2-62 Cross-section of a one-transistor DRAM cell built with a trench capacitor.

area. The bottom edge of the bottom plate makes the contact with the access transistor, saving additional area. The trench capacitor cell cross-section is shown in Figure 2-62 [Sun84]. A trench is etched into the chip, oxide is formed, and the trench is filled with polysilicon. This structure automatically connects the bottom plate to the grounded substrate; a contact is used to directly connect the polysilicon plate to the access transistor.

One should not expect one-transistor DRAMs that can be fabricated on a logic process to be equivalent to commodity DRAMs. The processing steps required to create a dense array of capacitors are not ideal for efficient logic transistors, so high-density DRAMs generally have lower-quality transistors. Since transistors make up a relatively small fraction of the circuitry in a commodity DRAM, those chips are optimized for the capacitors. However, in a process designed to implement large amounts of logic with some embedded DRAM, processing optimizations will generally be made in favor of logic, resulting in less-dense DRAM circuitry. In addition, the sorts of manufacturing optimizations possible in commodity parts are also not generally possible in logic-and-

DRAM processes, since more distinct parts will be manufactured, making it more difficult to measure the process. As a result, embedded DRAM will generally be larger and slower than what one would expect from evaluating commodity DRAM in a same-generation process.

2.7 Packages and Pads

Integrated circuits need to connect to the outside world but also be protected from environmental damage. The I/O pad is the circuit used to connect to the larger world beyond the chip. The package is designed to carry and protect the chip. In this section we will look at both since the electrical characteristics of the pads and package have some effect on system design.

2.7.1 Packages

A chip isn't very useful if you can't connect it to the outside world. You rarely see chips themselves because they are encased in **packages**. A complete discussion of packaging and its engineering problems fills several books [Bak90, Ser89]. The packaging problem most directly relevant to the designer is the pads which connect the chip's internals to the package and surrounding circuitry—the chip designer is responsible for designing the pad assembly. But first, we will briefly review packages and some of the system and electrical problems they introduce.

Chips are much too fragile to be given to customers in the buff. The package serves a variety of important needs: its pins provide manageable solder connections; it gives the chip mechanical support; it conducts heat away from the chip to the environment; ceramic packages in particular protect the chip from chemical damage.

Figure 2-63 shows a schematic of a generic package (though high-density packaging technologies often have very different designs). The chip sits in a cavity. The circuit board connects to the pins at the edge (or sometimes the bottom) of the package. Wiring built into the package (called traces) goes from the pins to the edge of the cavity; very fine **bonding wires** that connect to the package's leads are connected by robot machines to the chip's pads. The **pads** are metal rectangles large enough to be soldered to the leads. Figure 2-64 shows a photograph of a package before a chip has been bonded to it. The cavity is gold-plated to provide a connection to the chip's substrate for application of a bias voltage. Bonding pads connected to the package's pins surround the four

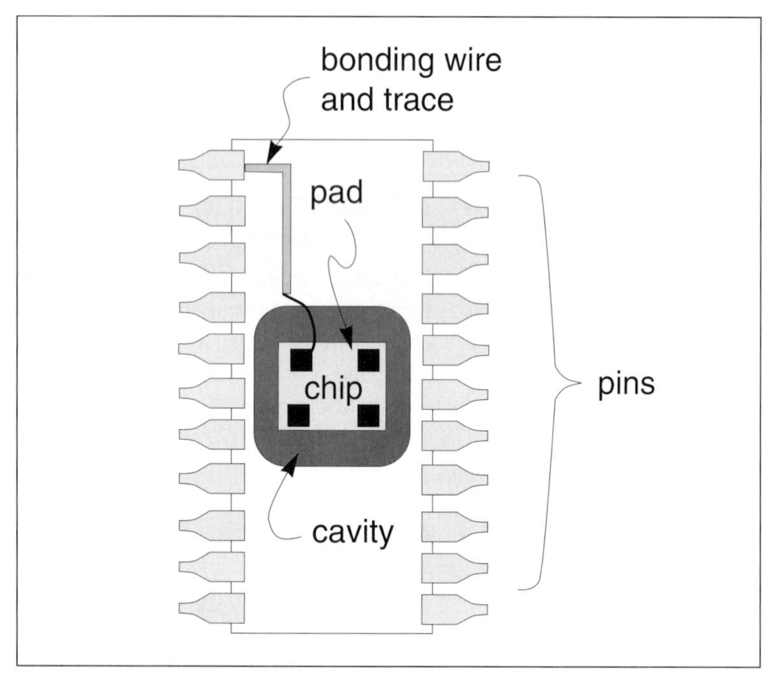

Figure 2-63 Structure of a typical package.

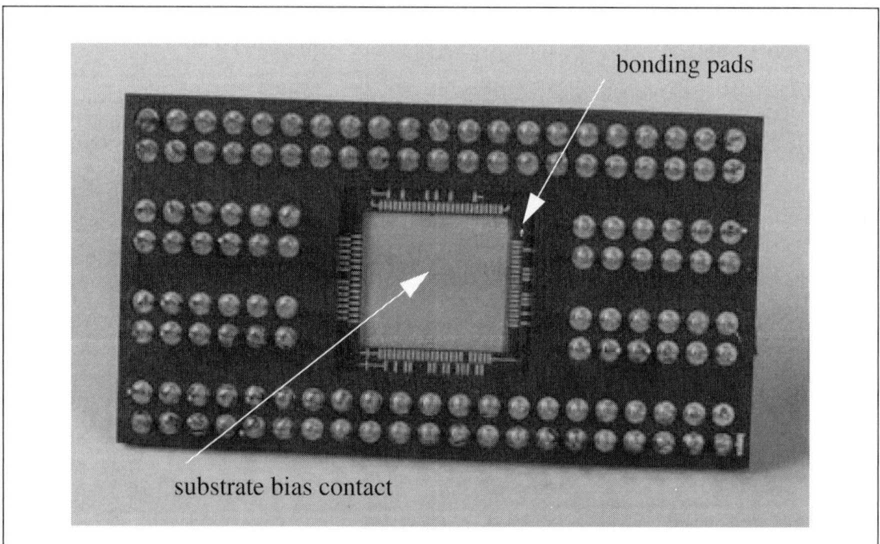

Figure 2-64 An empty package showing the substrate contact and bonding areas.

Figure 2-65 Common package types.

sides of the cavity. In a ceramic package, the cavity is sealed by a lid; to make a plastic package, the chip is soldered to a bare wiring frame, then the plastic is injection-molded around the chip-frame assembly. Ceramic packages offer better heat conductivity and environmental protection.

Figure 2-65 shows several varieties of packages. These packages vary in cost and the number of available pins; as you would expect, packages with more pins cost more money. The **dual in-line package** (**DIP**) is the cheapest and has the fewest number of pads, usually no more than 40. The **plastic leadless chip carrier** (**PLCC**) has pins around its four edges; these leads are designed to be connected to printed circuit boards without through-board holes. PLCCs typically have in the neighborhood of 128 pins. The **pin grid array** (**PGA**) has pins all over its bottom and can accommodate about 256 pins. The ball grid array (**BGA**) uses solder balls to connect to the package across the entire bottom of the package. The **plastic quad flat pack** (**PQFP**), which is not shown, resembles the PLCC in some respects, but has a different pin geometry.

Packages introduce system complications because of their limited pinout. Although 256 may seem like a lot of pins, consider a 32-bit microprocessor—add together 32-bit data and address ports for both instructions and data, I/O bus signals, and miscellaneous control signals, and those pins disappear quickly. Cost-sensitive chips may not be able to afford the more expensive packages and so may not get as many pins as they could use. Off-chip bandwidth is one of the most precious commodities in high-performance designs. It may be necessary to modify the chip's internal architecture to perform more work on-chip and conserve communication. It may also be necessary to communicate in units smaller than the natural data size, adding rate conversion circuitry to the chip and possibly slowing it down.

Packages also introduce electrical problems. The most common problems are caused by the inductance of the pins and the printed circuit board attached to them—on-chip wires have negligible inductance, but package inductance can introduce significant voltage fluctuations.

Example 2-6 Power line inductance

Inductance causes the most problems on the power line because the largest current swings occur on that pin. (Inductance can also cause problems for signals in very high-frequency parts.) Package inductance caused headaches for early VLSI designers, who didn't expect the package to introduce such serious problems. However, these problems can be easily fixed once they are identified.

The system's complete power circuit looks like this:

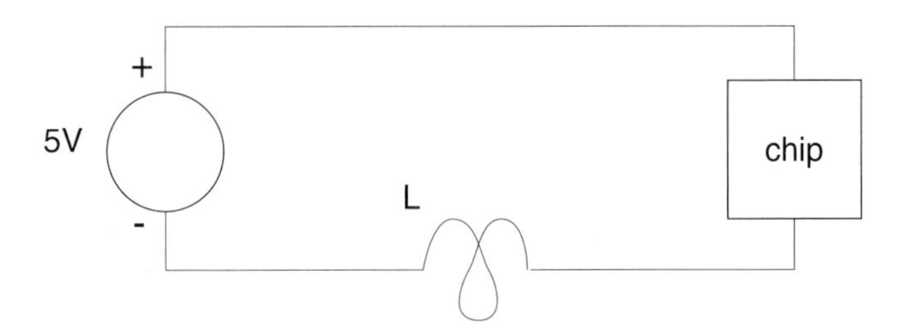

An off-chip power supply is connected to the chip. The inductance of the package and the printed circuit board trace is in series with the chip. (The chip also contributes capacitance to ground on the V_{DD} and V_{SS} lines which we are ignoring here.) The voltage across the inductance is

$$v_L = L\frac{di_L}{dt}.$$

In steady state there is no voltage drop across the inductance. But, if the current supplied by the power supply changes suddenly, v_L momentarily increases, and since the inductance and chip are in series, the power supply voltage seen at the chip decreases by the same amount. How much will the voltage supplied to the chip drop? Assume that the power supply current changes by 1 *A* in 1 *ns*, a large but not impossible value. A typical value for the package and printed circuit board total inductance is 0.5 *nH* [Ser89]. That gives a peak voltage drop of $v_L = 0.5\times10^{-9}H \cdot 1A/1\times10^{-9}s = 0.5V$, which may easily be large enough to cause dynamic circuits to malfunction. We can avoid this problem by introducing multiple power and ground pins. Running current through several pins in parallel reduces *di/dt* in each pin, reducing the total voltage drop. The first-generation Intel Pentium™ package, for example, has 497 V_{CC} pins and an equal number of V_{SS} pins [Int94].

2.7.2 Pads

A pad connects the chip to the outside world. The pad includes a large piece of metal that can be used to connect to a wire; in order to drive the large capacitive load of the pad itself, driver circuits are required for output. **Electrostatic discharge** (**ESD**) protection circuitry is added to input pads to eliminate damage caused from static electricity. A pad used for both input and output, sometimes known by its trade name of three-state pin[1], combines elements of both input and output pads.

input pad

The input pad may include circuitry to shift voltage levels, *etc*. The main job of an input pad is to protect the chip core from static electricity. People or equipment can easily deliver a large static voltage to a pin when handling a chip. MOS circuits are particularly sensitive to static discharge because they use thin oxides. The gate oxide, which may be a few hundred Angstroms thick, can be totally destroyed by a single static jolt, shorting the transistor's gate to its channel.

electrostatic discharge protection

An input pad puts protective circuitry between the pad itself and the transistor gates in the chip core. Electrostatic discharge can cause two types of problems: dielectric rupture and charge injection [Vin98]. When the dielectric ruptures, chip structures can short together. The charge injected by an ESD event can be sufficient to damage the small on-chip devices.

output pad

Electrostatic discharge protection is not needed for an output pad because the pad is not connected to any transistor gate. The main job of an output pad is to drive the large capacitances seen on the output pin. Within the chip, scaling ensures that we can use smaller and smaller transistors, but the real world doesn't shrink along with our chip's channel length. The output pad's circuitry includes a chain of drivers for the large off-chip load.

Figure 2-66 A three-state pad circuit.

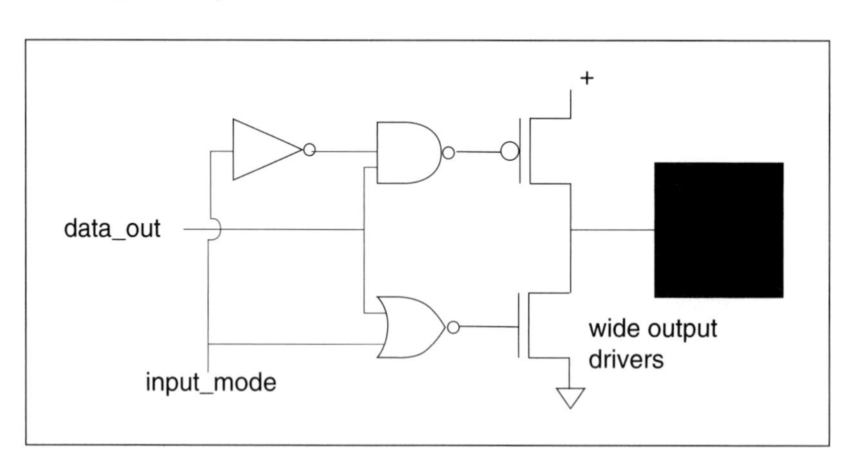

1.Tri-state is a trademark of National Semiconductor.

three-state pad

Three-state pads, used for both input and output, help solve the pin count crunch—if we don't need to use all combinations of inputs and outputs simultaneously, we can switch some pins between input and output. The pad cannot, of course, be used as an input and output simultaneously—the chip core is responsible for switching between modes. The pad requires electrostatic discharge protection for when the pad is used as an input, an output driver for when it is used as an output, plus circuitry to switch the pad between input and output modes. The circuit of Figure 2-66 can be used for mode switching. The n-type and p-type transistors are used to drive the pad when it is used as an output—the logic gates are arranged so that the output signal turns on exactly one of the two transistors when *input_mode* is 0. To use the pad as an input, *input_mode* is set to 1: the NOR gate emits a 0, turning off the pulldown, and the NAND gate emits a 1, turning off the pullup. Since both driver transistors are disabled, the pad can be used as an input. (The required ESD circuitry is not shown in the schematic.)

2.8 Summary

VLSI manufacturing processes determine the basic characteristics of the components used for digital design. The effort to make more complex chips with denser transistors has resulted in complex structures that can require careful analysis. We have seen how the characteristics of transistors and wires help determine delay and power consumption. We have also seen how transistors can be put together to form registers, flip-flops, and RAMs. In the next chapter we will use these elements to build FPGAs.

2.9 Problems

Use the technology parameters for transistors, wires, and other elements given in this chapter where appropriate.

Q2-1. Assuming that V_{gs} = 1V, compute the drain current through n-type transistors of these sizes at V_{ds} values of 0.5V, 0.75V, and 1V:

a. W/L = 5/2.

b. W/L = 8/2.

c. W/L =12/2.

d. W/L = 25/2.

Q2-2. Design the static complementary pullup and pulldown networks for these logic expressions:

a. $(a + b + c)$'.

b. $[(a + b)c]$'.

c. $(a + b)(c + d)$.

Q2-3. Write the defining logic equation and transistor topology for each complex gate below:

a. AOI-22.

b. OAI-22.

c. AOI-212.

d. OAI-321.

e. AOI-2222.

Q2-4. Size the transistors in a three-input, static complementary NOR gate such that the circuit's rise and fall times are approximately equal.

Q2-5. If the input to an inverter is a ramp and not a step, is the fall time calculation of Equation 2-11 optimistic or pessimistic? Explain your answer.

Q2-6. What is the difference in fall time of a two-input, static complementary NOR gate (assuming a minimum-size load capacitance) when one pulldown and when two pulldowns are activated?

Q2-7. Compute the low-to-high transition time and delay (at a power supply voltage of 1V) using the τ model through a two-input NAND gate which drives one input of a three-input NOR gate (both static complementary gates):

a. Compute the load capacitance on the NAND gate, assuming the NOR gate's transistors all have $W=6\lambda$, $L=2\lambda$.

b. Compute the equivalence resistance of appropriate transistors for the low-to-high transition, assuming the pulldown transistors have $W=6\lambda$, $L=2\lambda$ and the pullups have $W=6\lambda$, $L=2\lambda$.

c. Compute the transition time and delay.

Q2-8. Compute, using the τ model, the high-to-low transition time and delay (at a power supply voltage of 1V) of a two-input, static complementary NOR gate with minimum-size transistors driving these loads:

a. An inverter with minimum-size pullup and pulldown.

b. An inverter whose pullup and pulldown are both of size $W=10\lambda$, $L=10\lambda$.

c. A $2000\lambda \times 2\lambda$ poly wire connected to an inverter with minimum-sized pullup and pulldown (lump the wire and inverter parasitics into a single *RC* section).

d. Size the transistors in a two-input NOR gate such that the gate's rise time and fall time are approximately equal.

Q2-9. Draw a schematic of a switch logic network using transmission gates for a three-input multiplexer.

Q2-10. Draw a schematic of a switch logic network using pass transistors for a four-input multiplexer.

Q2-11. Using the capacitance and resistance values of Table 2-3, what length of minimum-resistance metal 1 wire has the same resistance as a minimum-width metal 3 wire that is 1 micron long?

Q2-12. Using the capacitance and resistance values of Table 2-3, what length of minimum-resistance metal 1 wire has the same RC product as a minimum-width metal 3 wire that is 1 micron long?

Q2-13. Plot the Elmore delay for a minimum-width metal 1 wire of length $2000\mu m$ using

a. 2 sections.

b. 4 sections.

c. 8 sections.

Q2-14. Plot the Elmore delay for a minimum-width metal 2 wire of length $3000\mu m$ using

a. 2 sections.

b. 4 sections.

c. 8 sections.

Q2-15. Compute the optimal number of buffers and buffer sizes for these minimum-width RC (non-inductive) wires when driven by a minimum-size inverter:

a. metal 1 $3000\mu m$.

b. metal 1 $5000\mu m$.

c. metal 2 $3000\mu m$.

d. metal 2 $4000\mu m$.

Q2-16. Design a transistor-level schematic for a dynamic D-type latch with an enable input.

Q2-17. Draw a transistor-level schematic for a three-port SRAM cell. The cell should have both bit and bit' lines for each port.

Q2-18. Draw a two-port version of the simplified SRAM cell of Figure 2-56.

3 FPGA Fabrics

Architectures of FPGAs.

SRAM-based FPGAs.

Antifuse-programmed FPGAs.

Programmable I/O pins.

FPGA circuits: logic and interconnect.

3.1 Introduction

In this chapter we will study the basic structures of FPGAs, known as **fabrics**. We will start with a brief introduction to the structure of FPGA fabrics. However, there are several fundamentally different ways to build an FPGA. Therefore, we will discuss combinational logic and interconnect for the two major styles of FPGA: SRAM-based and antifuse-based. The features of I/O pins are fairly similar among these two types of FPGAs, so we will discuss pins at the end of the chapter.

3.2 FPGA Architectures

ements of FPGAs

In general, FPGAs require three major types of elements:

- combinational logic;
- interconnect;
- I/O pins.

These three elements are mixed together to form an FPGA fabric.

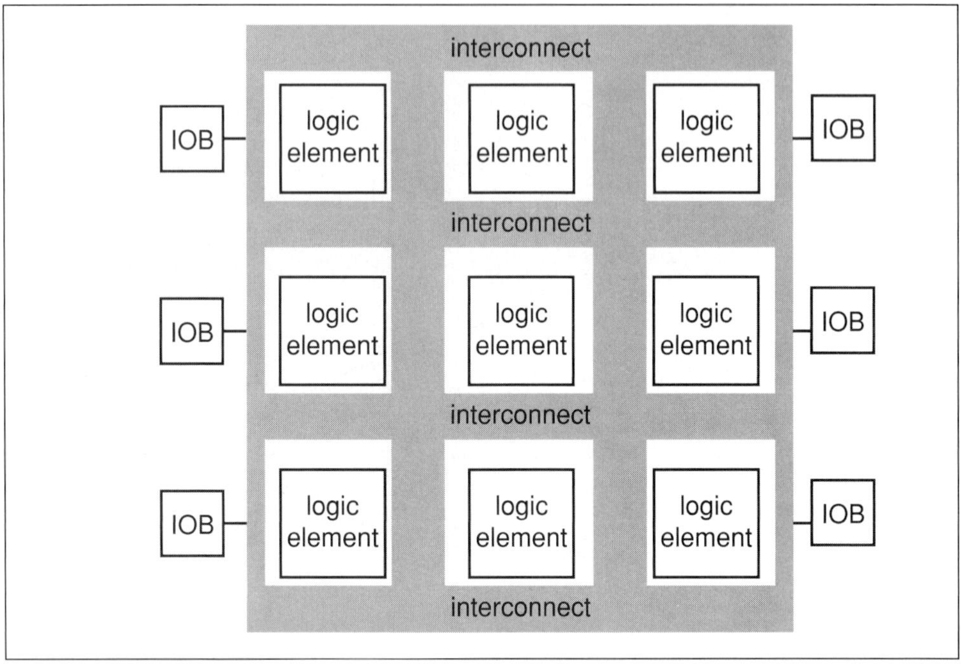

Figure 3-1 Generic structure of an FPGA fabric.

FPGA architectures

Figure 3-1 shows the basic structure of an FPGA that incorporates these three elements. The combinational logic is divided into relatively small units which may be known as **logic elements (LEs)** or **combinational logic blocks (CLBs)**. The LE or CLB can usually form the function of several typical logic gates but it is still small compared to the typical combinational logic block found in a large design. The interconnections are made between the logic elements using programmable interconnect. The interconnect may be logically organized into channels or other units. FPGAs typically offer several types of interconnect depending on the distance between the combinational logic blocks that are to be connected; clock signals are also provided with their own interconnection networks. The I/O pins may be referred to as **I/O blocks** (**IOBs**). They are generally programmable to be inputs or outputs and often provide other features such as low-power or high-speed connections.

FPGA interconnect

An FPGA designer must rely on pre-designed wiring, unlike a custom VLSI designer who can design wires as needed to make connections.

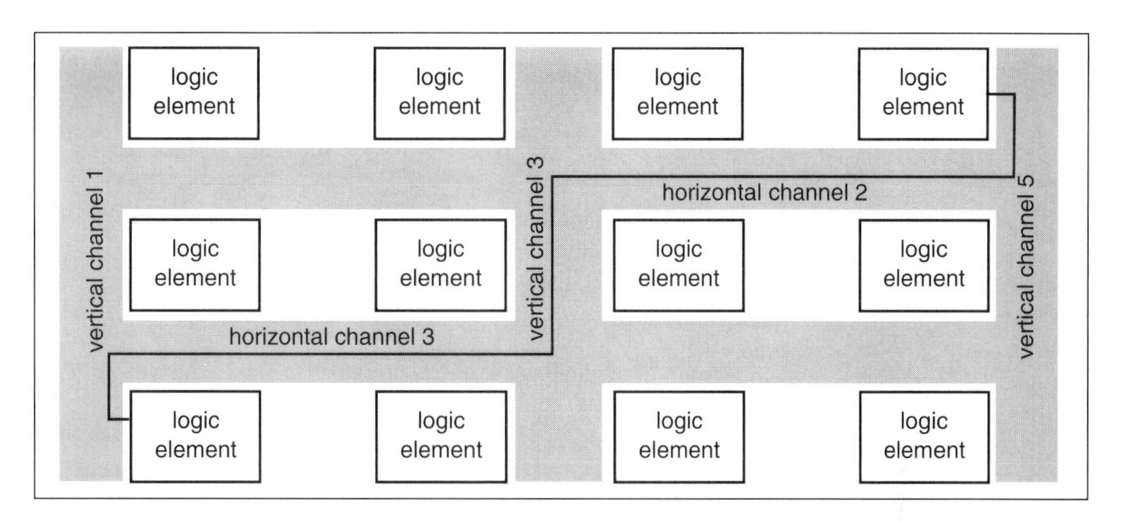

Figure 3-2 Interconnect may require complex paths.

The interconnection system of an FPGA is one of its most complex aspects because wiring is a global property of a logic design.

connection paths

Connections between logic elements may require complex paths since the LEs are arranged in some sort of two-dimensional structure as shown in Figure 3-2. We therefore need to make connections not just between LEs and wires but also between the wires themselves. Wires are typically organized in **wiring channels** or **routing channels** that run horizontally and vertically through the chip. Each channel contains several wires; the human designer or a program chooses which wire will be used in each channel to carry a signal. Connections must be made between wires in order to carry a signal from one point to another. For example, the net in the figure starts from the output of the LE in the upper-right-hand corner, travels down vertical channel 5 until it reaches horizontal channel 2, then moves down vertical channel 3 to horizontal channel 3. It then uses vertical channel 1 to reach the input of the LE at the lower-left-hand corner.

segmented wiring

In order to allow a logic designer to make all the required connections between logic elements, the FPGA channels must provide wires of a variety of lengths, as shown in Figure 3-3. Because the logic elements are organized in a regular array, we can arrange wires going from one LE to another. The figure shows connections of varying length as measured in units of LEs: the top signal of length 1 goes to the next LE, the second signal goes to the second LE, and so on. This organization is

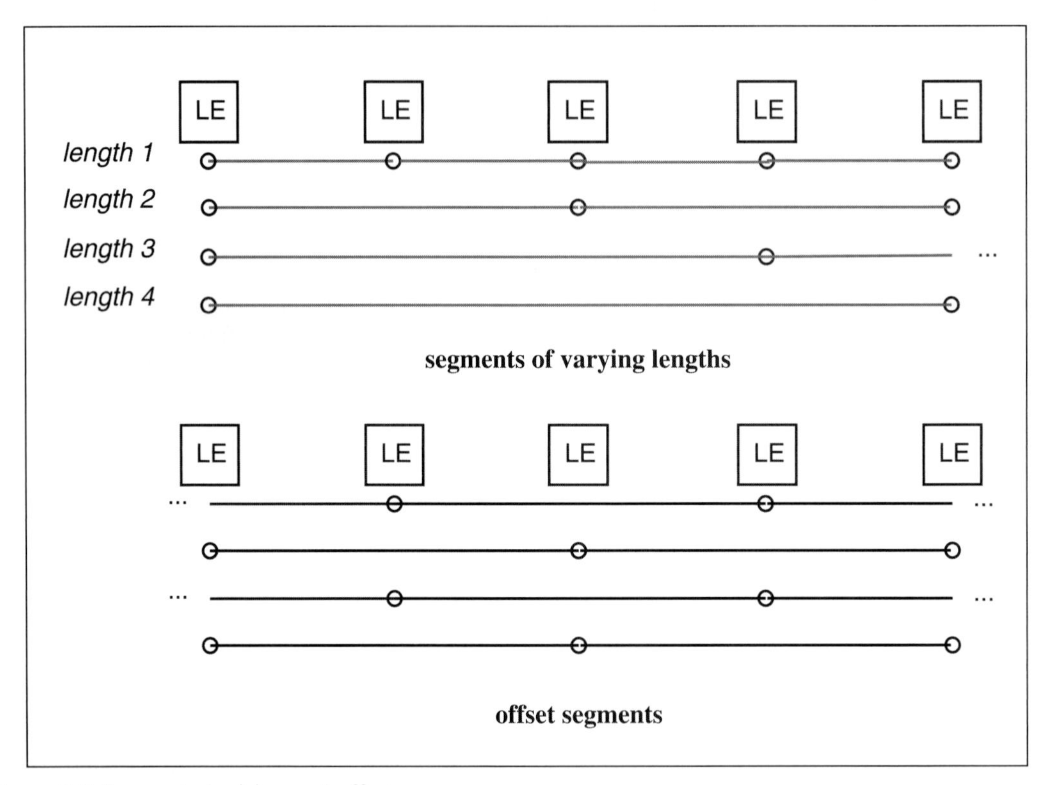

Figure 3-3 Segmented wiring and offsets.

known as a **segmented wiring** structure [ElG88] since the wiring is constructed of segments of varying lengths. The alternative to segmented wiring is to make each wire length 1. However, this would require a long connection to hop through many programmable wiring points, and as we will see in Section 3.6, that would lead to excessive delay along the connection. The segments in a group need not all end at the same point. The bottom part of Figure 3-3 shows segments of length 2 that are offset relative to each other.

FPGA configuration

All FPGAs need to be **programmed** or **configured**. There are three major circuit technologies for configuring an FPGA: SRAM, antifuse, and flash. No matter what circuits are used, all the major elements of the FPGA—the logic, the interconnect, and the I/O pins—need to be configured. The details of these elements vary greatly depending on how the FPGA elements are to be programmed. But FPGAs are very complex VLSI systems that can be characterized in many different ways.

Some of the characteristics of interest to the system designer who wants to use an FPGA include:

- How much logic can I fit into this FPGA?
- How many I/O pins does it have?
- How fast does it run?

While we can determine fairly easily how many I/O pins an FPGA has, determining how much logic can be fit into it and how fast that logic will run is not simple. As we will see in this chapter, the complex architecture of an FPGA means that we must carefully optimize the logic as we fit it into the FPGA. The amount of logic we can fit and how fast that logic runs depends on many characteristics of the FPGA architecture, the logic itself, and the logic design process. We'll look at the tools necessary to configure an FPGA in Chapter 4.

design of FPGA architectures

Some questions of interest to the person who designs the FPGA itself include:

- How many logic elements should the FPGA have?
- How large should each logic element be?
- How much interconnect should it have?
- How many types of interconnection structures should it have?
- How long should each type of interconnect be?
- How many pins should it have?

In Section 3.6 and Section 3.7 we will survey some of the extensive research results in the design of FPGA fabrics. A great deal of theory and experimentation has been developed to determine the parameters for FPGA architectures that best match the characteristics of typical logic that is destined for FPGA implementation.

fine-grain vs. coarse-grain

All of the FPGAs we will deal with in this chapter are **fine-grained** FPGAs—their logic elements can implement fairly small pieces of logic. Advances in VLSI technology are making possible **coarse-grained** FPGAs that are built from large blocks. We will look at some of these architectures in Chapter 7.

chapter outline

The next section looks at FPGAs based on static memory. Section 3.3 studies FPGAs built from permanently programmed parts, either antifuses or flash. Section 3.5 looks at the architecture of chip input and output, which is fairly similar in SRAM and antifuse/flash FPGAs. Section 3.6 builds on these earlier sections by studying in detail the circuits used to build FPGA elements, and Section 3.7 is a detailed study of the architectural characteristics of FPGAs.

3.3 SRAM-Based FPGAs

Static memory is the most widely used method of configuring FPGAs. In this section we will look at the elements of an FPGA: logic, interconnect, and I/O. In doing so we will consider both general principles and specific commercial FPGAs.

3.3.1 Overview

characteristics of SRAM-based FPGAs

SRAM-based FPGAs hold their configurations in static memory (though as we will see in Section 3.6 they don't use the same circuits as are used in commodity SRAMs). The output of the memory cell is directly connected to another circuit and the state of the memory cell continuously controls the circuit being configured.

Using static memory has several advantages:

- The FPGA can be easily reprogrammed. Because the chips can be reused, and generally reprogrammed without removing them from the circuit, SRAM-based FPGAs are the generally accepted choice for system prototyping.
- The FPGA can be reprogrammed during system operation, providing dynamically reconfigurable systems.
- The circuits used in the FPGA can be fabricated with standard VLSI processes.
- Dynamic RAM, although more dense, needs to be refreshed, which would make the configuration circuitry much more cumbersome.

SRAM-based FPGAs also have some disadvantages:

- The SRAM configuration memory burns a noticeable amount of power, even when the program is not changed.
- The bits in the SRAM configuration are susceptible to theft.

A large number of bits must be set in order to program an FPGA. Each combinational logic element requires many programming bits and each programmable interconnection point requires its own bit.

3.3.2 Logic Elements

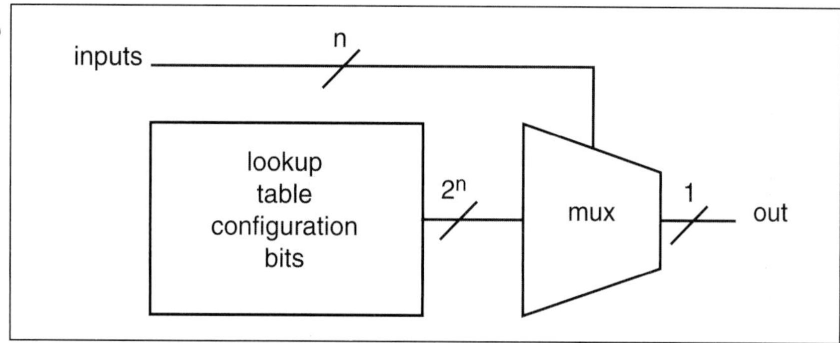

Figure 3-4 A lookup table.

lookup tables

The basic method used to build a **combinational logic block (CLB)**—also called a **logic element** or **LE**—in an SRAM-based FPGA is the **lookup table** (**LUT**). As shown in Figure 3-4, the lookup table is an SRAM that is used to implement a truth table. Each address in the SRAM represents a combination of inputs to the logic element. The value stored at that address represents the value of the function for that input combination. An n-input function requires an SRAM with 2^n locations. Because a basic SRAM is not clocked, the lookup table LE operates much as any other logic gate—as its inputs change, its output changes after some delay.

programming a lookup table

Unlike a typical logic gate, the function represented by the LE can be changed by changing the values of the bits stored in the SRAM. As a result, the n-input LE can represent 2^{2^n} functions (though some of these functions are permutations of each other). A typical logic element has four inputs. The delay through the lookup table is independent of the bits stored in the SRAM, so the delay through the logic element is the same for all functions. This means that, for example, a lookup table-based LE will exhibit the same delay for a 4-input XOR and a 4-input NAND. In contrast, a 4-input XOR built with static CMOS logic is considerably slower than a 4-input NAND. Of course, the static logic gate is generally faster than the LE.

Logic elements generally contain registers—flip-flops and latches—as well as combinational logic. A flip-flop or latch is small compared to the combinational logic element (in sharp contrast to the situation in custom VLSI), so it makes sense to add it to the combinational logic element. Using a separate cell for the memory element would simply take up routing resources. As shown in Figure 3-5, the memory element is

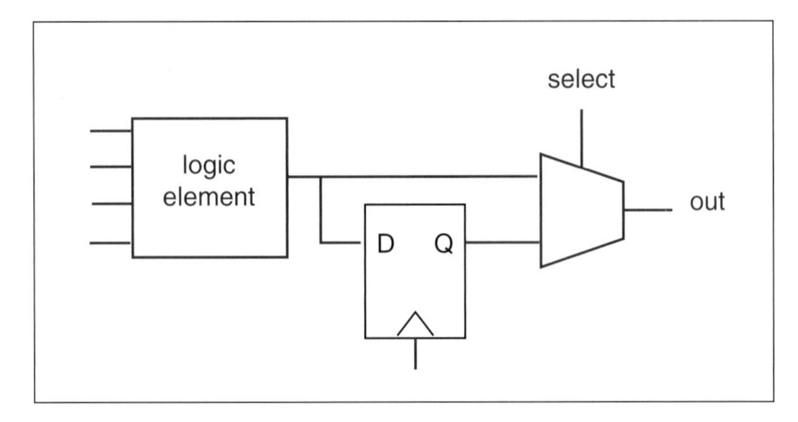

Figure 3-5 A flip-flop in a logic element.

connected to the output; whether it stores a given value is controlled by its clock and enable inputs.

complex logic elements

More complex logic blocks are also possible. For example, many logic elements also contain special circuitry for addition.

Many FPGAs also incorporate specialized adder logic in the logic element. The critical component of an adder is the carry chain, which can be implemented much more efficiently in specialized logic than it can using standard lookup table techniques.

The next two examples describe the logic elements in two FPGAs. They illustrate both the commonality between FPGA structures and the varying approaches to the design of logic elements.

Example 3-1 Xilinx Spartan-II combinational logic block

The Spartan-II combinational logic block [Xil01] consists of two identical slices, with each slice containing a LUT, some carry logic, and registers. Here is one slice:

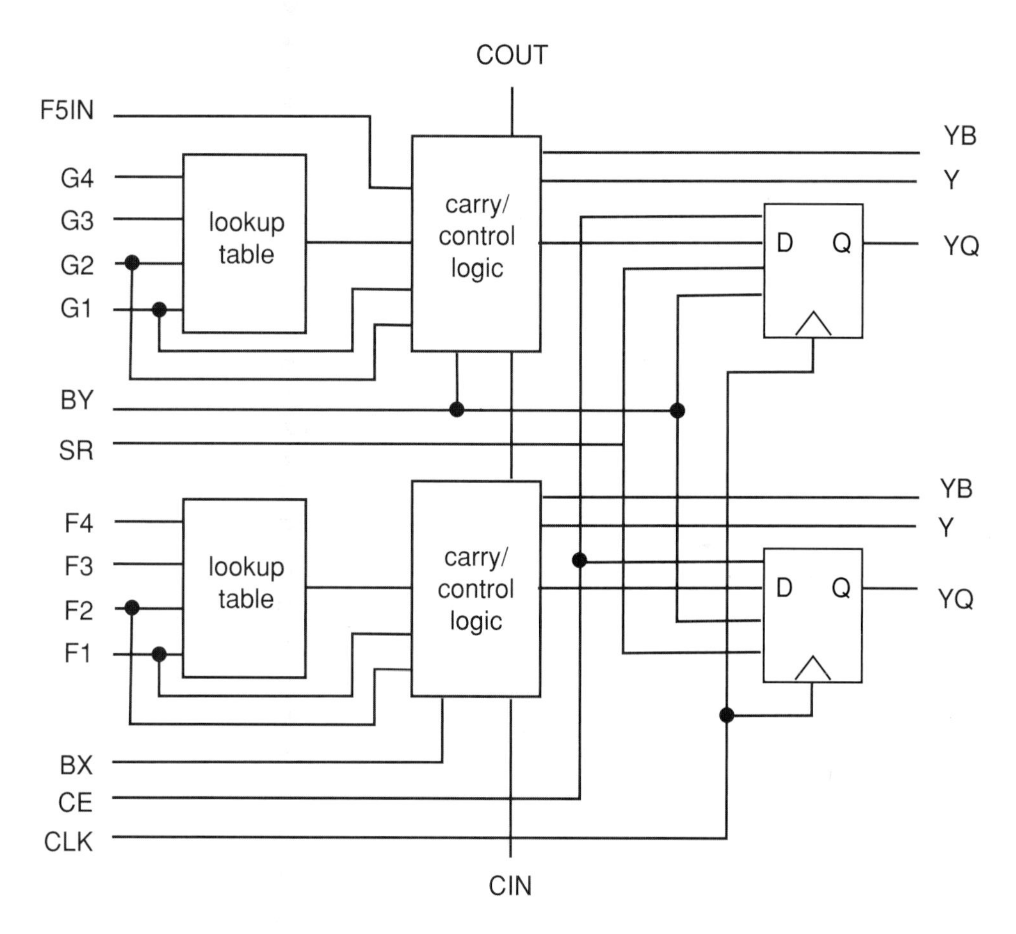

A slice includes two **logic cells** (**LCs**). The foundation of a logic cell is the pair of four-bit lookup tables. Their inputs are F1-F4 and G1-G4. Each lookup table can also be used as a 16-bit synchronous RAM or as a 16-bit shift register. Each slice also contains carry logic for each LUT so that additions can be performed. A carry in to the slice enters the CIN input, goes through the two bits of carry chain, and out through COUT. The arithmetic logic also includes an XOR gate. To build an adder, the

XOR is used to generate the sum and the LUT is used for the carry computation.

Each slice includes a multiplexer that is used to combine the results of the two function generators in a slice. Another multiplexer combines the outputs of the multiplexers in the two slices, generating a result for the entire CLB.

The registers can be configured either as D-type flip-flops or as latches. Each register has clock and clock enable signals.

Each CLB also contains two three-state drivers (known as **BUFTs**) that can be used to drive on-chip busses.

Example 3-2 Altera APEX II logic elements

The APEX II's logic [Alt02] is organized into **logic array blocks** (**LABs**). Each LAB includes 10 logic elements. Each logic element contains a lookup table, flip-flop, *etc*. The logic elements in an LAB share some logic, such as a carry chain and some control signal generation.

A single logic element looks like this:

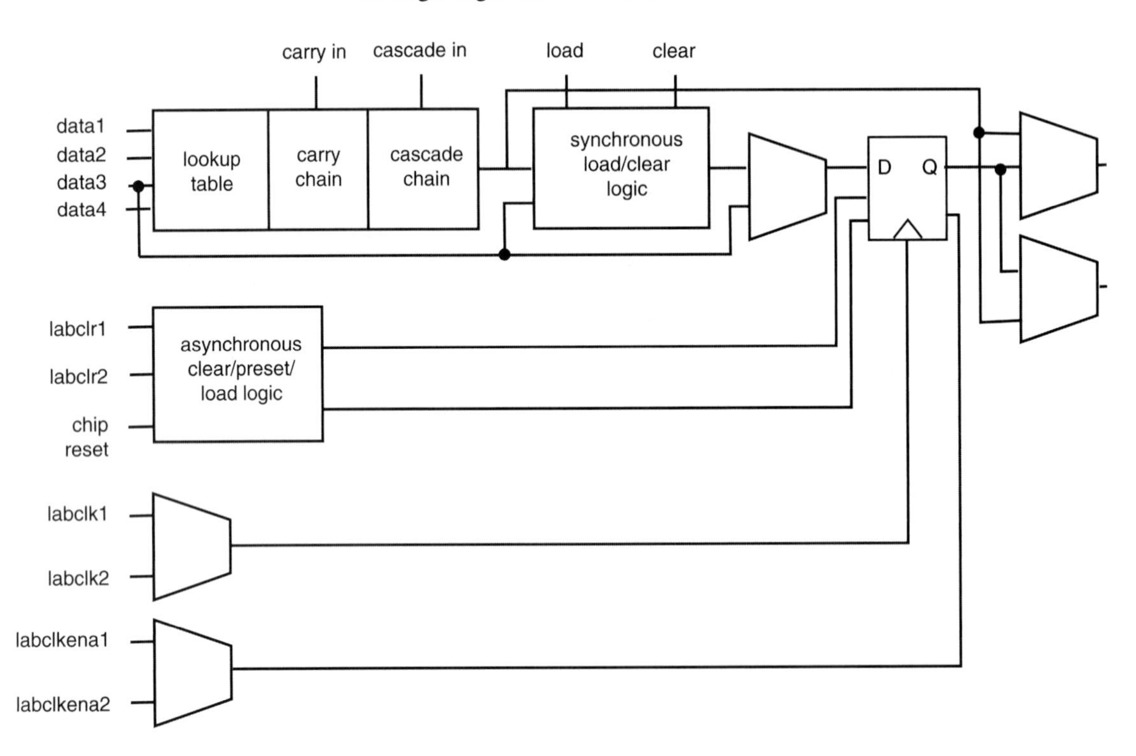

The main logic chain starts with a 4-input lookup table. The output of the LUT is fed to a carry chain. The cascade chain is used for cascading large fanin functions. For example, an AND function with a large number of inputs can be built using the cascade chain. The output of the cascade chain goes to a register. The register can be programmed to operate as a D, T, JK, or SR element.

To use all this logic, an LE can be operated in normal, arithmetic, or counter mode. In normal mode, the LE looks like this:

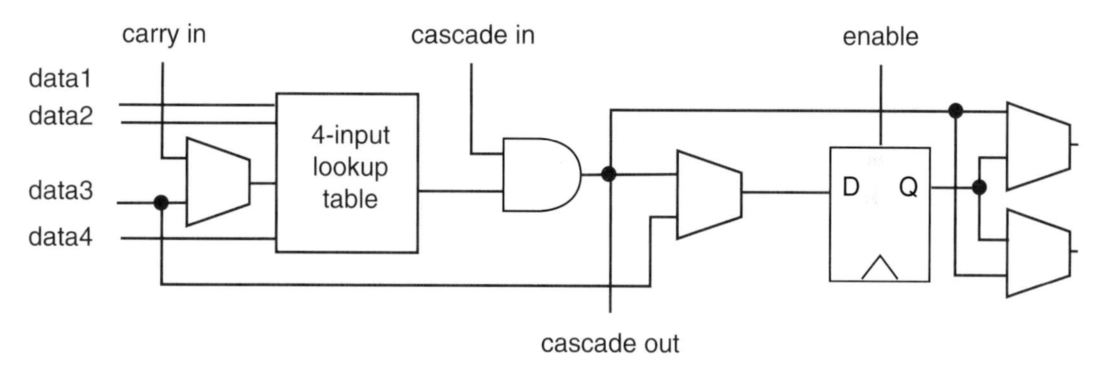

In arithmetic mode, the LE's configuration changes to take advantage of the carry chain logic:

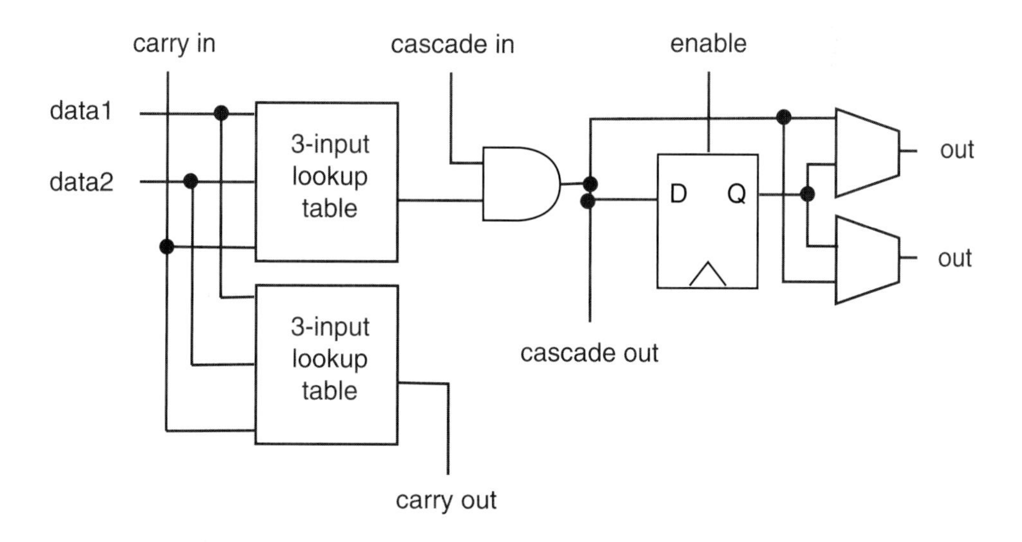

In counter mode, the configuration is altered slightly to provide a fast count:

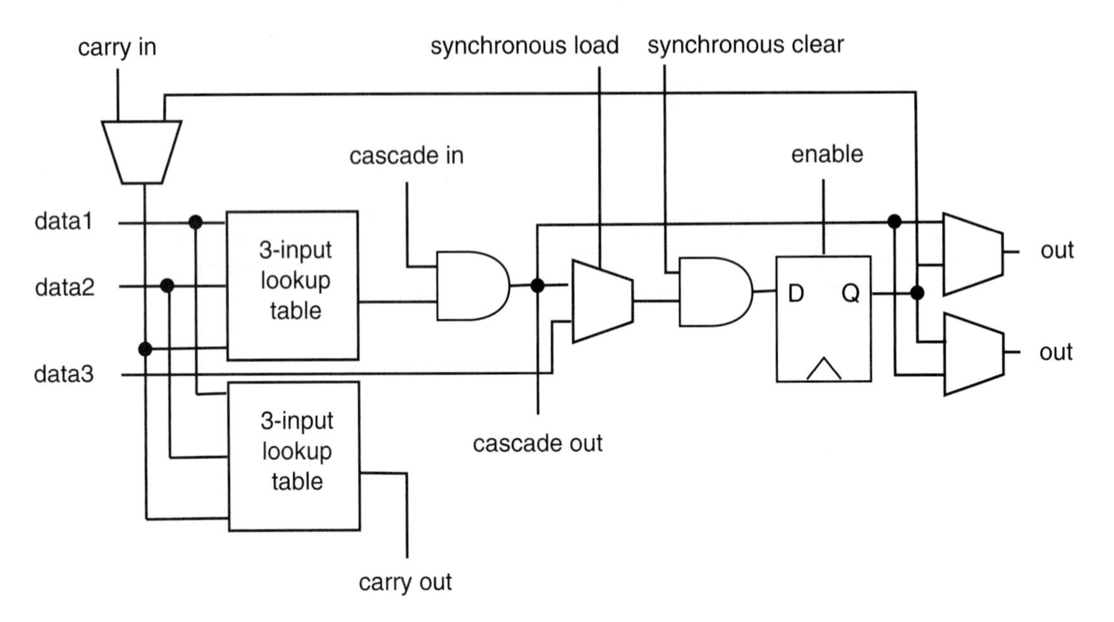

Each logic array block also includes some logic to generate control signals that can be distributed to all the LEs in the block:

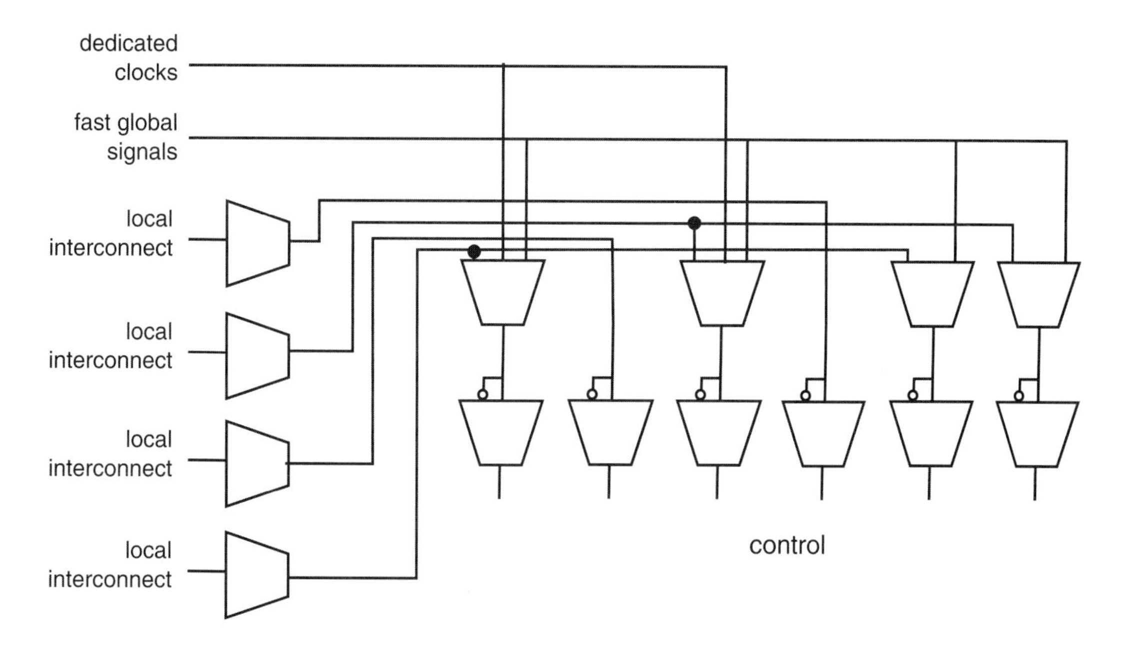

This block can take input signals from the rest of the chip and generate signals for the register: load, clear, enable, *etc.*

The LAB-wide control signals control the preset and clear signals in the LEs' registers. The APEX II also provides a chip-wide reset pin to clear all the registers in the device.

3.3.3 Interconnection Networks

Logic elements must be interconnected to implement complex machines. An SRAM-based FPGA uses SRAM to hold the information used to program the interconnect. As a result, the interconnect can be reconfigured, just as the logic elements can.

programmable interconnection points

Figure 3-6 shows a simple version of an **interconnection point,** often known as a **connection box**. A programmable connection between two wires is made by a CMOS transistor (a pass transistor). The pass transistor's gate is controlled by a static memory program bit (shown here as a

Figure 3-6 An interconnect point controlled by an SRAM cell.

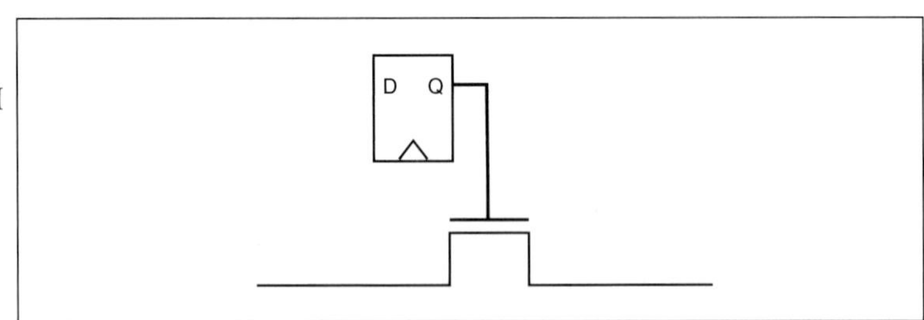

D register). When the pass transistor's gate is high, the transistor conducts and connects the two wires; when the gate is low, the transistor is off and the two wires are not connected. A CMOS transistor has a good off-state (though off-states are becoming worse as chip geometries shrink). In this simple circuit, the transistor also conducts bidirectionally—it doesn't matter which wire has the signal driver. However, the pass transistor is relatively slow, particularly on a signal path that includes several interconnection points in a row. As we will see in Section 3.6, there are several other circuits that can be used to build a programmable interconnection point, most of them unidirectional. These alternative circuits provide higher performance at the cost of additional chip area.

performance

FPGA wiring with programmable interconnect is slower than typical wiring in a custom chip for two reasons: the pass transistor and wire lengths. The pass transistor is not a perfect on-switch, so a programmable interconnection point is somewhat slower than a pair of wires permanently connected by a via. In addition, FPGA wires are generally longer than would be necessary for a custom chip. In a custom layout, a wire can be made just as long as necessary. In contrast, FPGA wires must be designed to connect a variety of logic elements and other FPGA resources. A net made of programmable interconnect may be longer, introducing extra capacitance and resistance that slows down the signals on the net.

types of programmable interconnect

An FPGA requires a large number of programmable wires in order to take advantage of the logic in the LEs. As we saw in Section 3.1, FPGAs use wires of varying lengths in order to minimize the delay through wires. Wiring is often organized into different categories depending on its structure and intended use:

- Short wires connect only local LEs. These wires don't take up much space and they introduce less delay. The carry chains through the LEs are one example of short interconnect.
- Global wires are specially designed for long-distance commu-

nication. As with high-speed highways with widely spaced exits, they have fewer connection points than local connections. This reduces their impedance. Global wires may also include built-in electrical repeaters to reduce the effects of delay.

- Special wires may be dedicated to distribute clocks or other register control signals.

connections and choice

In order to be able to select routes for connections, the FPGA fabric must provide choices for the interconnections. We must be able to connect a logic element's input or output to one of several different wires. Similarly, the ends of each wire must be able to connect to several different wires. Each of these choices requires its own connection box. This adds up to a large amount of circuitry and wiring that is devoted to programmable interconnection. If we add too many choices, we end up devoting too much of the chip to programmable interconnect and not enough to logic. If we don't have enough choice, we can't make use of the logic resources on the chip. One of the key questions in the design of an FPGA fabric is how rich the programmable interconnect fabric should be. Section 3.7 presents the results of several experiments that are designed to match the richness of FPGA fabrics to the characteristics of typical designs.

types of interconnect

One way to balance interconnect and logic resources is to provide several different types of interconnect. Connections vary both in length and in speed. Most FPGAs offer several different types of wiring so that the type most suited to a particular wire can be chosen for that wire. For example, the carry signal in an adder can be passed from LE to LE by wires that are designed for strictly local interconnections; longer connections may be made in a more general interconnect structure that uses segmented wiring.

The next examples describe the interconnect systems in the two FPGAs we discussed earlier.

Example 3-3 The Xilinx Spartan-II interconnect system

The Spartan-II includes several types of interconnect: local, general-purpose, I/O, global, and clock.

The local interconnect system provides several kinds of connections. It connects the LUTs, flip-flops, and general purpose interconnect. It also provides internal CLB feedback. Finally, it includes some direct paths for high-speed connections between horizontally adjacent CLBs. These paths can be used for arithmetic, shift registers, or other functions that need structured layout and short connections.

The general-purpose routing network provides the bulk of the routing resources. This network includes several types of interconnect:

- A **general routing matrix** (**GRM**) is a switch matrix used to connect horizontal and vertical routing channels as well as the connections between the CLBs and the routing channels.
- There are 24 single-length lines to connect each GRM to the four nearest GRMs, to the left, right, above, and below.
- Hex lines route GRM signals to the GRMs six blocks away. Hex lines provide longer interconnect. The hex lines include buffers to drive the longer wires. There are 96 hex lines, one third bidirectional and the rest unidirectional.
- 12 longlines provide interconnect spanning the entire chip, both vertically and horizontally.

The general routing matrices are related to the single-length and hex lines like this:

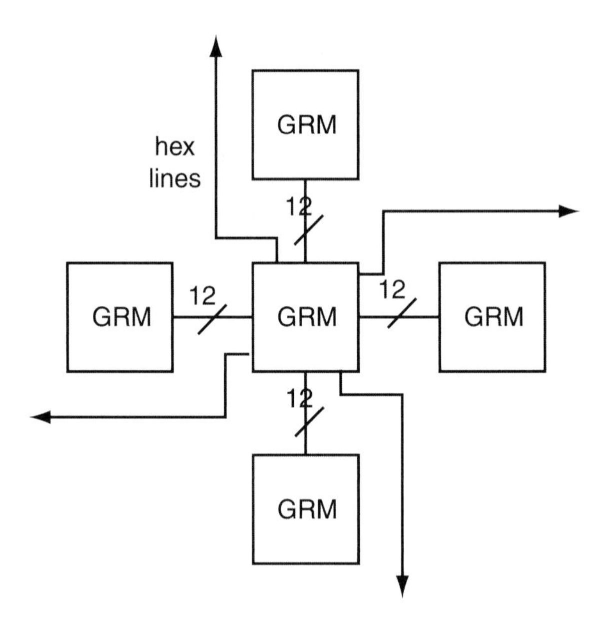

Some additional routing resources are placed around the edge of the chip to allow connections to the chip's pins.

One type of dedicated interconnect resource is the on-chip three-state bus:

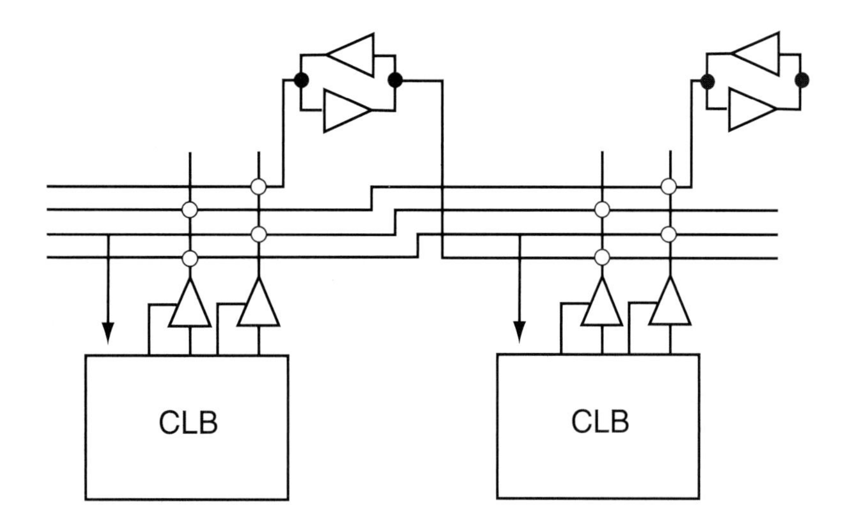

These busses run only horizontally. Four partitionable busses are available per CLB row.

Another type of dedicated routing resource is the wires connecting the carry chain logic in the CLBs.

The global routing system is designed to distribute high-fanout signals, including both clocks and logic signals. The primary global routing network is a set of four dedicated global nets with dedicated input pins. Each global net can drive all CLB, I/O register, and block RAM clock pins. The clock distribution network is buffered to provide low delay and low skew:

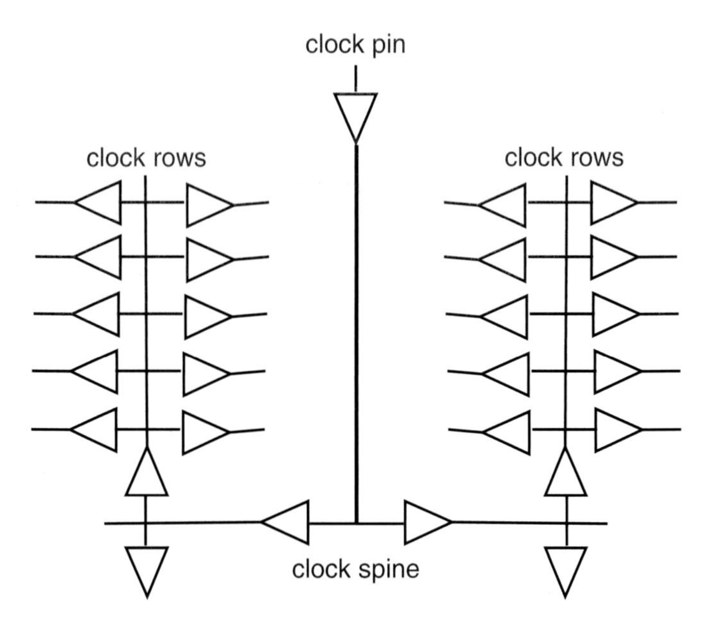

The secondary global routing network includes 24 backbone lines, half along the top of the chip and half along the bottom. The chip also includes a delay-locked loop (DLL) to regulate the internal clock.

Example 3-4 The Altera APEX II interconnect system

The APEX II uses horizontal and vertical interconnect channels to interconnect the logic elements and the chip pins. The interconnect structure looks like this:

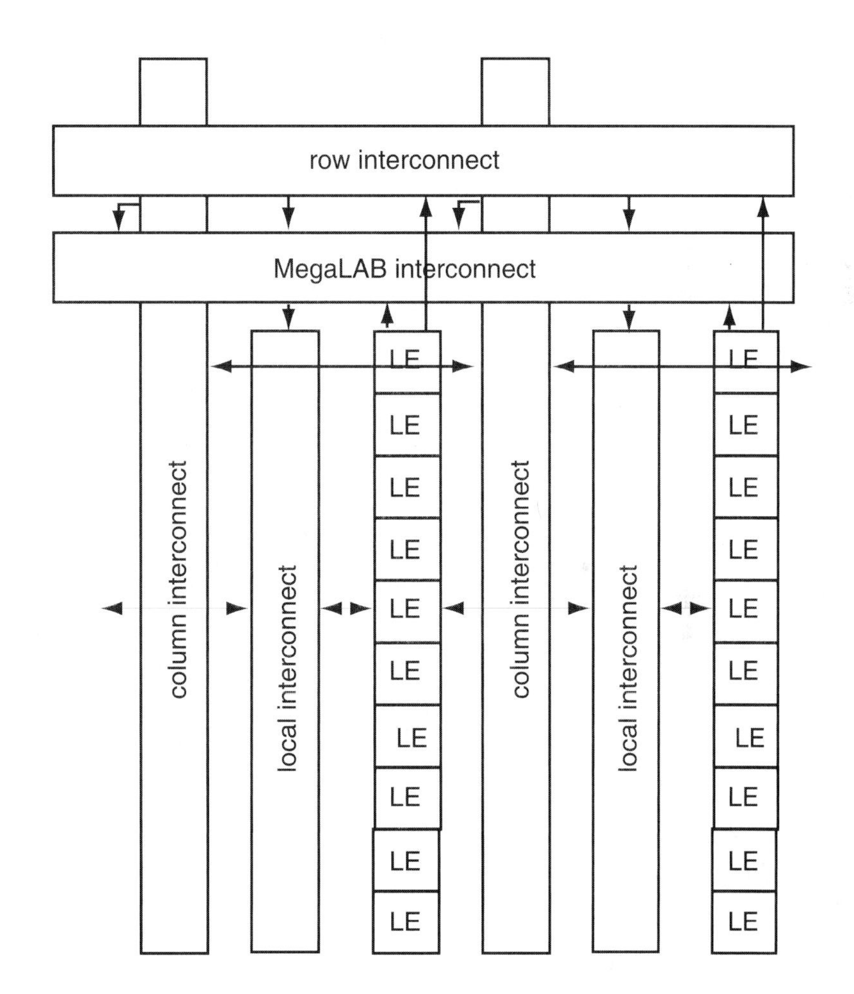

A row line can be driven directly by an LE, I/O element, or embedded memory in that row. A column line can also drive a row line; columns can be used to connect wires in two rows.

Some dedicated signals with buffers are provided for high-fanout signals such as clocks.

Column I/O pins can directly drive these interconnect lines. Each line traverses two MegaLAB structures, driving the four MegaLABs in the top row and the four MegaLABs in the bottom row of the chip.

3.3.4 Configuration

SRAM configuration

SRAM-based FPGAs are reconfigured by changing the contents of the configuration SRAM. A few pins on the chip are dedicated to configuration; some additional pins may be used for configuration and later released for use as general-purpose I/O pins. Because FPGAs are reconfigured relatively infrequently, configuration lines are usually bit-serial. However, it is possible to send several bits in parallel if configuration time is important.

During prototyping and debugging, we change the configuration frequently. A download cable can be used to download the configuration directly from a PC. When we move the design into production, we do not want to rely on a download cable and a PC. Specialized programmable read-only memories (PROMs) are typically used to store the configuration on the printed circuit board with the FPGA. The FPGA upon power-up runs through a protocol on its configuration pins. The EPROM has a small amount of additional logic to supply a clock signal and answer the FPGA's configuration protocol.

configuration time

When we start up the system, we can usually tolerate some delay to download the configuration into the FPGA. However, there are cases when configuration time is important. This is particularly true when the FPGA will be dynamically reconfigured—reconfigured on-the-fly while the system is operating, such as the Radius monitor described in the next example.

Example 3-5 Dynamic reconfiguration

The Radius monitors for the Apple Macintosh™ computer [Tri94] operated in horizontal (landscape) and vertical (portrait) modes. When the monitor was rotated from horizontal to vertical or vise versa, the monitor contents changed so that the display contents did not rotate. The Radius monitor used an SRAM-based FPGA to run the display. Because long shift registers to hold the display bits were easily built on the FPGA, the part made sense even without reconfiguration. However, the monitor's mode shift was implemented in part by reconfiguring the

FPGA. A mercury switch sensed the rotation and caused a new personality to be downloaded to the FPGA, implementing the mode switch.

---

configuration circuits

The configuration memory must be designed to be as immune as possible to power supply noise. Glitches in the power supply voltage can cause memory circuits to change state. Changing the state of a configuration memory cell changes the function of the chip and can even cause electrical problems if two circuits are shorted together. As a result, the memory cells used in configuration memory use more conservative designs than would be used in bulk SRAM. Configuration memory is slower to read or write than commodity SRAM in order to make the memory state more stable.

Although the configuration data is typically presented to the chip in serial mode in order to conserve pins, configuration is not shifted into the chip serially in modern FPGAs. If a shift register is directly connected to the programmable interconnection points and I/O pins, then the connections will constantly change as the configuration is shifted through. Many of these intermediate states will cause drivers to be shorted together, damaging the chip. Configuration bits are shifted into a temporary register and then written in parallel to a block of configuration memory [Tri98].

scan chains and JTAG

Many modern FPGAs incorporate their reconfiguration **scan chains** into their testing circuitry. Manufacturing test circuitry is used to ensure that the chip was properly manufactured and that the board on which the chip is placed is properly manufactured. The **JTAG** standard (JTAG stands for Joint Test Action Group) was created to allow chips on boards to be more easily tested. JTAG is often called **boundary scan** because it is designed to scan the pins at the boundary between the chip and the board. As shown in Figure 3-7, JTAG is built into the pins of the chip. During testing, the pins can be decoupled from their normal functions and used as a shift register. The shift register allows input values to be placed on the chip's pins and output values to be read from the pins. The process is controlled by the **test access port** (**TAP**) controller. The controller is connected to four pins: TDI, the shift register input; TDO, the shift register output; TCK, the test clock; and TMS, test mode select. (The standard also allows an optional test reset pin known as TRST.) The test access port includes an instruction register (IR) that determines what actions are taken by the TAP; the standard defines the state transition graph for the TAP's function. A bypass register (BP) allows bits to be either shifted into the IR or for the IR's contents to be left intact. Each pin on the chip is modified to include the JTAG shift register logic.

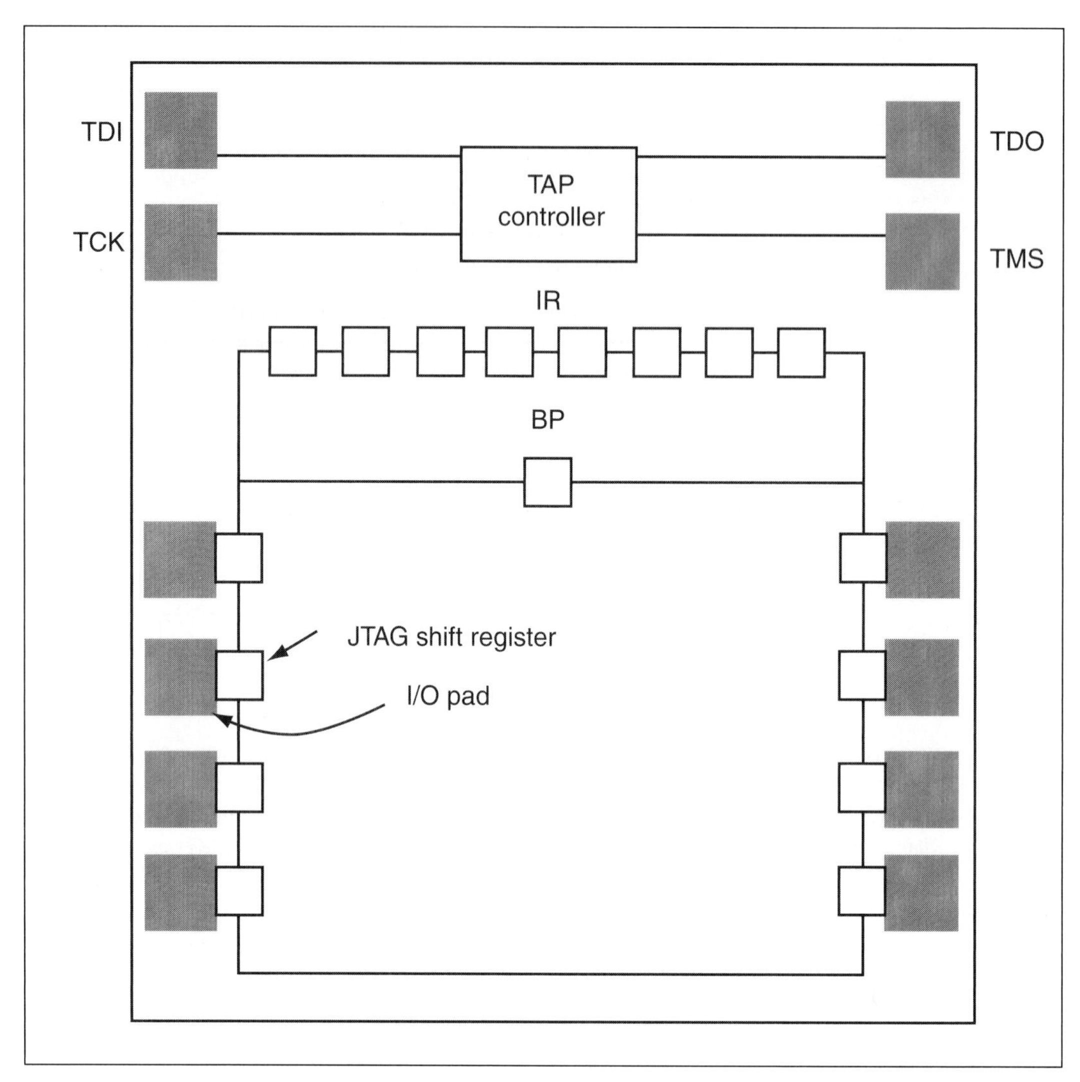

Figure 3-7 The JTAG architecture.

Using this relatively small amount of logic, an outside unit can control and observe all the pins on the chip.

The next two examples discuss the configuration systems for the Spartan-II and APEX-II.

Example 3-6 Xilinx Spartan-II configuration

The Spartan-II configuration requires, depending on the size of the chip, from about 200,000 to over 1.3 million bits.

The chip can be configured in one of several modes:

- Master serial mode assumes that the chip is the first chip in a chain (or the only chip). The master chip loads its configuration from an EPROM or a download cable.
- Slave serial mode gets its configuration from another slave serial mode chip or from the master serial mode chip in the chain.
- Slave parallel mode allows fast 8-bit-wide configuration.
- Boundary scan mode uses the standard JTAG pins.

Several pins are dedicated to configuration. The PROGRAM' pin carries an active-low signal that can be used to initiate configuration. The configuration mode is controlled by three pins M0, M1, and M2. The DONE pin signals when configuration is finished. The boundary scan pins TDI, TDO, TMS, and TCK can be used to configure the FPGA without using the dedicated configuration pins.

Example 3-7 Altera APEX-II configuration

The standard APEX-II configuration stream is one bit wide. The configuration can be supplied by a ROM or by a system controller such as a microprocessor. The APEX-II can be configured in less than 100 ms.

The APEX-II also supports a byte-wide configuration mode for fast configuration.

3.4 Permanently Programmed FPGAs

SRAM-based FPGAs have to be configured at power-up. There are two technologies used to build FPGAs that need to be configured only once: antifuses and flash. In this section we will survey methods for using both to build FPGAs.

3.4.1 Antifuses

An antifuse, as shown in Figure 3-8, is fabricated as a normally open disconnection. When a programming voltage is applied across the antifuse, it makes a connection between the metal line above it and the via to the metal line below. An antifuse has a resistance on the order of 100 Ω, which is more resistance than a standard via. The antifuse has several advantages over a fuse, a major one being that most connections in an FPGA should be open, so the antifuse leaves most programming points in the proper state.

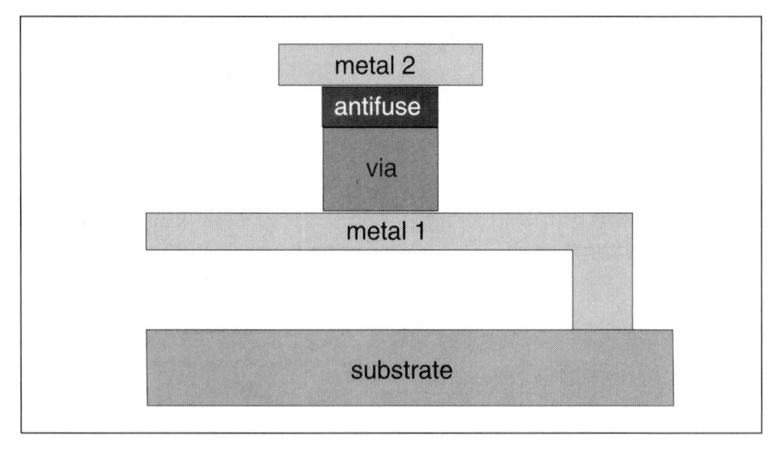

Figure 3-8 Cross-section of an antifuse.

An antifuse is programmed by putting a voltage across it. Each antifuse must be programmed separately. The FPGA must include circuitry that allows each antifuse to be separately addressed and the programming voltage applied.

3.4.2 Flash Configuration

Flash memory is a high-quality programmable read-only memory. Flash uses a floating gate structure in which a low-leakage capacitor holds a voltage that controls a transistor gate. This memory cell can be used to control programming transistors.

Figure 3-9 shows the schematic of a flash-programmed cell. The memory cell controls two transistors. One is the programmable connection point. It can be used for interconnect electrical nodes in interconnect or logic. The other allows read-write access to the cell.

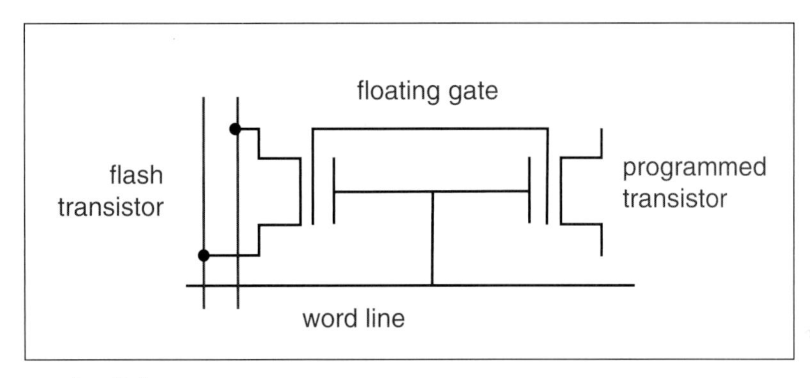

Figure 3-9 A flash programmed switch.

3.4.3 Logic Blocks

multiplexers and programming

The logic blocks in antifuse-programmed FPGAs are generally based upon multiplexing, since that function can be implemented by making or breaking connections and routing signals.

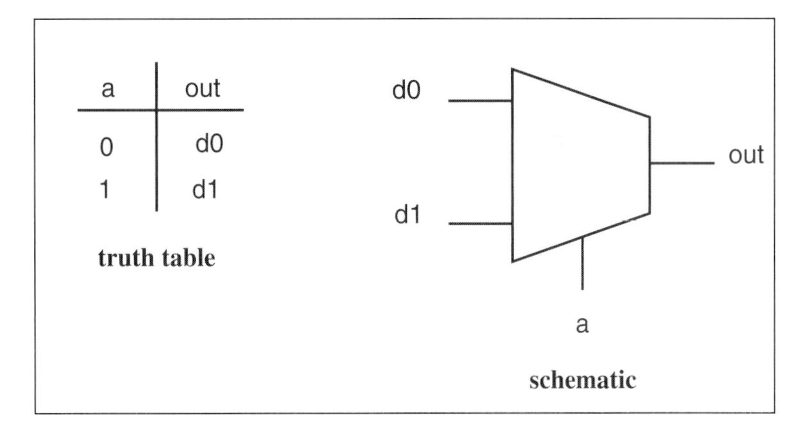

Figure 3-10 A single multiplexer used as a logic element.

To understand implementing logic functions with multiplexers, consider first the circuit of Figure 3-10. When the multiplexer control a is 0, the output is d0; when the control is 1, the output is d1. This logic element lets us configure which signal is copied to the logic element output. We can write the logic element's truth table as shown in the figure.

Now consider the more complex logic element of Figure 3-11. This element has two levels of multiplexing and four control signals. The con-

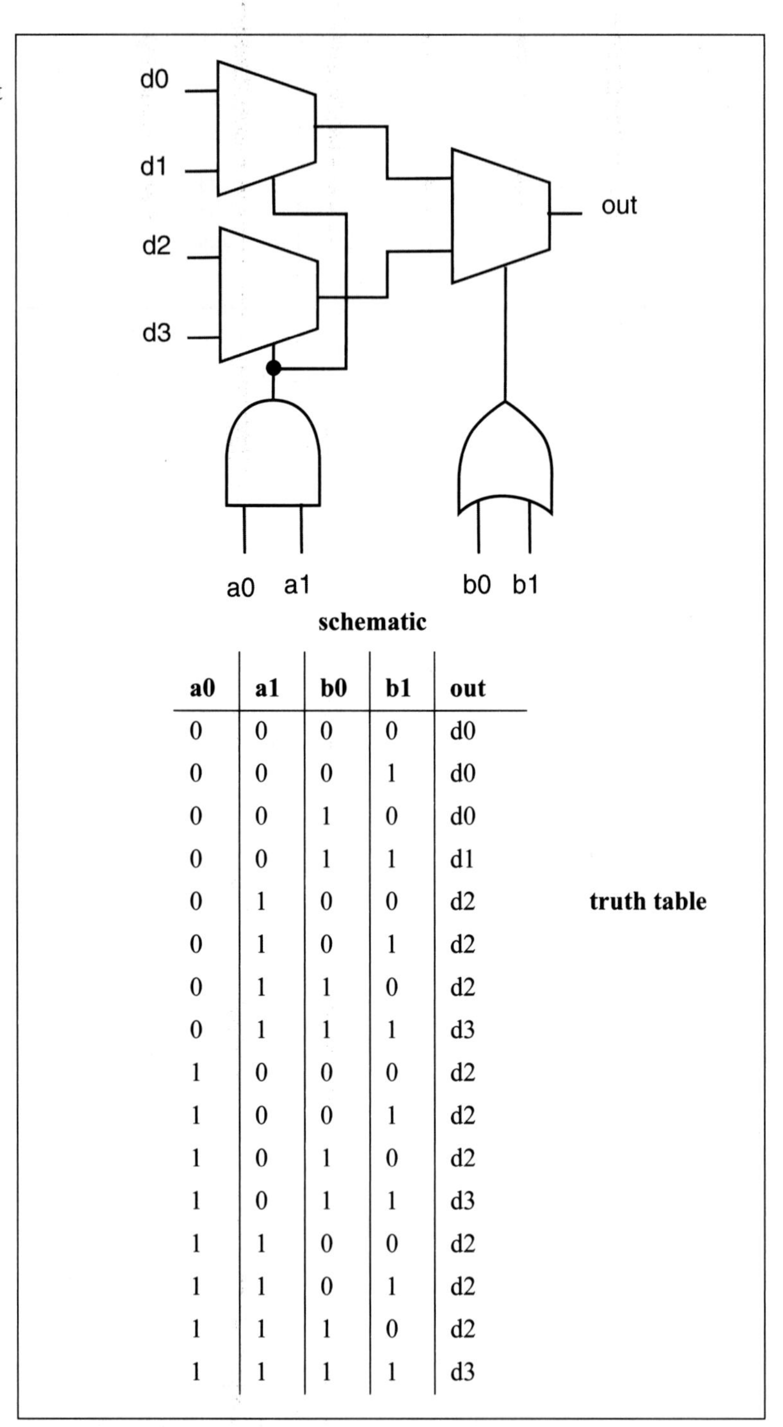

a0	a1	b0	b1	out
0	0	0	0	d0
0	0	0	1	d0
0	0	1	0	d0
0	0	1	1	d1
0	1	0	0	d2
0	1	0	1	d2
0	1	1	0	d2
0	1	1	1	d3
1	0	0	0	d2
1	0	0	1	d2
1	0	1	0	d2
1	0	1	1	d3
1	1	0	0	d2
1	1	0	1	d2
1	1	1	0	d2
1	1	1	1	d3

Figure 3-11 A logic element built from several multiplexers.

trol signals of both multiplexers in the first row are controlled by the same signal, which is the AND of two control signals. The final multiplexer stage is controlled by the OR of two other control signals. This provides a significantly more complex function.

The next example describes the logic blocks of the Actel Axcelerator family FPGA [Act02]. Members of this family range in capacity from 80,000 to 1 million gates.

Example 3-8 Actel Axcelerator family logic elements

The Actel Axcelerator family has two types of logic elements: the **C-cell** for combinational logic and the **R-cell** for registers. These cells are organized into SuperClusters, each of which has four C-cells, two R-cells, and some additional logic.

Here is the C cell:

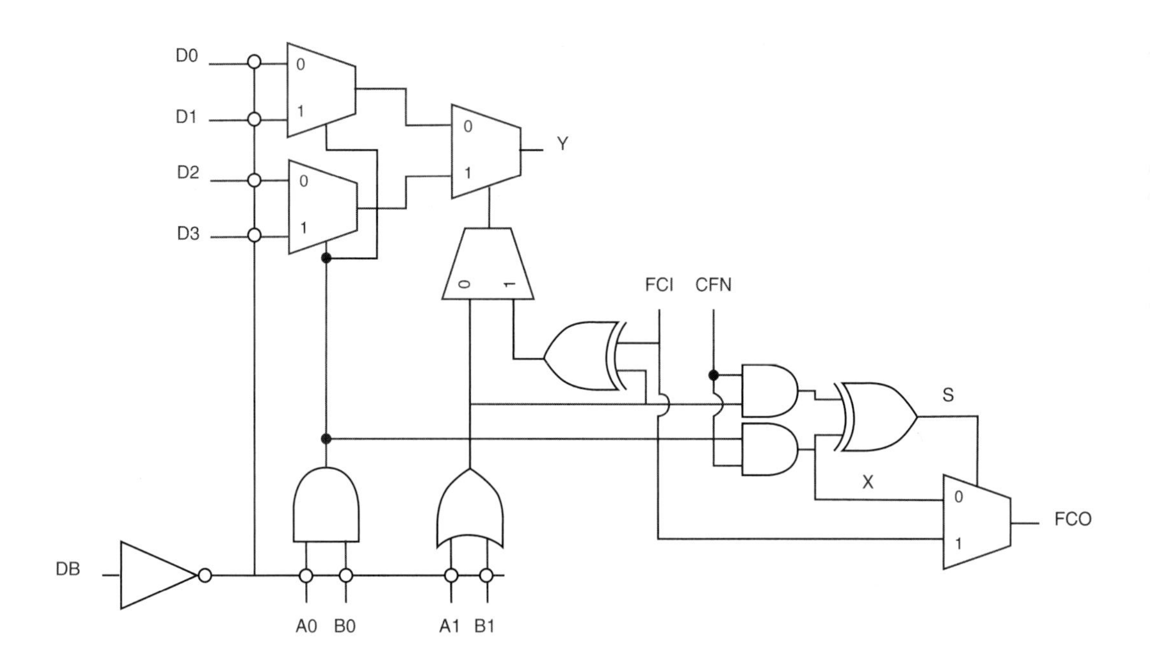

The core of the cell is the multiplexer, which has four data signals: *D0*, *D1*, *D2*, *D3;* and four select signals: *A0*, *A1*, *A2*, and *A3*. The *DB* input can be used to drive the inverted form of any of these signals to the multiplexer by using an antifuse to connect the signal to the *DB* input, then using one of the antifuses shown to connect the inverted signal to the desired multiplexer input. The signals can also be connected in their uncomplemented form. The *S* and *X* bits are used for fast addition and are not available outside the SuperCluster.

The cell includes logic for fast addition. The two bits to be added arrive at the *A0* and *A1* inputs. The cell has a carry input *FCI* and carry output *FCO*. The carry logic is active when the *CFN* signal is high. The carry logic operates in a group of two C-cells in a SuperCluster:

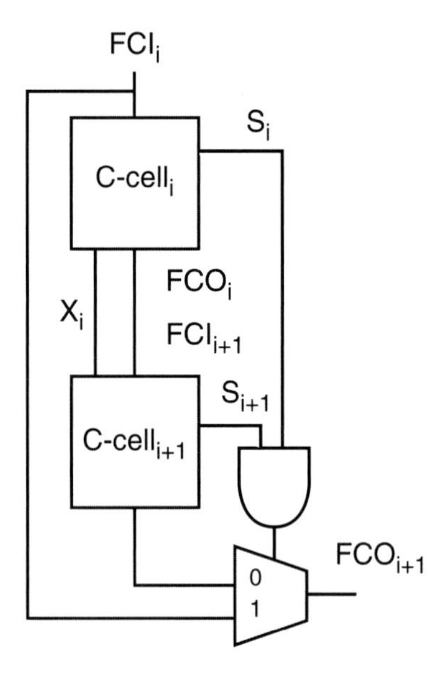

Within $C\text{-}cell_{i+1}$, the X_i bit is connected to the 1 input of the FCO multiplexer in this configuration. The *Y* output of $C\text{-}cell_{i+1}$ is used as the sum. This logic performs a carry-skip operation for faster addition.

The R-cell allows for various combinations of register inputs, clocking, *etc.*:

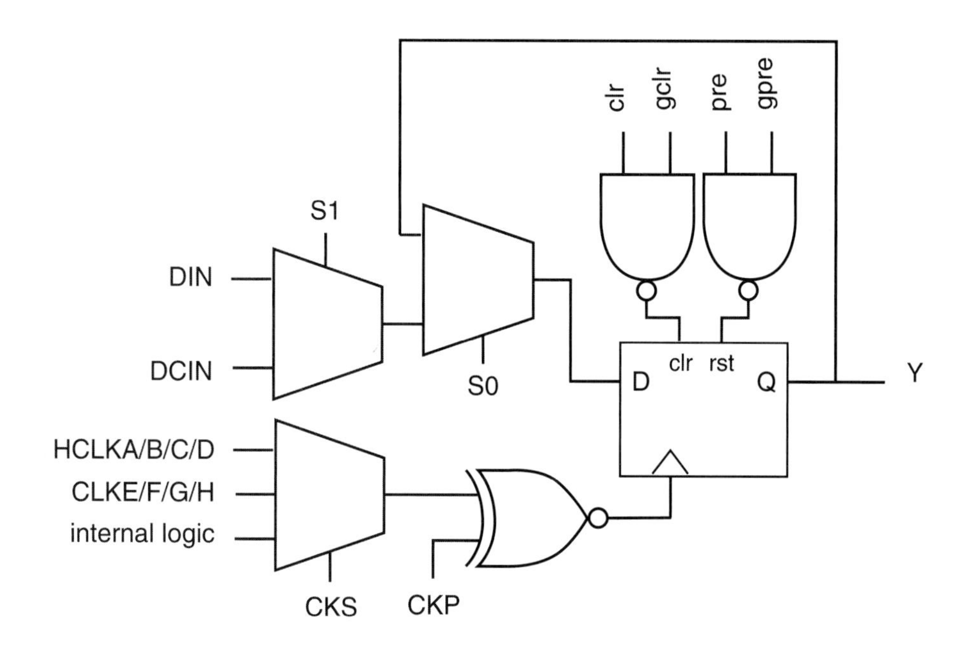

DCIN is a hardwired connection to the *DCOUT* signal of the adjacent *C-cell*; this connection is designed to have less than 0.1 ns of wire delay. The *S0* and S1 inputs act as data enables. The flip-flop provides active low clear and presets, with clear having higher priority. A variety of clock sources can be selected by the *CKS* control signal; the *CKP* signal selects the polarity of the clock used to control the flip-flop.

Each SuperCluster has two clusters. Each cluster has three cells in the pattern CCR.

logic elements for flash-based FPGAs

Flash-based FPGAs use switches for programmability. The next example describes the Actel ProASIC 400K family logic element [Act03b].

Example 3-9 ProASIC 500K logic element

The Actel ProASIC 500K logic element uses multiplexers to generate the desired logic function. The programmed switches are used to select alternate inputs to the core logic.

Here is the core logic tile:

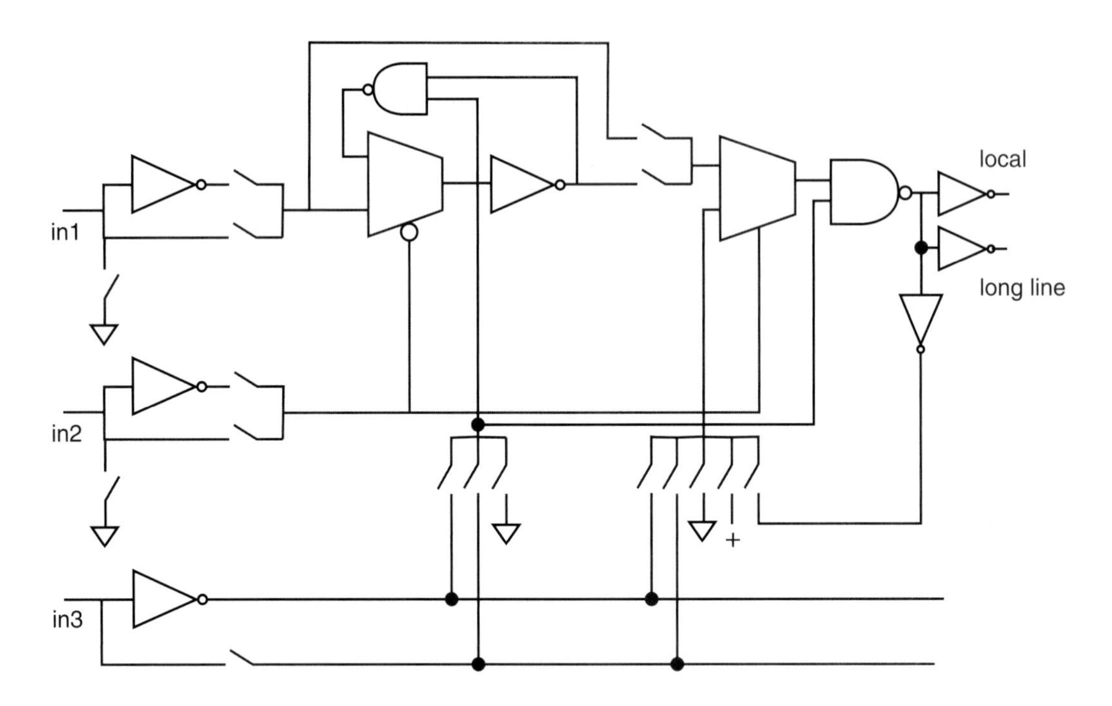

Each of the three inputs can be presented to the multiplexers in true or complement form. The multiplexer system can implement any function of three inputs except for three-input XOR. The feedback paths allow the logic element to be configured as a latch, in which case *in2* is used as the clock and *in3* as reset. The logic element provides two output drivers, one for local interconnect and a larger driver for long lines.

3.4.4 Interconnection Networks

Antifuses make it relatively easy to program interconnect. An antifuse also slows down the interconnect path less than a pass transistor in an SRAM-programmable FPGA.

The next example describes the wiring organization of the Actel Axcelerator family [Ac02].

Example 3-10 Actel Axcelerator interconnect system

The Axcelerator has three different local wiring systems. The FastConnect system provides horizontal connections between logic modules within a SuperCluster or to the SuperCluster directly below. CarryConnects route the carry signals between SuperClusters. DirectConnect connects entirely within a SuperCluster—it connects a C-cell to the neighboring R-cell. A DirectConnect signal path does not include any antifuses; because it has lower resistance it runs faster than programmable wiring.

Generic global wiring is implemented using segmented wiring channels. Routing tracks run across the entire chip both horizontally and vertically. Although most of the wires are segmented with segments of several different lengths, a few wires run the length of the chip.

The chip provides three types of global signals. Hardwired clocks (HCLK) can directly drive the clock input of each R-cell. Four routed clocks can drive the clock, clear, preset, or enable pin of an R-cell or any input of a C-cell. Global clear (GCLR) and global preset (GPSET) signals drive the clear and preset signals of R-cells and I/O pins.

interconnect in flash-based FPGAs

The next example describes the interconnect structure of the ProASIC 500K [Act02].

Example 3-11 Actel ProASIC 500K interconnect system

The ProASIC 500K provides local wires that allow the output of each tile to be directly connected to the eight adjacent tiles. Its general-purpose routing network uses segmented wiring with segments of length 1, 2, and 4 tiles. The chip also provides very long lines that run the length of the chip.

The ProASIC 500K provides four global networks. Each network can be accessed either from a dedicated global I/O pin or a logic tile.

3.4.5 Programming

antifuse programming

An antifuse is programmed by applying a large voltage sufficient to make the antifuse connection. The voltage is applied through the wires connected by the antifuse. The FPGA is architected so that all the antifuses are in the interconnect channels; this allows the wiring system to be used to address the antifuses for programming. In order to be sure

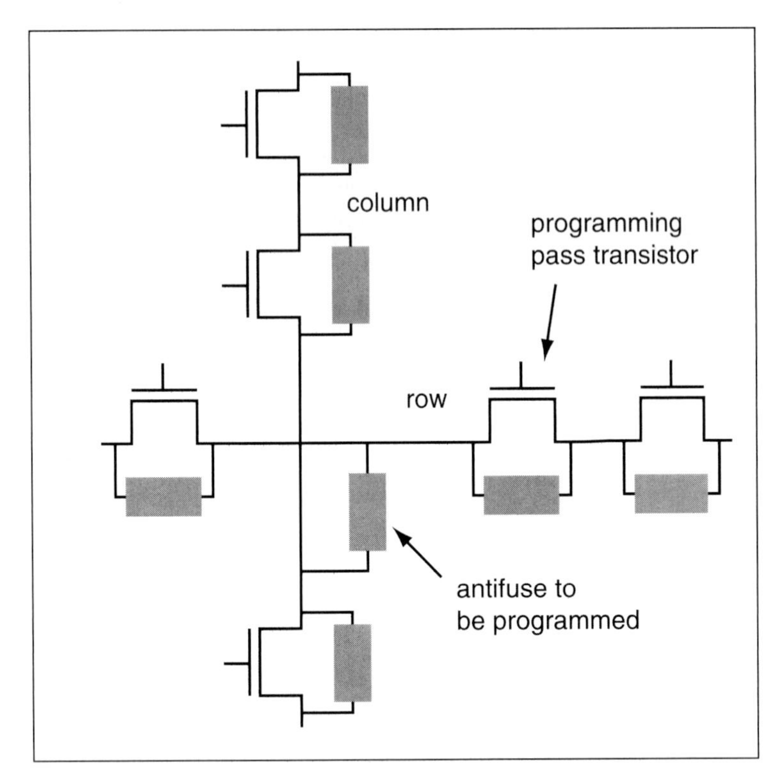

Figure 3-12 Selecting an antifuse to be programmed.

that every antifuse can be reached, each antifuse is connected in parallel with a pass transistor that allows the antifuse to be bypassed during programming. The gates of the pass transistors are controlled by programming signals that select the appropriate row and column for the desired antifuse, as shown in Figure 3-12. The programming voltage is applied across the row and column such that only the desired antifuse receives the voltage and is programmed [ElG98].

Because the antifuses are permanently programmed, an antifuse-based FPGA does not need to be configured when it is powered up. No pins need to be dedicated to configuration and no time is required to load the configuration.

3.5 Chip I/O

features of I/O pins

The I/O pins on a chip connect it to the outside world. The I/O pins on any chip perform some basic functions:

- Input pins provide electrostatic discharge (ESD) protection.
- Output pins provide buffers with sufficient drive to produce adequate signals on the pins.
- Three-state pins include logic to switch between input and output modes.

The pins on an FPGA must be programmable to accommodate the requirements of the configured logic. A standard FPGA pin can be configured as either an input, output, or three-state pin.

Pins may also provide other features. Registers are typically provided at the pads so that input or output values may be held. The slew rate of outputs may be programmable to reduce electromagnetic interference; lower slew rates on output signals generate less energetic high-frequency harmonics that show up as electromagnetic interference (EMI).

The next two examples describe the I/O pins of the Actel APEX-II and the Xilinx Spartan-II 2.5V FPGA.

Example 3-12 Altera APEX-II I/O pin

The Altera APEX-II I/O structure, known as an IOE, is designed to support SDRAM and double-data-rate (DDR) memory interfaces. It contains six registers and a latch as well as bidirectional buffers. The IOE supports two inputs and two outputs:

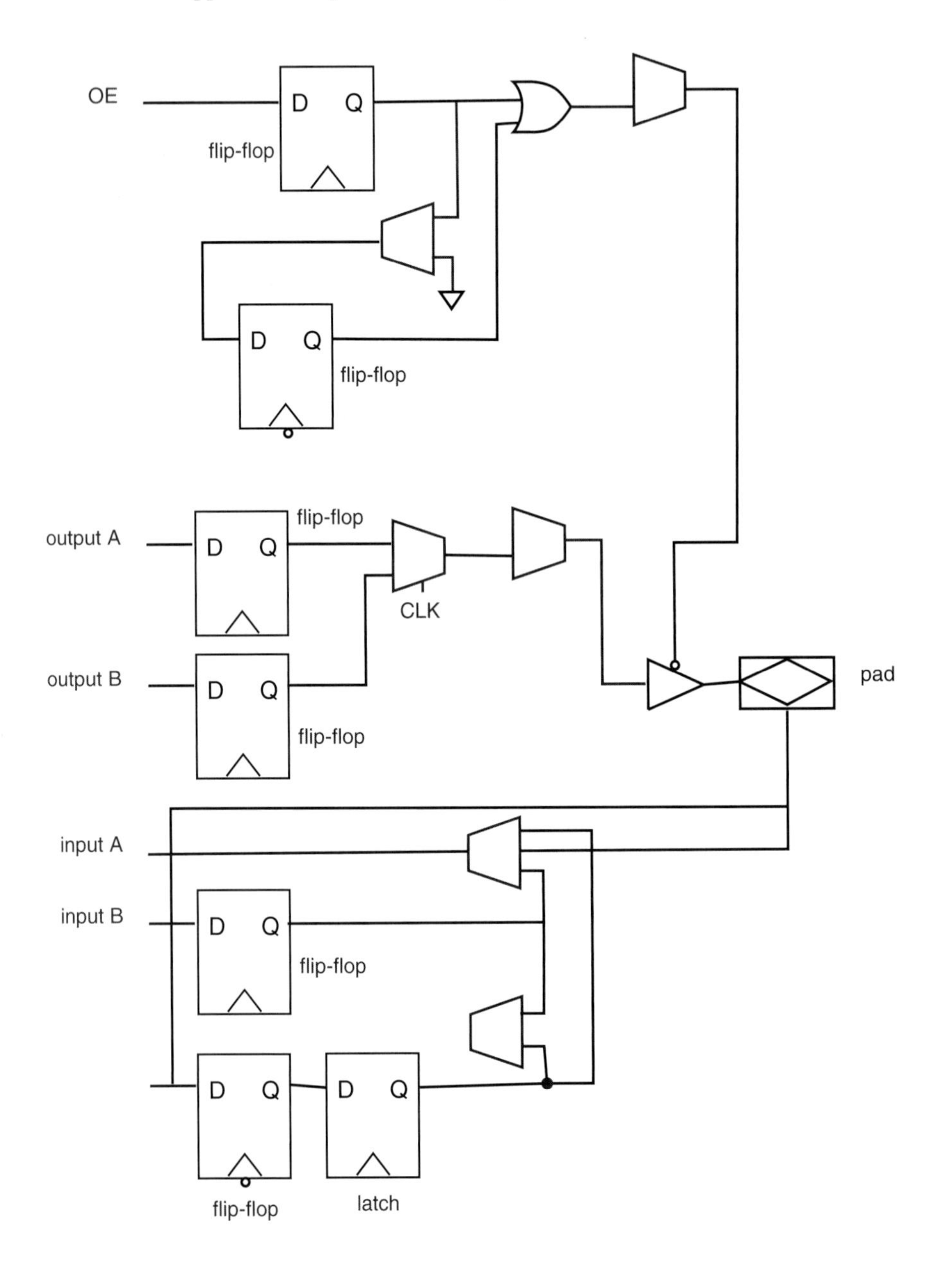

The OE signal controls the three-state behavior.

Example 3-13 The Xilinx Spartan-II 2.5V I/O pin

The Spartan-II 2.5V family is designed to support a wide variety of I/O standards:

I/O standard	Input reference voltage (V_{ref})	Output source voltage (V_{CCO})	Board termination voltage (V_{TT})
LVTTL	N/A	3.3	N/A
LVCMOS2	N/A	2.5	N/A
PCI	N/A	3.3	N/A
GTL	0.8	N/A	1.2
GTL+	1.0	N/A	1.5
HSTL Class I	0.75	1.5	0.75
HSTL Class III	0;9	1.5	1.5
HSTL Class IV	0.9	1.5	1.5
SSTL3 Class I and II	1.5	3.3	1.5
SSTL2 Class I and II	1.25	2.5	1.25
CTT	1.5	3.3	1.5
AGP-2X	1.32	3.3	N/A

Here is the schematic for the I/O block:

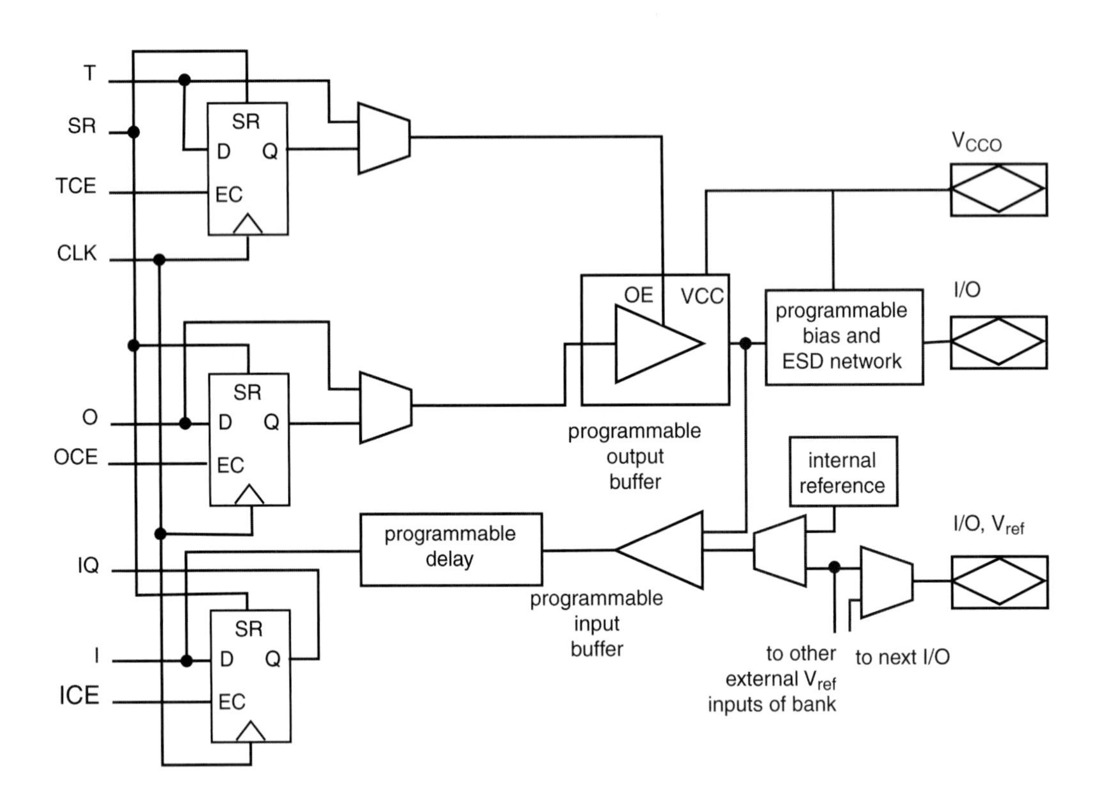

Much of the right-hand side of the schematic is devoted to handling the various I/O standards. Notice that pins are required for the various reference voltages as well as the I/O itself. The pins on the chip are divided into eight banks, with each bank sharing the reference voltage pins. Pins within a bank must use standards that have the same V_{CCO}.

The IOB has three registers, one each for input, output, and three-state operation. Each has its own enable (ICE, OCE, and TCE, respectively) but all three share the same clock connection. These registers in the IOB can function either as flip-flops or latches.

The programmable delay element on the input path is used to eliminate variations in hold times from pin to pin. Propagation delays within the FPGA cause the IOB control signals to arrive at different times, causing the hold time for the pins to vary. The programmable delay element is matched to the internal clock propagation delay and, when enabled, eliminates skew-induced hold time variations.

The output path has a weak keeper circuit that can be selected by programming. The circuit monitors the output value and weakly drives it to the desired high or low value. The weak keeper is useful for pins that are connected to multiple drivers; it keeps the signal at its last valid state after all the drivers have disconnected.

3.6 Circuit Design of FPGA Fabrics

Circuit design determines many of the characteristics of FPGA architectures. The size of a logic element determines how many can be put on a chip; the delay through a wire helps to determine the interconnection architecture of the fabric. In this section we will look at the circuit design of both logic elements and interconnections in FPGA fabrics, primarily concentrating on SRAM-based FPGAs, but with some notes on antifuse-based FPGAs. We will rely heavily on the results of Chapter 2 throughout this section.

3.6.1 Logic Elements

LEs vs. logic gates

The logic element of an FPGA is considerably more complex than a standard CMOS gate. A CMOS gate needs to implement only one chosen logic function. The logic element of an FPGA, in contrast, must be able to implement a number of different functions.

Antifuse-based FPGAs program their logic elements by connecting various signals, either constants or variables, to the inputs of the logic elements. The logic element itself is not configured as a SRAM-based logic element would be. As a result, the logic element for an antifuse-based FPGA can be fairly small. Figure 3-13 shows the schematic for a multiplexer-based logic element used in early antifuse-based FPGAs. Table 3-1 shows how to program some functions into the logic element by connecting its inputs to constants or signal variables. The logic element can also be programmed as a dynamic latch.

Example 3-14 compares lookup tables and static gates in some detail.

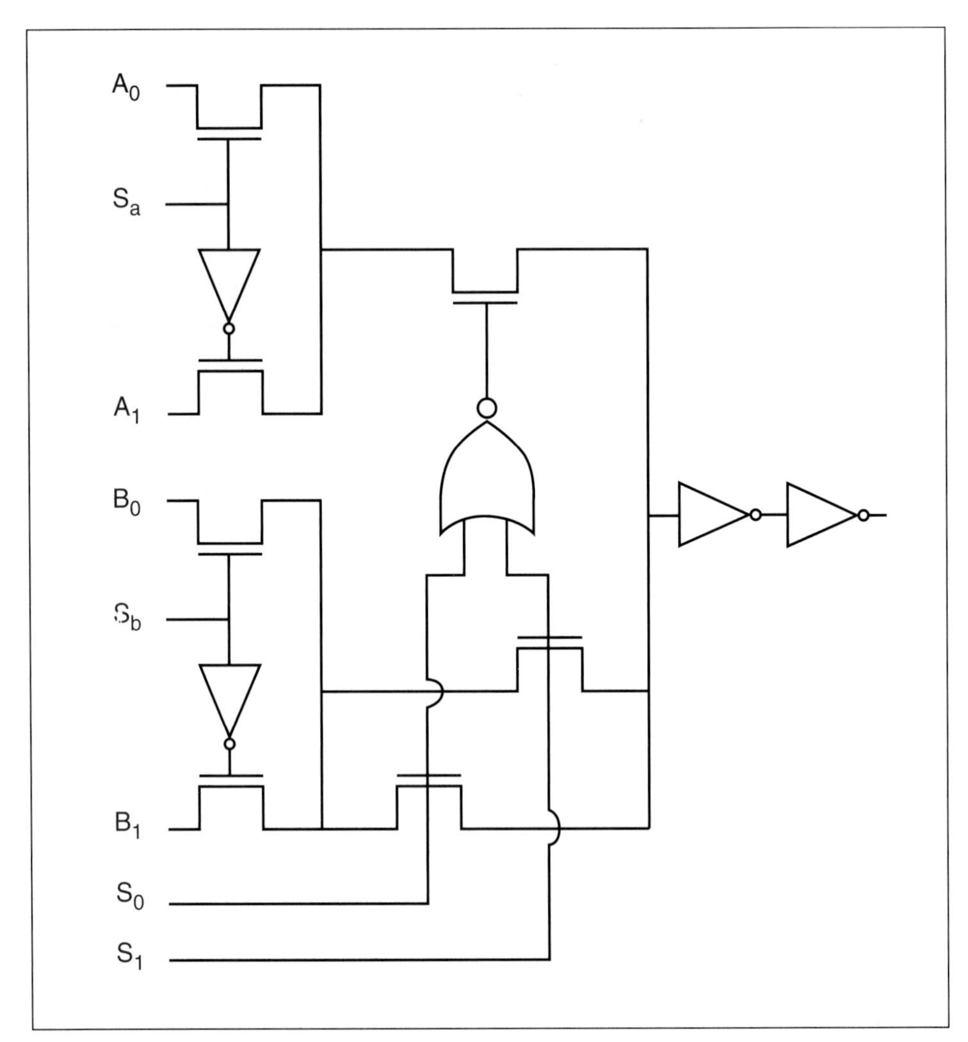

Figure 3-13 A logic element built from a multiplexer [ElG90].

equation	A_0	A_1	B_0	B_1	S_a	S_1	S_0	S_b
(AB)'	1	1	0	1	A	0	B	A
	0	1	0	1	0	0	B	A
	0	1	0	1	0	B	0	A
	0	1	0	1	0	0	A	B
A^B	1	0	0	1	A	0	B	A
	1	0	0	1	A	B	0	A
latch	Q	0	D	0	CLR	CLK	0	CLR
	Q	0	CLR	0	CLR	CLK	0	D

Table 3-1 Programming the mux-based logic element [ElG90].

Example 3-14 Lookup table *vs.* static CMOS gate

Let us compare the static CMOS gate and lookup table logic element in several respects: size, delay, and power consumption.

For our purposes, counting the number of transistors in a circuit gives us a sufficiently accurate estimate of the size of the circuit on-chip. The number of transistors in a static CMOS gate depend on both the number of inputs to the gate and the function to be implemented. A static NAND or NOR gate of n inputs has $2n$ transistors; more complex functions such as XOR can be more complicated. A single NAND or NOR gate with 16 inputs is impractical, but it would require 32 transistors.

In contrast, the SRAM cell in the lookup table requires eight transistors, including the configuration logic. For a four-input function, we would have $8 \times 16 = 128$ transistors just in the core cell. In addition, we need decoding circuitry for each bit in the lookup table. A straightforward decoder for the four-bit lookup table would be a multiplexer with 96 transistors, though smaller designs are possible.

The delay of a static gate depends not only on the number of inputs and the function to be implemented, but also on the sizes of transistors used. By changing the sizes of transistors, we can change the delay through the gate. The slowest gate uses the smallest transistors. Using logical effort theory [Sut99], we can estimate the delay of a chain of two four-input NAND gates that drives another gate of the same size as 9 τ units.

The delay of a lookup table is independent of the function implemented and dominated by the delay through the SRAM addressing logic. Logical effort gives us the decoding time as 21 τ units.

The power consumption of a CMOS static gate is, ignoring leakage, dependent on the capacitance connected to its output. The CMOS gate consumes no energy while the inputs are stable (once again, ignoring leakage). The SRAM, in contrast, consumes power even when its inputs do not change. The stored charge in the SRAM cell dissipates slowly (in a mechanism independent of transistor leakage); that charge must be replaced by the cross-coupled inverters in the SRAM cell.

As we can see, the lookup table logic element is considerably more expensive than a static CMOS gate.

Because the logic element is so complex, its design requires careful attention to circuit characteristics. In this section we will concentrate on lookup tables—most of the complexity of an antifuse-based logic element is in the antifuse itself, and the surrounding circuitry is highly dependent on the electrical characteristics of the antifuse material. The lookup table for an SRAM-based logic element incorporates both the memory and the configuration circuit for that memory.

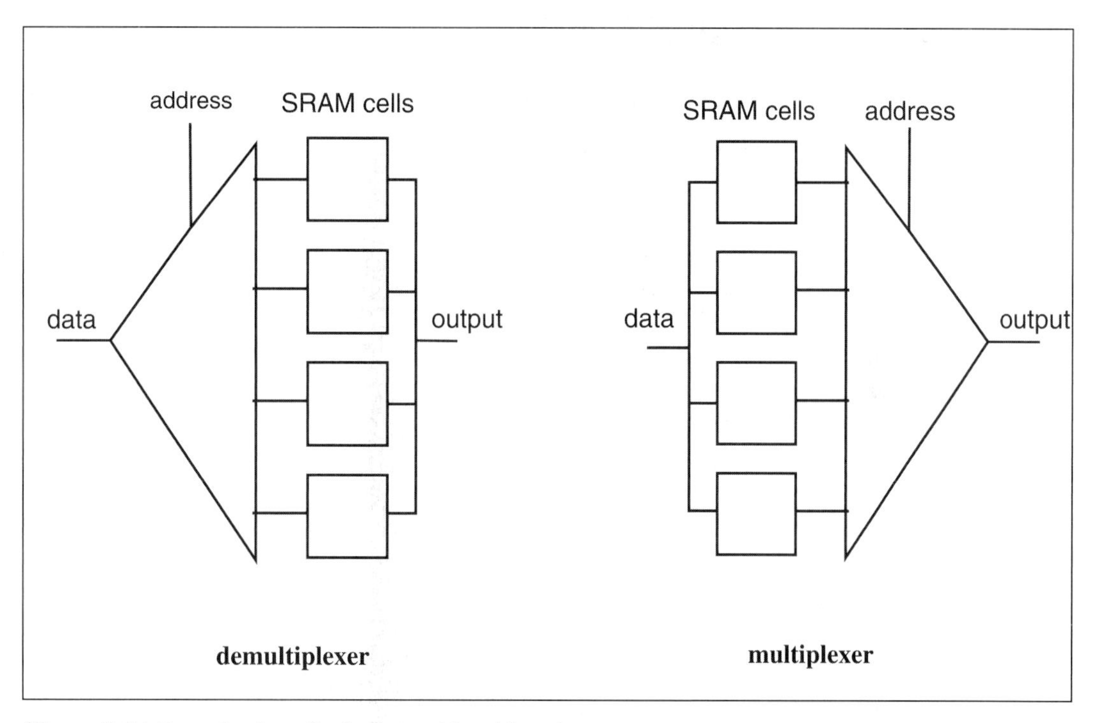

Figure 3-14 Organizations for lookup table addressing.

SRAMs for LEs

There are two possible organizations for the lookup table as shown in Figure 3-14: a demultiplexer that causes one bit to drive the output or a multiplexer that selects the proper bit. These organizations are logically equivalent but have different implications for circuitry. Bulk SRAMs generally use the demultiplexer architecture, as shown in Section 2.6.2. The demultiplexer selects a row to be addressed, and the shared bit lines are used to read or write the memory cells in that row. The shared bit line is very efficient in large memories but less so in small memories like those used in logic elements. Most FPGA logic elements use a multiplexer to select the desired bit.

Most large SRAMs use two bit lines that provide complementary values of the SRAM cell value. Using bit and bit-bar lines improves access times in large SRAMs but does not make any noticeable improvement in small SRAMs used in logic elements. As a result, logic element SRAM cells are generally read and written through only one side of the cell, not through both sides simultaneously.

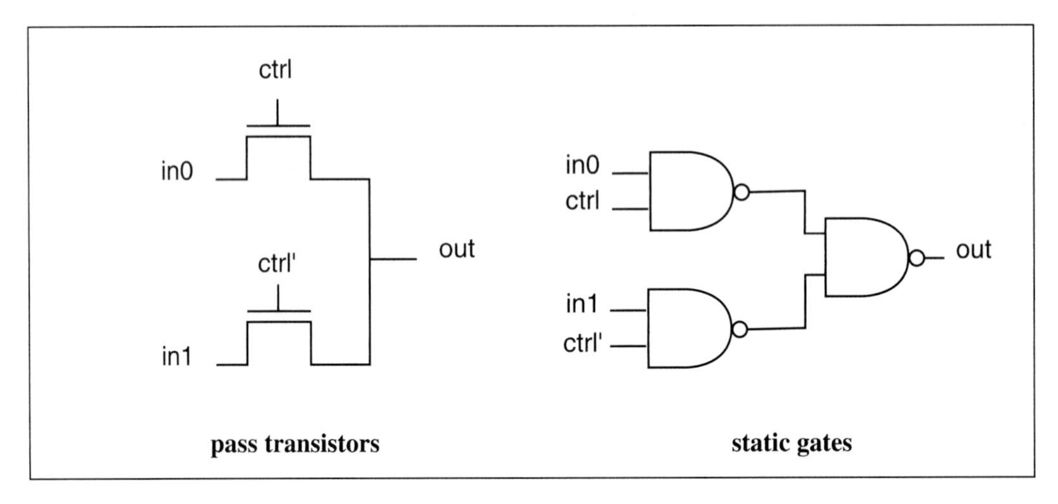

Figure 3-15 Alternative circuits for a multiplexer.

SRAM multiplexer design

Should that multiplexer be made of static gates or pass transistors? The alternatives for the case of a two-input multiplexer are shown in Figure 3-15. The pass transistor network is clearly smaller—each two-input NAND or NOR gate has four transistors. But as the number of series pass transistors grows the delay from the data input to the data output grows considerably. The delay through a series of pass transistors, in fact, grows as the square of the number of pass transistors in the chain, for reasons similar to that given by Elmore. The choice between static gates and pass transistors therefore depends on the size of the lookup table. The next example compares the delay through static gate and pass transistor multiplexers.

Example 3-15 Delay through multiplexer circuits

We want to build a b-input multiplexer that selects one of the b possible input bits. We will call the data input bits i_0, *etc.* and the select bits s_0, *etc.* In our drawings we will show four-input multiplexers; these are smaller than the multiplexers we want to use for lookup tables but they are large enough to show the form of the multiplexer.

Here is a four-input mux built from NAND gates:

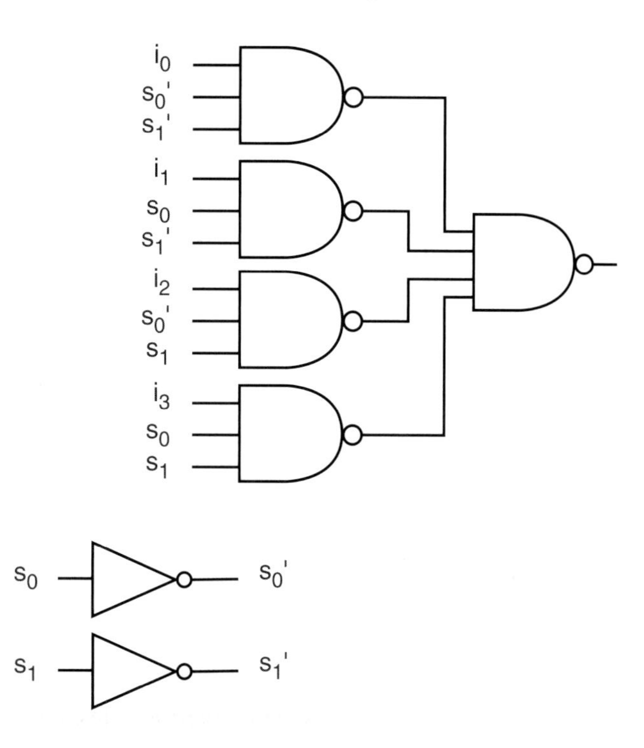

This multiplexer uses two levels of logic plus some inverters that form a third level of logic. Each of the NAND gates in the first level of logic have as inputs one of the data bits and true or complement forms of all the select bits. The inverters are used to generate the complement forms of the select bits. Each NAND gate in the first level determines whether to propagate the input data bit for which it is responsible; the second-level NAND sums the partial results to create a final output.

We can analyze the delay through a b-bit multiplexer as a function of b using logical effort [Sut99]. The delay through an n-input NAND gate is proportional to $(n+2)/3$. Each NAND gate in the first level of logic has lg b inputs for the select bits and one for the data input, giving a delay proportional to $(\lg b + 3)/3$. The second-level NAND gate has b inputs, for delay proportional to $(b+2)/3$. The delay through the inverters on the select bits is proportional to 1. This means that the total delay through the b-bit static gate multiplexer grows as $b \lg b$.

Here is one form of multiplexer built from pass transistors:

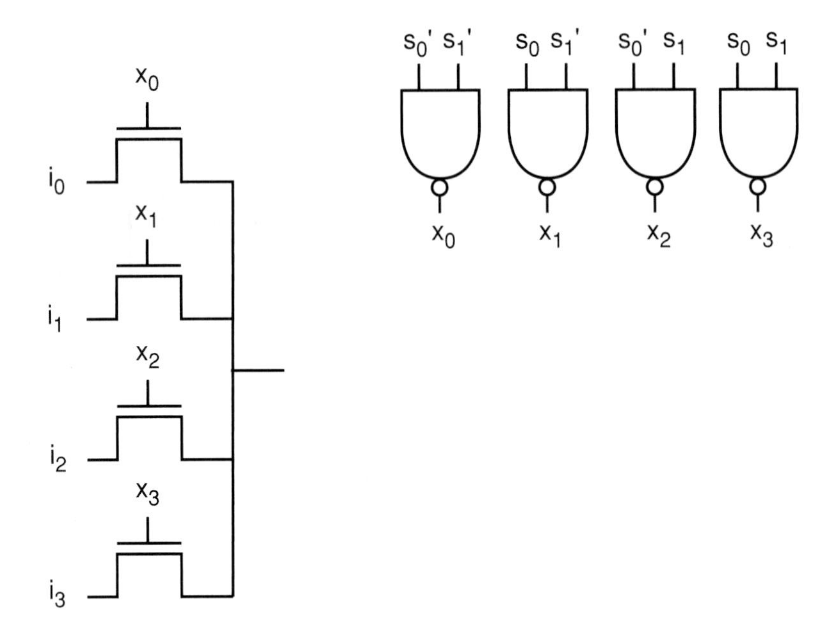

While this form may seem simple and fast, the pass transistors are not the only transistors in the multiplexer. The gates must be driven by decoded address signals generated from the select bits. This circuit is not good for large multiplexers because it combines the worst aspects of static gate and pass transistor circuits.

A better form of circuit built from pass transistors is a tree:

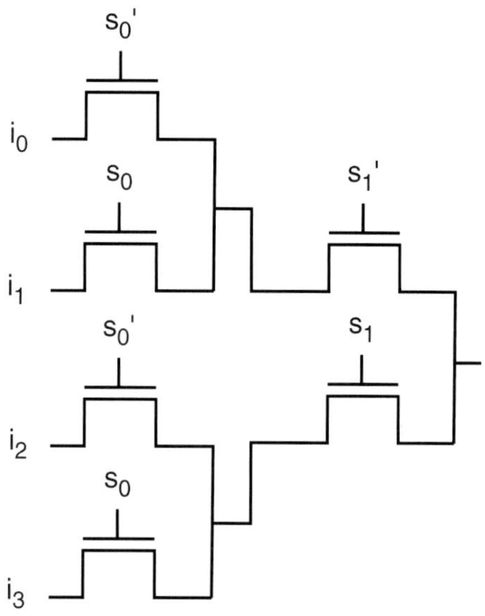

The gates of these pass transistors are driven directly by select bits or the complements of the select bits (generated by inverters), eliminating the decoder NAND gates. However, because the pass transistors can be (roughly) modeled as resistors, the delay through a chain of pass transistors is proportional to the square of the number of switches on the path. (We analyzed delay through RC chains in Section 2.5.4.) The tree for a b-input multiplexer has lg b levels of logic, so the delay through the tree is proportional to lg b^2.

One question that can be asked is whether transmission gates built from parallel n-type and p-type transistors are superior to pass transistors (that is, single n-type transistors). While transmission gates are more egalitarian in the way they propagate logic 0 and 1 signals, their layouts are also significantly larger. Chow et al. [Cho99] found that pass transistors were the better choice for multiplexers.

It is possible to build a mux from a combination of pass transistors and static gates, using switches for some of the select stages and static gates for the remaining stages.

LE output drivers

The output of the logic element must be designed with drivers sufficient to drive the interconnect at the LE's output. The transistors must be sized properly based upon a given load on the output and a desired rise/fall time. Once the characteristics of the interconnect are known, sizing the output transistors is straightforward. However, because the LE may need to drive a long wire that also includes logic used to make programmable connections, the output buffer must be powerful and large.

3.6.2 Interconnect

varieties of interconnect

The first question we need to ask about FPGA interconnect is why an FPGA has so many different kinds of interconnect. A typical FPGAs has short wires, general-purpose wires, global interconnect, and specialized clock distribution networks. The reason that FPGAs need different types of wires is that wires can introduce a lot of delay, and wiring networks of different length and connectivity need different circuit designs. We saw some uses for different types of interconnect when we studied existing FPGAs, but the rationale for building several types of interconnect becomes much clearer when we study the circuit design of programmable interconnect.

Figure 3-16 A generic signal path between two logic elements.

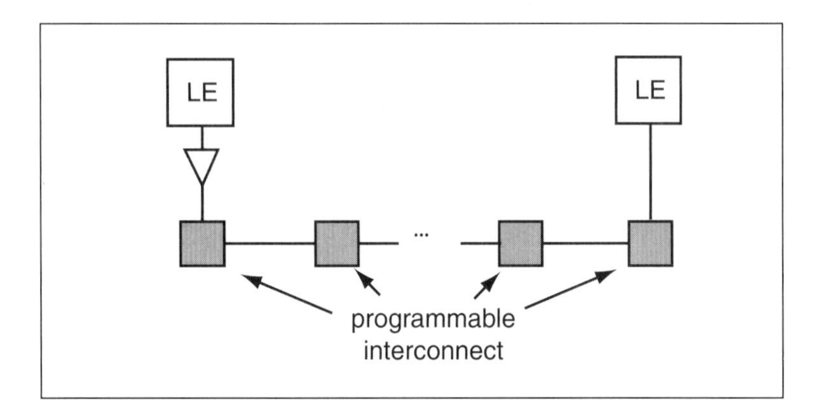

In Example 2-4 we compared the delay through logic gates and the delay through wires. We saw that a relatively short wire—a wire that is much shorter than the size of the chip—has a delay equal to the delay through a logic gate. Since many connections on FPGAs are long, thanks to the relatively large size of a logic element, we must take care to design circuits that minimize wire delay. A long wire that goes from one point to another needs a chain of buffers to minimize the delay through the wire. We studied optimal buffer sizing and buffer insertion in Section 2.5.5. Now we must apply that general knowledge to the interconnect circuits in FPGAs.

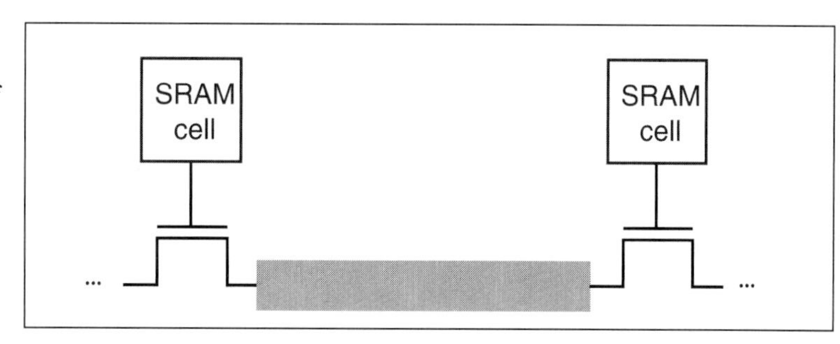

Figure 3-17 Organization of a pass-transistor-based interconnection point.

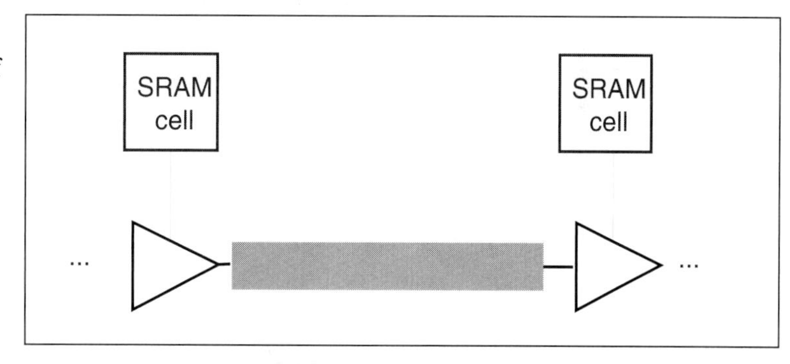

Figure 3-18 Organization of a three-state buffer-based interconnection point.

Figure 3-16 shows the general form of a path between two logic elements. A signal leaves a logic element, goes through a buffer, enters the routing channel through a programmable interconnect block, passes through several more programmable interconnect blocks, then passes through a final programmable interconnect block to enter the destination LE. We have studied the circuit design of logic elements; now we need to consider the circuits for programmable interconnect blocks. Brown et al. [Bro92] concluded that most of the area in an SRAM-based FPGA is consumed by the routing switches, so we must carefully design the programmable interconnect to make best use of the available area.

programmable interconnection circuit

Antifuses provide low-impedance connections between wires. However, the design of a programmable interconnection point for an SRAM-based or flash-based FPGA requires more care because the circuitry can introduce significant delay as well as cost a significant amount of area. The circuit design of a pass-transistor-based programmable interconnection point is shown in Figure 3-17. If we use pass transistors at the programmable interconnection points, we have two parameters we can use to minimize the delay through the wire segment: the width of the pass transistor and the width of the wire. As we saw in Section 2.3, the current through a transistor increases proportionately with its width. The increased current through the transistor reduces its effective resistance, but at the cost of a larger transistor. Similarly, we can increase the width of a wire to reduce its resistance, but at the cost of both increased capacitance and a larger wire. Rather than uniformly change the width of the wire, we can also taper the wire as described in Section 2.5.3.

We can also ask ourselves whether we should use three-state buffers rather than pass transistors at the programmable interconnection points. The use of three-state buffers in programmable interconnect is illustrated in Figure 3-18. The three-state buffer is larger than a pass transistor but it provides amplification that the pass transistor does not.

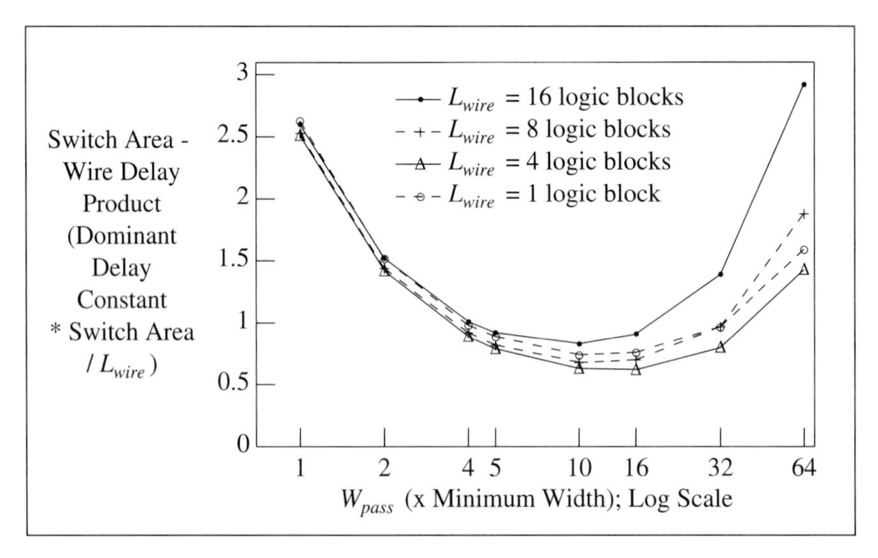

Figure 3-19 Switch area * wire delay *vs*. routing pass transistor width (from Betz and Rose [Bet99], © 1999 IEEE).

pass transistor and wire sizing

Betz and Rose [Bet99] considered the effects of pass transistor and wire sizing as well as the merits of three-state buffers as programmable interconnection points; their studies used a 0.35 μm technology. They use the product of area and wire delay as a metric for the cost-effectiveness of a given circuit design. Figure 3-19 compares the product of switch area and wire delay as a function of the width of the pass transistor at a programmable interconnection point. The plot shows curves for wires of different lengths, with the length of a wire being measured in multiples of the size of a logic element. The plot shows a clear minimum in the area-delay product when the pass transistor is about ten times the minimum transistor width.

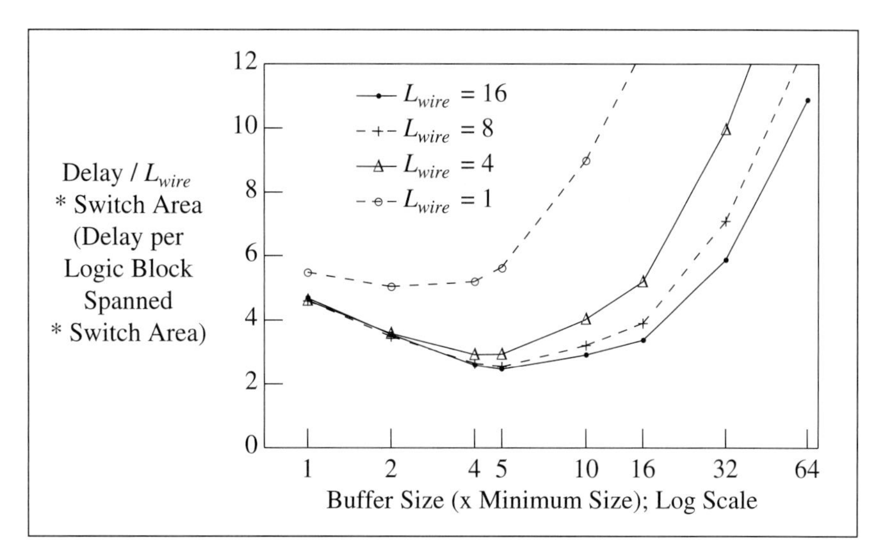

Figure 3-20 Switch area * wire delay *vs*. routing three-state buffer size (from Betz and Rose [Bet99], © 1999 IEEE).

Figure 3-20 shows the area-delay curve for a three-state buffer. The minimum area-delay product occurs when the three-state driver's transistors are about five times the minimum transistor size; the three-state requires smaller transistors than does the pass transistor because it provides amplification. Betz and Rose also found that increasing the width of the wire uniformly gave very little improvement in delay: doubling the wire width reduced delay by only 14%. Uniformly increasing the wire width has little effect because the wire capacitance is much larger, swamping the effects of reduced resistance.

simultaneous driver and pass transistor sizing

Chandra and Schmit [Cha02] studied the effect of simultaneously optimizing the sizes of the drivers at the LE outputs and the pass transistors in the interconnect. Figure 3-21 shows how delay through the wire varies as the driving buffer and routing switch size change; these curves were generated for a 0.18 μm technology. Each curve shows the delay for a given size of driver. The curves are U shaped—as the routing switch increases in size, delay first decreases and then increases. The initial drop in delay is due to decreasing resistance in the switch; the ultimate increase in delay happens when the increases in capacitance overwhelm the improvements obtained from lower resistance. This plot

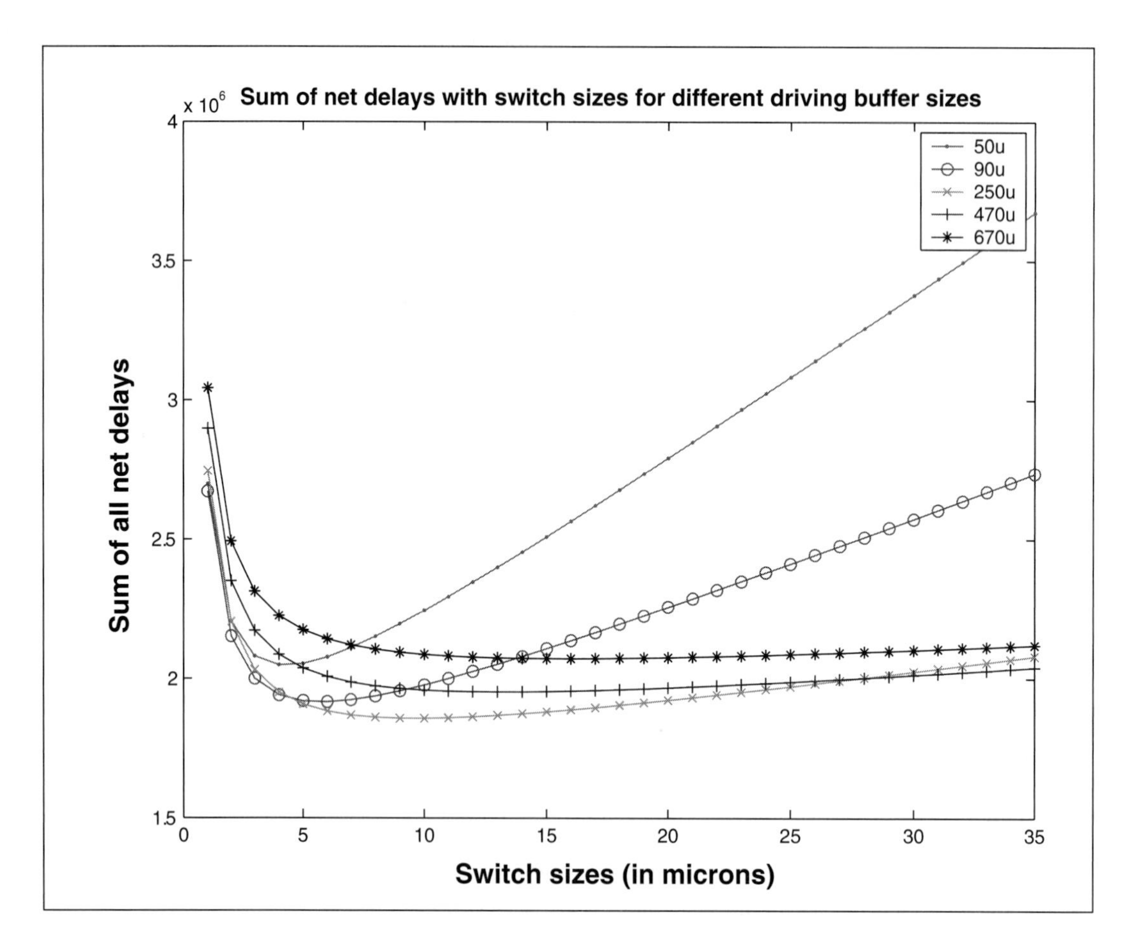

Figure 3-21 Variation of delay with routing switch and driving buffer sizes (from Chandra and Schmit [Cha02], ©2002 IEEE).

shows that there is a best size for the pass transistor routing switch for any given choice of driver size.

Figure 3-22 A clock driver tree.

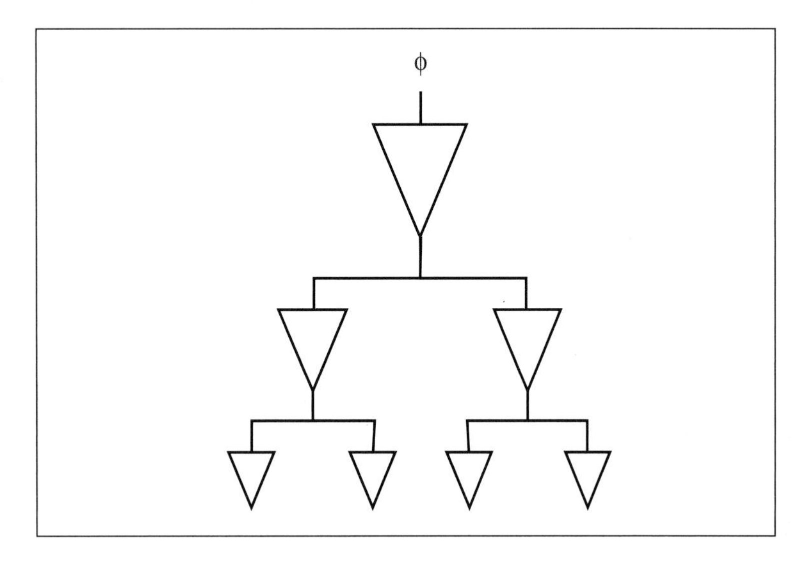

clock networks

FPGAs have specialized clock wiring because the clock signal combines a large number of destinations (all the registers in use in the system) with low delays and skew between the arrival of the clock at different points on the chip. A clock network is particularly difficult because it must go to many different places. As illustrated in Figure 3-22, clock signals are often distributed by trees of drivers, with larger transistors on the drivers near the clock source and smaller transistors on the drivers close to the flip-flops and latches. This structure presents a much larger capacitive load than does a point-to-point wire. Buffers must be distributed throughout the clock tree in order to minimize delay.

3.7 Architecture of FPGA Fabrics

issues in fabric architecture

In addition to designing the circuits of the FPGA, we need to design the FPGA's architecture—the characteristics of the logic elements and interconnection that form the FPGA fabric.

We need to answer quite a few questions to move from the concept of an FPGA to a specific FPGA fabric:

- How many logic elements on the FPGA? Logic elements and interconnect are, to some extent, mutually exclusive, since we have only a limited amount of area on chip. Wires do exist on several levels but the transistors for the interconnection points and amplifiers take up area that could be devoted to logic elements.
- What functions should the logic element perform? How many inputs should it have? Should it provide dedicated logic for addition or other functions?
- What different types of interconnect do we need? Do we need global interconnect, local interconnect, and other types? How much of each type do we need?
- How long should interconnect segments be? Longer segments provide shorter delay but less routing flexibility.
- How should we distribute the interconnect elements? Interconnect may be distributed uniformly or in various patterns.

Figure 3-23 Methodology for evaluating FPGA fabrics.

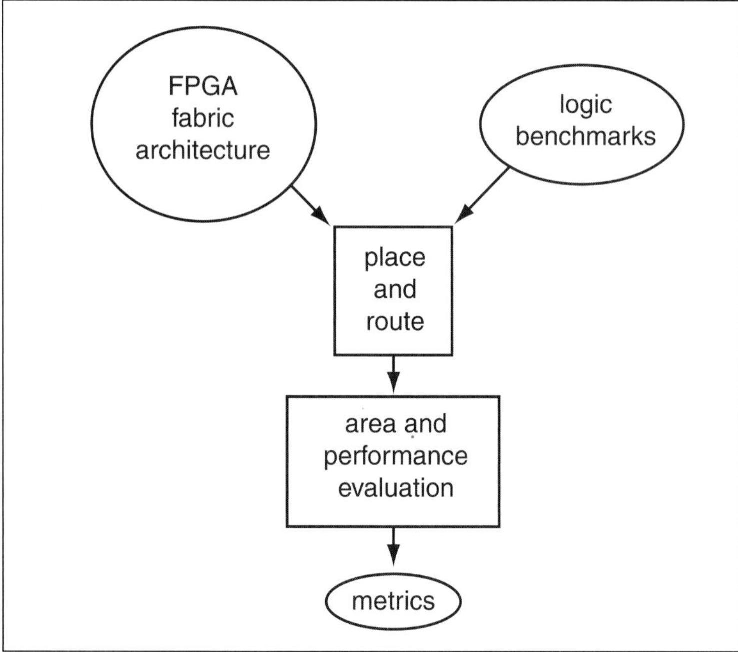

We can answer all these questions and more using the same basic methodology shown in Figure 3-23: choose an FPGA fabric to evaluate; select a set of benchmark designs; implement the benchmarks on the test fabric; evaluate the resulting metrics. An FPGA fabric is different from a custom chip in that it is intended to be used for many different logic designs. As a result, the standard by which the fabric should be judged is the quality of implementation of a typical set of logic designs. Companies that design FPGAs usually do not use them to build systems, so they may collect benchmarks from customers or from public sources. By implementing a set of designs and then measuring how well they fit onto the fabric, FPGA designers can get a better sense of what parts of their fabric work well and what need improvement.

We can measure the quality of a result in several ways:

- logic utilization;
- size of the logic element;
- interconnect utilization;
- area consumed by the connection boxes for the interconnect;
- worst-case delay.

This methodology works because we use computer-aided design (CAD) tools to implement logic designs on FPGAs. We will discuss CAD tools in more detail in later chapters; at the moment it is only necessary to know that we can map logic designs onto FPGAs automatically.

3.7.1 Logic Element Parameters

factors in LE design

The most basic question we can ask about a lookup table logic element is the number of inputs it should have. This choice is subject to delicate trade-offs:

- If the LE has too few inputs, each lookup table and the associated interconnect have a higher proportion of overhead circuitry. Fewer transistors are devoted to logic.
- If the LE has too many inputs, the logic designs that are mapped into the LEs may not be large enough to make use of all the capacity of the lookup tables, thus wasting the capacity of the LEs.

The choice of a size for the lookup table therefore depends on both the circuit design of lookup tables and the characteristics of the designs to be implemented in the FPGAs.

Many experiments [Bro92] have found that a lookup table with four inputs (and therefore 16 entries) is the best choice.

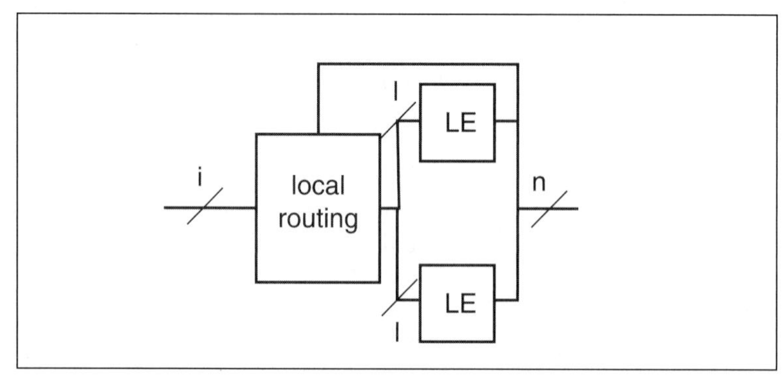

Figure 3-24 A logic element cluster.

LE clusters

Betz and Rose [Bet98] studied logic element clusters. As shown in Figure 3-24, a logic cluster has several logic elements and some dedicated interconnect. The *i* inputs to the cluster are routed to the LEs and to the *n* cluster outputs by a local routing network. Each LE has *l* inputs. Because this local network does not provide for full connectivity between the cluster inputs and LEs, it requires much less chip area than programmable interconnect, but it provides only certain connections. This leads one to ask whether clusters are better than monolithic logic elements and what sort of cluster configuration is best.

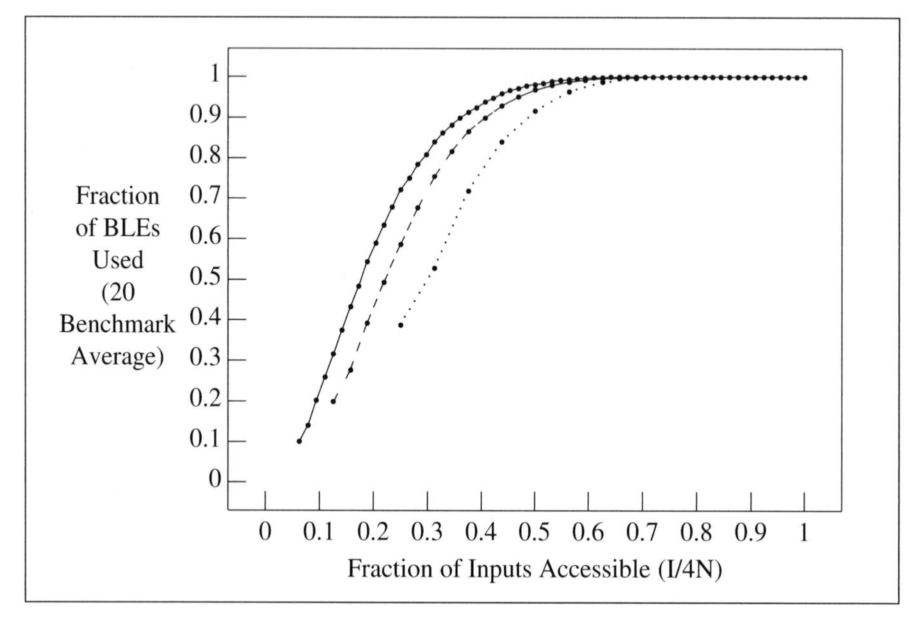

Figure 3-25 Logic utilization *vs.* number of logic cluster inputs (from Betz and Rose [Bet98], ©1998 IEEE).

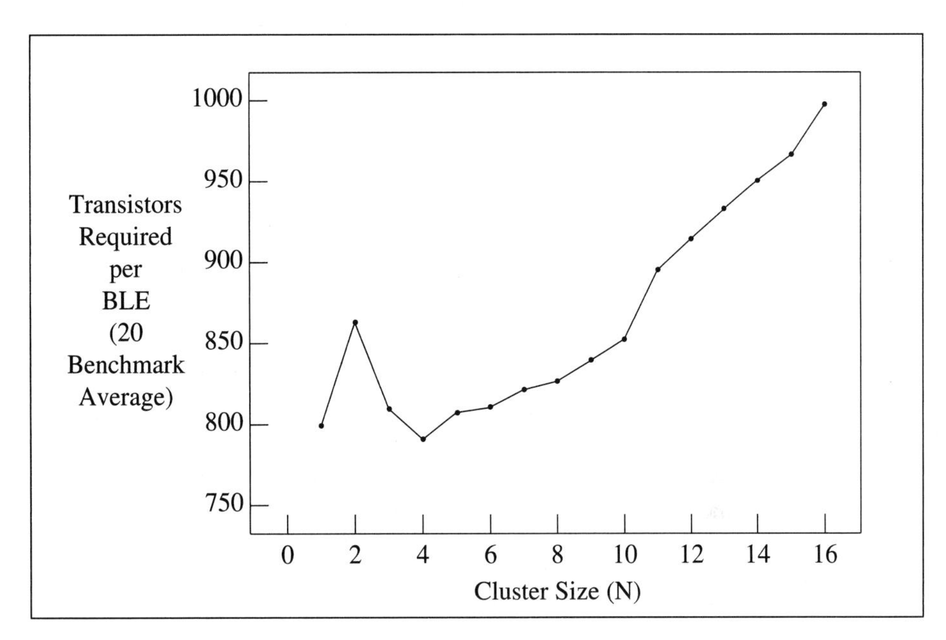

Figure 3-26 Area efficiency *vs*. cluster size (from Betz and Rose [Bet98], ©1998 IEEE).

LUT utilization

Figure 3-25 shows how well the lookup tables are used as a function of the number of accessible inputs. Ideally, we would like to be able to use all of the bits in the lookup tables, but the fixed interconnect in the cluster may not match well with the logic functions to be implemented. However, the figure shows that utilization reaches 100% when only 50% to 60% of the lookup table inputs are accessible; the common inputs and locally generated outputs do not cause a problem.

Betz and Rose also studied the number of routing tracks to which each pin should connect. When a cluster has a large number of pins, it is not necessary that all the pins be able to connect to all of the tracks. They found that an organization that allows each of the routing tracks to be driven by one output pin on each logic block was sufficient to ensure high utilization.

area efficiency vs. cluster size

Figure 3-26 shows how area efficiency varies with cluster size. Clusters in the 1 to 8 size range showed good area efficiency.

3.7.2 Interconnect Architecture

interconnect channel considerations

The design of the interconnect architecture is distinct from the design of the interconnect circuits. The circuit designs for drivers and amplifiers that we discussed in the last section are for a single wire; the interconnect architecture, in contrast, describes the entire set of wires that are used to make all the connections in the FPGA.

interconnect structures

The connection between two LEs must in general be built through three types of connections:

- the logic element must connect to the wiring channel;
- the wire segments in the wiring channel must be connected together to reach an intermediate point or destination;
- a connection must be made between wiring channels.

The connections between wire segments in a channel can generally be made to adjacent wires. The connections into the wiring channel or between wiring channels should be richer but that richness comes at the cost of additional programming circuitry. The logic elements in an antifuse-based FPGA are generally connected to one wire segment in the channel. SRAM-based FPGAs generally allow an input or output of a logic element to connect to any of several wires in the wiring channel. Connections between wiring channels would ideally be crossbars that allow any wire in one channel to be connected to any wire in the other channel. However, few FPGAs offer such rich inter-channel interconnections due to the cost of the programming circuitry.

segmented wiring

As described in Section 3.2, one of the key questions in the design of interconnect is the length of the routing segment in a general-purpose interconnection network. An individual segment may run for several LEs before it stops at a programmable interconnection point. The segment will not be able to connect to some LEs but it will provide less delay than a segment that is broken at every LE by an interconnection point. Segmented routing channels may also use offset wiring segments so that not all logic elements connect to the same wire segments.

routing segment length vs. delay

Brown et al. [Bro96] studied the effects of routing segment length on delay. Figure 3-27 shows routing delays for a number of different fabrics that have different proportions of segment lengths. In this graph, the vertical axis shows the percentage of tracks that were of length 2 while the horizontal axis shows the percentage of tracks that were of length 3; for data points where the sum of length 2 and length 3 tracks was less than 100%, the remaining tracks were of length 1. The figure shows that

	0	10	20	30	40	50	60	70	80	90	100
100	8.3										
90	8.4	8.2									
80	8.4	8.1	8.0								
70	8.6	8.3	8.0	7.9							
60	9.0	8.5	8.2	7.9	7.7						
50	9.5	8.9	8.4	7.9	7.7	7.6					
40	10.3	9.5	8.7	8.2	7.9	7.5	7.5				
30	11.1	10.1	9.4	8.7	8.2	7.7	7.5	7.4			
20	11.9	11.0	10.2	9.3	8.6	8.0	7.7	7.4	7.5		
10	12.4	11.7	10.9	10.1	9.2	8.3	7.9	7.6	7.4	7.5	
0	12.8	12.5	11.8	10.8	9.9	9.0	8.4	7.9	7.5	7.4	7.5

Figure 3-27 Routing delays of segmentation schemes (from Brown et al. [Bro96], © 1996 IEEE).

the sweet spot in the design space is centered around a point where more of the tracks are of length 3 and most of the remainder are of length 2.

Betz and Rose [Bet98] studied FPGAs with varying routing channel widths. Their study was motivated by commercial FPGAs which had larger routing channels with more wires in the center of the FPGA and smaller routing channels with fewer wires toward the edge of the chip. They found that, in fact, using the same size routing channels throughout the chip made the FPGA more routable, although if the design had the positions of its I/O positions constrained by external requirements, making the routing channels that feed the I/O pins 25% larger aided routability.

3.7.3 Pinout

Another concern is how many pins to provide for the FPGA. In a custom design the number of pins is determined by the application. But since an FPGA provides uncommitted logic we must find some other way to provide the right number of pins. If we provide too many pins we will drive up the cost of the chip unnecessarily (the package is often more expensive than the chip itself). If we don't provide enough pins then we may not be able to use all the logic in the FPGA.

Rent's Rule

The best characterization of the relationship between logic and pins was provided by E. F. Rent of IBM in 1960. He gathered data from several designs and plotted the number of pins versus the number of components. He showed that the data fit a straight line on a log-log plot. This gives the relationship known as **Rent's Rule**:

$$N_p = K_p N_g^{\beta} \quad \textbf{(EQ 3-1)}$$

where N_p is the number of pins and N_g is the number of logic gates. The formula includes two constants: β is Rent's constant while K_p is a proportionality constant. These parameters must be determined empirically by measuring sample designs. The parameters vary somewhat depending on the type of system being designed. For example, Rent measured the parameters on early IBM mainframes as $\beta = 0.6$ and $K_p = 2.5$; others have measured the parameters for modern microprocessors as $\beta = 0.45$ and $K_p = 0.82$.

3.8 Summary

FPGA fabrics are complex so that they can support realistic system designs. They contain several different components: logic elements, multiple types of interconnection networks, and I/O elements. The characteristics of FPGA fabrics are determined in part by VLSI technology and in part by the applications for which we want to use FPGAs.

3.9 Problems

Q3-1. You have a two-input lookup table with inputs a and b. Write the lookup table contents for these Boolean functions:

a. a AND b.

b. NOT a.

c. a XOR b.

Q3-2. You have a three-input lookup table with inputs a, b, and c. Write the lookup table contents for these Boolean functions:

a. a AND b.

b. a AND b AND c.

c. a XOR b XOR c.

d. a + b + c (arithmetic).

Q3-3. You have a logic element with two lookup tables, each with three inputs. The output of the first lookup table in the pair can be connected to the first input of the second lookup table using an extra configuration bit. Show how to program this logic element to perform:

a. a + b + c (arithmetic, sum only not carry).

b. a - b (arithmetic, difference only not borrow).

Q3-4. Redesign the logic element of Figure 3-11 to be controlled by a_0 OR a_1 in the first stage and b_0 AND b_1 on the second stage. Draw the schematic and write the truth table.

Q3-5. Design each of these functions using a tree of multiplexers:

a. a | ~b.

b. a & (b | c).

c. (a & ~b) | (c & d).

Q3-6. Program the logic element of Figure 3-13 to perform these functions:

a. a & b.

b. a | b.

c. a NOR b.

d. ab + bc + ac.

Q3-7. How many two-input LUTs would be required to implement a four-bit ripple-carry adder? How many three-input LUTs? How many four-input LUTs?

Q3-8. Prove that the fast arithmetic circuitry of the Actel Axcelerator performs a correct two-bit addition with carry.

Q3-9. Draw a transistor-level schematic diagram for the programmable interconnection point shown in Figure 3-6. The interconnection point should be controlled by a five-transistor SRAM cell.

Q3-10. Populate the array of logic elements in Figure 3-2 with wires and programmable interconnection points. Each wiring channel should have two wires. Assume that each logic element has two inputs and one output; each logic element should be able to connect its inputs to the channel on its left and its output to the channel on its right. When two wiring channels cross, you should be able to make connections between all the crossing wires.

Q3-11. Redo your routing design of Question Q3-10 but add local connection wires. Each local wire should be the output of an LE to one of the inputs of the LE on its right. Each end of the local connection should be controlled by a programmable interconnection point.

Q3-12. Your FPGA has 128 logic elements, each with four inputs and one output. There are 128 vertical and 128 horizontal routing channels with four wires per channel. Each wire in the routing channel can be connected to every input of the LE on its right and the output of the LE on its left. When two routing channels intersect, all possible connections between the intersecting wires can be made. How many configuration bits are required for this FPGA?

Q3-13. Draw a transistor schematic for two programmable interconnection points that are connected in a scan chain. Each programmable interconnection point should be implemented by a five-transistor SRAM cell.

Q3-14. Draw a schematic for a four-input multiplexer that uses a combination of pass transistors and static gates. The first stage of multiplexing should be performed by pass transistors while the remaining multiplexing should be performed by static gates.

Q3-15. Draw a transistor-level schematic for a programmable interconnection point implemented using a three-state buffer.

Q3-16. Draw eight LEs and a routing channel with eight wires. The routing channel should have two sets of length 1 segments, two sets of length 2 segments, and four sets of length 3 segments. Each LE should be able to be connected to at least one length 2 and one length 3 segment.

Q3-17. Draw a block diagram for a logic element cluster with two two-input LEs and four inputs to the cluster. The local interconnect network in the cluster should be able to connect the two inputs of an LE to two of the cluster inputs.

4 Combinational Logic

Hardware description languages.

Combinational network design.

Delay modeling and optimization.

Power consumption.

Logic for arithmetic.

Placement and routing.

4.1 Introduction

This chapter takes a broad view of logic design and studies the design of combinational logic networks for implementation in FPGA fabrics. There are many ways to implement any combinational logic function. Logic design is guided by the requirements imposed on the implementation, such as performance and power. To design fast, energy-efficient logic, we need to understand some basic principles about logic design in general, FPGA logic in particular, and the languages and tools we use to design logic.

The next section uses a simple design problem to survey the logic design process. Section 4.3 talks about hardware description languages and simulation. Section 4.4 describes how to analyze the delay through combinational logic networks. Section 4.5 talks about power and energy optimization. Section 4.5 concentrates on the design of arithmetic logic circuits, an important class of combinational logic. Section 4.7 talks about logic synthesis for FPGAs while Section 4.8 talks about physical design for FPGAs. At the end of the chapter we will take another look at the logic design process using a larger example.

The Verilog sources for many of the design examples in this chapter are available on the book's Web site; these files should help you follow along with the examples as you use the tools on your computer. We will use the Xilinx Integrated Software Environment (ISE) to illustrate design principles. We don't have room to cover every feature of this complex system, but hopefully these examples will give you some idea of how to use the system as well as illustrate the principles we discuss throughout the chapter.

4.2 The Logic Design Process

In this section we will survey the logic design process; in the rest of the chapter we will study the topics introduced here in much more detail. We may use some terms here without explaining them in too much detail. If you are unfamiliar with some of the terms here, they will be covered in later sections of this chapter.

logic expressions and gates

First, it is important to distinguish between *combinational logic expressions* and *logic gate networks*. A combinational logic expression is a mathematical formula which is to be interpreted using the laws of Boolean algebra: given the expression $a + b$, for example, we can compute its truth value for any given values of a and b; we can also evaluate relationships such as $a + b = c$. A logic gate computes a specific Boolean function, such as $(a + b)'$.

logic optimization

The goal of logic design or optimization is to find a network of logic gates which together compute the combinational logic function we want. Logic optimization is interesting and difficult for several reasons:

- We may not have a logic gate for every possible function, or even for every function of n inputs. It therefore may be a challenge to rewrite our combinational logic expression so that each term represents a gate.
- Not all gate networks that compute a given function are alike—networks may differ greatly in their area and speed. We want to find a network that satisfies our area and speed requirements,

which may require drastic restructuring of our original logic expression.

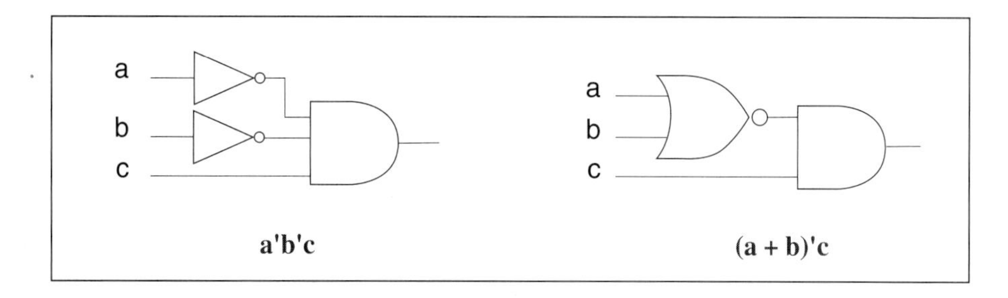

Figure 4-1 Two logic gate implementations of a Boolean function.

Figure 4-1 illustrates the relationship between logic expressions and gate networks. The two expressions are logically equivalent: $(a + b)'c = a'b'c$. We have shown a logic gate network for each expression which directly implements each function—each term in the expression becomes a gate in the network. The two logic networks have very different structures. As we discussed at the start of this chapter, which is best depends on the requirements—the relative importance of area and delay—and the characteristics of the technology. But we must work with both logic expressions and gate networks to find the best implementation of a function, keeping in mind the relationships:

- combinational logic expressions are the specification;
- logic gate networks are the implementation.

requirements

Our goals in designing logic networks are correctness, performance, size, and power consumption. Clearly, the logic we design must perform the desired function. The functions tell us what output to expect for each possible input. We refer to the main logic inputs as **primary inputs** and the main outputs as **primary outputs** to distinguish them from the outputs of gates buried in the logic network.

performance, size, and power

Performance, size, and power consumption are all **non-functional requirements**, in that they do not relate to the logical function being performed. Each imposes its own constraints:

- Performance is clearly a major design goal in most digital systems. The propagation delays through the combinational logic are a major determinant of clock speed. Performance is usually expressed as a lower bound—the logic must run at least at a given rate, but running faster often accrues no advantage.

- Size is directly related to manufacturing cost. The amount of circuitry required to implement a logical function determines chip size, which in turn is a major component of chip cost. In some instances size constraints are coarsely quantized—if your logic doesn't fit in one size FPGA, you must use the next larger FPGA. However, in a larger system design built from several subsystems there may be more subtle trade-offs.
- Power and energy consumption are important in all sorts of digital systems. Even if your design isn't run off a battery, lower energy consumption will lead to less heat dissipation. Higher heat dissipation could force you to use a chip that comes in a more expensive package that is better able to dissipate heat; it could also force you into more extreme measures such as cooling fans.

Since we want to write complex logic functions, we need a way to ensure that the logic we build meets all the non-functional requirements while maintaining its logical correctness. We do this today by using tools that help us optimize a logic description. We first write the logic functions as directly as possible. We then rely on optimization tools to transform that logical description to meet our other goals without breaking the original functions. The combination of translation plus optimization is a stark change from the early days of logic design, when designers were responsible for manually optimizing designs without introducing bugs.

In the next sections we will consider how to optimize combinational logic. First, we will study the basic sources of delay in combinational logic and what can be done about them. We will then consider power consumption. All along the way, we will keep size in mind as a criterion. Many of these concepts are common to all forms of logic design; some are specific to FPGAs. We will then study logic optimization methods. Although our primary goal is to be able to use these CAD tools, not to be able to write CAD tools from scratch, it is important to have a basic understanding of how these tools work in order to make best use of them. Logic synthesis is not sufficiently automated that we can simply press a button and expect to obtain ideal results. Some knowledge of how the tools work helps us figure out how to get the most from these tools.

mapping to an FPGA

We clearly have to select an FPGA at some point in the design so that we can map our design into it. Knowing the brand of FPGA we will use is not enough to be able to transform the logic into a configuration. Even knowing the family of FPGA we will use is not enough. FPGA families are offered in several different sizes: different numbers of logic elements (which of course changes the amount of interconnect on the chip); different numbers of I/O pins. The size of the FPGA is generally specified by the number of pins on the chip since putting too many LEs on the chip relative to the number of pins means that the LEs will be hard to use effectively. But the same number of pins may be offered in more than one type of package. The chip package houses the chip and provides structure for the pins. Packages vary in the type of material used (plastic or ceramic) and the style of pin used. Therefore, both the package type and number of pins must be specified. And for a given package and pin count the chip is usually available in several **speed grades**. Speed variations are a natural result of the VLSI manufacturing process; simply fabricating the same chip over time yields a Gaussian distribution of speeds. The speed grade is an element of the FPGA part number. The next example gives an example part number and what it means.

Example 4-1 Parsing an FPGA part number

Consider the Xilinx Spartan-IIE. There are several different parts in this family that have different numbers of CLBs, etc. Some of the part numbers include XC2S50E, XC2S100E, etc. The next parameter to consider is speed grade. The Spartan-IIE family provides two speed grades, each with its own part code: -6 for standard performance and -7 for high performance. There are several different packages available, each with its own code: TQ144 for a plastic thin quad flat pack, PQ208 for a plastic quad flat pack, etc. Finally, there is a code for the temperature range: C for commercial and I for industrial. (The industrial temperature range is larger than the commercial range.)

hardware description

We can specify the logical function we want to build in several different ways. The two major methods in use today are **hardware description languages** (**HDL**s) and **schematics**. A hardware description language is a textual description. It can contain elements of function, such as an I/O relation; it can also include structural information, such as how components are connected by wires. A schematic is a graphical description. Schematics are generally structural descriptions of the components and wires in a design. While schematics are often used to illustrate a design, HDLs are the most common way to describe a logic design for CAD tools.

logic optimization

Writing the logical function that we want to build in hardware is only a first step in digital design. We need to take into account the characteristics of the logic gates that we will use to build the system. That requires rewriting the logical function in terms of the primitive operations supplied by those logic gates. As we rewrite our original logical function, we must be sure that we do not change the basic input/output relation defined by that function. The rewrite must also ensure that we meet the non-functional requirements such as performance and power.

Logic optimization (also known as **logic synthesis**) is a rewriting process that is designed to improve the non-functional characteristics of logic—size, performance, and/or power—while maintaining the logical function. Logic optimization tools use several techniques to improve the logic design. The tools may use the laws of Boolean algebra to simplify the logic. They may use don't-cares to improve the logic without changing the essential requirements of the input/output relation. They may also add common factors to reduce size or eliminate levels of logic to increase performance.

macros

Neither CAD tools nor human designers are perfect—they may miss opportunities to fit the logic into the primitive gates of the FPGA or other target technology. One way to help ensure a good result is to specify the function using **macros**. A macro is a pre-defined primitive that has a known implementation in the target technology. For instance, many FPGAs provide specialized logic for addition. A macro for addition provides a simple way for the designer to write the addition—simpler, for example, than using XORs, NANDs, *etc*. The CAD tools identify the macro and map it into the specialized addition logic. If the logic is later targeted to a different device that doesn't provide this specialized logic, it is still possible to rewrite the macro in terms of generic logic. Macros provide a fairly painless way for designers to make use of specialized features.

physical design

Once we have rewritten the logical functions for our device, we must still choose the physical implementation. In general, physical design is divided into **placement** and **routing**; FPGAs add the step of **configuration generation**. Placement chooses the locations of the gates on the chip. On a custom chip, placement decides where to build gates, while in FPGAs it decides what existing gates will be used to implement each piece of the logic. Routing makes the connections between the gates. Once again, custom chip design requires creating wires while FPGA routing chooses among the existing wires on the chip. Generating the configuration for an FPGA chooses the bits that will configure the logic and wiring on the chip; the order of those bits in the configuration is determined by the placement and routing.

The next two examples will go through the design and verification of a simple logic design. The example should help to illustrate the steps in

FPGA-based logic design. The second example simulates the design to verify its functional correctness. It may help you to run the tools yourself as you follow along with the example.

Example 4-2 Implementing a logic design

We will implement an even parity function to illustrate the various steps in logic design. A four-input parity function can be described as

$$p = a_0 \oplus a_1 \oplus a_2 \oplus a_3$$

where $\oplus$ is the symbol for the exclusive OR (XOR) operation.

We will use the Xilinx ISE (version 6.1.03i to be exact) to work through the process. Here is the main screen for the Project Navigator, which we have opened to an existing version of the parity project to illustrate the ISE's major features:

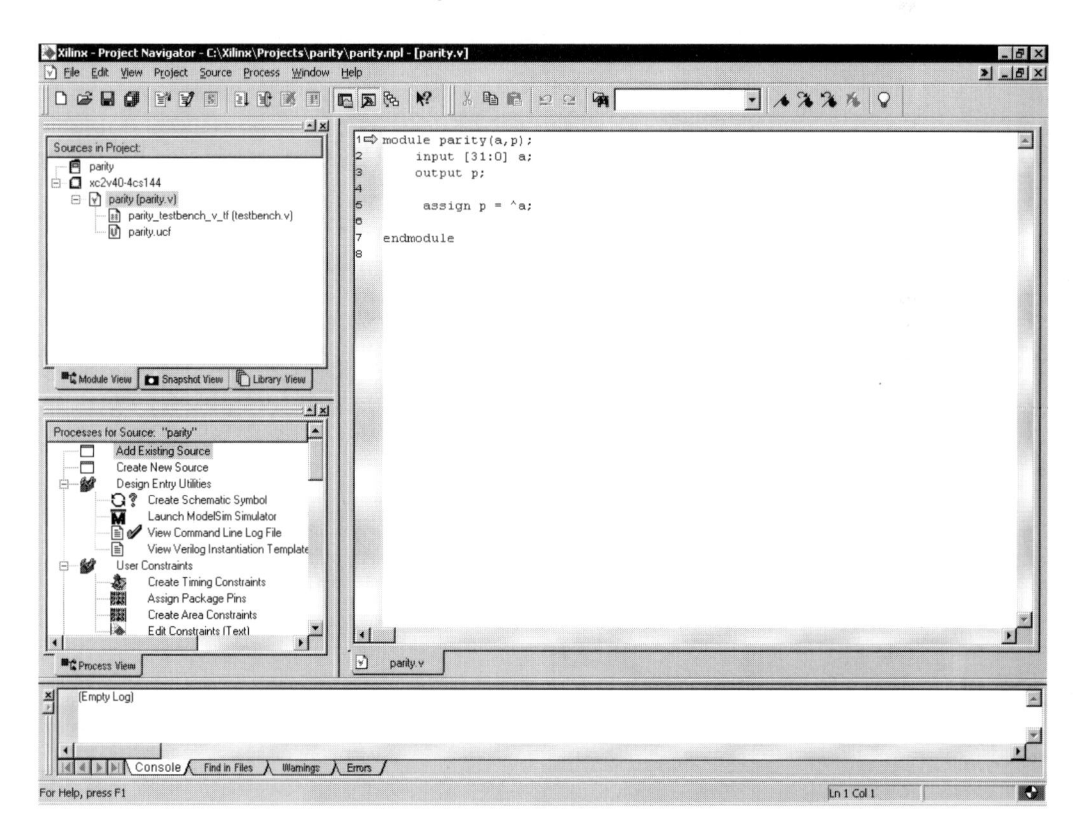

The main window is divided into several panes:

- At the top left, the pane labeled *Sources in Project* shows the files used in the project as well as the type of Xilinx part being targeted.

- At the bottom left, the pane labeled *Processes for Source "parity"* shows the design flow and the steps that can be performed at this time. A step with a check mark next to it has been successfully completed, while a step with a red X needs to be finished. We can run a step (including all the precursor steps that need to be run) by double-clicking on that step.

- The pane at the top right shows the sources that we are viewing. We opened *parity.v* by double-clicking on that name in the *Sources in Project* window.

- The pane at the bottom is the console window that shows the output from the various programs we run. This pane has several tabs: console for regular output, find in files, warnings, and errors.

Let's use Xilinx ISE to create, design, and compile the parity project. We will use the *File->New Project* command to create a new project directory. This command will show us a series of windows that prompt us for various pieces of information about the project. The first window asks us for a name and location for the project:

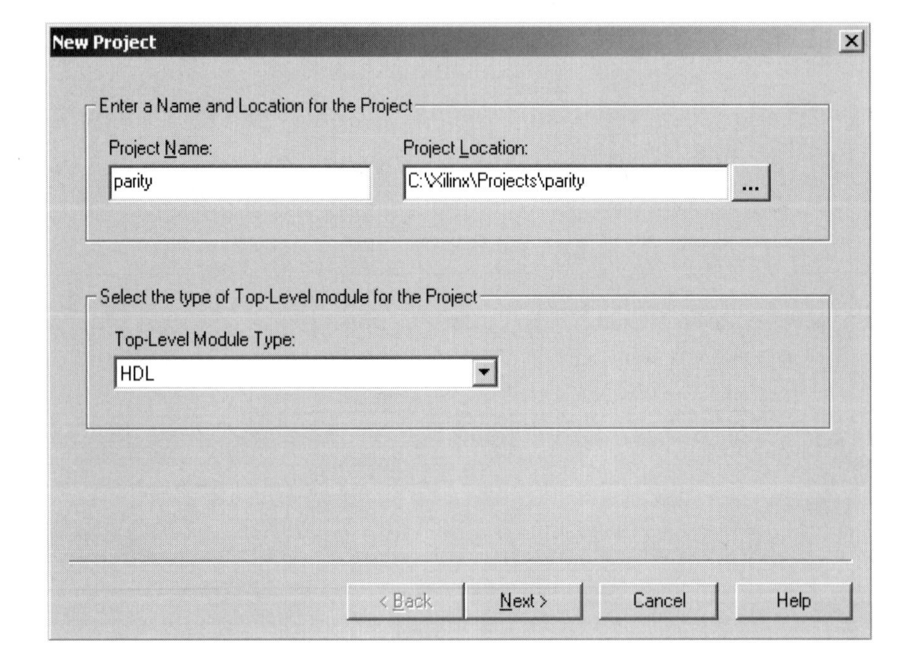

If we enter *parity* into the name field, the Project Navigator will create a directory named *parity* in *C:\Xilinx\Projects* (a subdirectory of *Xilinx* that we created). We could put it into another directory, but this version of Project Navigator cannot use directory names that include spaces, such as "My Directory." We also tell it that the top-level module is an HDL description, not a schematic or other representation.

The next window asks us which type of part we want to use:

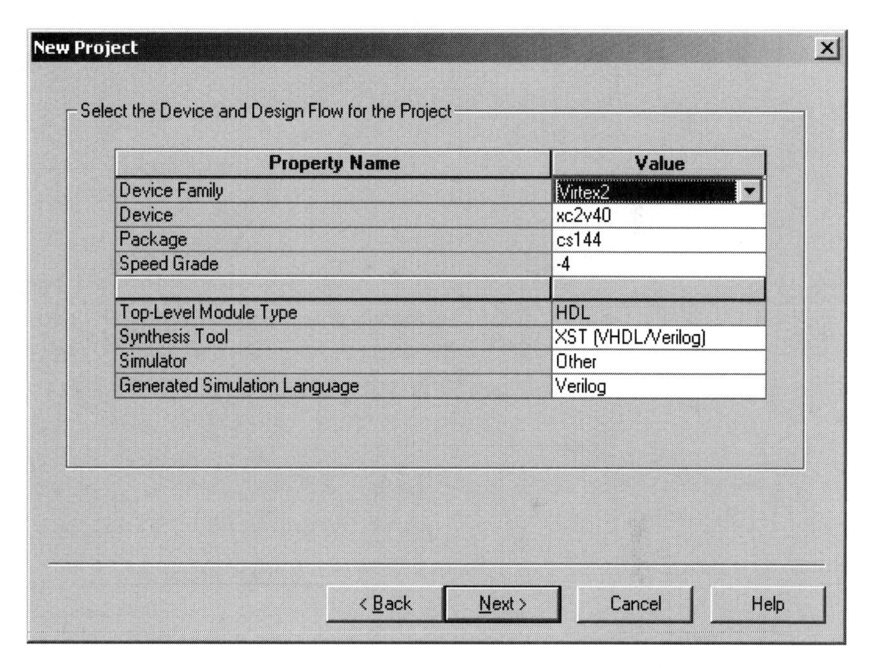

The synthesis tools look up a description of the part in a database; that description tells the tools how the logic elements work, how many logic elements are on the chip, *etc*. There are several fields to take into account: the device family (which determines the overall architecture); the device number (which determines the exact number of LEs, *etc*.); the package type; and the speed grade, which specifies the maximum rated speed of the part (faster parts are more expensive). We can also specify what synthesis tool we want to use, *etc*.

The next window asks us whether we want to create a new HDL file:

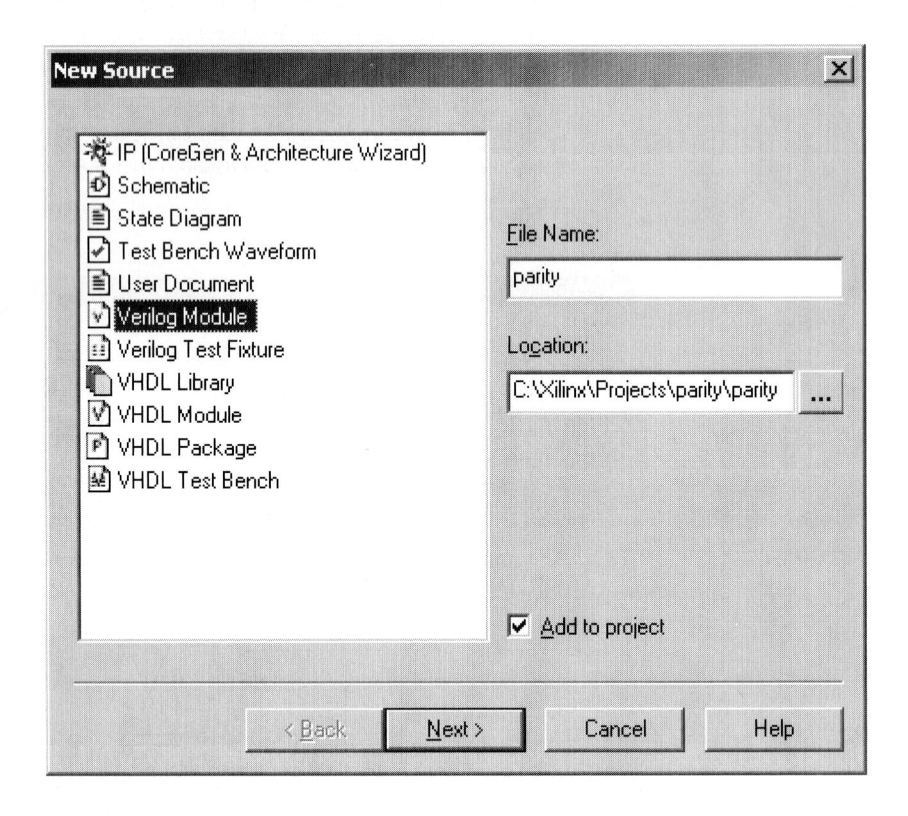

When we tell the system that we want to add one new Verilog module file called *parity*, it gives us a new window that asks us to specify the inputs and outputs of the module:

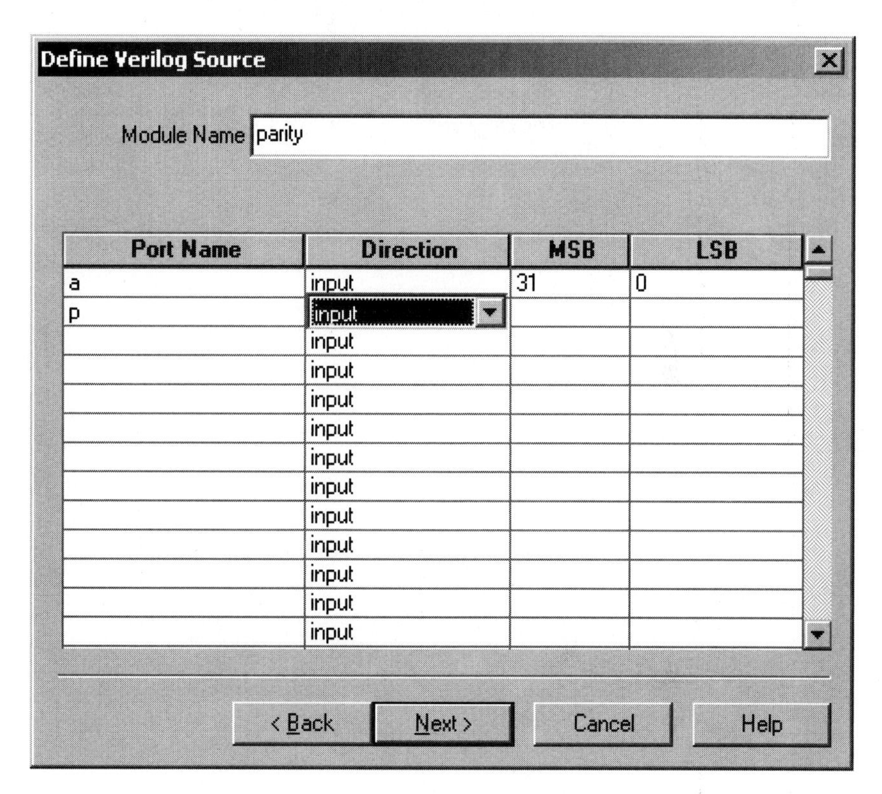

We will tell the system to create one input and one output. The input is *a* and we tell the system that the most-significant bit (MSB) is 31 and the least-significant bit (LSB) is 0 to create a 32-input parity checker. The output is *p* and we do not specify the MSB or LSB to create a scalar output. The system will then create a skeleton Verilog file that includes declarations for these ports.

The system then gives us a window that reports on the description it is about to create:

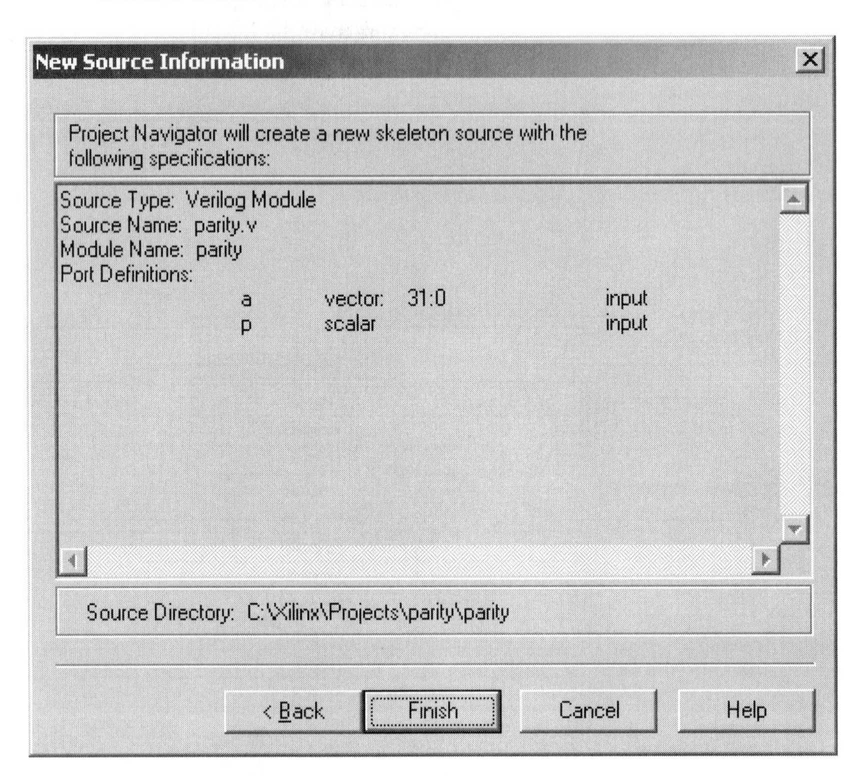

The Project Navigator next asks us about existing files that we want to add to the project. In this case, we don't have any, but we might have existing files from other team members, previous projects, Web sites, or other sources.

The *New Project* process ends with a status window that tells us about the project that is being created:

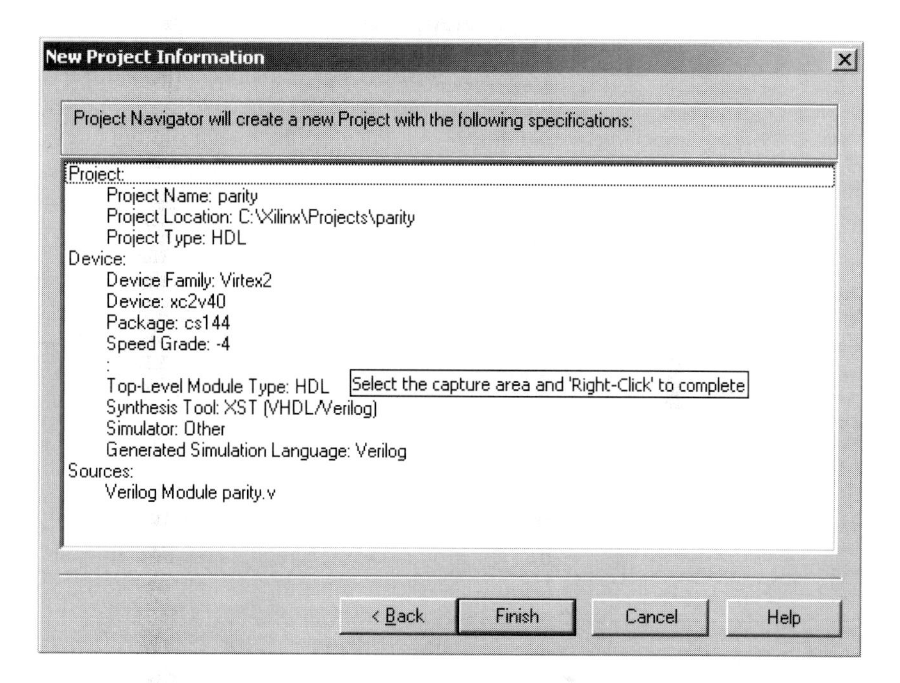

At this point the system shows us the Verilog file, *parity.v*, that was created:

```
module parity(a,p);
   input [31:0] a;
   output p;

endmodule
```

The editor window shows us the Verilog shell that *New Project* created, including the input array and the single output. All we need to do now is to add the statements to compute the parity output from the inputs. The added code is fairly simple:

```
assign p = ^a; /* compute XOR of all the bits */
```

We can simply type this code into the appropriate point in the Verilog shell:

```
module parity(a,p);
   input [31:0] a;
   output p;
```

```
    assign p = ^a;

endmodule
```

This description is longer than the original function we wrote because the Verilog code includes additional information. We define a name for the function and declare the inputs and outputs of the function. Because Verilog functions can contain many statements, Verilog uses a block syntax similar to that used in programming languages to delimit the function definition.

The *Processes for Source* window shows us the design flow for the FPGA design. This design flow specifies the steps we need to follow to synthesize the final configuration file that we can load into the FPGA. As we complete each phase in the implementation process, the system will put a check mark next to that phase. If we want to rerun a phase that has successfully completed, we can select the phase and then choose *Process->Rerun*.

The *User Constraints* item allows us to specify timing and pinout. Timing constraints are not very interesting in a design this small but we often want to determine the pins on which signals will be visible. If we click on *Assign Package Pins* we are presented with a window that lets us assign signals in the design to pins on the FPGA. The window shows us a map of the pin locations on the chip. In addition to location, we can specify the other configurable attributes of the pin, such as voltage and slew rate.

Although we have described the function in terms of a network of logic elements we cannot yet configure the function into an FPGA. We must first optimize the logic to fit into logic elements, next place the logic elements in the FPGA fabric, then route the wires to connect the logic elements, and finally generate a configuration file.

Logic synthesis will optimize the function from the HDL description. Because this function is simple there is little optimization work per se, but we must still map it into logic elements. By clicking on the *Synthesize* item in the *Processes for Source* window we can run the logic synthesis tool. The *Console* window gives us a long report on the results. Here is what it tells us about the logic synthesis process:

```
Started process "Synthesize".
```

```
=================================================
========================
*                    HDL Compilation                    *
=================================================
========================
Compiling source file "parity.v"
Module <parity> compiled
No errors in compilation
Analysis of file <parity.prj> succeeded.


=================================================
========================
*                    HDL Analysis                    *
=================================================
========================
Analyzing top module <parity>.
Module <parity> is correct for synthesis.


=================================================
========================
*                    HDL Synthesis                    *
=================================================
========================

Synthesizing Unit <parity>.
   Related source file is parity.v.
   Found 1-bit xor32 for signal <p>.
   Summary:
     inferred   1 Xor(s).
Unit <parity> synthesized.


=================================================
========================
HDL Synthesis Report

Macro Statistics
# Xors                          : 1
  1-bit xor32                   : 1
```

```
=========================================================
======================

=========================================================
======================
*                Advanced HDL Synthesis              *
=========================================================
======================


=========================================================
======================
*                  Low Level Synthesis               *
=========================================================
======================

Optimizing unit <parity> ...
Loading device for application Xst from file '2v40.nph' in environ-
ment C:/Xilinx.

Mapping all equations...
Building and optimizing final netlist ...
Found area constraint ratio of 100 (+ 5) on block parity, actual
ratio is 2.

=========================================================
======================
*                     Final Report                   *
=========================================================
======================

Device utilization summary:
---------------------------

Selected Device : 2v40cs144-4

 Number of Slices:                  6  out of   256     2%
 Number of 4 input LUTs:           11  out of   512     2%
 Number of bonded IOBs:            33  out of    88    37%


=========================================================
======================
```

```
TIMING REPORT


Clock Information:
------------------
No clock signals found in this design

Timing Summary:
---------------
Speed Grade: -4

   Minimum period: No path found
   Minimum input arrival time before clock: No path found
   Maximum output required time after clock: No path found
   Maximum combinational path delay: 8.573ns

=========================================================
=========================
Completed process "Synthesize".
```

This report gives us a blow-by-blow description of the steps used to compile the Verilog file into optimized logic; the exact steps called out depend on the synthesis tool you use and may even change from version to verison of the same tool. The *HDL Compilation* and *HDL Analysis* steps parse the Verilog file and perform some simple checks. The *HDL Synthesis* step starts the process of translating the Verilog into logic in earnest. This step gives us a report about an XOR function that it found in the logic; this step will often tell us about other functions that it finds. The *Advanced HDL Synthesis* step continues to optimize the logic. The *Low Level Synthesis* step maps the logic into the logic elements on the FPGA; the tool tells us that it loads a description of the device that includes the number of logic elements, *etc.*

The *Final Report* section tells us how much of the on-chip logic was used; since this was a simple design it used only a small fraction of the chip. The timing report is mainly used for synchronous machines. Since this parity function is combinational it doesn't give us much information but in general it would tell us about the longest path between registers. The *Errors* tab on the console window shows us error messages for syntax, *etc.* In this case, we didn't have any syntax errors and synthesis completed.

We can look at the schematic of the logic that was generated by selecting "View RTL Schematic" on this menu:

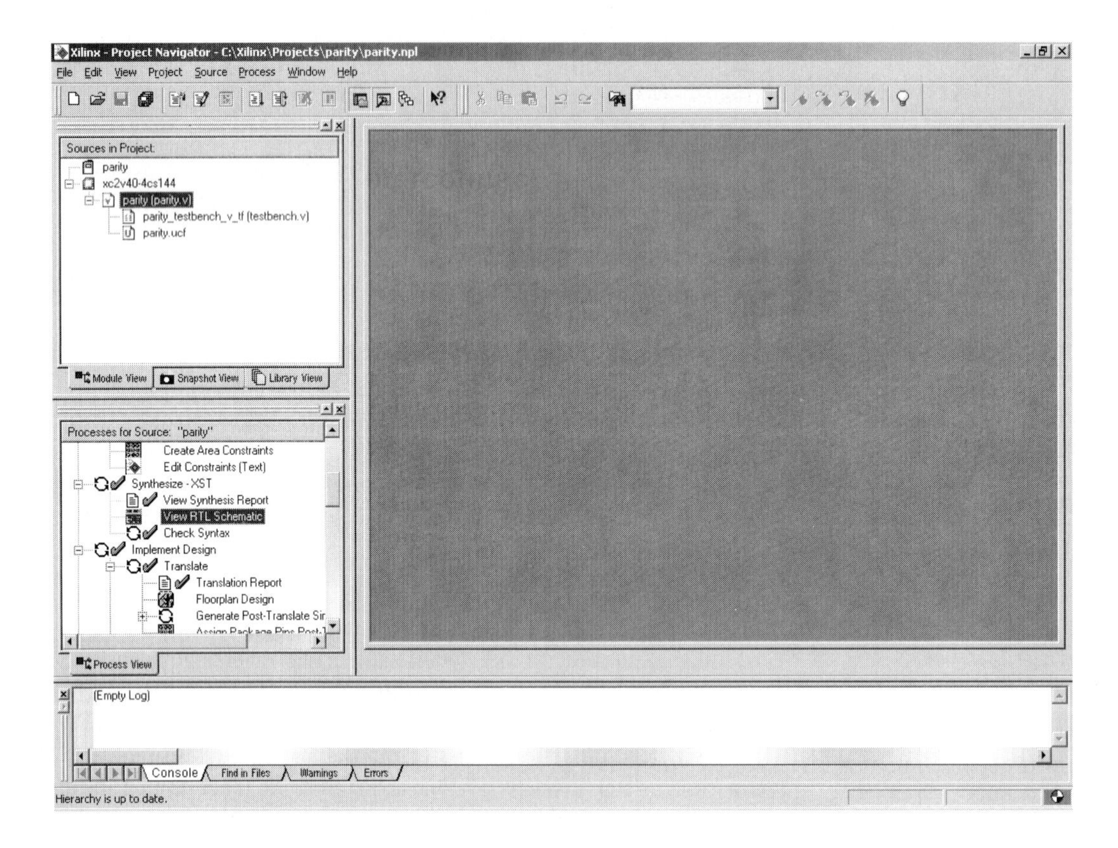

The first schematic shown is the black box view that would be used when you instantiate this design as a component:

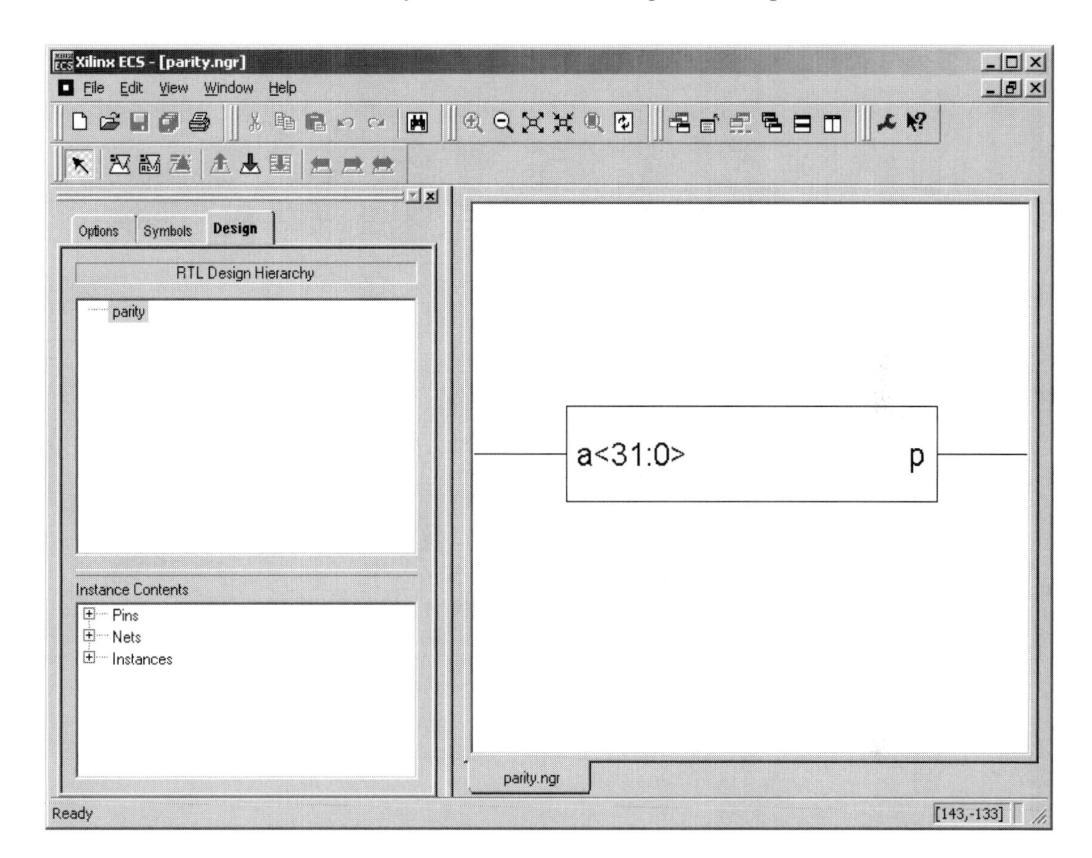

This view shows the inputs and outputs of the module. By double-clicking on this symbol we can look inside to see the logic that implements the module:

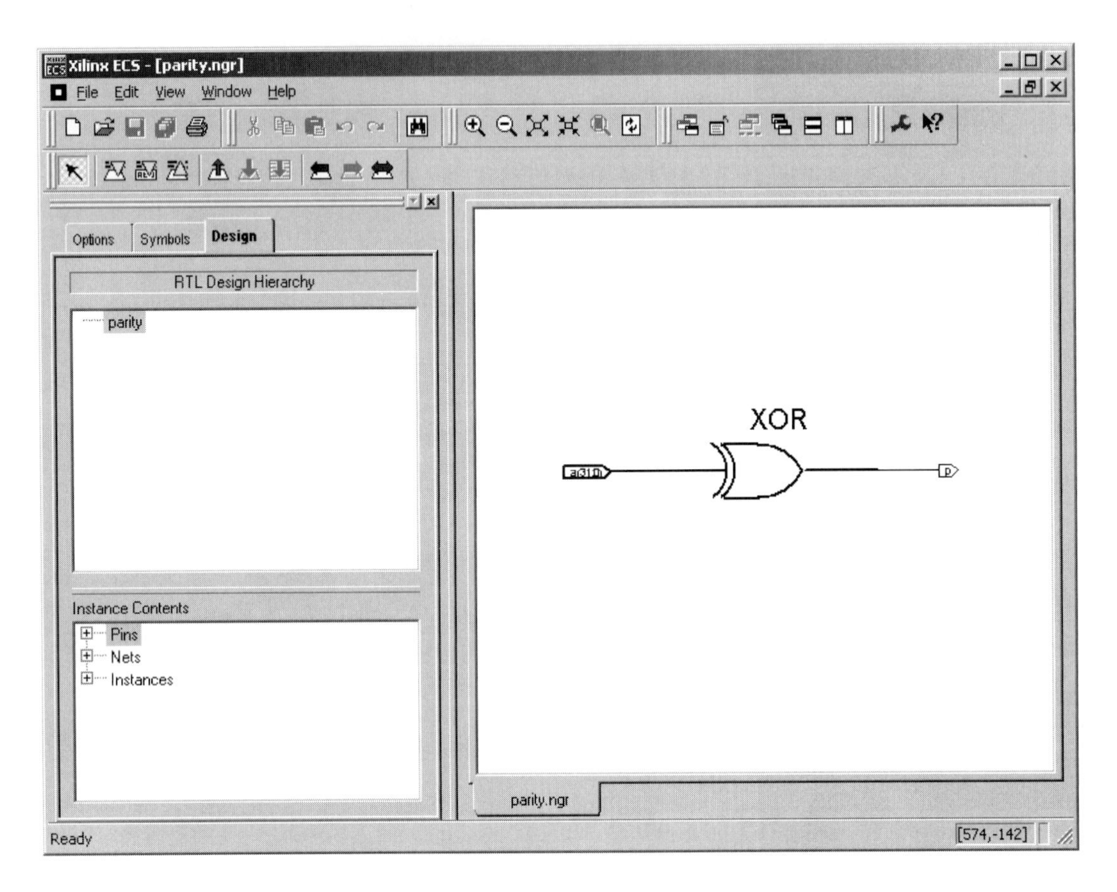

This schematic is very simple but it does show the array of signals and logic used to build the parity function.

The *Implement* phase maps the logic into the FPGA and performs placement and routing. Translation and mapping fit the logic into pieces that will fit into the logic elements. Placement chooses which logic elements will hold each of those pieces. Routing chooses which wire segments are to be used to make the necessary connections between the logic elements. In this case routing is simple because the function occupies a small part of the FPGA. If the function were considerably larger, routing would be more complex; the router may need to rip up certain connections and reroute them in a different place in order to make room for all the connections. Here are the results from our *Implement* step:

```
Started process "Translate".


Command Line: ngdbuild -intstyle ise -dd c:\xilinx\projects\parity/
_ngo -uc
parity.ucf -p xc2v40-cs144-4 parity.ngc parity.ngd

Reading NGO file "C:/Xilinx/Projects/parity/parity.ngc" ...
Reading component libraries for design expansion...

Annotating constraints to design from file "parity.ucf" ...

Checking timing specifications ...
Checking expanded design ...

NGDBUILD Design Results Summary:
  Number of errors:     0
  Number of warnings:   0

Total memory usage is 35564 kilobytes

Writing NGD file "parity.ngd" ...

Writing NGDBUILD log file "parity.bld"...

NGDBUILD done.
Completed process "Translate".


Started process "Map".

Using target part "2v40cs144-4".
Removing unused or disabled logic...
Running cover...
Running directed packing...
Running delay-based LUT packing...
Running related packing...

Design Summary:
Number of errors:     0
Number of warnings:   0
Logic Utilization:
  Number of 4 input LUTs:           11 out of     512    2%
```

```
Logic Distribution:
  Number of occupied Slices:           9 out of    256    3%
  Number of Slices containing only related logic:      9 out of      9
100%
  Number of Slices containing unrelated logic:        0 out of      9
0%
      *See NOTES below for an explanation of the effects of unre-
lated logic
Total Number 4 input LUTs:            11 out of    512    2%

  Number of bonded IOBs:              33 out of     88   37%

Total equivalent gate count for design:  66
Additional JTAG gate count for IOBs:  1,584
Peak Memory Usage:  60 MB

NOTES:

   Related logic is defined as being logic that shares connectivity -
   e.g. two LUTs are "related" if they share common inputs.
   When assembling slices, Map gives priority to combine logic
that
   is related.  Doing so results in the best timing performance.

   Unrelated logic shares no connectivity.  Map will only begin
   packing unrelated logic into a slice once 99% of the slices are
   occupied through related logic packing.

   Note that once logic distribution reaches the 99% level through
   related logic packing, this does not mean the device is com-
pletely
   utilized.  Unrelated logic packing will then begin, continuing until
   all usable LUTs and FFs are occupied.  Depending on your tim-
ing
   budget, increased levels of unrelated logic packing may
adversely
   affect the overall timing performance of your design.


Mapping completed.
See MAP report file "parity_map.mrp" for details.
Completed process "Map".
```

Mapping Module parity . . .
MAP command line:
map -intstyle ise -p xc2v40-cs144-4 -cm area -pr b -k 4 -c 100 -tx off -o parity_map.ncd parity.ngd parity.pcf
Mapping Module parity: DONE

tarted process "Place & Route".

Constraints file: parity.pcf

Loading device database for application Par from file "parity_map.ncd".
"parity" is an NCD, version 2.38, device xc2v40, package cs144, speed -4
Loading device for application Par from file '2v40.nph' in environment
C:/Xilinx.
The STEPPING level for this design is 1.
Device speed data version: PRODUCTION 1.116 2003-06-19.

Resolving physical constraints.
Finished resolving physical constraints.

Device utilization summary:

Number of External IOBs 33 out of 88 37%
Number of LOCed External IOBs 0 out of 33 0%

Number of SLICEs 9 out of 256 3%

Overall effort level (-ol): Standard (set by user)
Placer effort level (-pl): Standard (set by user)
Placer cost table entry (-t): 1
Router effort level (-rl): Standard (set by user)

```
Phase 1.1
Phase 1.1 (Checksum:9896d3) REAL time: 0 secs

Phase 2.2
Phase 2.2 (Checksum:1312cfe) REAL time: 0 secs

Phase 3.3
Phase 3.3 (Checksum:1c9c37d) REAL time: 0 secs

Phase 4.5
Phase 4.5 (Checksum:26259fc) REAL time: 0 secs

Phase 5.8
.
Phase 5.8 (Checksum:98d8b5) REAL time: 0 secs

Phase 6.5
Phase 6.5 (Checksum:39386fa) REAL time: 0 secs

Phase 7.18
Phase 7.18 (Checksum:42c1d79) REAL time: 0 secs

Phase 8.24
Phase 8.24 (Checksum:4c4b3f8) REAL time: 0 secs

Phase 9.27
Phase 9.27 (Checksum:55d4a77) REAL time: 0 secs

Writing design to file parity.ncd.

Total REAL time to Placer completion: 0 secs
Total CPU time to Placer completion: 1 secs


Phase 1: 43 unrouted;    REAL time: 0 secs

Phase 2: 43 unrouted;    REAL time: 0 secs

Phase 3: 15 unrouted;    REAL time: 0 secs

Phase 4: 0 unrouted;    REAL time: 0 secs
```

```
Total REAL time to Router completion: 0 secs
Total CPU time to Router completion: 1 secs

Generating "par" statistics.

Generating Pad Report.

All signals are completely routed.

Total REAL time to PAR completion: 2 secs
Total CPU time to PAR completion: 1 secs

Peak Memory Usage:  47 MB

Placement: Completed - No errors found.
Routing: Completed - No errors found.

Writing design to file parity.ncd.


PAR done.
Completed process "Place & Route".


Started process "Generate Post-Place & Route Static Timing".


Analysis completed Mon Sep 29 14:43:12 2003
--------------------------------------------------------------------------------

Generating Report ...

Completed process "Generate Post-Place & Route Static Tim-
ing".

Place & Route Module parity . . .
PAR command line: par -w -intstyle ise -ol std -t 1 parity_map.ncd
parity.ncd parity.pcf
PAR completed successfully
```

This step also gives us some reports about the quality of our design. It starts with a summary of the logic resources utilized by this design, similar to the report given at the end of logic synthesis. It then calls out various phases in placement and routing. It ends with a timing report that gives delays between clocked registers. This report is much more accurate than the delay estimates given at the end of logic synthesis. Logic synthesis doesn't know where the logic will go on chip, so it has no idea how long the wires are and what delays will be incurred by signals traveling along those wires. The post-place and route static timing report is based on the actual placement of logic and delays through wires, so the accuracy of this report is limited only by the basic accuracy of the underlying timing models for logic and interconnect.

The placement and routing information tells us where everything is on the chip but that information isn't in the proper format for configuration. The configuration process examines the placement and routing results and generates a final file that can be used to configure the FPGA. In the case of an SRAM-based FPGA the configuration is downloaded into the FPGA; the configuration is fed to a device programmer to burn an antifuse-based FPGA. The configuration file is generally different for each different type of part (though speed grades of a part generally use the same configuration file). The configuration file contains all the bits for configuring the logic elements, wiring, and I/O pins. It usually contains some additional information that is specific to the part, such as security information.

When we select *Generate Programming File* the system will generate a configuration file for us. Here are the console results for *parity*:

```
Started process "Generate Programming File".

Completed process "Generate Programming File".
```

We can now download the configuration into the FPGA.

Example 4-3 But does it work?

The fact that our Verilog file synthesizes does not mean that it implements the function we need. One of the key tools for verifying design correctness is simulation. In this example we will use the ModelSim simulator (ModelSim XE II 5.7c to be exact) to test the correctness of our design.

Simulation requires that we apply inputs to our model and observe its outputs. Those inputs are often called the **stimulus** or **test vectors**. The output is sometimes called the **response** or **output vectors**. For each input we need to determine whether the model produced the correct output. Determining the proper output is fairly simple for our parity function but in a large system it can be much more difficult to determine what output to expect.

In order to stimulate our model and observe its response we need to create another Verilog module known as the **testbench** or **test fixture**. The model we want to simulate is called the **unit under test** or UUT. The testbench connects to the unit under test—it feeds the stimulus into the UUT's inputs and catches the results at the output. The relationship between the testbench and the UUT looks something like this:

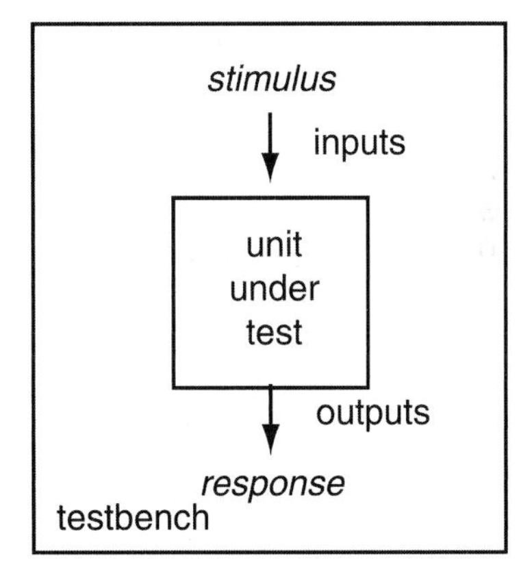

The testbench connects to the UUT's inputs and outputs; its own simulation code generates the inputs and receives the outputs. However, the testbench itself has no inputs or outputs. We use the simulator's tools to see what is going on inside the testbench.

We can use the Xilinx tools to generate a testbench for our module by creating a new file of type Verilog test fixture. When we tell the system what Verilog file we want to use as the unit under test, it reads the Verilog description to determine the UUT's inputs and outputs, then builds the testbench to match. Here is the basic testbench created for our parity module:

```
module parity_testbench_v_tf();

// DATE:     11:48:13 11/07/2003
// MODULE:   parity
// DESIGN:   parity
// FILENAME: testbench.v
// PROJECT:  parity
// VERSION:
```

```
// Inputs
   reg [31:0] a;

// Outputs
   wire p;

// Bidirs

// Instantiate the UUT
   parity uut (
      .a(a),
      .p(p)
      );

// Initialize Inputs
   'ifdef auto_init

      initial begin
         a = 0;
      end

   'endif

endmodule
```

This code starts with some declarations of registers and wires to connect to the UUT's inputs and outputs. It then instantiates the UUT itself, giving it the name uut. The instantiation connects the registers and wires it just declared to the UUT's pins. For example, the declaration .a(a) connects the a pin on uut to the a wire in the testbench. It then includes some code to initialize the inputs at the start of the simulation.

We can then add our own code to generate stimuli and report responses. This code goes after the 'endif and before the endmodule:

```
initial begin
  $monitor("a = %b, parity=%b\n",a,p);
  #10 a = 0;
  #10 a = 1;
```

```
  #10 a = 2'b10;
  #10 a = 2'b11;
  #10 a = 3'b100;
  #10 a = 3'b101;
  #10 a = 3'b110;
  #10 a = 3'b111;
  #10 a = 32;
  #10 a = 33;
  #10 a = 34;
  #10 a = 255;
  #10 a = 256;
  #10 a = 257;
  #10 a = 1023;
  #10 a = 1024;
  #10 a = 1025;
  #10 a = 16'b1010101010101010;
  #10 a = 17'b11010101010101010;
  #10 a = 17'b10010101010101010;
  #10 a = 32'b10101010101010101010101010101010;
  #10 a = 32'b11101010101010101010101010101010;
  #10 a = 32'b10101010101010101010101010101011;
  $finish;
end
```

This part of the code is declared as initial, meaning that it is executed at the start of the simulation. An initial block may also be used to initialize a regular simulation model, but in this case it is used to generate the stimulus and observe the response.

The block begins with a $monitor statement which is similar to a C printf. This statement is executed every time the inputs or outputs change. This is a simple way to observe what is going on in the system. To check the correctness of our design we must read the output generated by the $monitor statement and compare the response for each input to the response we expected.

The next lines cause the simulator to apply a series of inputs to the UUT. Each line begins with #10, which tells the simulator to advance the time by 10 units. In this case, the amount of time between inputs is arbitrary since we are only functionally simulating the design; if we were worried about delays within the UUT we would have to be sure that we satisfied the timing constraints of the module. The rest of the line then assigns a value to the input signal. (If the UUT has several inputs we can put several of them after one # statement but we can have only one # per line.)

During execution, as the inputs change and new outputs are computed, the simulator will use the $format statement to print out the new input and output results. We could also write code that would compare the outputs to some expected inputs, but in this simple example we just want to observe the inputs and outputs directly.

The initial block ends with a $finish statement that causes the simulation to terminate. The stimuli in this testbench are probably not thorough enough to fully exercise the parity module. We don't want to exhaustively simulate the design even in something this small, but we need to make sure that we exercise various data ranges, boundary conditions, etc. to adequately test the correctness of the design.

In order to run the simulation, we must first create a project file for ModelSim. This project file is similar to the Xilinx project file in that it helps the simulator organize the files relating to simulation but ModelSim uses its own project file format. The *File->New->Project...* command presents this menu that we use to create the project:

The default library is used to organize the code; we need to remember that our library is named *work*. We will next add our Verilog files to the simulation project. First we add the *parity.v* file:

We also add the testbench file using the same procedure.

This process tells the simulator what files need to be compiled into the simulation using the *Compile->Compile All* command. We next tell the simulator to compile the simulation. After doing so we tell the simulator to load our compiled Verilog into the simulator using the

Simulate->Simulate... command. This command will present us with this menu:

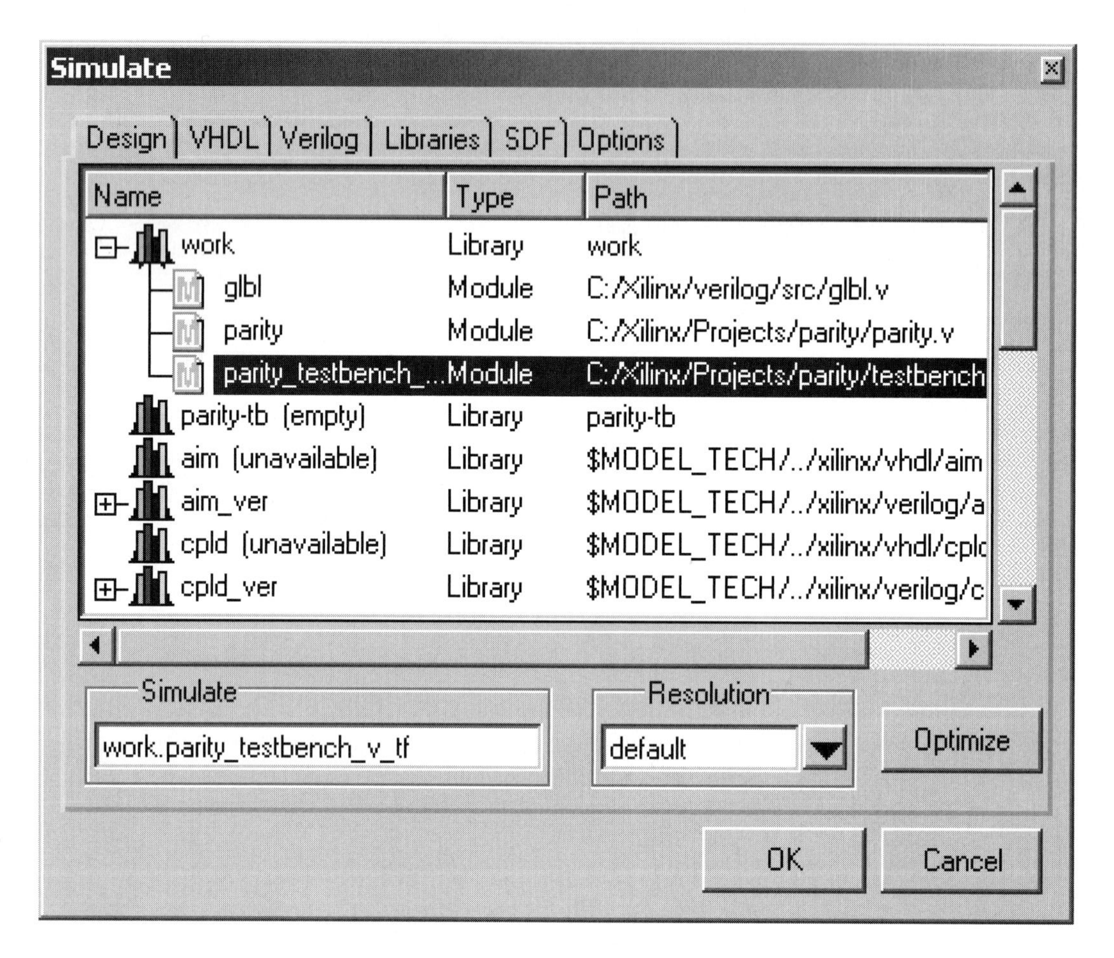

We must use this menu to tell which one of all the Verilog modules in the simulation to execute first. In our case, we want to simulate the test-bench, which will in turn simulate the parity module. We select the test-bench by name, remembering that our code was put in the library named *work*.

Once the simulation has been loaded we can then execute it. The simulator lets us choose the amount of time that we want to simulate. In this simple design, we will use the *Simulate->Run->Run -All* command to run the simulation to completion.

This simulation simply prints stimulus/response pairs into the simulator's standard output window. The result looks like this:

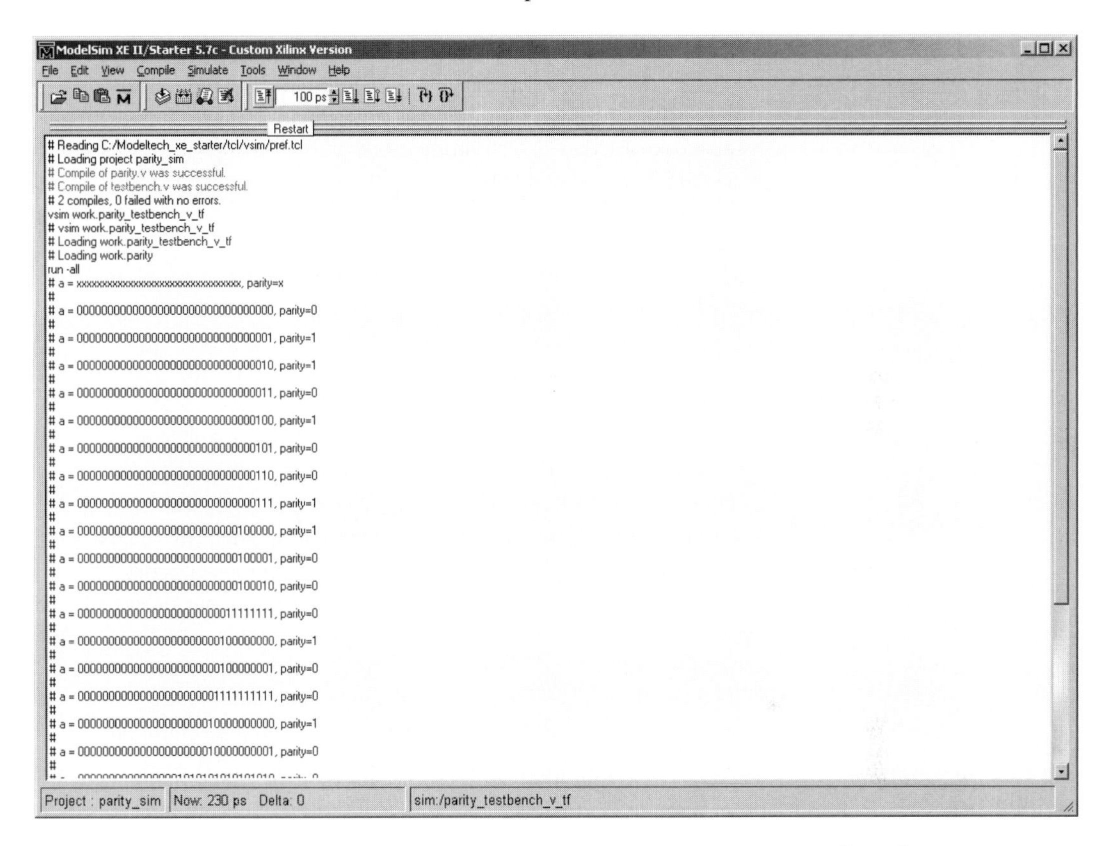

We can read the inputs/outputs printed by the $monitor statement and check that the model always generates the proper result. The simulator also provides waveform and other displays that we can use to look at the activity in the design.

4.3 Hardware Description Languages

Hardware description languages (HDLs) are the most important modern tools used to describe hardware. HDLs become increasingly important as we move to higher levels of abstraction. While schematics can convey some information very clearly, they are generally less dense than textual descriptions of languages. Furthermore, a textual HDL description is much easier for a program to generate than is a schematic with pictorial information such as placement of components and wires.

Verilog and VHDL

In this section we will introduce the use of the two most widely used hardware description languages, **Verilog** [Tho98,Smi00] and **VHDL** [IEE93,Bha95]. Since both these languages are built on the same basic framework of event-driven simulation, we will start with a description of the fundamental concepts underlying the two languages. We will then go on to describe the details of using VHDL and Verilog to model hardware. We don't have room to discuss all the details of Verilog and VHDL modeling, but this brief introduction should be enough to get you started with these languages.

4.3.1 Modeling with HDLs

Both Verilog and VHDL started out as simulation languages—they were originally designed to build efficient simulations of digital systems. Some other hardware description languages, such as **EDIF**, were designed to describe the structure of nets and components used to build a system. Simulation languages, on the other hand, are designed to be executed. Simulation languages bear some resemblance to standard programming languages. But because they are designed to describe the parallel execution of hardware components, simulation languages have some fundamental differences from sequential programming languages.

HDLs vs. programming languages

There are two important differences between simulation and sequential programming languages. First, statements are not executed in sequential order during simulation. When we read a sequential program, such as one written in C, we are used to thinking about the lines of code being executed in the order in which they were written. In contrast, a simulation may describe a series of logic gates, all of which may change their outputs simultaneously. If you have experience with a parallel programming language, you may be used to this way of thinking. Second, most simulation languages must support some notion of real time in order to provide useful results. Even parallel programming languages usually do not explicitly support real time. Time may be measured in nanoseconds for more realistic simulation or in some more abstract unit such as gate delays or clock cycles in faster, more abstract simulators. One important job of the simulator is to determine how long it takes to compute a given value. Delay information determines not only clock speed but also proper operation: glitches caused by unbalanced delays may, for example, cause a latch to be improperly clocked. Simulating functional behavior in the absence of time can be relatively easy; however, the simulator must go to a great deal of effort to compute the time at which values are computed by the simulated hardware.

simulation algorithms

Simulation languages serve as specialized programming languages for the **simulation engines** that execute simulations. Both VHDL and Verilog use a combination of **event-driven simulation** and **compiled simulation**. Event-driven simulation allows us to efficiently simulate large designs in which the amount of activity in the system varies over time; compiled simulation helps us quickly execute smaller chunks of the simulation.

event-driven simulation

Event-driven simulation is a very efficient algorithm for hardware simulation because it takes advantage of the activity levels within the hardware simulation. In a typical hardware design, not all the nets change their values on every clock cycle: having fewer than 50% of the nets in a system keep their value on any given clock cycle is not unusual. The most naive simulation algorithm for a clocked digital system would scan through all the nets in the design for every clock cycle. Event-driven simulation, in contrast, ignores nets that it knows are not active.

simulation events

An **event** has two parts: a value and a time. The event records the time at which a net takes on a new value. During simulation, a net's value does not change unless an event records the change. Therefore, the simulator can keep track of all the activity in the system simply by recording the events that occur on the nets. This is a sparse representation of the system's activity that both saves memory and allows the system activity to be computed more efficiently.

Figure 4-2 illustrates the event-driven simulation of gates; the same principle can be applied to digital logic blocks at other levels of abstraction as well. The top part of the figure shows a NAND gate with two inputs: one input stays at 0 while the other changes from a 0 to a 1. In this case, the NAND gate's output does not change—it remains 1. The simulator determines that the output's value does not change. Although the gate's input had an event, the gate itself does not generate a new event on the net connected to its output. Now consider the case shown on the bottom part of the figure: the top input is 1 and the bottom input changes from 0 to 1. In this case, the NAND gate's output changes from 1 to 0. The activity at the gate's input in this case causes an event at its output.

simulation timewheel

The event-driven simulator uses a **timewheel** to manage the relationships between components. As shown in Figure 4-3, the timewheel is a list of all the events that have not yet been processed, sorted in time. When an event is generated, it is put in the appropriate point in the time-

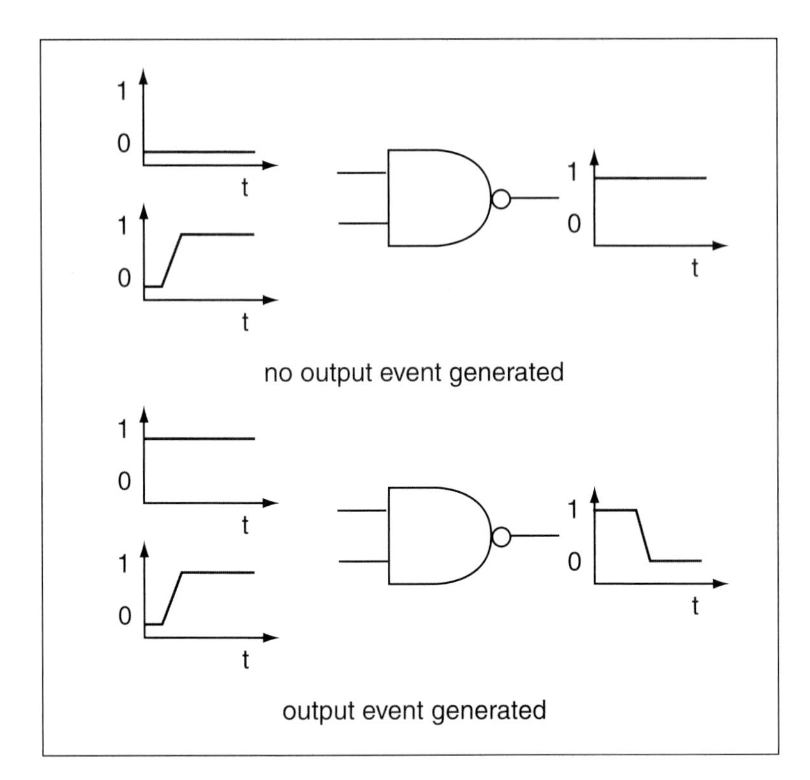

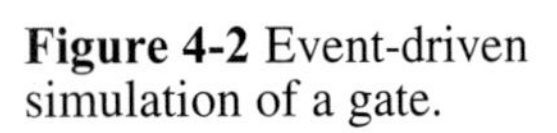
Figure 4-2 Event-driven simulation of a gate.

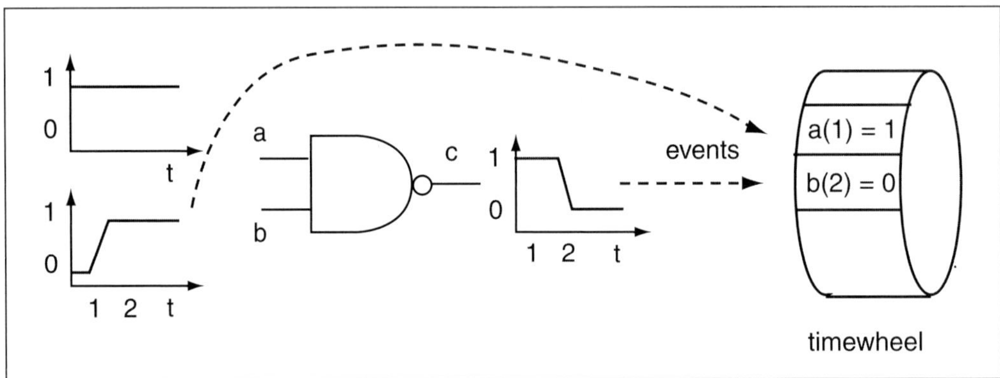

Figure 4-3 The event-driven timewheel.

wheel's list. The simulator therefore processes events in the time order in which they occur by pulling events in order from the head of the timewheel's list. Because a component with a large internal delay can generate an event far in the future from the event that caused it, operations during simulation may occur in a very different order than is apparent from the order of statements in the HDL program.

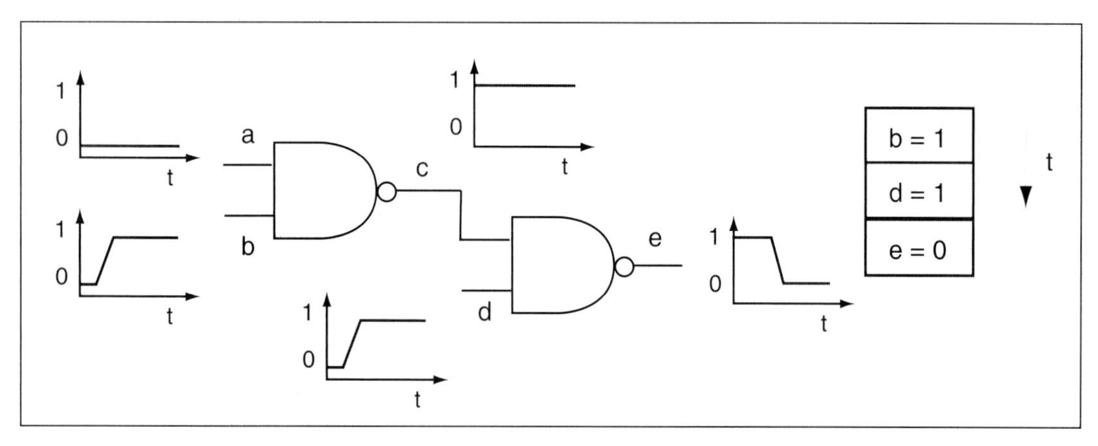

Figure 4-4 Order of evaluation in event-driven simulation.

As shown in Figure 4-4, an event caused by the output of one component causes events to appear at the inputs of the components being driven by that net. As events are put into the timewheel, they are ordered properly to ensure causality, so that the simulator events are processed in the order in which they occur in the hardware. In the figure, the event at the input causes a cascade of other events as activity flows through the system.

compiled simulation

Compiled simulation is used to speed up chunks of simulation in which we do not care about the detailed structure of events and when there is enough activity in the block. When designing at the register-transfer level, for example, we do not care about the events within a block of combinational logic. Generating events for all the internal logic is very time consuming. Although event-driven simulation is designed to take advantage of idleness in the system being simulated, it is sometimes more efficient to simply compute everything, including recomputing some values.

Compiled simulation generates code directly to execute a block of the design under simulation. Simulation systems often generate high-level language code that is then compiled into machine code for execution. Consider a C function whose arguments are the inputs and outputs of a hardware block. The body of the function takes the input values presented to it and computes the outputs corresponding to those inputs. The internals of that function don't make use of the timewheel and the rest of the event-driven simulation mechanism.

compiled/event-driven interface

A block of compiled code may interact with the event-driven simulation even if it does not use events internally. When an event arrives at the input of a compiled block, the simulator can schedule the compiled code for execution. The compiled block will generate new events at its outputs; those events go back into the event queue. The combination of event-driven and compiled simulation allows the simulator to take advantage of activity patterns in the overall design while efficiently simulating the pieces of the design.

structural vs. behavioral modeling

There are two ways to describe a design for simulation: **structural** or **behavioral modeling**. A structural model for a component is built from smaller components. The structural model specifies the external connections, the internal components, and the nets that connect them. The behavior of the model is determined by the behavior of the components and their connections. A behavioral model is more like a program—it uses functions, assignments, *etc.* to describe how to compute the component's outputs from its inputs. However, the behavioral model deals with events, not with variables as in a sequential programming language. Simulation languages define special constructs for recognizing events and for generating them.

component types

Whether a component is described structurally or behaviorally, we must define it and use it. As in a programming language, a hardware description language has separate constructs for the type definition of a component and for instantiations of that component in the design of some larger system. In C, the statement `struct { int a; char b; } mydef;` defines a data structure called `mydef`. However, that definition does not allocate any instances of the data structure; memory is committed for an instance of the data structure only by declaring a variable of type `mydef`. Similarly, in order to use a component, we must have a definition of that component available to us. The module that uses the component does not care whether the component is modeled behaviorally or structurally. In fact, we often want to simulate the system with both behavioral and structural models for key components in order to verify the correctness of the behavioral and structural models. Modern hardware description languages provide mechanisms for defining components and for choosing a particular implementation of a component model.

testbenches

We can use the simulator to exercise a design as well as describe it. Testing a design often requires complex sequences of inputs that must be compared to expected outputs. If you were testing a physical piece of hardware, you would probably wire up a test setup that would supply the necessary input vectors and capture and compare the resulting output vectors. We can do the equivalent during simulation. We build components to generate the inputs and test the outputs; we then wire them together with the component we want to test and simulate the entire system. This sort of simulation setup is known as a **testbench**.

HDL simulation subset

Both VHDL and Verilog were originally designed as simulation languages. However, one of their principal uses today is as a synthesis language. A VHDL or Verilog model can be used to define the functionality of a component for logic synthesis (or for higher levels of abstraction such as behavioral synthesis). The synthesis model can also be simulated to check whether it is correct before going ahead with synthesis. However, not all simulatable models can be synthesized. Synthesis tools define a **synthesis subset** of the language that defines the constructs they know how to handle. The synthesis subset defines a modeling style that can be understood by the synthesis tool and also provides reasonable results during simulation. There may in fact be several different synthesis subsets defined by different synthesis tools, so you should understand the synthesis subset of the tool you plan to use.

register-transfer synthesis

The most common mode of synthesis is **register-transfer synthesis**. RT synthesis uses logic synthesis on the combinational logic blocks to optimize their implementations, but the registers are placed at the locations specified by the designer. The result is a sequential machine with optimized combinational logic. The signals in the combinational section of the model may or may not appear in the synthesized implementation, but all the registers appear in the implementation. (Some register-transfer synthesis tools may use state assignment algorithms to assign encodings to symbolic-valued symbols.) The most commonly used RT synthesis subset is the one defined for the Synopsys Design Compiler™, which is accepted by that tool and several others.

One item to look out for in RT synthesis is the **inferred storage element**. A combinational logic block is defined by a set of assignments to signals. If those signals form a cycle, many synthesis tools will insert a storage element in the loop to break the cycle. Consider this Verilog fragment:

```
assign a = ~i1 | c;
assign b = a & y;
assign c = b | d;
```

This code contains a cycle *a* –> *b* –> *c*. While the inferred storage element can be handy if you want it to appear, it can cause confusion if the combinational cycle was caused by a bug in the synthesis model. The synthesis tool will emit a warning message when an inferred storage element is inserted into an implementation; that warning is generally used to warn of an unintended combinational cycle in the design.

4.3.2 Verilog

Figure 4-5 A structural Verilog model.

```
// this is a comment
module adder(a,b,cin,sum,cout);
    input a, b, cin;
    output sum, cout;

    // sum
    xor #2
        s(sum,a,b,cin);
    // carry out
    and #1
        c1(x1,a,b);
        c2(x2,a,cin);
        c3(x3,b,cin);
    or #1
        c4(cout,x1,x2,x3);
endmodule
```

Verilog basics

Verilog is in many respects a very different language from VHDL. Verilog has much simpler syntax and was designed for efficient simulation (although it now has a synthesis subset). Figure 4-5 gives a simple Verilog structural model of an adder. The module statement and the succeeding input and output statement declare the adder's inputs and outputs. The following statements define the gates in the adder and the connections between them and the adder's pins. The first gate is an XOR; the #2 modifier declares that this gate has a delay of two time units. The XOR's name is s and its parameters follow. In this case, all the XOR's pins are connected directly to other pins. The next statements define the AND and OR gates used to compute the carry out. Each of these gates is defined to have a delay of one time unit. The carry out requires internal wires c1, c2, and c3 to connect the AND gates to the OR gate. These names are not declared in the module so Verilog assumes that they are wires; wires may also be explicitly declared using the wire statement at the beginning of the module.

Verilog expressions

Table 4-1 summarizes some basic syntax for Verilog expressions. We can use the 'define compiler directive (similar to the C #define preprocessor directive) to define a constant:

a & b	**Boolean AND**
a | b	**Boolean OR**
~a	**Boolean NOT**
a = b	**assignment**
a <= b	**concurrent assignment, less than or equal to**
a >= b	**greater than or equal to**
==	**equality**
2'b00	**two-bit binary constant with value 00**
#1	**time**

Table 4-1 Some elements of Verilog syntax.

	0	**1**	**x**	**z**
0	0	1	x	x
1	1	1	1	1
x	x	1	x	x
z	x	1	x	x

Table 4-2 Truth table for OR in four-valued logic.

```
'define aconst 2'b00
```

x and z values

Verilog uses a four-valued logic that includes the value *x* for unknown and *z* for high-impedance. Table 4-2 and Table 4-3 show the truth tables for four-valued OR and AND functions. These additional logic values help us better simulate the analog behavior of digital circuits. The high-impedance z captures the behavior of disabled three-state gates. The unknown value is a conservative, pessimistic method for dealing with unknown values and is particularly helpful in simulating initial conditions. A circuit often needs a particular set of initial conditions to behave properly, but nodes may come up in unknown states without initializa-

```
module testbench;
    // this testbench has no inputs or outputs
    wire awire, bwire, cinwire, sumwire, coutwire;

    // declare the adder and its tester
    adder a(awire,bwire,cinwire,sumwire,coutwire);
    adder_teser at(awire,bwire,cinwire,sumwire,coutwire);
endmodule

module adder(a,b,cin,sum,cout);
    input a, b, cin;
    output sum, cout;

    // sum
    xor #2
        s(sum,a,b,cin);
    // carry out
    and #1
        c1(x1,a,b);
        c2(x2,a,cin);
        c3(x3,b,cin);
    or #1
        c4(cout,x1,x2,x3);
endmodule

module adder_tester(a,b,cin,sum,cout);
    input sum, cout;
    output a, b, cin;
    reg a, b, cin;

    initial
        begin
            $monitor($time,,"a=%b, b=%b, cin=%cin, sum=%d,
cout=%d",
                    a,b,cin,sum,cout);
            // waveform to test the adder
            #1 a=0; b=0; cin=0;
            #1  a=1; b=0; cin=0;
            #2  a=1; b=1; cin=1;
            #2 a=1; b=0; cin=1;
        end
endmodule
```

Figure 4-6 A Verilog testbench.

	0	**1**	**x**	**z**
0	0	0	0	0
1	0	1	x	x
x	0	x	x	x
z	0	x	x	x

Table 4-3 Truth table for AND in four-valued logic.

tion. If we assumed that certain nodes were 0 or 1 during simulation, we may optimistically assume that the circuit works when, in fact, it fails to operate in some initial conditions.

adder testbench

Figure 4-6 shows a testbench for the adder. The testbench includes three modules. The first is the testbench itself. The testbench wires together the adder and the adder tester; it has no external inputs or outputs since none are needed. The second module is the adder, which is unchanged from Figure 4-5. The third module is the adder's tester. adder_test generates a series of inputs for the adder and prints the results. In order to hold the adder's inputs for easier observation of its behavior, adder_test's outputs are declared as registers with the reg statement. The initial statement allows us to define the initial behavior of the module, which in this case is to apply a stimulus to the adder and print the results. The $monitor statement is similar to a C printf statement. The succeeding statements define the changes to signals at different times. The first #1 line occurs one time unit after simulation starts; the second statement occurs one time unit later; the last two each occur two time units apart. At each time point, the $monitor statement is used to print the desired information.

4.3.3 VHDL

VHDL basics

VHDL is a general-purpose programming language as well as a hardware description language, so it is possible to create VHDL simulation programs ranging in abstraction from gate-level to system. VHDL has a rich and verbose syntax that makes its models appear to be long and verbose. However, VHDL models are relatively easy to understand once you are used to the syntax. We will concentrate here on register-transfer simulation in VHDL, using the sequencer of the traffic light controller of Section 5.4.2 as an example. The details of running the simulator will vary substantially depending on which VHDL simulator you use and your local system configuration. We will concentrate here on basic techniques for coding VHDL simulation models.

Figure 4-7 Abstract types in VHDL.

```
package lights is
    -- this package defines constants used by the
    -- traffic light controller light encoding
    subtype light is bit_vector(0 to 1);
    constant red : light := B"00";
    constant green : light := B"01";
    constant yellow : light := B"10";
end lights;
```

types in VHDL

VHDL provides extensive type-definition facilities: we can create an abstract data type and create signals of that data type, rather than directly write the simulation model in terms of ones and zeroes. Abstract data types and constants serve the same purposes in hardware modeling that they do in programming: they identify design decisions in the source code and make changing the design easier. Figure 4-7 shows a set of type definitions for the traffic light controller. These data types are defined in a package, which is a set of definitions that can be used by other parts of the VHDL program. Since we encode traffic light values in two bits, we define a data type called light to hold those values. We also define the constants red, green, and yellow and their encoding; the syntax B"00" defines a constant bit vector whose value is 00. If we write our program in terms of these constants, we can change the light encoding simply by changing this package. (VHDL is case-insensitive: yellow and YELLOW describe the same element.) When we want to use this package in another section of VHDL code, the use statement imports the definitions in the package to that VHDL code.

VHDL entities

VHDL requires an entity declaration for each model; Figure 4-8 shows the entity declaration for the traffic light controller. The entity declaration defines the model's primary inputs and outputs. We can define one or more bodies of code to go with the entity; that allows us to, for example, simulate once with a functional model, then swap in a gate-level model by changing only a few declarations.

Figure 4-8 A VHDL entity declaration.

```
-- define the traffic light controller's pins
entity tlc_fsm is
    port( CLOCK: in BIT; -- the machine clock
    reset : in BIT; -- global reset
    cars : in BIT; -- signals cars at farm road
    short, long : in BIT; -- short and long timeouts
    highway_light : out light := green; -- light values
    farm_light : out light := red ;
    start_timer : out BIT -- signal to restart timer
    );
end;
```

Figure 4-9 A process in a VHDL model.

```
combin : process(state, hg)
begin
highway_light <= green;
end process combin;
```

VHDL processes

The basic unit of modeling in VHDL is the **process**. A process defines the actions that are taken whenever any input to the process is activated by an event. As shown in Figure 4-9, a process starts with the name of the process and a **sensitivity list**. The sensitivity list declares all the signals to which the process is sensitive: if any of these signals change, the process should be evaluated to update its outputs. In this case, the process proc1 is sensitive to a, b, and c. Assignment to a signal are defined by the <= symbol. The first assignment is straightforward, assigning the output *x* to the or of inputs a and *b*.

Figure 4-10 Conditional assignment in VHDL.

```
if (b or c) = '1' then
    y <= '1';
else
    y <= '0';
```

assignment to y

```
if (b or c) = '1' then
    y <= '1'
else
    z <= a or b;
```

assignment to y or z

conditional assignments

The second statement in the example defines a conditional assignment to y. The value assigned to y depends on the value of the conditional's test. Figure 4-10 shows the combinational logic that could be used to implement this statement.

semantics of conditionals

What if a signal is not assigned to in some case of a conditional? Consider, for example, the conditional of Figure 4-10. If (b or c) = '1' then y is assigned a value; if not, then z is assigned a value. This statement illustrates some subtle differences between the semantics of simulation and synthesis:

- During simulation, the simulator would test the condition and execute the statements in the selected branch of the conditional. The signal referred to in the branch not taken would retain its value since no event is generated for that signal. This case is somewhat similar to sequential software.
- Synthesis may interpret this statement as don't-care conditions for both y and z: y's value is a don't-care if (b or c) is not '1', while z is a don't-care if (b or c) is '1'. However, unlike in software, both y and z are always evaluated. Although a C program with this sort of conditional would assign to either y or z but not both, the logic shown in the figure makes it clear that both y and z are combinational logic signals.

These differences are minor, but they do highlight the differences between simulation and logic synthesis. A simulation run results in a single execution of the machine; with different inputs, the simulation would have produced different outputs. Don't-care values could be used in simulation, but they can cause problems for later stages of logic that may not know what value to produce. Logic synthesis, in contrast, results in the structure of the machine that can be run to produce desired values. Don't-care values are very useful to logic synthesis during minimization.

Table 4-4 Some elements of VHDL syntax.

a and b	**Boolean AND**
a or b	**Boolean OR**
not a	**Boolean NOT**
a <= b	**signal assignment, less than or equal to**
a = b	**equality**
a = b	**equality**
after 5 ns	**time**

VHDL expressions

Table 4-4 shows the syntax of a few typical VHDL expressions. VHDL modelers can build complex signals with arrays of signals and bundles of different signals. Signals also need not carry binary values. By defining a series of VHDL functions, one can create a signal definition that works on a variety of logical systems: three-valued logic (0, 1, x) or symbolic logic such as the states of a state machine (s1, s2, s3). VHDL defines a basic bit type that provides two values of logic, '0' and '1'. The library IEEE.std_logic_1164 defines a nine-valued signal type known as std_ulogic.

traffic light controller

Here is a complete, simple VHDL model of a traffic light controller:

```
Library IEEE;
use IEEE.std_logic_1164.all;
use work.lights.all; -- use the traffic light controller data types

-- define the traffic light controller's pins
entity tlc_fsm is
    port( CLOCK: in BIT; -- the machine clock
    reset : in BIT; -- global reset
    cars : in BIT; -- signals cars at farm road
    short, long : in BIT; -- short and long timeouts
    highway_light : out light := green; -- light values
    farm_light : out light := red ;
    start_timer : out BIT -- signal to restart timer
    );
end;

-- define the traffic light controller's behavior
architecture register_transfer of tlc_fsm is

    -- internal state of the machine
    -- first define a type for symbolic control states,
    -- then define the state signals
    type ctrl_state_type is (hg,hy,fg,fy);
    signal ctrl_state, ctrl_next : ctrl_state_type := hg;

begin

-- the controller for the traffic lights
ctrl_proc_combin : process(ctrl_state, short, long, cars)
begin
if reset = '1' then
    -- reset the machine
    ctrl_next <= hg;
 else
    case ctrl_state is
    when hg =>
        -- set lights
        highway_light <= green;
        farm_light <= red;
```

```
            -- decide what to do next
            if (cars and long) = '1' then
                ctrl_next <= hy;
                start_timer <= '1';
            else -- state doesn't change
                ctrl_next <= hg;
                start_timer <= '0';
            end if;
        when hy =>
            -- set lights
            highway_light <= yellow;
            farm_light <= red;
            -- decide what to do next
            if short = '1' then
                ctrl_next <= fg;
                start_timer <= '1';
            else
                ctrl_next <= hy;
                start_timer <= '0';
            end if;
        when fg =>
            -- set lights
            highway_light <= red;
            farm_light <= green;
            -- decide what to do next
            if (not cars or long) = '1' then -- sequence to yellow
                ctrl_next <= fy;
                start_timer <= '1';
            else
                ctrl_next <= fg;
                start_timer <= '0';
            end if;
        when fy =>
            -- set lights
            highway_light <= red;
            farm_light <= yellow;
            -- decide what to do next
            if short = '1' then
                ctrl_next <= hg;
                start_timer <= '1';
            else
                ctrl_next <= fy;
                start_timer <= '0';
            end if;
    end case; -- main state machine
end if; -- not a reset
end process ctrl_proc_combin;

-- the sync process updates the present state of the controller
sync: process(CLOCK)
begin
    wait until CLOCK'event and CLOCK = '1';
```

```
        ctrl_state <= ctrl_next;
    end process sync;

    end register_transfer;
```

VHDL model elements

The description has several parts. The first statements declare the libraries needed by this model. The VHDL simulator or synthesis tool gathers the declarations and other information it needs from these libraries. The next statement is the entity declaration. After the entity declaration, we can have one or more architecture statements. An architecture statement actually describes the component being modeled for simulation or synthesis. We may want to have several architecture description, for a component at different levels of abstraction or to have faster simulation models for some purpose. The architecture of this model is named register_transfer; this name has no intrinsic meaning in VHDL and is used only to identify the model. After the architecture declaration proper, we can define signals, required type definitions, *etc.*

combinational and sequential processes

This model has two processes, one for the combinational behavior and another for the sequential behavior. Each process begins with its sensitivity list—the signals that should cause this process to be reevaluated when they change. The **combinational process** first uses an if to check for reset, then uses a case statement to choose the right action based on the machine's current state. In each case, we may examine primary inputs that help determine the proper action in this state, then set outputs and the next state as appropriate. There may be several combinational processes in a register-transfer model, which would correspond to a system partitioned into several communicating machines.

4.4 Combinational Network Delay

In this section we will study the delay through combinational logic networks and ways to optimize logic to minimize delay. Combinational delay is measured along paths in the logic, generally paths from a primary input to a primary output. We need to analyze the sources of delay along a path as well as how to efficiently analyze the various paths through the logic.

After considering the types of delay measurements we want to make, we will build a bottom-up view of delay through combinational logic. We will start with the basic model for delay from one gate to another. Next, we will consider paths through combinational logic blocks. We will close with some notes on how physical design—placement and routing—affect delay.

4.4.1 Delay Specifications

We are interested in delay not just for a single gate but for an entire block of logic. For example, when designing a sequential machine, we often care about the delay from any primary input to any primary output, since the clock period is determined by the maximum delay through the logic. In other cases we may care about the delay from one or more primary inputs to one or more primary outputs.

data-dependent delay

The delay through a block of logic may depend on what values we present at its inputs. Some logic families have different rise and fall delays; the structure of the logic itself may cause the delay to vary based upon the input values. We often care about the worst-case delay through logic but we may be interested in the delay for a particular input vector. This is particularly true when we want to improve the logic delay—knowing the worst-case input may help us optimize the logic.

slack and criticality

The term **slack** is commonly used in optimization to refer to the non-critical part of a measurement. In delay analysis, the timing slack along a path is the difference between the longest delay through the logic block and the actual delay along that path. If the slack is zero, then that path is **critical** and helps determine the maximum delay. If the slack is greater than zero, it refers to the extra time between when the signal finishes propagating and the longest delay for the block.

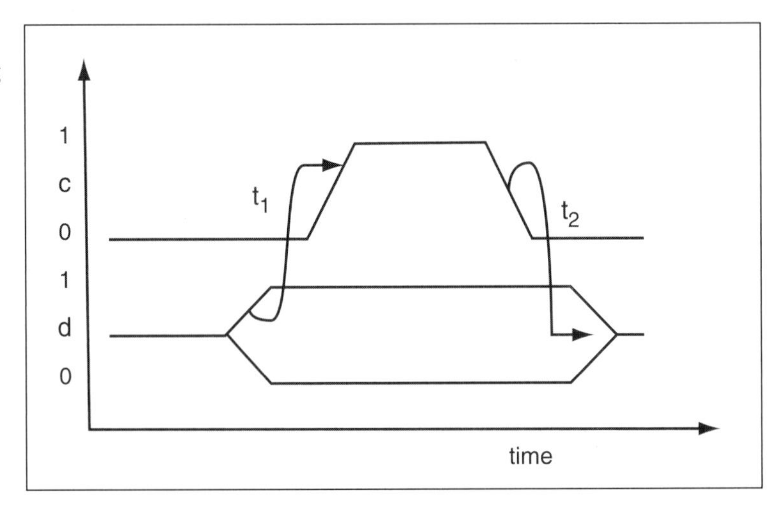

Figure 4-11 A simple timing diagram.

timing diagrams

Logic behavior is sometimes specified as **timing diagrams**. An example timing diagram is shown in Figure 4-11. The system described by this diagram shows a communication from *d* to *c* to signal and later a response from c back to d. The d line is specified as either **unknown**, **changing,** or **stable**: the unknown value is the horizontal line through the signal value; the changing value are the diamond-shaped points and the stable value is shown as a pair of low and high horizontal lines. The data line can take on different values; unlike the control signal, we do not want to tie the timing specification to a given data value. An unknown value indicates that the data value on the bus is not useful, perhaps because no component on the bus is driving the data lines. A changing value indicates a transition between unknown and stable states.

Figure 4-11 also shows timing constraints between the d and c signals. The head and tail of the arrow tell us what is constrained while the label tells us the value of the timing constraint. The arrow represents an inequality relating the times of events on the d and c signals. If we denote the time of d becoming stable as t_d and the time of c rising to 1 as t_c, then the arrow tells us that

$$t_c \geq t_d + t_1 . \quad \text{(EQ 4-1)}$$

4.4.2 Gate and Wire Delay

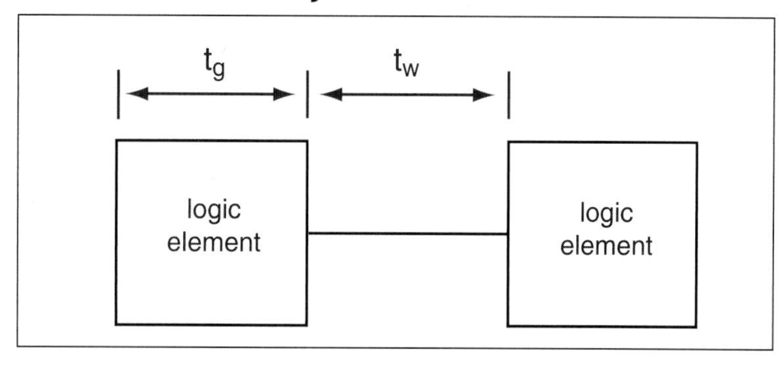

Figure 4-12 A model for delay between two logic elements.

delay models

In order to understand paths through complex logic networks, we first need to consider the basic building blocks of a path—the connection from one logic element to another. This model will be somewhat abstract because logic elements are themselves complex. FPGA manufacturers do not often provide detailed delay parameters directly to customers; we must instead rely on CAD tools that analyze delay for us. However, a basic understanding of delay mechanisms helps us design logic networks.

Figure 4-12 shows a logic element being driven by another logic element. We will break the total delay from the input of the first logic element to the input of the second one into two components:

- **Gate delay** t_g. This is the delay through the circuits in the logic elements themselves. This delay is usually substantial because FPGAs use complex logic elements.
- **Wire delay** t_w. The delay through the wire connecting the two gates. This delay clearly depends on the length and other characteristics of the wire.

The total measurements is made from the input of one logic element to the input of the next logic element. By measuring t_g and t_w for each segment of a path, we can determine the total delay along that path.

We can use several different definitions of the endpoints of a delay measurement: we can start and stop at the 50% voltage point; from 10% to 90%; etc. There are some technical reasons to prefer one over the other in certain circumstances. However, when we measure delays along a path we must be sure to use the same measurement standard along the entire path.

delay optimizations

There is little we can do about the first delay component in FPGAs but much we can do about the second. We can change the gate delay only by selecting a different FPGA with a different logic element. However, we can select the path of the wire through the FPGA interconnect system, as well as selecting buffered paths.

We have to consider two cases in wire delay: short and long wires. Both are defined relative to the drive capability of the driving gate. A short wire presents a load sufficiently small that it can be modeled as a single capacitor. Calculating the delay through short wires is relatively simple. The delays through short wires are small compared to the possible delays through long wires. And since the wires are short, there is relatively little we can do to improve the delay.

The circuits for the logic element and the interconnect must be designed to meet some delay specification. Unlike a custom chip design, we have no control over the circuits used in an FPGA. However, we may be able to select between several available circuits when configuring an FPGA. In particular, we may be able to choose between several different styles of interconnect that offer different delays. We saw in Section 3.6 and Section 3.7 why FPGAs offer different types of interconnect—they provide different levels of performance at varying costs in silicon. Since FPGA resources are fixed, we may need to carefully choose which logic paths are allocated to high-performance interconnect to be sure that those resources are not consumed by low-priority signals.

4.4.3 Fanout

driving multiple gates

When we generalize our delay model, the next problem we run into is **fanout**, or driving multiple logic elements from a single logic element. As the number of logic elements being driven increases, the load on the output gates increases, slowing it down. That capacitance may come from the transistor gates or from the wires to those gates.

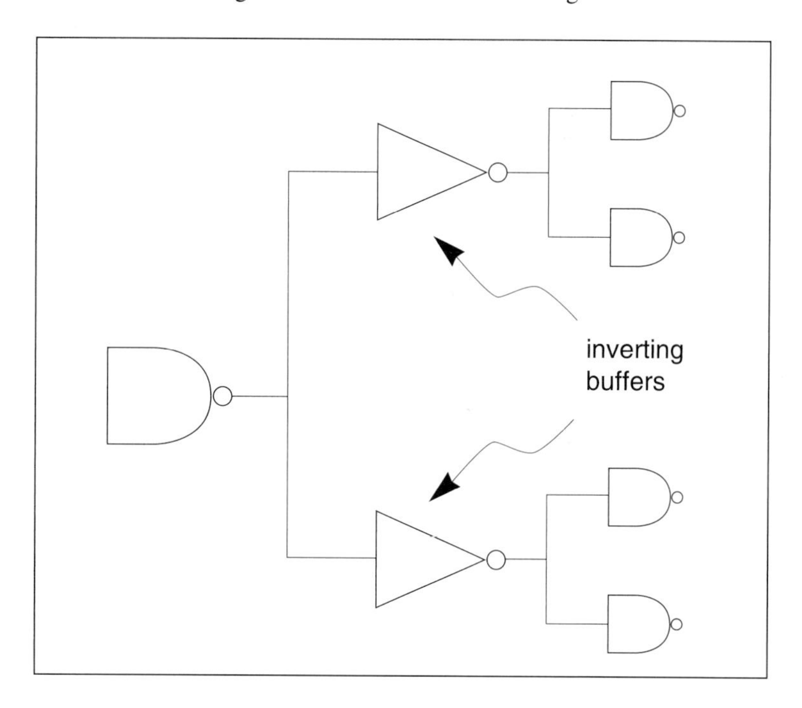

Figure 4-13 Fanout reduction by buffer insertion.

Example 4-4 Driving fanout

A simple model of a logic gate helps make clear the problem with fanout. Consider the logic gate's output as a current source. Model the input of the next logic gate as a capacitor:

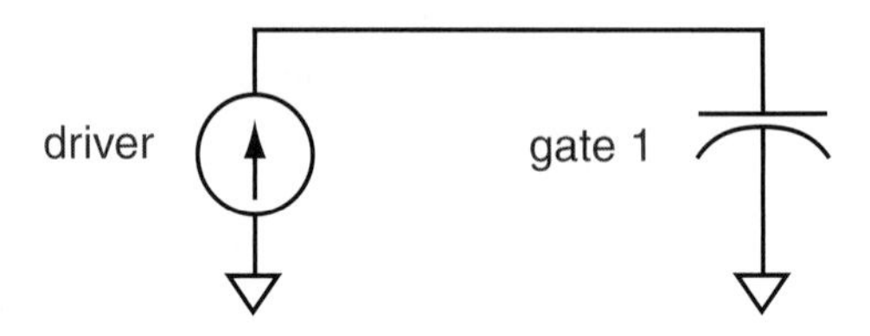

The current source must supply current to drive the capacitor's voltage to the power supply value. When we add a gate to the driving gate's fanout, we add another capacitor:

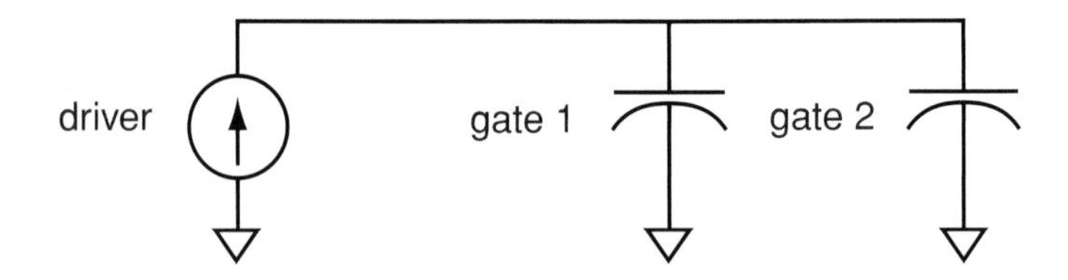

The same current source must now drive twice as much capacitance. This will slow down the signal seen by both the gates.

In custom logic we can enlarge a transistor to increase the amount of current it supplies when we fan out to multiple gates. FPGAs do not provide so much flexibility. If fanout delay becomes a problem, we may need to restructure the logic to reduce the number of fanout gates at that point. This will not fundamentally change the function being computed but it will redistribute the fanout gates to be driven by several other gates.

4.4.4 Path Delay

paths and delay

In general, performance may be limited not by a single gate, but by a path through a number of gates. To understand how this can happen and what we can do about it, we need a concise model of the combinational logic that considers only delays. We can model the logic network and its delays as a directed graph that is sometimes called a **timing graph**. Each logic gate and each primary input or output is assigned its own node in the graph. When one gate drives another, an edge is added from the driving gate's node to the driven gate's node; the number assigned to the edge is the delay required for a signal value to propagate from the driver to the input of the driven gate. (The delay for $0 \rightarrow 1$ and $1 \rightarrow 0$ transitions will in general be different; since the wires in the network may be changing arbitrarily, we will choose the worst delay to represent the delay along a path.)

In building the graph we need to know the gate along each edge in the graph. We use our delay model to estimate the delay from one gate's input through the gate and its interconnect to the next gate's input.

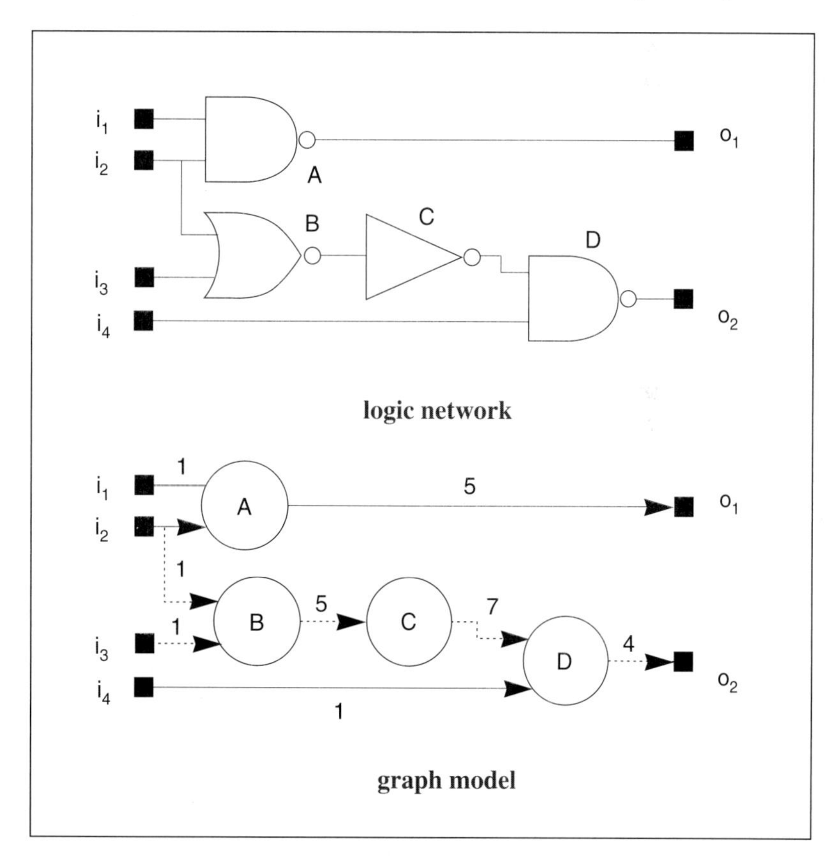

Figure 4-14 A graph model of delays through a logic network.

The simplest delay problem to analyze is to change the value at only one input and determine how long it takes for the effect to be propagated to a single output. (Of course, there must be a path from the selected input to the output.) That delay can be found by summing the delays along all the edges on the path from the input to the output.

static timing analysis

We could use a logic simulator which models delays to compute the delays through various paths in the logic. However, system performance is determined by the *maximum* delay through the logic—the longest delay from any input to any output for any possible set of input values. To determine the maximum delay by simulation, we would have to simulate all 2^n possible input values to the combinational logic. It is possible, however, to find the logic network's maximum delay without exhaustive simulation. **Timing analysis** [McW80,Ost83] builds a graph which models delays through the network and identifies the longest delay path. Timing analysis is also known as **static timing analysis** because it determines delays statically, independent of the values input to the logic gates.

critical path

The longest delay path is known as the **critical path** since that path limits system performance. We know that the graph has no cycles, or paths from a node back to itself—a cycle in the graph would correspond to feedback in the logic network. As a result, finding the critical path isn't too difficult. In Figure 4-14 there are two paths of equal length: $i2 \rightarrow B \rightarrow C \rightarrow D \rightarrow o2$ and $i3 \rightarrow B \rightarrow C \rightarrow D \rightarrow o2$. Both have total delays of 17 ns. Any sequential system built from this logic must have a total delay of 17 ns, plus the setup time of the latches attached to the outputs, plus the time required for the driving latches to switch the logic's inputs (a term which was ignored in labeling the graph's delays).

There are many possible paths through a timing graph. We can use longest path algorithms, well known in computer science, to find the critical path or paths through the timing graph. These algorithms will give us both the length of the critical path and all the nodes and edges on the critical path.

optimizing for delay

The critical path not only tells us the system cycle time, it points out what part of the combinational logic must be changed to improve system performance. Speeding up a gate off the critical path, such as A in the example, won't speed up the combinational logic. The only way to reduce the longest delay is to speed up a gate on the critical path. That can be done by increasing transistor sizes or reducing wiring capacitance. It can also be done by redesigning the logic along the critical path to use a faster gate configuration.

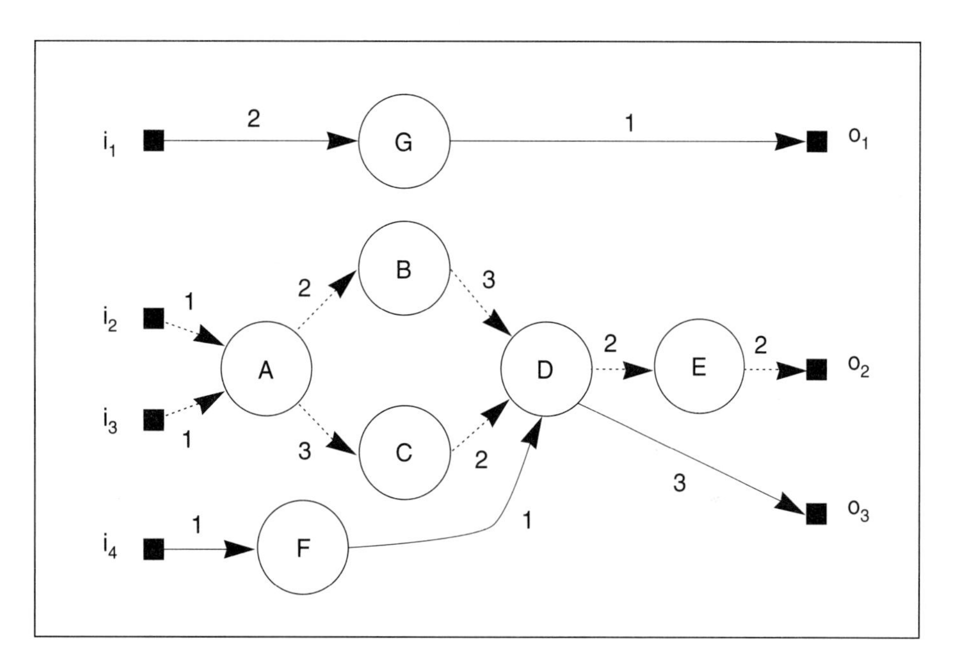

Figure 4-15 A cutset through a critical timing path.

Speeding up the system may require modifying several sections of logic since the critical path can have multiple branches. The circuit in Figure 4-15 has a critical path with a split and a join in it. Speeding up the path from *B* to *D* will not speed up the system—when that branch is removed from the critical path, the parallel branch remains to maintain its length. The system can be improved only by speeding up both branches [Sin88]. A **cutset** is a set of edges in a graph that, when removed, break the graph into two unconnected pieces. Any cutset that separates the primary inputs and primary outputs identifies a set of speedups sufficient to reduce the critical delay path. The set *b-d* and *c-d* is one such cutset; the single edge *d-e* is another. We probably want to speed up the circuit by making as few changes to the network as possible. It may not be possible, however, to speed up every connection on the critical path. After selecting a set of optimization locations identified by a cutset, you must analyze them to be sure they can be sped up, and possibly alter the cutset to find better optimization points.

4.4.5 Delay and Physical Design

Physical design is, strictly speaking, a design step that comes after combinational logic design. But the characteristics of the placement in particular are a major contributor to the overall delay characteristics of the final logic. We will study algorithms for placement and routing in more detail in Section 4.8 but we can understand the major challenges presented by physical design using some simple models.

delay modeling

Physical design is important in FPGAs because of interconnect. The wires that connect the logic elements introduce delay between the logic elements: they present a load on the output circuits of the logic elements and they have their own transmission line delay. The delay characteristics of the interconnect should be taken into account during the design of the logic network. However, we can only estimate the characteristics of the interconnect when designing the logic network. Furthermore, physical design tools are good but have their limitations.

Inaccuracies in the models for interconnect come from several sources. First, we do not know the placement of the logic around the fabric. Second, we do not know the details of the routing of the interconnections between the logic elements. Placement is the most critical piece of information. Given a placement we can estimate the length of the interconnections by measuring the distance between the logic elements to be connected. However, different placements can give very different wiring patterns and therefore very different delays.

Accuracy is important because we really want to be able to identify the critical path through the logic. In order to improve delay, we must improve the critical path(s). Optimizations applied to the wrong part of the logic network will be wasted. It is difficult to know the exact critical path without a detailed placement and routing. In the early stages of logic optimization rough estimates may be sufficient, but as the design moves closer to its maximum performance value, performance estimates must become increasingly accurate to be useful.

algorithmic limitations

Placement and routing algorithms are essential to successful FPGA design. Many designs were manually placed and routed in the early days of FPGAs but only at the cost of huge amounts of labor. However, physical design algorithms aren't perfect. In particular, placement algorithms are not nearly as good as people at dealing with regular designs, such as data paths, multipliers, *etc*. Placement algorithms do not recognize the regularity that is apparent to the human designer and so do not place elements with regular connections in a regular pattern. This makes for longer, slower wires.

timing constraints

Placement and routing algorithms can take **timing constraints** that guide their efforts. Most algorithms let us set either overall timing constraints on any path or a constraint on a particular input-output pair. The algorithms use these constraints as goals. If they need to speed up one path with a tight constraint, they may do so at the expense of another path with a looser constraint. They will also not improve the logic too much more than the constraint requires since timing improvements may come at the expense of other characteristics. However, timing-directed physical design does not always solve all our problems.

The next example illustrates the limitations of automatic placement.

Example 4-5 Automatic placement of an adder

Let us use an eight-bit ripple-carry adder as an example of a regular design. This logic is very regular: each bit is identical and carry signals link adjacent bits. A good placement for the adder should look like this:

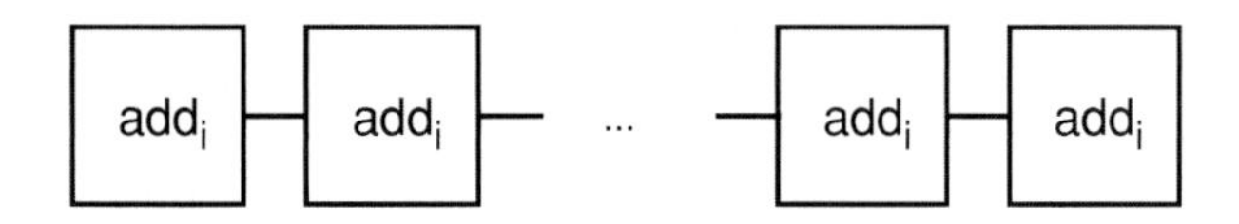

This layout reduces the length of the carry wire, which we know is on the critical delay path.

Here is the result of a placement of the adder in which the placer has no delay constraint:

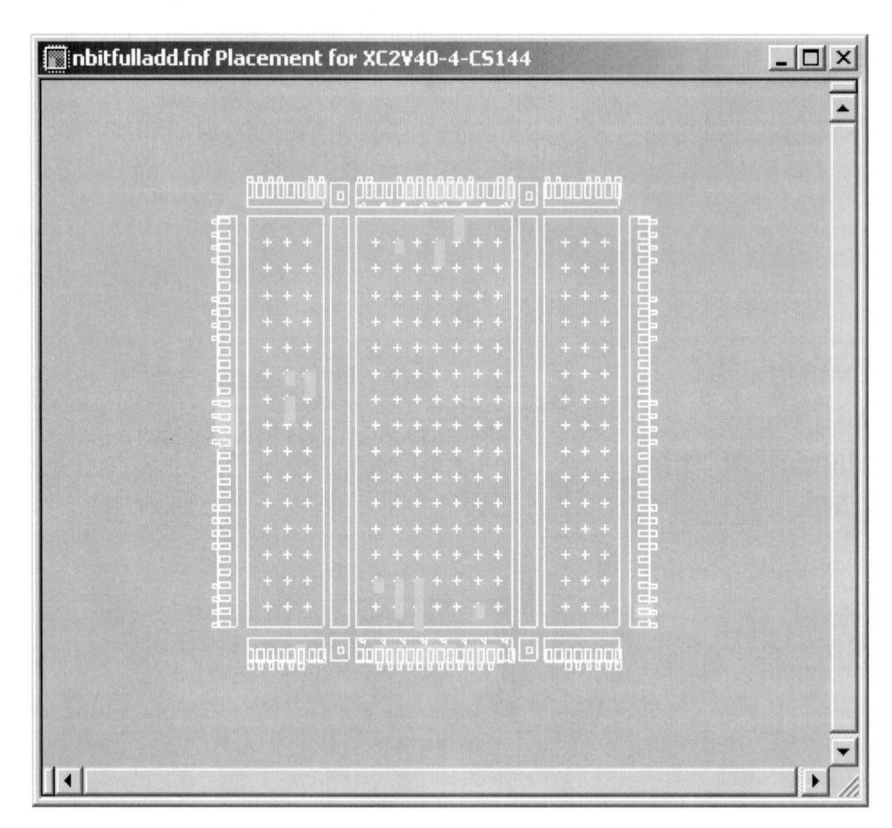

The shaded sections are the logic elements assigned to the various pieces of the adder. It is clear from this plot that the adder has been dispersed throughout the FPGA, which will make its wires longer than necessary. The plot also shows that the input and output pins for the adder have been spread around the chip, further increasing the length of wires.

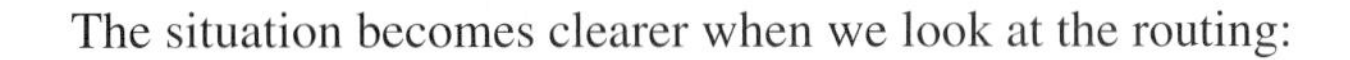
The situation becomes clearer when we look at the routing:

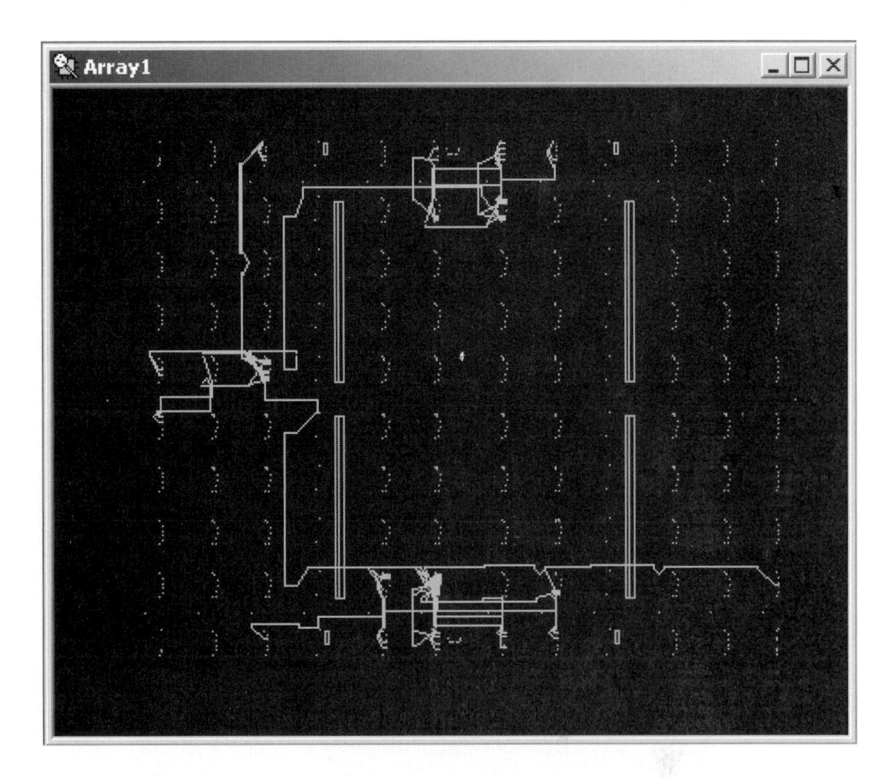

This plot clearly shows that, although there are several clusters of logic, the wires between them are very long. The physical design algorithms did not catch on to the regularity inherent in this design.

We can rerun the design with a timing constraint. In this case we choose a delay of 12 ns. Here is the placement:

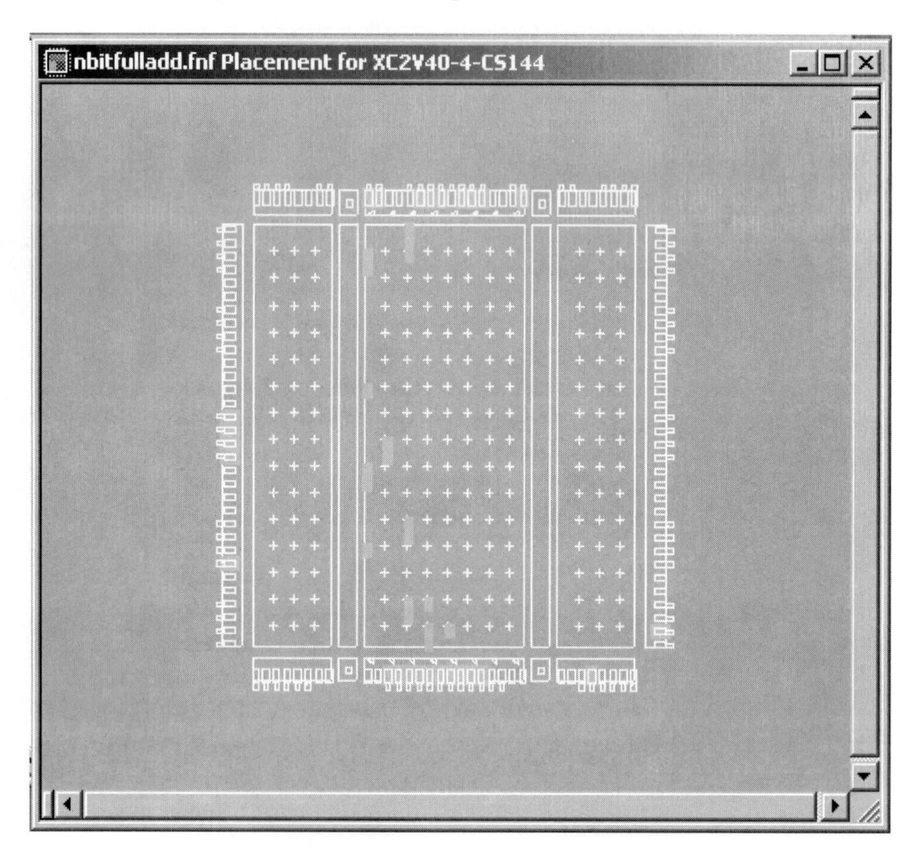

This placement is somewhat better but it is still far from perfect. The logic is less spread out horizontally but still very spread out vertically.

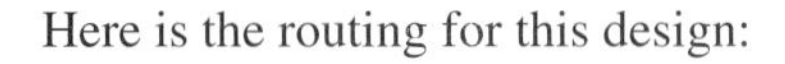

Here is the routing for this design:

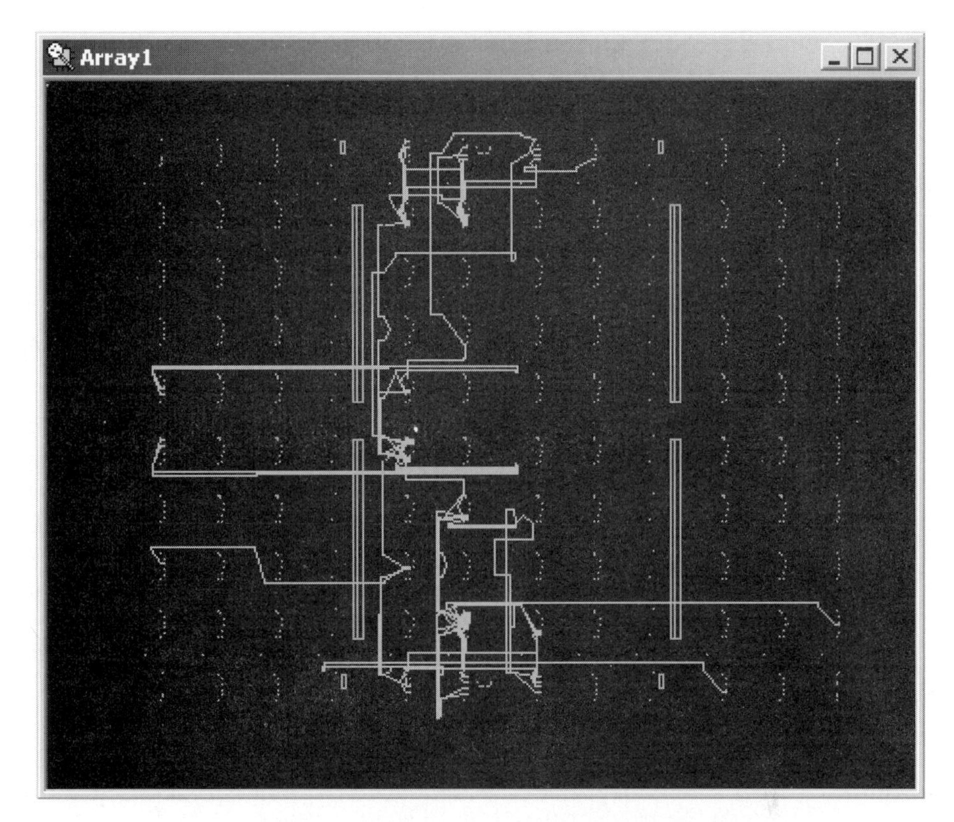

Once again, this routing looks better than the unconstrained design but it is still far from perfect. The placement algorithm has clustered the logic for the adder closer together but it has not exploited the regularity in the wiring. We will see how to add constraints directly to the placement in Example 4-15; placement constraints allow us to more closely control the placement of objects.

---

ways to improve placement

Macros are one way to improve the results of placement. A macro can provide placement information that will constrain the relative placement of the elements in the macro. Most placement tools also allow the designer to constrain or hand-place some components.

4.5 Power and Energy Optimization

In this section we consider power and energy consumption of combinational networks. Power consumption is important whenever heat dissipation is a problem; energy consumption is critical in battery-operated and other low-power systems.

energy consumption mechanisms

CMOS gates exhibit several energy consumption mechanisms; each suggests its own mechanisms for minimizing energy consumption. The first and largest source of energy consumption is caused by transitions on gate outputs. Driving a gate's load consumes considerable energy. The energy consumption can be reduced by reducing the number of times the gate's output changes. Proper logic design can, in some cases, reduce glitching.

Second, leakage contributes to energy consumption in more advanced VLSI fabrication processes. The traditional CMOS transistor has a very high off resistance and very little current leaked through the channel. However, as transistor geometries increase, leakage current increases. Leakage occurs even when the logic is idle. The only way to eliminate leakage current is to disconnect the power supply from the logic. This can be done when it is known that the logic will not be needed for some time. It generally takes a considerable period—longer than a clock cycle—to reconnect power and let the circuits stabilize.

4.5.1 Glitching Analysis and Optimization

glitches and energy

One important way to reduce a gate's energy consumption is to make it change its output as few times as possible. While the gate would not be useful if it never changed its output value, it is possible to design the logic network to reduce the number of *unnecessary* changes to a gate's output as it works to compute the desired value.

Figure 4-16 shows an example of energy-consuming glitching in a logic network. Glitches are more likely to occur in multi-level logic networks because the signals arrive at gates at different times. In this example, the NOR gate at the output starts at 0 and ends at 0, but differences in arrival times between the gate input connected to the primary input and the output of the NAND gate cause the NOR gate's output to glitch to 1.

glitches and unused outputs

Often the most effective way to minimize glitches is to keep unnecessary transitions from propagating. A logic block's inputs may change even though the output of the logic is not used. Consider, for example, an ALU in a CPU. The ALU output is not used on every cycle. If the

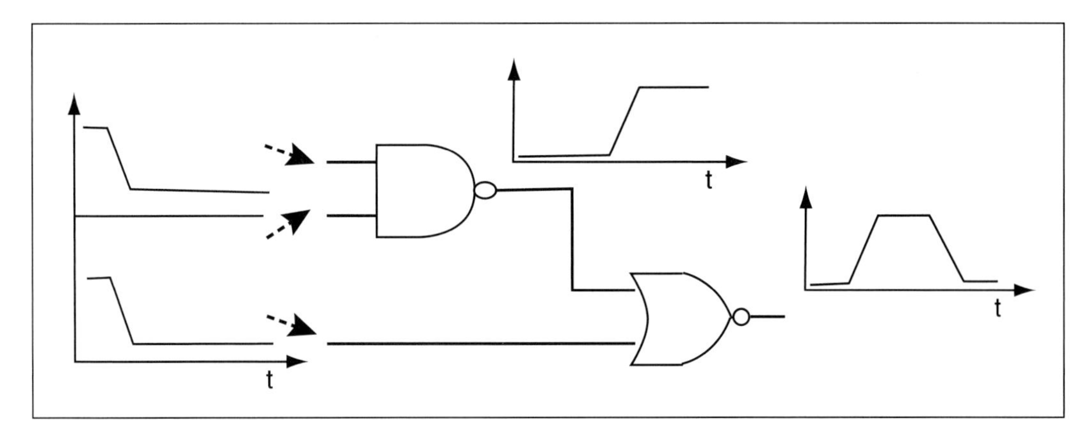

Figure 4-16 Glitching in a simple logic network.

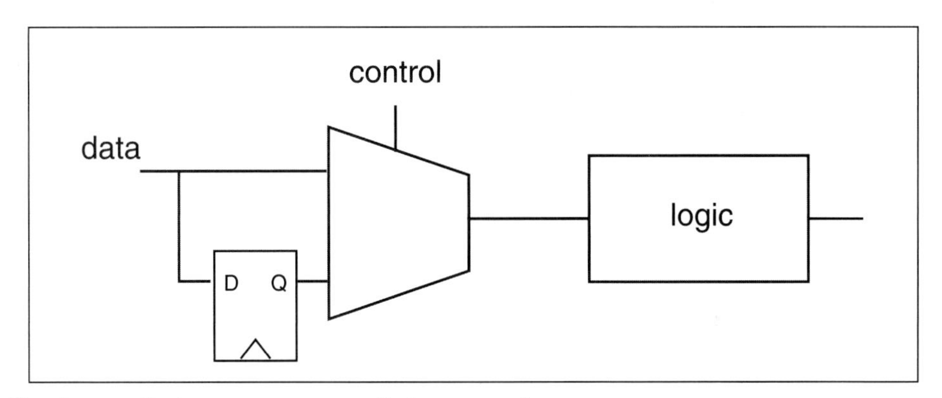

Figure 4-17 Circuitry to eliminate unnecessary glitch propagation.

ALU's inputs change when the ALU is not in use, then energy is needlessly consumed in the ALU logic. As shown in Figure 4-17, we can use gating logic to guard against unnecessary transitions. The control signal selects whether data is allowed to pass to the logic or whether the previous value is held to avoid transitions in the logic block.

4.6 Arithmetic Logic

Arithmetic circuits are both important in real applications and challenging to design. Luckily, a large body of knowledge has been developed on the design of arithmetic logic. We will start with a brief review of number representations. We will then consider several types of arithmetic circuits: shifters, adders, ALUs, and multipliers.

4.6.1 Number Representations

Before looking at logic for arithmetic operations, we will briefly review some basic facts about the representation of numbers (specifically integers) in computers. The most commonly used number representation in computers is **twos-complement**. We traditionally write the base of a particular number representation as a subscript, so $5_{10} = 101_2$. Let us assume that we have an n-bit binary word for representing numbers. We often refer to the **most significant bit** (**MSB**), bit n-1 in this case, and the **least significant bit** (**LSB**), bit 0 in our notation.

We want to put the number N in binary form. Then if N is positive, we have

$$N = \sum_{i=0}^{n-1} 2^i b_i . \qquad \text{(EQ 4-2)}$$

For example, if $N = 5$, then the twos-complement representation 101_2 satisfies this equality since $2^2 + 2^0 = 5$. If N is negative, we can form its twos-complement representation from the twos-complement representation for the absolute value of N (its positive form): we first take the **ones-complement** by complementing each bit; we then add one to the ones-complement version. If $n = 4$ (we use a four-bit number representation), then we can write -5 as 1011_2 since the ones-complement of 5 is 1010_2 and adding 1 to that value gives us 1011_2. A twos-complement number of n bits can represent numbers in the range

$$-2^{n-1} \leq N \leq 2^{n-1} - 1 . \qquad \text{(EQ 4-3)}$$

Twos-complement form has the useful property that there is a unique representation for zero. (The ones-complement representation, in contrast, has two zeroes, 0000 and 1111.)

Most of the basic operations we want to perform are fairly straightforward:

- We can test for zero easily because there is only one representation for zero—we check that every bit in the number is zero.

- We can test for a negative number by testing the most significant bit, with an MSB of 1 indicating a negative number.
- We can form the ones-complement of a number by taking the Boolean complement of each bit.
- We can perform $a - b$ by negating b and then adding. If we write the ones-complement as *ones*(), then we must compute $a + (\text{ones(b)}+1)$.

4.6.2 Combinational Shifters

shifter operation

A **shifter** is most useful for arithmetic operations since shifting is equivalent to multiplication by powers of two. Shifting is necessary, for example, during floating-point arithmetic. The simplest shifter is the shift register, which can shift by one position per clock cycle. However, that machine isn't very useful for most arithmetic operations—we generally need to shift several bits in one cycle and to vary the length of the shifts.

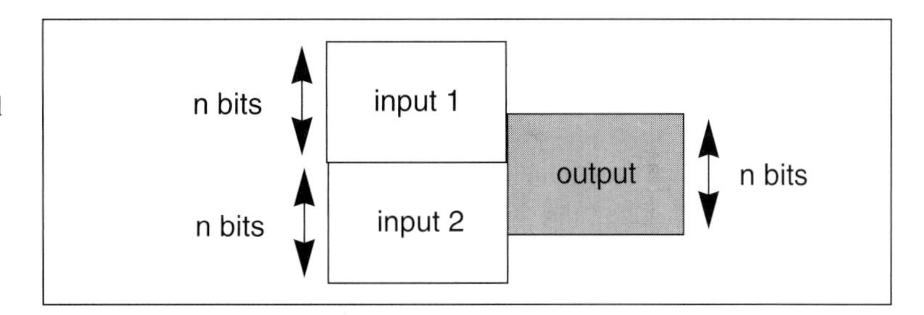

Figure 4-18 How a barrel shifter performs shifts and rotates.

A **barrel shifter** [Mea80] can perform n-bit shifts in a single combinational function, and it has a very efficient layout. It can rotate and extend signs as well. Its architecture is shown in Figure 4-18. The barrel shifter accepts $2n$ data bits and n control signals and produces n output bits. It shifts by transmitting an n-bit slice of the $2n$ data bits to the outputs. The position of the transmitted slice is determined by the control bits; the exact operation is determined by the values placed at the data inputs. Consider two examples:

- Send a data word d into the top input and a word of all zeroes into the bottom input. The output is a right shift (imagine standing at the output looking into the barrel shifter) with zero fill. Setting the control bits to select the top-most n bits is a shift of zero, while selecting the bottom-most n bits is an n-bit shift that pushes the entire word out of the shifter. We can shift with a ones fill by sending an all-ones word to the bottom input.

- Send the same data word into both the top and bottom inputs. The result is a rotate operation—shifting out the top bits of a word causes those bits to reappear at the bottom of the output.

Example 4-6 Shifters

Here is the Verilog for an N-bit shifter that shifts up (to more significant bits) by b bits and fills with zeroes:

```
module shifter(data,b,result);
    parameter Nminus1 = 31;  /* 32-bit shifter */
    input [Nminus1:0] data; /* compute parity of these bits */
    input [3:0] b; /* amount to shift */
    output [Nminus1:0] result; /* shift result */

    assign result = data << b;
    endmodule
```

4.6.3 Adders

The adder is probably the most studied digital circuit. There are a great many ways to perform binary addition, each with its own area/delay trade-offs. A great many tricks have been used to speed up addition: encoding, replication of common factors, *etc*. We will first review several well-known algorithms for addition; at the end of this section we will compare the speed of several of these methods when implemented in FPGA logic.

full adder

The basic adder is known as a **full adder**. It computes a one-bit sum and carry from two addends and a carry-in. The equations for the full adder's functions are simple:

$$s_i = a_i \oplus b_i \oplus c_i$$
$$c_{i+1} = a_i b_i + a_i c_i + b_i c_i \cdot \qquad \text{(EQ 4-4)}$$

In these formulas, s_i is the sum at the ith stage and c_{i+1} is the carry out of the ith stage.

half adder

A **half adder** accepts two input bits rather than the three inputs of a full adder. It produces the sum output $a_i \oplus b_i$ and the carry output $a_i b_i$.

serial adder

Serial adders present one approach to high-speed arithmetic—they require many clock cycles to add two n-bit numbers, but with a very short cycle time. Serial adders can work on nybbles (four-bit words) or bytes, but the bit-serial adder [Den85], shown in Figure 4-19, is the most extreme form. The data stream consists of three signals: the two numbers to be added and a least-significant bit signal that is high when the current data bits are the least significant bits of the addends. The addends appear LSB first and can be of arbitrary length—the end of a pair of numbers is signaled by the LSB bit for the next pair. The adder itself is simply a full adder with a register for the carry. The LSB signal clears the carry register. Subsequently, the two input bits are added with the carry-out of the last bit. The serial adder is small and has a cycle time equal to that of a single full adder plus the register delay.

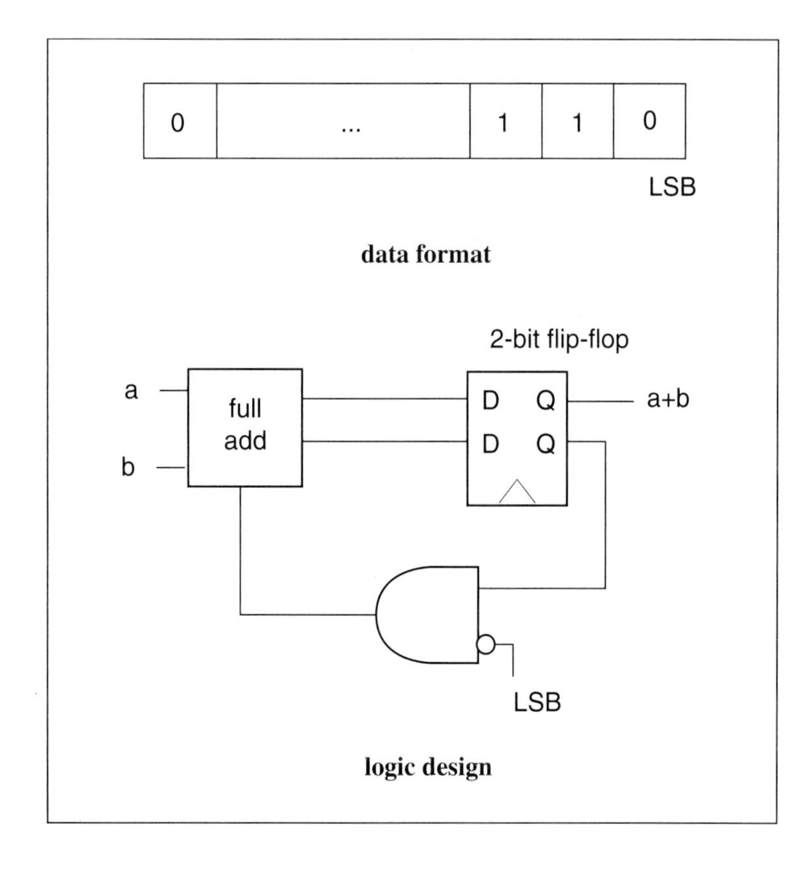

Figure 4-19 A bit-serial adder.

ripple-carry adder

The n-bit adder built from n one-bit full adders is known as a **ripple-carry adder** because of the way the carry is computed. The addition is not complete until the $n\text{-}1^{th}$ adder has computed its s_{n-1} output; that result depends on c_i input, and so on down the line, so the critical delay path goes from the 0-bit inputs up through the c_i's to the n-1 bit. (We can find the critical path through the n-bit adder without knowing the exact logic in the full adder because the delay through the n-bit carry chain is so much longer than the delay from a and b to s.) The ripple-carry adder is area efficient and easy to design.

Example 4-7 Ripple-carry adder

Here is the Verilog for a one-bit full adder:

```
module fulladd(a,b,carryin,sum,carryout);
    input a, b, carryin; /* add these bits*/
    output sum, carryout; /* results */

    assign {carryout, sum} = a + b + carryin;
          /* compute the sum and carry */
    endmodule
```

The notation *{carryout, sum}* on the left-hand side of the assignment forms a two-bit array from the one-bit signals; since the result of the right-hand side expression is two bits in size, this allows to easily capture the sum and carry. Using the + operator makes it easier for the tools to recognize the addition and to map it efficiently onto the FPGA's resources. We use the one-bit adder to build an N-bit adder. This is a structural description in which we wire together the full adder components:

```
module nbitfulladd(a,b,carryin,sum,carryout)
    input [7:0] a, b; /* add these bits */
    input carryin; /* carry in*/
    output [7:0] sum; /* result */
    output carryout;
    wire [7:1] carry; /* transfers the carry between bits */

    fulladd a0(a[0],b[0],carryin,sum[0],carry[1]);
    fulladd a1(a[1],b[1],carry[1],sum[1],carry[2]);
    fulladd a2(a[2],b[2],carry[2],sum[2],carry[3]);
    fulladd a3(a[3],b[3],carry[3],sum[3],carry[4]);
    fulladd a4(a[4],b[4],carry[4],sum[4],carry[5]);
    fulladd a5(a[5],b[5],carry[5],sum[5],carry[6]);
    fulladd a6(a[6],b[6],carry[6],sum[6],carry[7]);
```

```
    fulladd a7(a[7],b[7],carry[7],sum[7],carryout]);
endmodule
```

Of course, we could build an N-bit adder more simply by defining *a*, *b*, *sum*, and *carry* as vectors and using the + operator. However, this form gives us practice in mixed functional/structural designs that will come in handy in more complex logic designs.

The **carry-lookahead adder** is an alternate way to formulate the carry computation. The carry-lookahead adder breaks the carry computation into two steps, starting with the computation of two intermediate values. The adder inputs are once again the a_i's and b_i's; from these inputs, P (propagate) and G (generate) are computed:

$$\begin{aligned} P_i &= a_i \oplus b_i \\ G_i &= a_i \cdot b_i \end{aligned} \quad \text{(EQ 4-5)}$$

(Some authors define the propagate signal as $P_i = a_i + b_i$. The XOR-based definition of P is most useful if you also want to make use of a carry annihilate signal.) If $G_i = 1$, there is definitely a carry out of the ith bit of the sum—a carry is generated. If $P_i = 1$, then the carry from the $i\text{-}1^{th}$ bit is propagated to the next bit. The sum and carry equation for the full adder can be rewritten in terms of P and G:

$$\begin{aligned} s_i &= c_i \oplus P_i \oplus G_i \\ c_{i+1} &= G_i + P_i c_i \end{aligned} \quad \text{(EQ 4-6)}$$

The carry formula is smaller when written in terms of P and G, and therefore easier to recursively expand:

$$\begin{aligned} c_{i+1} &= G_i + P_i \cdot (G_{i-1} + P_{i-1} \cdot c_{i-1}) \\ &= G_i + P_i G_{i-1} + P_i P_{i-1} \cdot (G_{i-2} + P_{i-2} \cdot c_{i-2}) \\ &= G_i + P_i G_{i-1} + P_i P_{i-1} G_{i-2} + P_i P_{i-1} P_{i-2} c_{i-2} \end{aligned} \quad \text{(EQ 4-7)}$$

The c_{i+1} formula of Equation 4-7 depends on c_{i-2}, but not c_i or c_{i-1}. After rewriting the formula to eliminate c_i and c_{i-1}, we eliminated parentheses, which substitutes larger gates for long chains of gates. There is a limit beyond which the larger gates are slower than chains of smaller gates; typically, four levels of carry can be usefully expanded.

A depth-4 carry-lookahead unit is shown in Figure 4-20. The unit takes the P and G values from its four associated adders and computes four carry values. Each carry output is computed by its own logic. The logic for c_{i+3} is slower than that for c_i, but the flattened c_{i+3} logic is faster than the equivalent ripple-carry logic.

There are two ways to hook together depth-b carry-lookahead units to build an n-bit adder. The carry-lookahead units can be recursively connected to form a tree: each unit generates its own P and G values, which are used to feed the carry-lookahead unit at the next level of the tree. A simpler scheme is to connect the carry-ins and carry-outs of the units in a ripple chain. This approach is most common in chip design because the wiring for the carry-lookahead tree is hard to design and area-consuming.

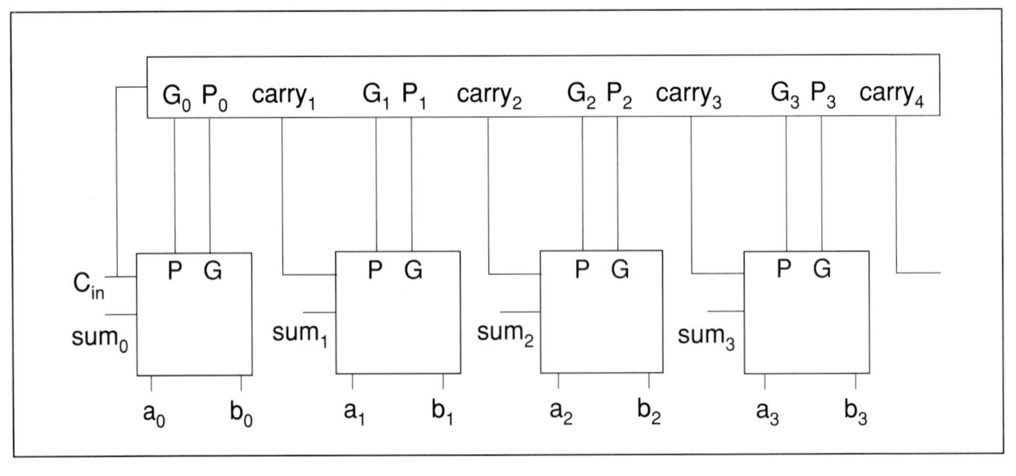

Figure 4-20 Structure of a carry lookahead adder.

Example 4-8 Carry lookahead adder

Here is the Verilog for a four-bit carry-lookahead unit:

```
module carry_block(a,b,carryin,carry);
    input [3:0] a, b; /* add these bits*/
    input carryin; /* carry into the block */
    output [3:0] carry; /* carries for each bit in the block */
    wire [3:0] g, p; /* generate and propagate */

    assign g[0] = a[0] & b[0]; /* generate 0 */
    assign p[0] = a[0] ^ b[0]; /* propagate 0 */
    assign g[1] = a[1] & b[1]; /* generate 1 */
```

```
    assign p[1] = a[1] ^ b[1]; /* propagate 1 */
    assign g[2] = a[2] & b[2]; /* generate 2 */
    assign p[2] = a[2] ^ b[2]; /* propagate 2 */
    assign g[3] = a[3] & b[3]; /* generate 3 */
    assign p[3] = a[3] ^ b[3]; /* propagate 3 */
    assign carry[0] = g[0] | (p[0] & carryin);
    assign carry[1] = g[1] | p[1] & (g[0] | (p[0] & carryin));
    assign carry[2] = g[2] | p[2] &
        (g[1] | p[1] & (g[0] | (p[0] & carryin)));
    assign carry[3] = g[3] | p[3] &
        (g[2] | p[2] & (g[1] | p[1] & (g[0] | (p[0] & carryin))));

endmodule
```

We will use a module called sum to perform the XOR since the carry is computed separately:

```
module sum(a,b,carryin,result);
    input a, b, carryin; /* add these bits*/
    output result; /* sum */

    assign result = a ^ b ^ carryin;
            /* compute the sum */
endmodule
```

And here is the Verilog for a 16-bit carry-lookahead adder. This is a structural description that uses the carry block and the sum unit we designed before:

```
module carry_lookahead_adder(a,b,carryin,sum,carryout);
    input [15:0] a, b; /* add these together */
    input carryin;
    output [15:0] sum; /* result */
    output carryout;
    wire [16:1] carry; /* intermediate carries */

    assign carryout = carry[16]; /* for simplicity */
    /* build the carry-lookahead units */
    carry_block b0(a[3:0],b[3:0],carryin,carry[4:1]);
    carry_block b1(a[7:4],b[7:4],carry[4],carry[8:5]);
    carry_block b2(a[11:8],b[11:8],carry[8],carry[12:9]);
    carry_block b3(a[15:12],b[15:12],carry[12],carry[16:13]);
    /* build the sum */
    sum  a0(a[0],b[0],carryin,sum[0]);
    sum  a1(a[1],b[1],carry[1],sum[1]);
```

```
    sum  a2(a[2],b[2],carry[2],sum[2]);
    sum  a3(a[3],b[3],carry[3],sum[3]);
    sum  a4(a[4],b[4],carry[4],sum[4]);
    sum  a5(a[5],b[5],carry[5],sum[5]);
    sum  a6(a[6],b[6],carry[6],sum[6]);
    sum  a7(a[7],b[7],carry[7],sum[7]);
    sum  a8(a[8],b[8],carry[8],sum[8]);
    sum  a9(a[9],b[9],carry[9],sum[9]);
    sum a10(a[10],b[10],carry[10],sum[10]);
    sum a11(a[11],b[11],carry[11],sum[11]);
    sum a12(a[12],b[12],carry[12],sum[12]);
    sum a13(a[13],b[13],carry[13],sum[13]);
    sum a14(a[14],b[14],carry[14],sum[14]);
    sum a15(a[15],b[15],carry[15],sum[15]);
endmodule
```

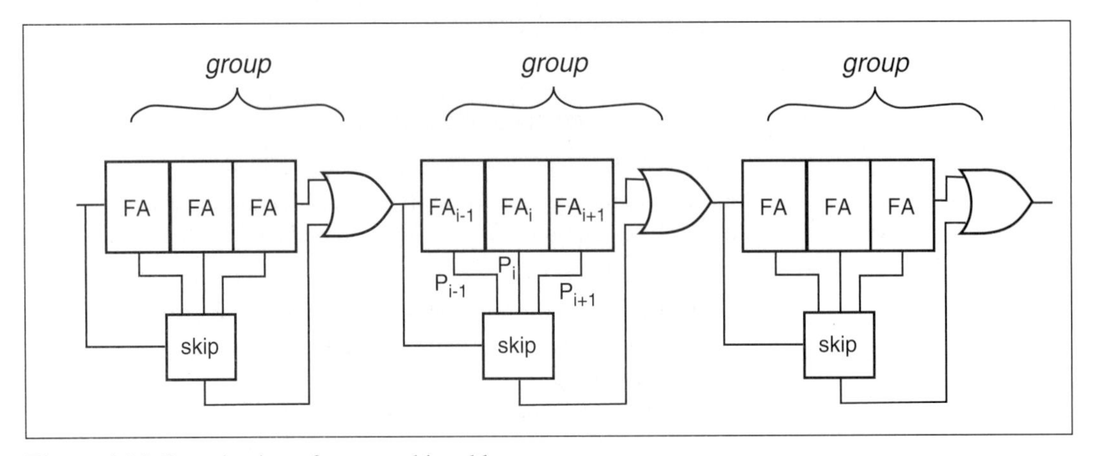

Figure 4-21 Organization of a carry-skip adder.

carry-skip adder

The **carry-skip adder** [Leh61] looks for cases in which the carry out of a set of bits is the same as the carry in to those bits. This adder makes a different use of the carry-propagate relationship. A carry-skip adder is typically organized into m-bit *groups*; if the carry is propagated for every bit in the stage, then a bypass gate sends the stage's carry input directly to the carry output.

The structure of the carry chain for a carry-skip adder divided into groups of bits is shown in Figure 4-21. A true carry into the group and

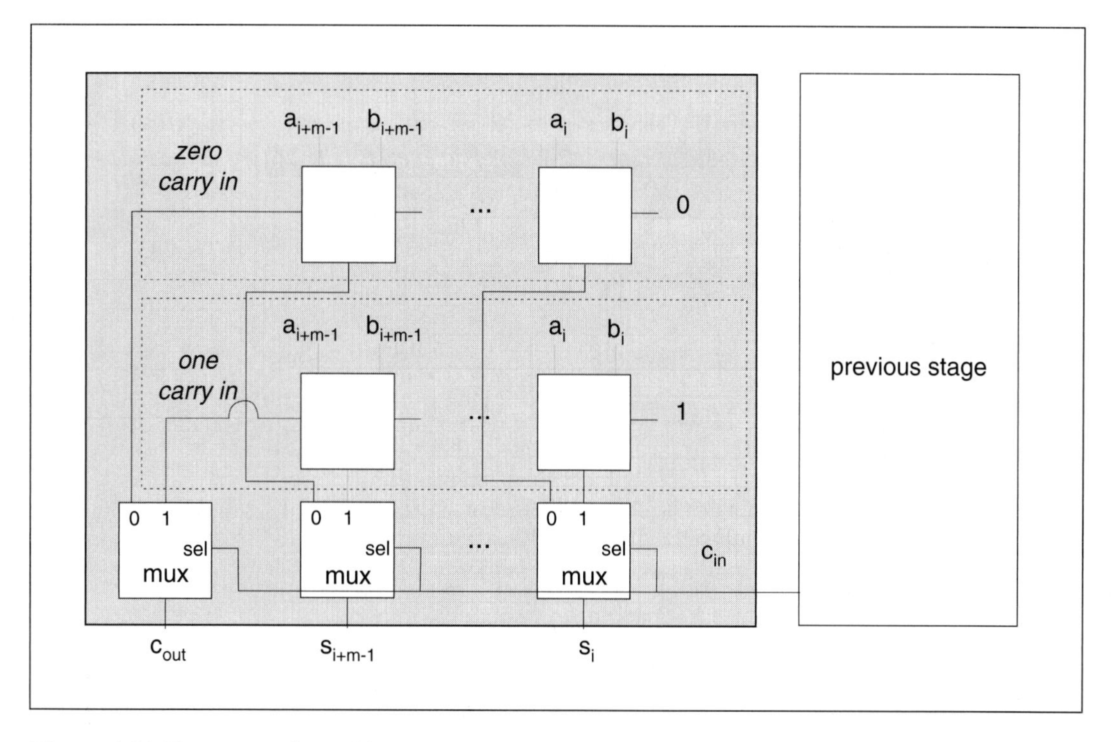

Figure 4-22 The carry-select adder.

true propagate condition P at every bit in the group is needed to cause the carry to skip. It is possible to determine the optimum number of bits in a group [Kor93]. The worst case for the carry signal occurs when there is a carry propagated through every bit, but in this case P_i will be true at every bit. Therefore, the longest path for the carry begins when the carry is generated at the bottom bit of the bottom group (rippling through the remainder of the group), is skipped past the intermediate groups, and ripples through the last group; the carry must out of necessity ripple through the first and last groups to compute the sum. Using some simple assumptions about the relative delays for a ripple through a group and skip, Koren estimates the optimum group size for an n-bit word as

$$k_{opt} = \sqrt{n/2}\,. \qquad \text{(EQ 4-8)}$$

Since the carry must ripple through the first and last stages, the adder can be further speeded up by making the first and last groups shorter than this length and by lengthening the middle groups.

Example 4-9 Carry-skip adder

To build the carry-skip adder more easily, we can modify the one-bit full adder to produce the propagate signal used for skipping the carry:

```
module fulladd_p(a,b,carryin,sum,carryout,p);
    input a, b, carryin; /* add these bits*/
    output sum, carryout, p; /* results including propagate */

    assign {carryout, sum} = a + b + carryin;
            /* compute the sum and carry */
   assign p = a | b;
endmodule
```

Here is the code for a carry-skip adder that uses this augmented full adder:

```
module carryskip(a,b,carryin,sum,carryout);
    input [7:0] a, b; /* add these bits */
    input carryin; /* carry in*/
    output [7:0] sum; /* result */
    output carryout;
    wire [8:1] carry; /* transfers the carry between bits */
    wire [7:0] p; /* propagate for each bit */
    wire cs4; /* final carry for first group */

    fulladd_p a0(a[0],b[0],carryin,sum[0],carry[1],p[0]);
    fulladd_p a1(a[1],b[1],carry[1],sum[1],carry[2],p[1]);
    fulladd_p a2(a[2],b[2],carry[2],sum[2],carry[3],p[2]);
    fulladd_p a3(a[3],b[3],carry[3],sum[3],carry[4],p[3]);
    assign cs4 = carry[4] | (p[0] & p[1] & p[2] & p[3] & carryin);
    fulladd_p a4(a[4],b[4],cs4,    sum[4],carry[5],p[4]);
    fulladd_p a5(a[5],b[5],carry[5],sum[5],carry[6],p[5]);
    fulladd_p a6(a[6],b[6],carry[6],sum[6],carry[7],p[6]);
    fulladd_p a7(a[7],b[7],carry[7],sum[7],carry[8],p[7]);
    assign carryout = carry[8] | (p[4] & p[5] & p[6] & p[7] & cs4);
endmodule
```

carry-select adder

The **carry-select adder** computes two versions of the addition with different carry-ins, then selects the right one. Its structure is shown in Figure 4-22. As with the carry-skip adder, the carry-select adder is typically divided into m-bit stages. The second stage computes two values: one assuming that the carry-in is 0 and another assuming that it is 1. Each of these candidate results can be computed by your favorite adder structure. The carry-out of the previous stage is used to select which version is correct: multiplexers controlled by the previous stage's carry-out choose the correct sum and carry-out. This scheme speeds up the addition because the ith stage can be computing the two versions of the sum in parallel with the $i\text{-}1^{th}$'s computation of its carry. Once the carry-out is available, the ith stage's delay is limited to that of a two-input multiplexer.

adders in FPGAs

All these adders were originally developed for logic families in which we have gates with different numbers of inputs, costs, and delays. How do different adders compare when implemented in FPGAs? The answer depends in part on the FPGA fabric you use, but we can make some general conclusions. Xing and Yu compared several different adder designs in FPGAs [Xin98]. Assume that an n-bit adder is divided into x blocks. Then the ripple-carry delay at the kth block is

$$R(y_k) = \lambda_1 + \delta y_k, \qquad \text{(EQ 4-9)}$$

where λ_1 is a constant, δ is the incremental delay through a stage, and y_k is the number of stages in block k. The carry-generate delay is

$$G(y_k) = \lambda_2 + \delta(y_k - 1), \qquad \text{(EQ 4-10)}$$

where λ_2 is a constant. The carry-terminate or carry-kill delay is $T(y_k) = G(y_k)$.

As we have seen, the carry-skip adder is built from ripple-carry adders. Interconnect delay dominates the logic element delay, and wire delay is proportional to the square of the length of the wire. However, the longest wire length in a carry-skip adder is not obvious; Xing and Wu found the length l to be

$$l = \gamma \log_4(1 + 3N) \cong \gamma \log_4(4 + y_k), \qquad \text{(EQ 4-11)}$$

where γ is a constant. This gives a total carry-skip delay of

$$S(y_k) = \lambda_3 + \alpha \log_4^2(4 + y_k), \qquad \text{(EQ 4-12)}$$

where λ_3 is the delay of carry-in and carry-out logic and $\alpha = \beta\gamma^2$ is a constant. As with custom carry-skip adders, we can use our delay models to optimize the sizes of the carry-skip blocks.

The delay of a basic carry-select adder, in which ripple-carry adders are used in every block, is

$$T = (x-1)\mu + R(y_1) = \frac{1}{x}(\delta n - \mu) + \frac{\mu}{2}x + \left(\lambda_1 + \frac{\mu}{2}\right). \qquad \text{(EQ 4-13)}$$

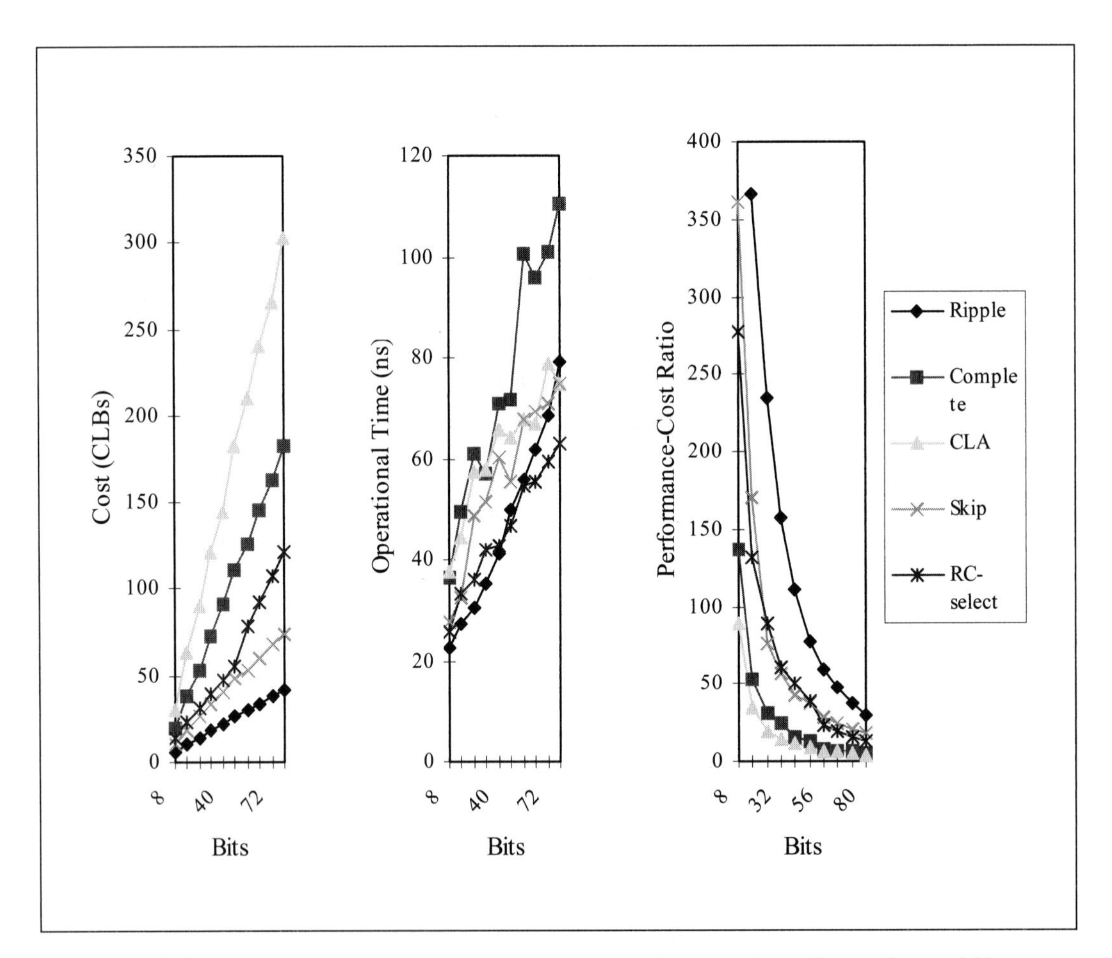

Figure 4-23 FPGA adders compared for cost, speed, and performance/cost (from Xing and Yu [Xin98], © 1998 IEEE).

Figure 4-23 compares ripple-carry, carry-skip, and carry-select adders using these formulas with delay parameters fitted from a Xilinx 4010 FPGA. These results show that ripple-carry adders were faster than either carry-skip or carry-select adders for any adder smaller than 48-bits. The carry-skip adder was second fastest, but did not gain an advantage until additions of 56 bits.

4.6.4 ALUs

ALU requirements

The **arithmetic logic unit**, or **ALU**, is a modified adder. While an ALU can perform both arithmetic and bit-wise logical operations, the arithmetic operations' carry chain usually shapes the design.

A basic ALU takes two data inputs and a set of control signals, also called an **opcode**. The opcode, together with the ALU's carry-in, determine the ALU's function. For example, if the ALU is set to add, then $c_0 = 0$ produces $a+b$ while $c_0 = 1$ produces $a+b+1$.

ALU operations

We saw in Section 4.6.1 how to perform some basic arithmetic operations. Many other operations are relatively easy to recast in a form that fits into the operations performed by an ALU:

- We can perform $a = b$ by computing $a - b$ and testing whether the result is zero.
- We can perform $a \geq b$ by computing $a - b$ and then testing the result for non-negativity (sign bit is zero).

There are 16 possible functions of two inputs. Not all ALUs must implement the full set of logical functions. If the ALU need implement only a few functions, the function block scheme may be overkill. An ALU that, for example, must implement only addition, subtraction, and one or two bit-wise functions may be smaller than a full-function ALU.

If we need to perform a small number of operations, we may be able to squeeze the logic into fewer LEs. For example, if we want to perform only addition and subtraction using four-input, two-output logic elements, we can squeeze the opcode and data inputs for one bit into a single LE, as shown in this truth table:

a	b	carry/ borrow in	opcode (0=sum, 1=difference)	a +/- b	carry/ borrow
0	0	0	0	0	0
0	0	1	0	1	0
0	1	0	0	1	0
0	1	1	0	0	1
1	0	0	0	1	0
1	0	1	0	0	1
1	1	0	0	0	1
1	1	1	0	1	1
0	0	0	1	0	0
0	0	1	1	1	1
0	1	0	1	1	1
0	1	1	1	0	1
1	0	0	1	1	0
1	0	1	1	0	0
1	1	0	1	0	0
1	1	1	1	1	1

When choosing the functions to go into the ALU, it makes sense to keep in mind the capacity of the logic elements. A set of functions that will fit into one, or perhaps two, LEs would provide a much more efficient physical design.

The next example introduces a simple ALU design.

Example 4-10 ALU

Here is an ALU that performs several basic operations. It includes a register for the output value:

```
'define PLUS 0
'define MINUS 1
'define AND 2
'define OR 3
```

```
'define NOT 4

module alu(fcode,op0,op1,result,oflo);
    parameter n=16, flen=3;
    input [flen-1:0] fcode; // operation code
    input [n-1:0] op0, op1; // operands
    output [n-1:0] result; // operation result
    output oflo; // overflow

assign
    {oflo,result} =
        (fcode == 'PLUS) ? (op0 + op1) :
          (fcode == 'MINUS) ? (op0 - op1) :
            (fcode == 'AND) ? (op0 & op1) :
              (fcode == 'OR) ? (op0 | op1) :
                (fcode == 'NOT) ? (~op0) : 0;
endmodule
```

The case statement does not synthesize into purely combinational logic, so the ? operator allows us to clearly express choice in the code.

4.6.5 Multipliers

multiplication algorithms

Multiplier design starts with the elementary school algorithm for multiplication. Consider the simple example of Figure 4-24. At each step, we multiply one digit of the multiplier by the full multiplicand; we add the result, shifted by the proper number of bits, to the partial product. When we run out of multiplier digits, we are done. Single-digit multiplication is easy for binary numbers—binary multiplication of two bits is performed by the AND function. The computation of partial products and their accumulation into the complete product can be optimized in many ways, but an understanding of the basic steps in multiplication is important to a full appreciation of those improvements.

serial-parallel multiplier

One simple, small way to implement the basic multiplication algorithm is the **serial-parallel** multiplier of Figure 4-25, so called because the n-bit multiplier is fed in serially (LSB first) while the m-bit multiplicand is held in parallel during the course of the multiplication. The multiplier is fed in least-significant bit first and is followed by at least m zeroes. The result appears serially at the end of the multiplier chain. A one-bit multiplier is simply an AND gate. The sum units in the multiplier include a combinational full adder and a register to hold the carry. The chain of

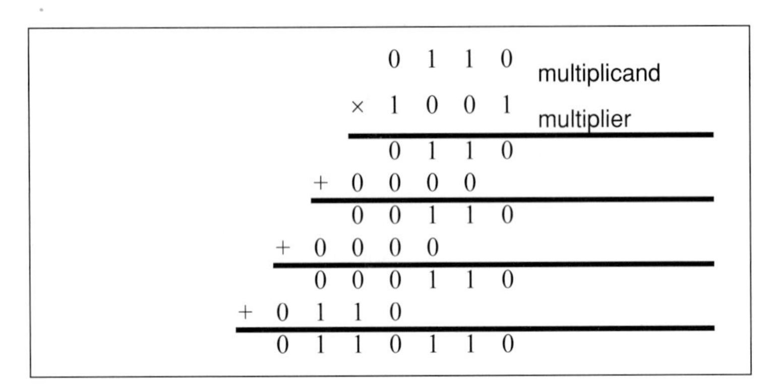

Figure 4-24 Multiplication using the elementary school algorithm.

summation units and registers performs the shift-and-add operation—the partial product is held in the shift register chain, while the multiplicand is successively added into the partial product.

bit-serial multiplication

We can use the serial-parallel multiplier to build a **bit-serial multiplier**. Rather than present the bits of the parallel multiplicand in parallel, we shift them in serially and perform the multiplication. This multiplier requires a controller in order to sequence the shift-in, multiply-add steps, and shift-out.

Baugh-Wooley multiplier

One important complication in the development of efficient multiplier implementations is the multiplication of two's-complement signed numbers. The **Baugh-Wooley multiplier** [Bau73] is the best-known algorithm for signed multiplication because it maximizes the regularity of the multiplier logic and allows all the partial products to have positive sign bits. The multiplier X can be written in binary as

$$X = x_{n-1}2^{n-1} + \sum_{i=0}^{n-2} x_i 2^i, \qquad \text{(EQ 4-14)}$$

where n is the number of bits in the representation. The multiplicand Y can be written similarly. The product P can be written as

$$P = p_{2n-1}2^{2n-2} + \sum_{i=0}^{2n-2} p_i 2^i. \qquad \text{(EQ 4-15)}$$

When this formula is expanded to show the partial products, it can be seen that some of the partial products have negative signs:

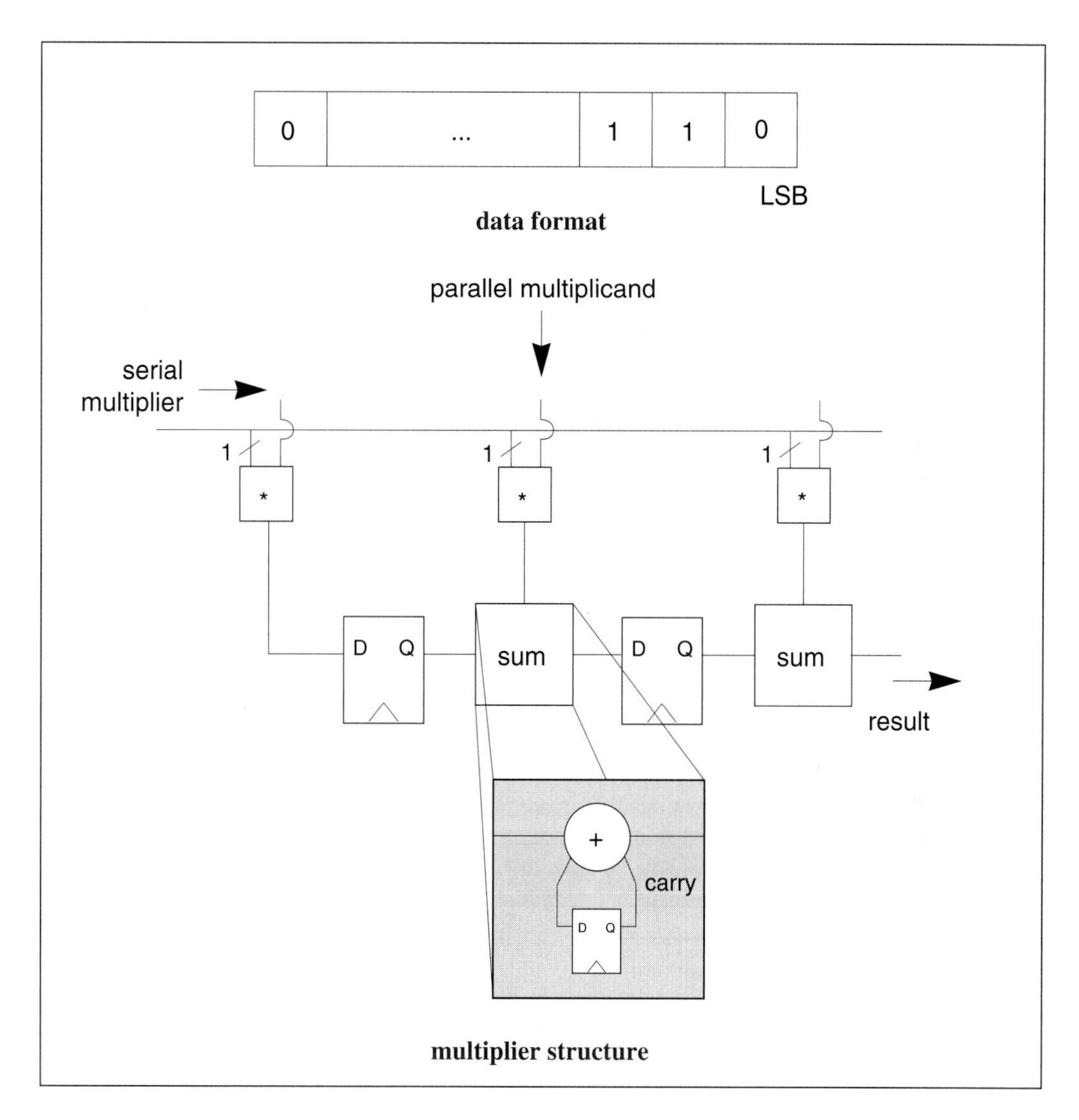

Figure 4-25 Basic structure of a serial-parallel multiplier.

$$P = \left[x_{n-1}y_{n-1}2^{2n-2} + \sum_{i=0}^{n-2}\sum_{j=0}^{n-2} x_i y_j 2^{i+j}\right] - \left[\sum_{i=0}^{n-2} (x_{n-1}y_i + y_{n-1}x_i)2^{n-1+i}\right]. \quad \text{(EQ 4-16)}$$

The formula can be further rewritten, however, to move the negative-signed partial products to the last steps and to add the negation of the partial product rather than subtract. Further rewriting gives this final form:

$$P = 2^{n-1}\left(-2^n + 2^{n-1} + \bar{x}_{n-1}2^{n-1} + x_{n-1} + \sum_{i=0}^{n-2} x_{n-1}\bar{y}_i 2^i\right). \quad \text{(EQ 4-17)}$$

Each partial product is formed with AND functions and the partial products are all added together. The result is to push the irregularities to the end of the multiplication process and allow the early steps in the multiplication to be performed by identical stages of logic.

array multiplier

The elementary school multiplication algorithm (and the Baugh-Wooley variations for signed multiplication) suggest a logic and layout structure for a multiplier which is surprisingly well-suited to VLSI implementation—the **array multiplier**. The structure of an array multiplier for unsigned numbers is shown in Figure 4-26. The logic structure is shown in parallelogram form both to simplify the drawing of wires between stages and also to emphasize the relationship between the array and the basic multiplication steps shown in Figure 4-24. As when multiplying by hand, partial products are formed in rows and accumulated in columns, with partial products shifted by the appropriate amount. In layout, however, the y bits generally would be distributed with horizontal wires since each row uses exactly one y bit.

Notice that only the last adder in the array multiplier has a carry chain. The earlier additions are performed by full adders which are used to reduce three one-bit inputs to two one-bit outputs. Only in the last stage are all the values accumulated with carries. As a result, relatively simple adders can be used for the early stages, with a faster (and presumably larger and more power-hungry) adder reserved for the last stage. As a

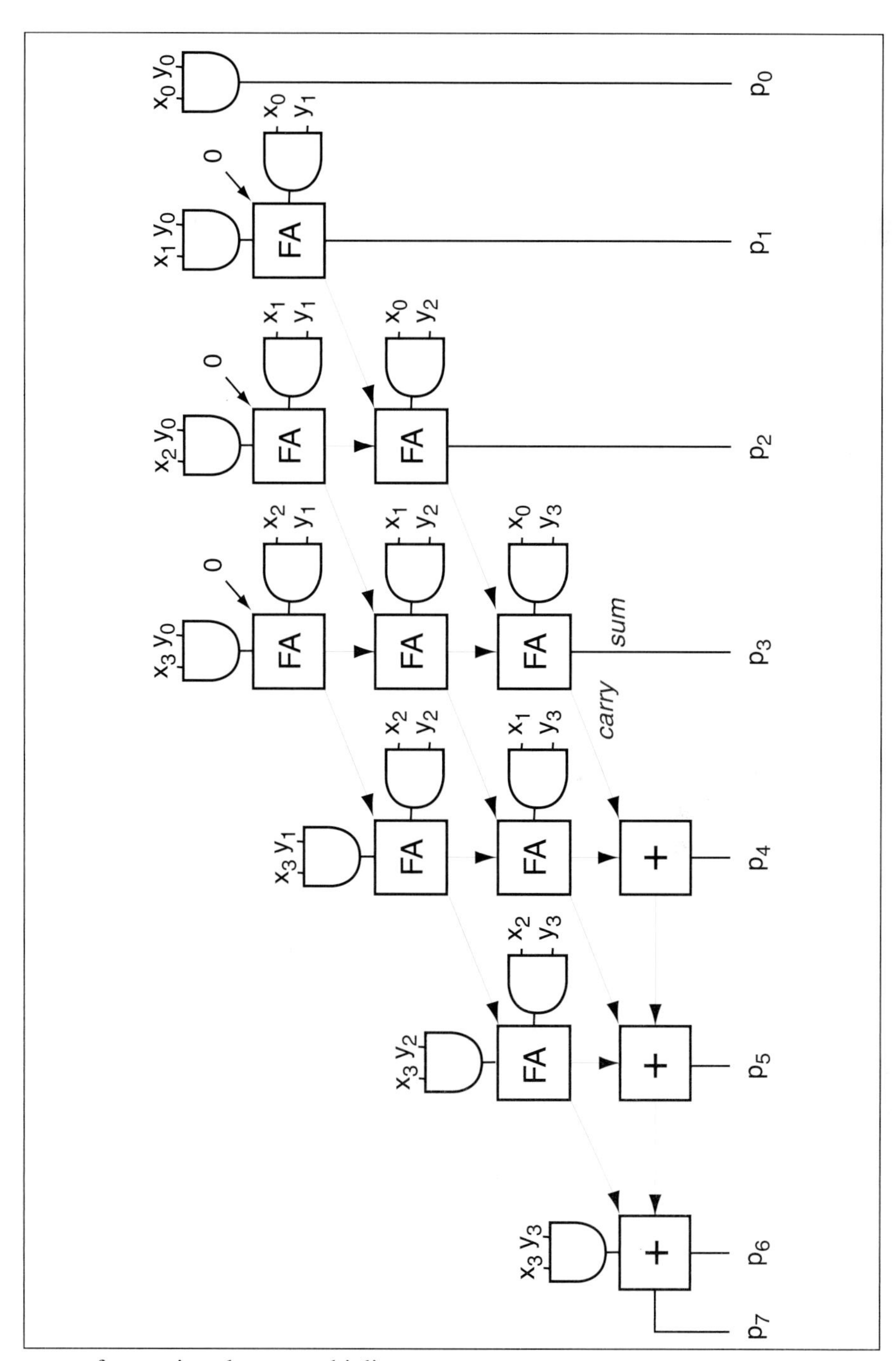

Figure 4-26 Structure of an unsigned array multiplier.

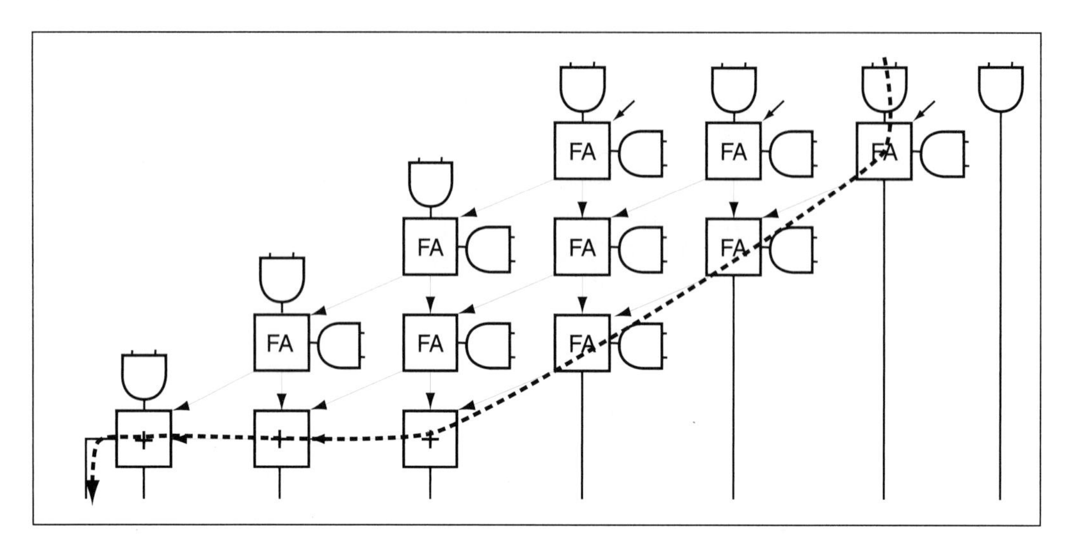

Figure 4-27 The critical delay path in the array multiplier.

result, the critical delay path for the array multiplier follows the trajectory shown in Figure 4-27.

The next example shows the design of an array multiplier in Verilog.

Example 4-11 Array multiplier

We will build a 4×4 multiplier; a large multiplier can be built along these same lines.

The first step in building the multiplier is to create a cell that contains a row of the add/1-bit multiply array:

```
module multrow(part,x,ym,yo,cin,s,cout);
   /* A row of one-bit multiplies */
   input [2:0] part;
   input [3:0] x;
   input ym, yo;
   input [2:0] cin;
   output [2:0] s;
   output [2:0] cout;

  assign {cout[0],s[0]} = part[1] + (x[0] & ym) + cin[0];
  assign {cout[1],s[1]} = part[2] + (x[1] & ym) + cin[1];
  assign {cout[2],s[2]} = (x[3] & yo) + (x[2] & ym) + cin[2];
endmodule
```

Here is the code for the last row of the multiplier, which is somewhat different:

```
module lastrow(part,cin,s,cout);
    /* Last row of adders with full carry chain. */
  input [2:0] part;
  input [2:0] cin;
  output [2:0] s;
  output cout;
  wire [1:0] carry;

  assign {carry[0],s[0]} = part[0] + cin[0];
  assign {carry[1],s[1]} = part[1] + cin[1] + carry[0];
  assign {cout,s[2]} = part[2] + cin[2] + carry[1];
endmodule
```

We can now build the full multiplier, building the last row and some other logic as special cases:

```
module array_mult(x,y,p);
  input [3:0] x;
  input [3:0] y;
  output [7:0] p;
  wire [2:0] row0, row1, row2, row3, c0, c1, c2, c3;

  /* generate first row of products */
  assign row0[2] = x[2] & y[0];
  assign row0[1] = x[1] & y[0];
  assign row0[0] = x[0] & y[0];
  assign p[0] = row0[0];
  assign c0 = 3'b000;
  multrow p0(row0,x,y[1],y[0],c0,row1,c1);
  assign p[1] = row1[0];
  multrow p1(row1,x,y[2],y[1],c1,row2,c2);
  assign p[2] = row2[0];
  multrow p2(row2,x,y[3],y[2],c2,row3,c3);
  assign p[3] = row3[0];
  lastrow l({x[3] & y[3],row3[2:1]},c3,p[6:4],p[7]);
endmodule
```

constant-coefficient multipliers

In many FPGA designs, the multiplier is used to multiply a value by a constant coefficient. In this case, the constant can be used to simplify the multiplier logic either by hand or using logic synthesis. The savings gained by using a **constant-coefficient multiplier** can be substantial.

Booth encoding

One way to speed up general multiplication is **Booth encoding** [Boo51], which performs several steps of the multiplication at once. Booth's algorithm takes advantage of the fact that an adder-subtractor is nearly as fast and small as a simple adder. In the elementary school algorithm, we shift the multiplicand x, then use one bit of the multiplier y if that shifted value is to be added into the partial product. The most common form of Booth's algorithm looks at three bits of the multiplier at a time to perform two stages of the multiplication.

Consider once again the two's-complement representation of the multiplier y:

$$y = (-2)^n y_n + 2^{n-1} y_{n-1} + 2^{n-2} y_{n-2} + \dots \quad \text{(EQ 4-18)}$$

We can take advantage of the fact that $2^a = 2^{a+1} - 2^a$ to rewrite this as

$$y = 2^n(y_{n-1} - y_n) + 2^{n-1}(y_{n-2} - y_{n-1}) + 2^{n-2}(y_{n-3} - y_{n-2}) + \dots . \quad \text{(EQ 4-19)}$$

Now, extract the first two terms:

$$2^n(y_{n-1} - y_n) + 2^{n-1}(y_{n-2} - y_{n-1}) . \quad \text{(EQ 4-20)}$$

Each term contributes to one step of the elementary-school algorithm: the right-hand term can be used to add x to the partial product, while the left-hand term can add $2x$. (In fact, since y_{n-2} also appears in another term, no pair of terms exactly corresponds to a step in the elementary school algorithm. But, if we assume that the y bits to the right of the decimal point are 0, all the required terms are included in the multiplication.) If, for example, $y_{n-1} = y_n$, the left-hand term does not contribute to the partial product. By picking three bits of y at a time, we can determine whether to add or subtract x or $2x$ (shifted by the proper amount, two bits per step) to the partial product. Each three-bit value overlaps with its neighbors by one bit. Table 4-5 shows the contributing term for each three-bit code from y.

Let's try an example to see how this works: x = 011001 (25_{10}), y = 101110 (-18_{10}). Call the ith partial product P_i. At the start, P_0 = 00000000000 (two six-bit numbers give an 11-bit result):

1. $y_1 y_0 y_{-1} = 100$, so $P_1 = P_0 - (10 \cdot 011001) = 11111001110$.

Table 4-5 Actions during Booth multiplication.

$y_i\ y_{i-1}\ y_{i-2}$	increment
0 0 0	0
0 0 1	x
0 1 0	x
0 1 1	2x
1 0 0	-2x
1 0 1	-x
1 1 0	-x
1 1 1	0

2. $y_3y_2y_1 = 111$, so $P_2 = P_1 + 0 = 11111001110$.
3. $y_5y_4y_3 = 101$, so $P_3 = P_2\text{-}0110010000 = 11000111110$.

In decimal, $y_1y_0y_{-1}$ contribute $-2x\cdot1$, $y_3y_2y_1$ contribute $0\cdot4$, and $y_5y_4y_3$ contribute $-x\cdot16$, giving a total of -18x. Since the multiplier is -18, the result is correct.

Figure 4-28 shows the detailed structure of a Booth multiplier. The multiplier bits control a multiplexer which determines what is added to or subtracted from the partial product. Booth's algorithm can be implemented in an array multiplier since the accumulation of partial products still forms the basic trapezoidal structure. In this case, a column of control bit generators on one side of the array analyzes the triplets of y bits to determine the operation in that row.

Wallace tree

Another well-known multiplication method is the **Wallace tree** [Wal64], although this method is infrequently used in FPGAs. It speeds up multiplication by using more adders to speed the accumulation of partial products. This adder is built from a tree of carry-save adders which, given three n-bit numbers a, b, c, computes two new numbers y, z such that $y + z = a + b + c$. The Wallace tree performs the three-to-two reductions; at each level of the tree, i numbers are combined to form $\lceil 2i/3 \rceil$ sums. When only two values are left, they are added with a high-speed adder. Figure 4-29 shows the structure of a Wallace tree. The partial products are introduced at the bottom of the tree. Each of the z outputs is shifted left by one bit since it represents the carry out.

A Wallace tree multiplier is considerably faster than a simple array multiplier because its height is logarithmic in the word size, not linear.

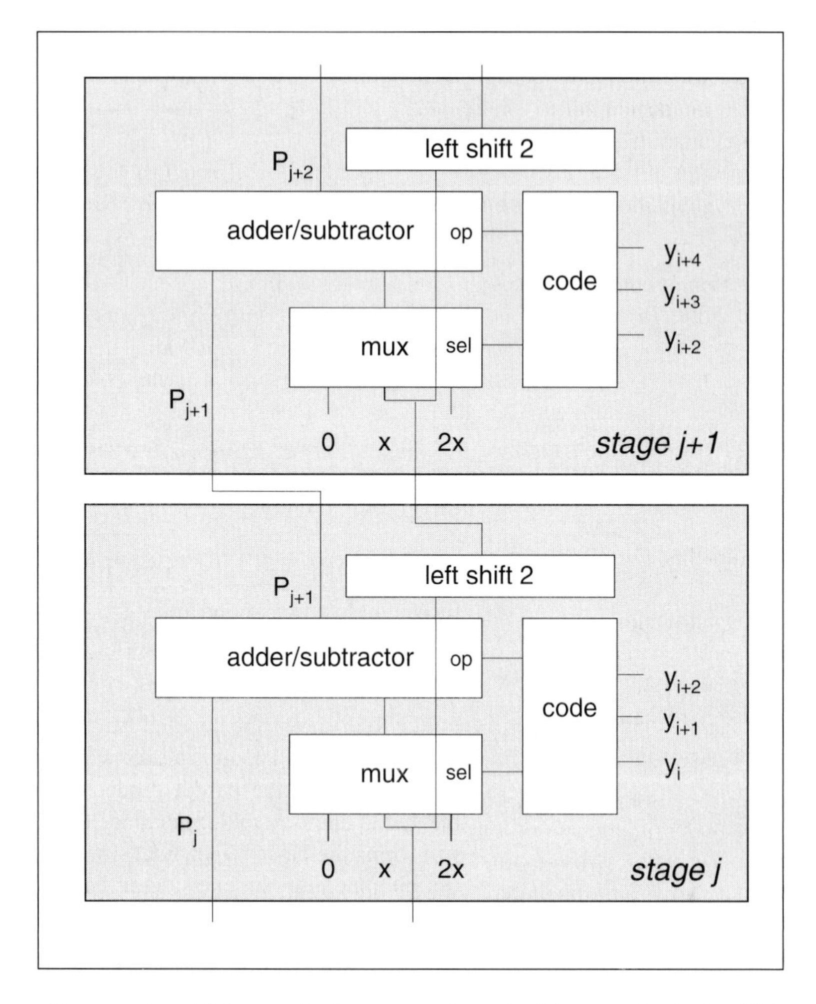

Figure 4-28 Structure of a Booth multiplier.

However, in addition to the larger number of adders required, the Wallace tree's wiring is much less regular and more complicated. As a result, Wallace trees are often avoided by designers who do not have extreme demands for multiplication speed and for whom design complexity is a consideration.

Callaway and Schwartzlander [Cal96] also evaluated the power consumption of multipliers. They compared an array multiplier and a Wallace tree multiplier (both without Booth encoding) and found that the Wallace tree multiplier used significantly less power for bit widths between 8 and 32, with the advantage of the Wallace tree growing as word length increased.

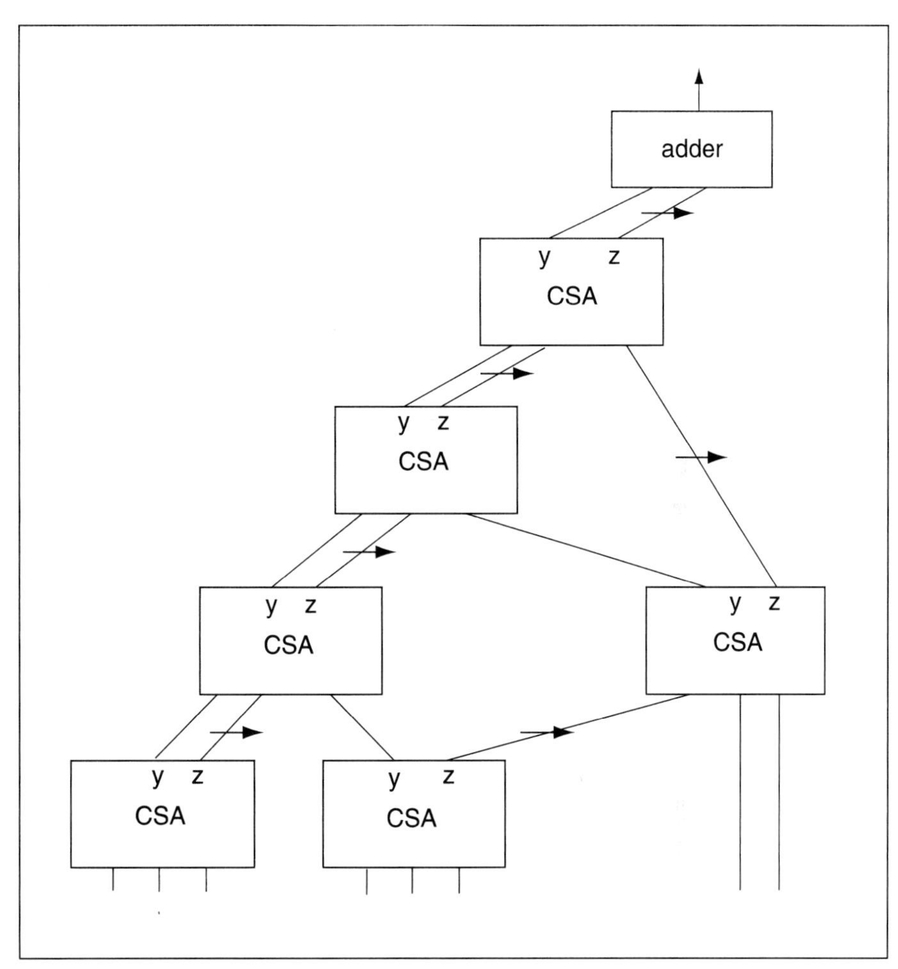

Figure 4-29 Structure of a Wallace tree.

4.7 Logic Implementation for FPGAs

logic optimization strategies

There are several ways to transform a logic description—whether it was written in a hardware description language or was drawn as a schematic—into a logic implementation optimized for whatever goals are appropriate to our project. **Logic synthesis** uses optimization algorithms to rewrite the logical description in order to improve some non-functional metric such as performance, size, or power consumption. **Macros** allow us to map a design more directly into pre-designed components that generally take advantage of characteristics of our target FPGA. Finally, **hand tuning** relies on people to massage the design.

We may have several different goals when designing logic: small size, high performance, low power or energy consumption. In general, we are interested in several of these goals. Unfortunately, the goals often conflict (though high performance designs often use less energy). Tools can help us optimize designs but some amount of engineering judgment may be required to achieve all our goals.

In this section we will look in detail at both logic synthesis and macro-based designs as means for turning our logic description into a usable implementation. First, we need to briefly describe ways in which we can capture the logic description from our HDL model.

4.7.1 Syntax-Directed Translation

translating HDLs into logic

Syntax-directed translation is a standard method in compilers for the early stages of compilation. It is used to extract the logical functions from an HDL description. As we will see, the functions extracted in this way may not be well-suited to direct implementation and requiring logic synthesis. However, syntactic hints are the major means for identifying where macros may be used.

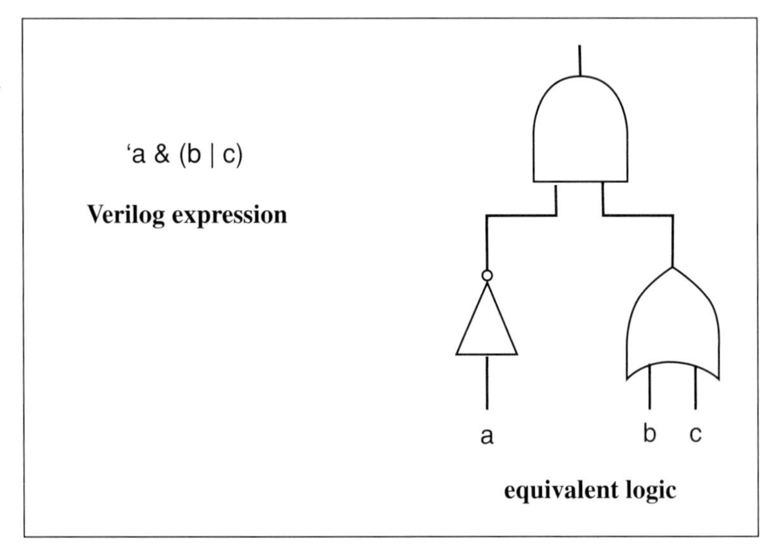

Figure 4-30 Translating a logical expression in an HDL description.

Some parts of an HDL description are in fact logical expressions. These are fairly easy to extract. Figure 4-30 shows a simple Verilog expression

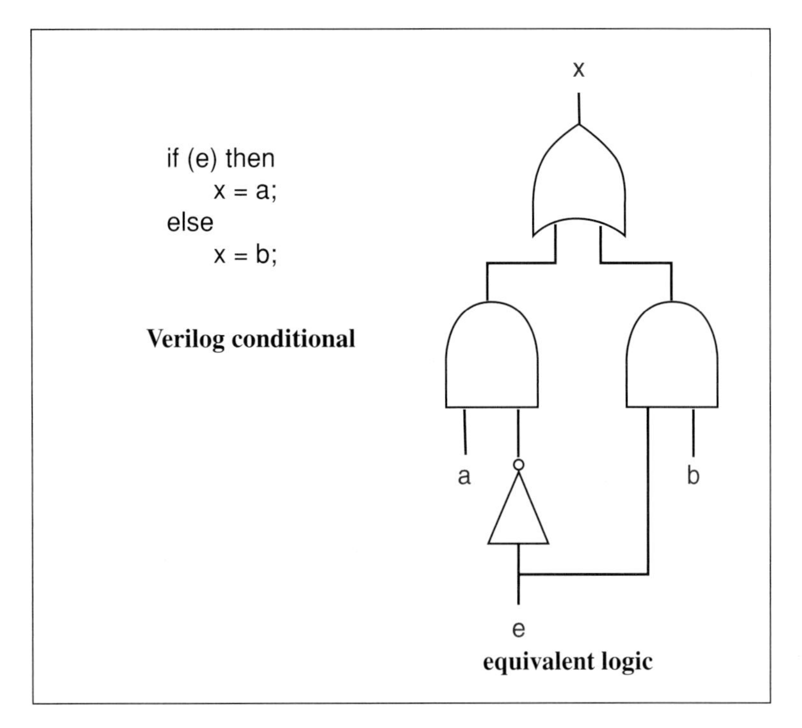

Figure 4-31 Translating a conditional in an HDL description.

and a schematic for equivalent gates. We can build the logic from the expression by traversing the expression and assigning gates at each operator and wires from the gate's inputs to the subexpressions that feed the operator.

Other parts of an HDL map less directly onto Boolean functions. Figure 4-31 shows an *if* statement in VHDL and some equivalent logic for it. These gates implement a multiplexer that selects which value is assigned to x. Of course, a conditional need not always assign to the same signals in both the true and false branches. In such cases, the logic may be more involved and a signal may not always be driven. Various synthesis subsets may take actions to handle the cases in which the signal is not driven.

4.7.2 Logic Implementation by Macro

FPGAs provide a rich structure to map onto logic. Most modern FPGAs offer features that help implement common or important functions. A key example is the adder. Addition-oriented structures are a prime example. Adders and counters are very common in digital logic and the carry chain determines the speed of the adder for even a small number of bits in the addend. Many FPGA fabrics offer specialized logic and interconnect for faster carry chains. However, the fabric must be properly configured to use these features, which means that the logic that maps into the features must be identified.

macros

Macros give a name to a piece of an implementation of a design. For example, a macro for an adder would describe how to configure the adder into the FPGA fabric using the appropriate features. Macros are a form of design using **intellectual property (IP)**. The macro is a predesigned element that can be incorporated into a larger design.

Macros (as with many other forms of intellectual property) may be **hard macros** or **soft macros**, depending on whether they include layout information. A hard macro describes where elements are placed—for example, that bit $n+1$ of the adder is in the logic element directly above bit n. A soft macro only describes logic and interconnections. Hard macros generally provide the highest performance because they describe physical relationships designed to minimize wire lengths. Some macros may specify placement of logic but not interconnections. However, physical restrictions may limit the applicability of use of the hard macro. Soft macros may provide less dramatic improvements but may be useful in more situations.

4.7.3 Logic Synthesis

logic synthesis

Logic synthesis (or logic optimization) creates a logic gate network which computes the functions specified by a set of Boolean functions, one per primary output. While we could get a gate implementation for a function by directly constructing a two-level network from the sum-of-products specification, using one logic gate per product plus one for the sum, that implementation almost certainly won't meet our area or delay requirements. Logic synthesis is challenging because, while there are many gate networks which implement the desired function, only a few of them are small enough and fast enough. Logic synthesis helps us explore more of the design space to find the best gate implementation, particularly for relatively unstructured random logic functions.

logic optimization phases

Synthesizing a logic gate network requires balancing a number of very different concerns: simplification from don't-cares; common factors; fanin and fanout; and many others. To help manage these complex decisions, logic synthesis is divided into two phases: first, **technology-independent** logic optimizations operate on a model of the logic that does not directly represent logic gates; later, **technology-dependent** logic optimizations improve the network's gate-level implementation. The transformation from the technology-independent to the technology-dependent gate network is called **library binding**.

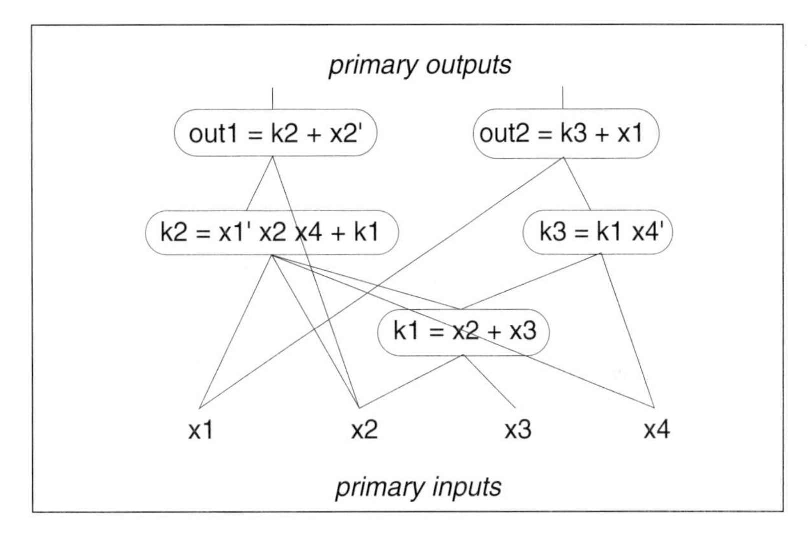

Figure 4-32 A Boolean network.

Boolean networks

Figure 4-32 shows a **Boolean network** [Bra90], which is the standard technology-independent model for a logic network. Nodes in the network are primary inputs, primary outputs, or functions; the function nodes can be thought of as sum-of-product expressions. Edges show on which variables a node depends (each node is referred to by a variable name). The Boolean network has a structure reminiscent of a logic network, but the function nodes are more general than logic gates: they can have an unbounded number of input variables and express an arbitrary Boolean function. The real set of logic gates we have to work with may be limited in several ways: we may not have gates for every function of n variables; we may have several different gates for a function, each with different transistor sizes; and we definitely want to limit the fanin and fanout of a gate. We can create the overall structure of the logic network using the Boolean network, ignoring some details of the gate design, then fine-tune the implementation using technology-dependent optimizations on the gate network.

terminology

A few definitions are helpful: a function's **support** is the set of variables used by a function; the **transitive fanin** of a node is all the primary inputs and intermediate variables used by the function; the transitive fanin is also known as a **cone** of logic; the **transitive fanout** of a node is all the primary outputs and intermediate variables which use the function.

4.7.4 Technology-Independent Logic Optimization

technology-independent optimization methods

Technology-independent optimizations can be grouped into categories based on how they change the Boolean network:

- **Simplification** rewrites (hopefully simplifying) the sum-of-products representation of a node in the network; most of the improvement usually comes from taking advantage of don't-cares.
- **Network restructuring** creates new function nodes that can be used as common factors and collapses sections of the network into a single node in preparation for finding new common factors.
- **Delay restructuring** changes the factorization of a subnetwork to reduce the number of function nodes through which a delay-critical signal must pass.

How do we judge whether an operation has improved the network? We can estimate area by counting literals (true or complement forms of variables) in the Boolean network. Counting the literals in all the functions of the Boolean network gives a reasonable estimate of the final layout area [Lig90]—each literal corresponds to a transistor if the functions in the network are implemented directly. Delay can be roughly estimated by counting the number of functions from primary input to primary output.

simplification

Simplification rewrites a single function in the network to reduce the number of literals in that network. In Figure 4-33, we have shown the function as a set of points: each variable is represented by a dimension with the coordinates 0 and 1; the function $x_1'x_2'x_3' + x_1x_2'x_3' + x_1'x_2x_3' + x_1x_2x_3$ is therefore represented by a three-dimensional space. A fully specified function is defined by two sets: the **on-set**, or the set of points (black in the figure) for which the function's value is 1; and the **off-set**, the points for which the function is 0. In the figure, the point $x_1 = 1, x_2 = 1, x_3 = 1$ is in the on-set and $x_1 = 1, x_2 = 1, x_3 = 0$ is in the off-set.

There are many ways to write this function as a sum of products. Each representation is called a **cover** because it must cover all the points in the function's on-set. Each member of the cover is a subspace of the function space, which is written algebraically as a product (or **cube**). For example, $x_1 = 1, x_3 = 0$ or $x_1 x_3'$ is a one-dimensional subspace of the function, and $x_2 = 0$ or x_2' is a two-dimensional subspace. The cubes in the cover must include all the points in the on-set and none of the

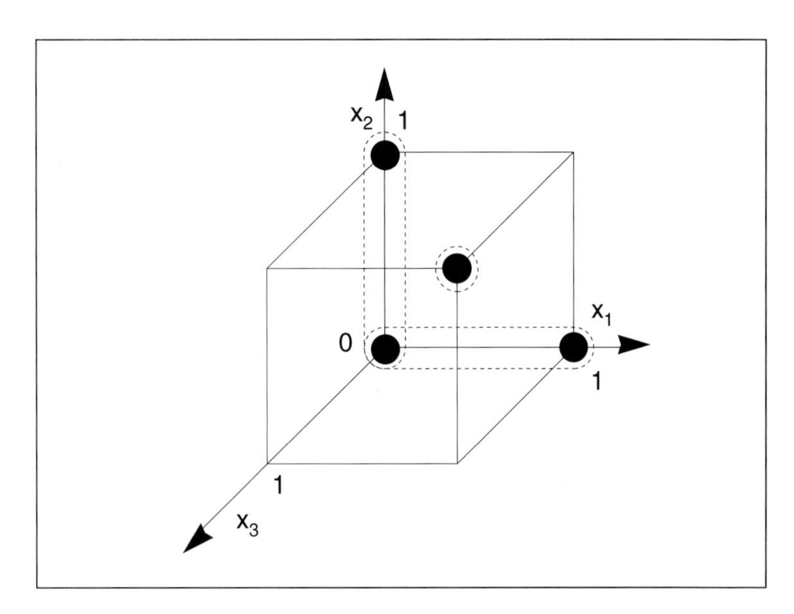

Figure 4-33 A three-input function shown in three-space and a cover of the function.

points in the off-set. The number of literals is counted in the cube representation—the larger the subspace a cube covers, the fewer literals in the cube.

A function has many covers, some of which have more literals than others. The simplest and largest cover of a function is just the sum of all the points in the on-set, for example $x_1'x_2'x_3' + x_1x_2'x_3' + x_1'x_2x_3' + x_1x_2x_3$. A cover with both fewer cubes and fewer literals is $x_2'x_3' + x_1'x_3' + x_1x_2x_3$; the point $x_1'x_2'x_3'$ is covered by two cubes.

don't-cares and simplification

Simplification is important largely because most function specifications include don't-care values. Such a function is called a **partially specified** function because, for some input values, the function value is not specified. Of course, the logic implementation must produce a 0 or 1 value for all inputs, but the specification states that we don't care which value the output assumes. Don't-cares let us put a point in the on-set or off-set, depending on what produces the smallest cover.

In Figure 4-34, the don't-care set consists of a single point $x_1x_2'x_3'$. If we assume that point is in the off-set, the cover is the same as before; if we put that point in the on-set, we can use the smaller cover $x_3' + x_1x_2x_3$ and save three literals.

two-level optimization

Espresso [Bra84] is the best-known two-level logic optimizer. The algorithm starts with a cover given by the user (simply the cover used to

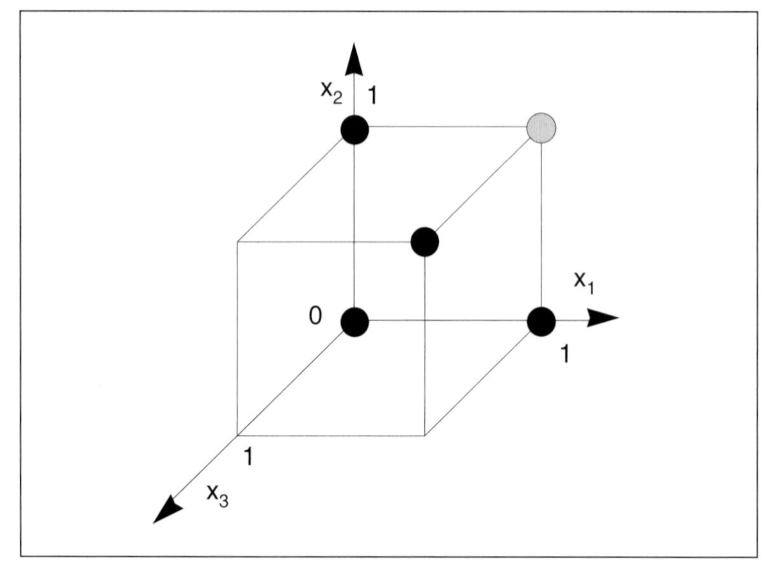

Figure 4-34 A partially specified Boolean function.

describe the function) and iteratively reduces it in size using an **expand-irredundant-reduce** loop. An example is shown in Figure 4-35. The first step is to make each cube as large as possible without covering a point in the off-set. This step increases the number of literals in the cover, but sets the stage for finding a new and possibly better solution. The next step is to throw out redundant cubes—points may be covered by many cubes after expansion, so the algorithm throws out smaller cubes whose points are covered by larger cubes. Finally, the cubes in the cover are reduced in size, reducing the number of literals in the cover. In general, the new cover will be different than the starting cover because the *expand* and *minimize* steps find a new way to cover the points in the on-set. Hopefully, this new cover will also be smaller. The algorithm successively generates new covers until the cover can no longer be improved. While this solution is not guaranteed to be the global optimum, experience shows that the solution is near optimal for typical functions.

structural don't-cares

When simplifying a function in a Boolean network, we must consider more than the output don't-cares specified by the user. Figure 4-36 shows where different varieties of don't-cares occur in Boolean networks. **Structural don't-cares** [Bra90] are created by the internal structure of the Boolean network. A single function has no structural don't-cares because it has no internal structure, but a logic optimization system for Boolean networks must be able to extract and use structural don't-cares. There are two types of structural don't-cares: **satisfiability**

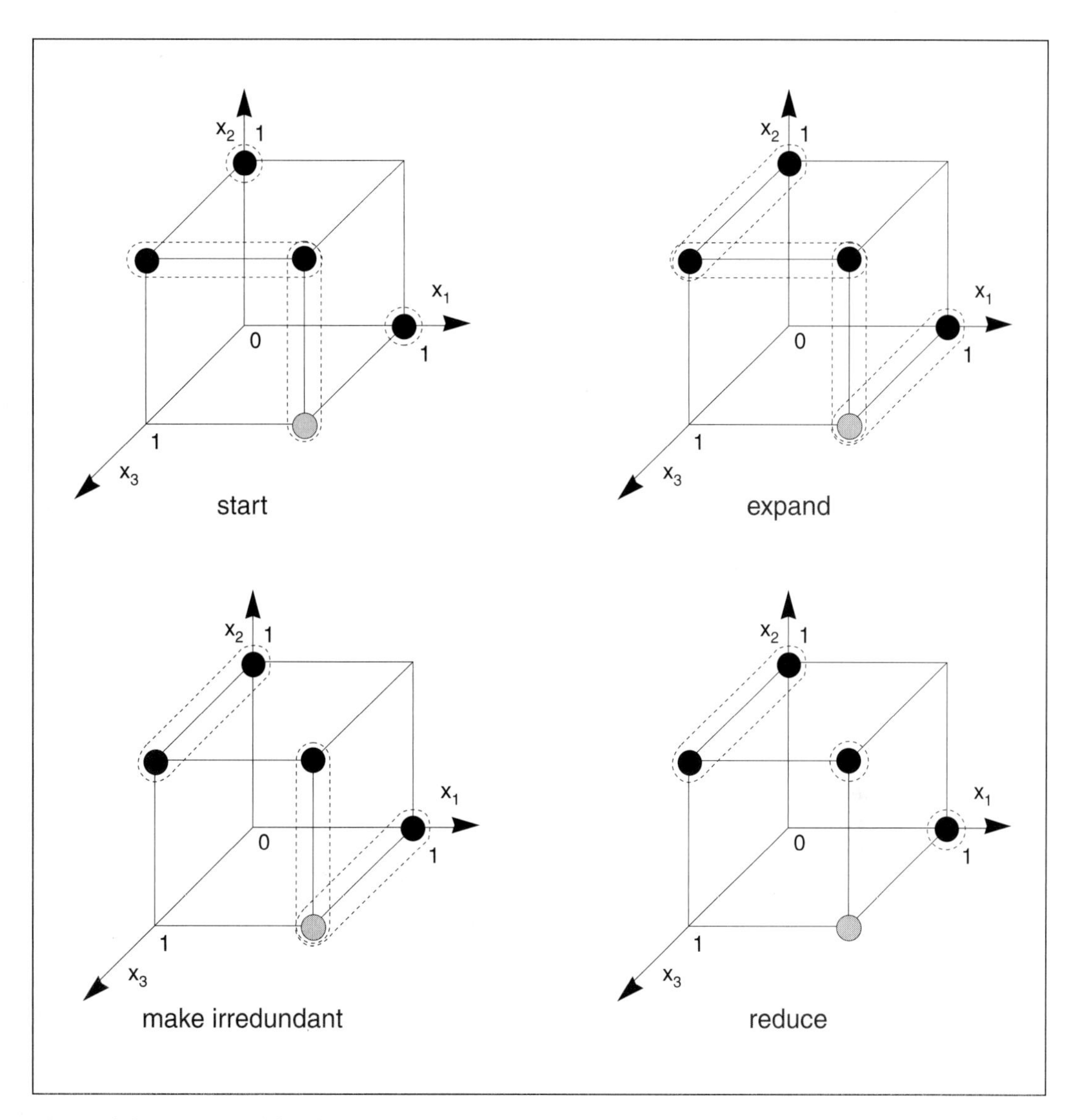

Figure 4-35 An expand-irredundant-reduce cycle.

don't-cares occur when an intermediate variable value is inconsistent with its function inputs; **observability don't-cares** occur when an intermediate variable's value doesn't affect the network's primary outputs.

An example of satisfiability don't-care optimization is shown in Figure 4-37. The *g* function defines the relationship between the primary inputs

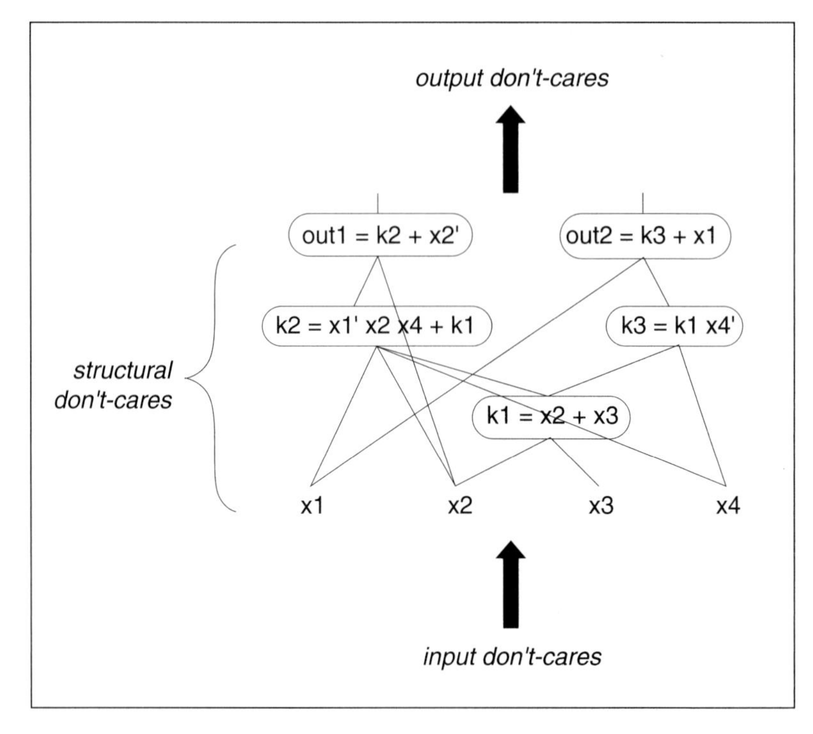

Figure 4-36 Varieties of don't-cares in Boolean networks.

a and *b* and the intermediate variable *y*. We know that the value of *y* is always equal to the value of the *g* function, so for example, we know that the case $a = b = 0$, $y = 1$ can never occur in the network. The case $y'g + yg'$ can be used as a don't-care condition for simplifying the *f* function.

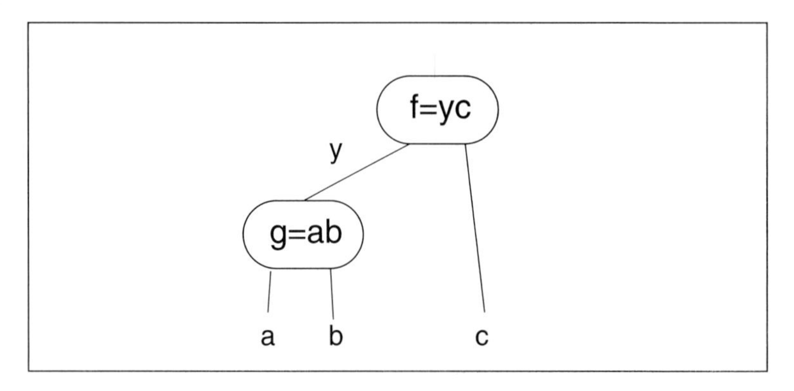

Figure 4-37 Satisfiability don't-cares.

Observability don't-cares are more complex to compute than are satisfiability don't-cares. As a simple example [Hac96], consider an OR gate with two inputs that are intermediate variables in the Boolean network.

If the gate's a input is 1, then the gate's output is 1 independent of the value of its b input; the symmetric situation holds for the b input. These conditions define observability don't-cares for this simple case.

Simplification can have a minor influence on the structure of the Boolean network—simplifying a function may eliminate a variable from a function, deleting an edge from the network. However, we need optimizations that can radically restructure the network. We may want to decompose a network for two reasons. The parity function is an excellent example of the savings provided by recursively decomposing a function into subfunctions. Each function is used only once in the network, yet the multilevel network has many fewer literals than the flat, sum-of-products form, which has an exponential number of literals. In less regular functions, common factors are the predominant form of intermediate node. Particularly in networks with several primary outputs, a subfunction may appear in several different functions.

logic factorization methods

Factorization algorithms extract new nodes to add to the Boolean network. Factorization algorithms are based on **division**—they first choose candidate functions for the intermediate nodes, then divide the candidate functions into the functions already in the network and measure whether the candidate function makes the network smaller. Division is the mechanism we use to determine how a function f splits into two functions: the candidate function c and a cofactor function g. If we find that $g = f/c$, we can use c as an intermediate function to compute f.

We can develop surprisingly powerful algorithms for factorization using relatively weak **algebraic division** [Bra90]. Algebraic division relies only on the algebraic identities: commutativity ($a + b = b + a$); associativity ($(a+b) + c = a + (b+c)$); and distributivity ($a(b+c) = ab + ac$). In contrast, more general Boolean division uses algebraic identities plus the Boolean identities: $a \cdot 0 = 0$; $a \cdot 1 = a$; $a + 0 = a$; $a + 1 = 1$. Algebraic division is important because, even though it is less powerful than **Boolean division**, algebraic division requires much less CPU time. That means we can test many more candidate functions. Only a subset of the possible candidate common factors are algebraic divisors, but algebraic divisors include many useful intermediate functions for typical networks. We generally use algebraic factorization to find the general structure of the Boolean network, then use Boolean factorization to fine-tune the design.

Whether potential new factors are generated using algebraic or Boolean division, factoring the network takes three steps:

1. generate all potential common factors and compute how many literals each saves when substituted into the network;

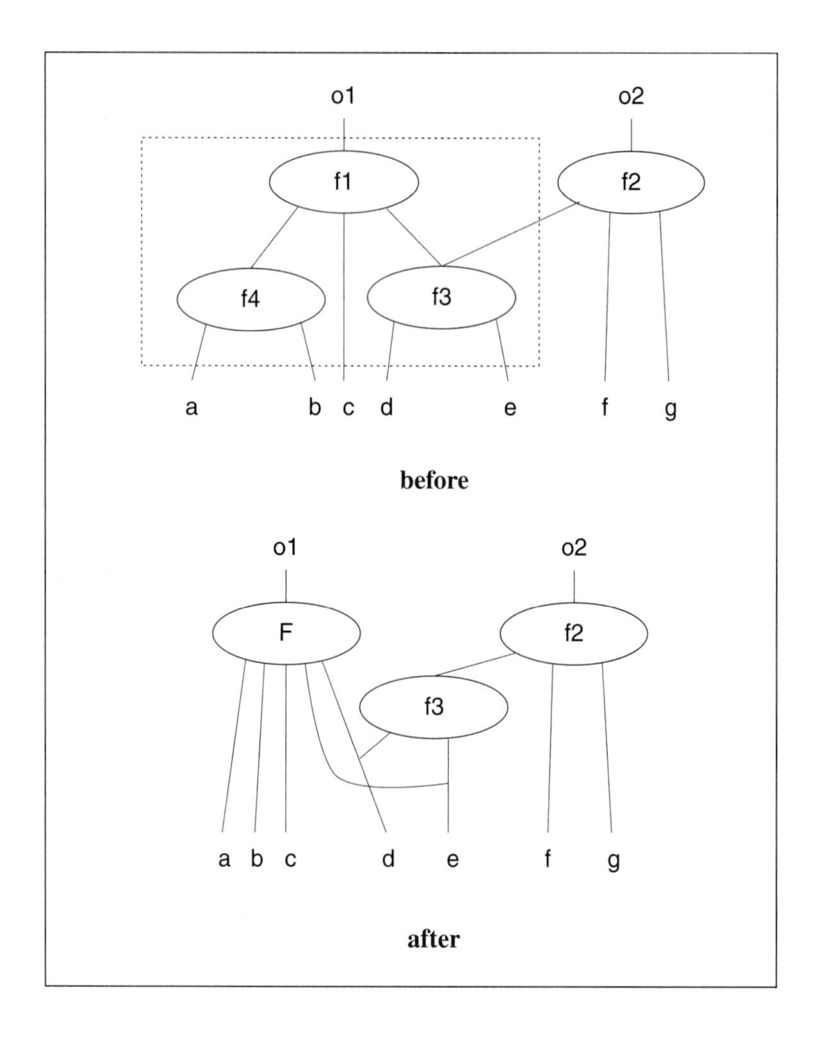

Figure 4-38 Partial collapsing of a Boolean network.

2. choose which factors to substitute into the network;
3. restructure the network by adding the new factors and rewriting the other functions to use those factors.

Boolean networks can also be optimized for delay. While delay estimates made on the network are rough, technology-dependent delay optimization can avoid creating logic with very long delay paths. As shown in Figure 4-39, a node can be re-factored to compensate for late-arriving signals. In the example, the latest-arriving signal must go through both *f1* and *f2* [Sin88]. We can collapse those two nodes into a single function, then extract new factors. But rather than choose a factor for its lit-

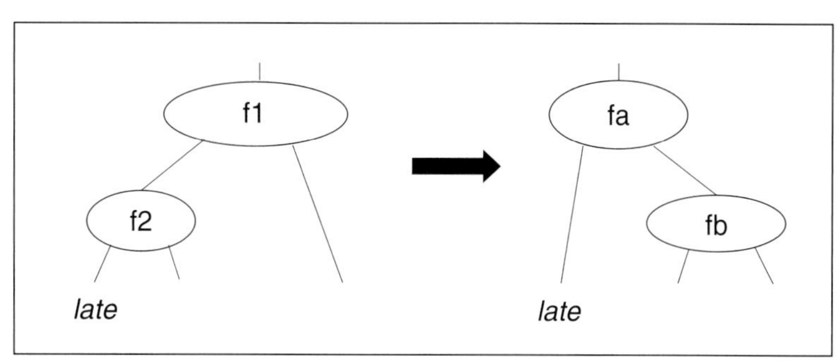

Figure 4-39 Factorization for delay.

eral savings, we choose factors that are functions of early-arriving signals.

collapsing

Unfortunately, our factorization algorithms are not perfect, so it may be possible to **collapse** several nodes, then apply node extraction to the larger node to find a better structure for the network. Figure 4-38 shows a network that has been partially collapsed, leaving f_2 intact. We can work from the collapsed function to find a new structure for F that could not be found starting from the given multilevel structure.

4.7.5 Technology-Dependent Logic Optimizations

Technology mapping transforms the Boolean network into a network of logic elements. We can optimize for both area and delay during technology mapping. An example is shown in Figure 4-40. The logic network is divided into three logic elements. Each LE has four or fewer inputs so each block of logic will fit into a typical LE. The NAND gate at the top of the logic network feeds two other gates and its functionality is replicated in both LEs. This costs nothing so long as the total function fits into the logic element.

technology mapping for LUTs

One method for technology mapping into lookup table architectures is a two-step approach [San93]. First, the technology-independent network is decomposed into nodes with no more than k inputs. This provides a feasible mapping onto lookup tables. The number of nodes is then reduced; various methods can be used to map nodes onto special features of the logic elements.

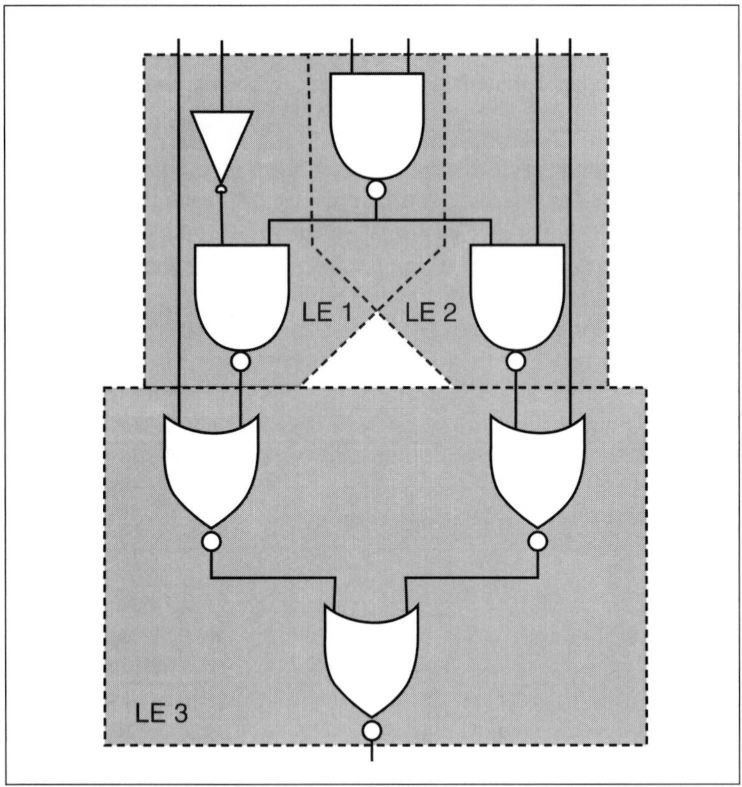

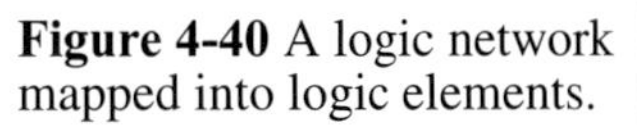
Figure 4-40 A logic network mapped into logic elements.

4.7.6 Logic Synthesis for FPGAs

Logic synthesis tools for FPGAs must be able to fit the logic into the primitives provided by the FPGA fabric. Murgai *et al.* [Mur90] and Francis *et al.* [Fra90] both observed that lookup table architectures can implement any function of n variables, where n is the number of inputs to the LUT, but that each function in the Boolean network can have no more than n inputs. The problem, therefore, is to cover the Boolean network with the minimum number of nodes, each of which has no more than n inputs. Murgai *et al.* used Roth-Karp decomposition along with other algorithms to solve this problem; Francis *et al.* based their technique on tree matching. Chen *et al.* [Che92] used mapping algorithms that did not require the network to be first broken into fanout-free trees.

FlowMap

The FlowMap algorithm [Con94] is a technology-mapping algorithm for LUT-based FPGA fabrics that optimizes for delay. The key step in

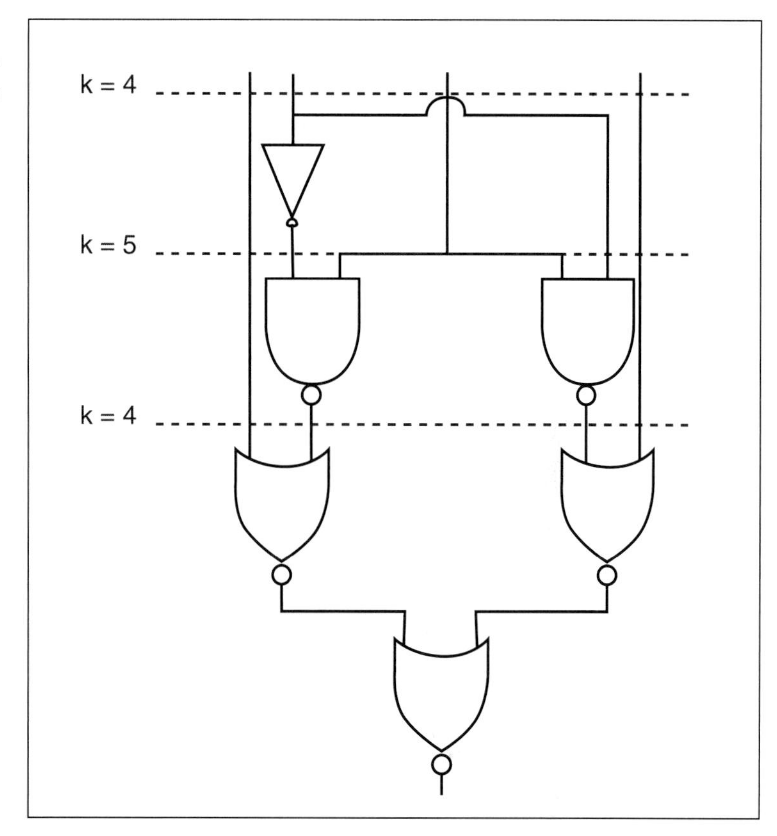

Figure 4-41 The number of inputs to a subnetwork does not change monotonically.

FlowMap is to break the network into LE-sized blocks by using an algorithm that finds a minimum-height k-feasible cut in the network. This algorithm uses network flow algorithms to optimally find this minimum-height cut. As the example of Figure 4-41 shows, the number of inputs to a block does not monotonically increase as we add logic to the block; this makes the computation of the feasible cut challenging. By finding a series of these cuts, FlowMap divides the network into blocks that can be fit into logic elements. It uses heuristics to maximize the amount of logic fit into each cut in order to reduce the number of logic elements required.

4.8 Physical Design for FPGAs

To be able to configure an FPGA we must place the logic into logic elements in the fabric and choose paths for each of the nets that must be implemented. Clearly, the way we do these steps helps determine whether our logic will fit into the FPGA. It also has a big effect on the delay and energy consumption of our logic.

placement and routing

Physical design is broken down into two major phases:

- **placement** determines the positions of the logic elements and I/O pads;
- **routing** selects the paths for connections between the logic elements and I/O pads.

These two phases interact—one placement of the logic may not be routable whereas a different placement of the same logic can be routed. However, this division allows us to make the physical design problem more tractable.

evaluating physical design quality

We are interested in using several different metrics to judge the quality of a placement or routing. Size is the most obvious metric. Ultimately, we are concerned about whether we can fit our complete design into the chip. We may also be interested in placing and routing a subsystem, in which case we want its placement and routing to be compact; we may also want the piece to conform to some shape. In FPGAs, size is closely tied to routing. The number of logic elements required is determined by logic synthesis. If we cannot find a legal routing for a given placement, we may need to change the placement. This won't change the number of LEs used, but we may have to spread out some LEs to find a less-crowded wiring area.

Delay is also a critical measure in most designs. Delay is measured relative to the specifications—slack is the actual measure of interest. A long delay that is not critical is not important, whereas a relatively short delay path that is critical must be carefully considered.

Detailed delay characteristics are somewhat expensive to compute, so tools generally use surrogates to estimate delay. Simpler measures are used in placement—since we don't yet have a routing, we can't expect very detailed delay measurements, but simple estimates such as distance help us compare different candidate placements. Increasingly detailed measures are used during various routing phases.

In FPGAs, design time may be an important metric for the placement and routing process. You may want to know whether a design will fit into a particular part, but not require an optimized routing at that point. You may also be interested in a quick-and-dirty physical design when building a prototype, in order to get some basic debugging information before refining the design. Some new placement and routing techniques have been developed to minimize design time.

randomness in physical design

Placement and routing tools need to make a great many decisions: which component to choose next, which direction to take for a wire, *etc.* Sometimes these decisions are guided by random number generators. We can take advantage of randomness to generate several different implementations of our logic. For example, the multi-pass place and route feature in the Xilinx ISE runs the physical design tools several times with different random seeds.

4.8.1 Placement

judging a placement

The separation of placement and routing immediately raises an important problem: how do we judge the quality of a placement? Because the wires are not routed at the end of placement, we cannot judge the placement quality by the two obvious measures: area of the layout and delay. We cannot afford to execute a complete routing for every placement to judge its quality; we need some metric which estimates the quality of a routing that can be made from a given placement. Different placement algorithms use different metrics, but a few simple metrics suggest important properties of placement algorithms.

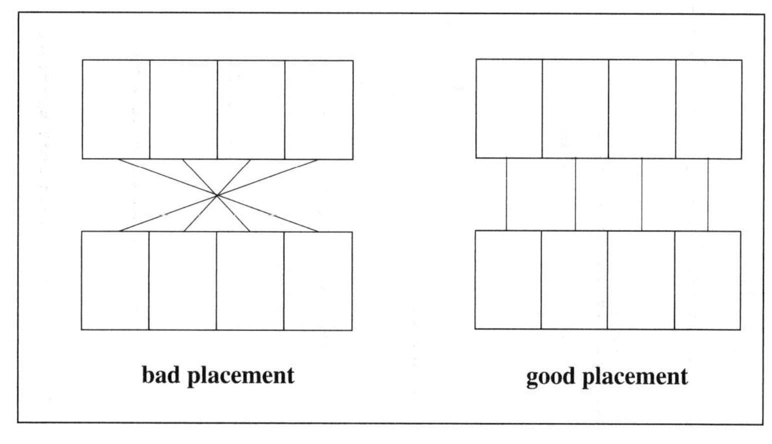

Figure 4-42 Total distance between connected pins as a placement metric.

One way to judge the area of a layout before routing is to measure the total distance between interconnections. Of course, we can't measure total wire length without routing the wires, but we can estimate the length of a wire by measuring the distance between the pins it connects. When straight lines between connected pins are plotted, the result is known as a **rat's nest** plot. Figure 4-42 shows two placements, one with a longer total rat's nest interconnection set and one with shorter total interconnections. While rat's nest distance is not absolutely correlated to

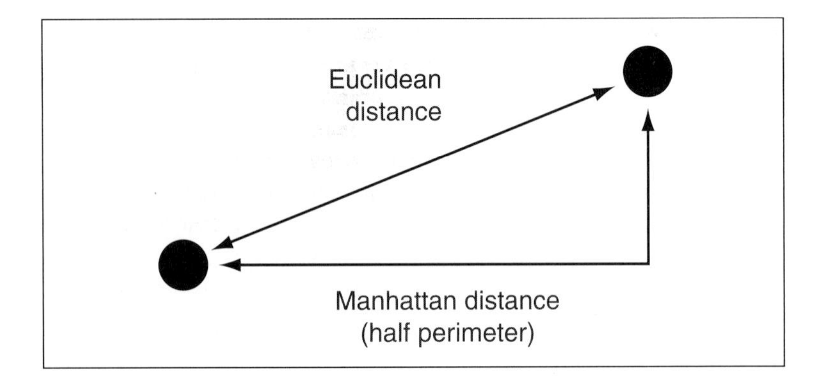

Figure 4-43 Wire distance metrics.

chip area, we would expect that a placement which gave much longer rat's nest connections would eventually result in a larger layout.

There are several ways to measure that distance as shown in Figure 4-43: the Euclidean distance of the direct line between the two; we can also measure the **Manhattan distance**, also known as the **half-perimeter**. Manhattan distance is more reflective of final wire length; it is also easier to compute because unlike Euclidean distance it does not require computing a square root.

clustering vs. partitioning

There are many algorithms for partitioning but these algorithms can generally be divided into two different approaches: bottom-up and top-down. Bottom-up methods are generally referred to as **clustering** methods—they cluster together nodes to create partitions. The growth of a cluster is shown in Figure 4-44. Top-down methods divide the nodes into groups that are then further divided. These methods are known as **partitioning** methods. Figure 4-45 shows how a placement can be constructed by recursively partitioning netlist (the connections between components are not shown for clarity).

placement by partitioning

The partitioning process for a **min-cut**, **bisecting** criterion is illustrated in Figure 4-46 [Bre77,Dun85]. The circuit is represented by a graph where nodes are the components and edges are connections between components. The goal is to separate the graph's nodes into two partitions that contain equal (or nearly equal) number of nodes and which have the minimum number of edges (wires) that cross the partition boundary to connect nodes (components) in separate partitions. With *A* and *B* in their positions in the **before** configuration, there are five wires crossing the partition boundary. Swapping the two nodes reduces the net cut count to one, a significant reduction. However, moves cannot be considered in isolation—it may require moving several nodes before it is clear whether a new configuration is better.

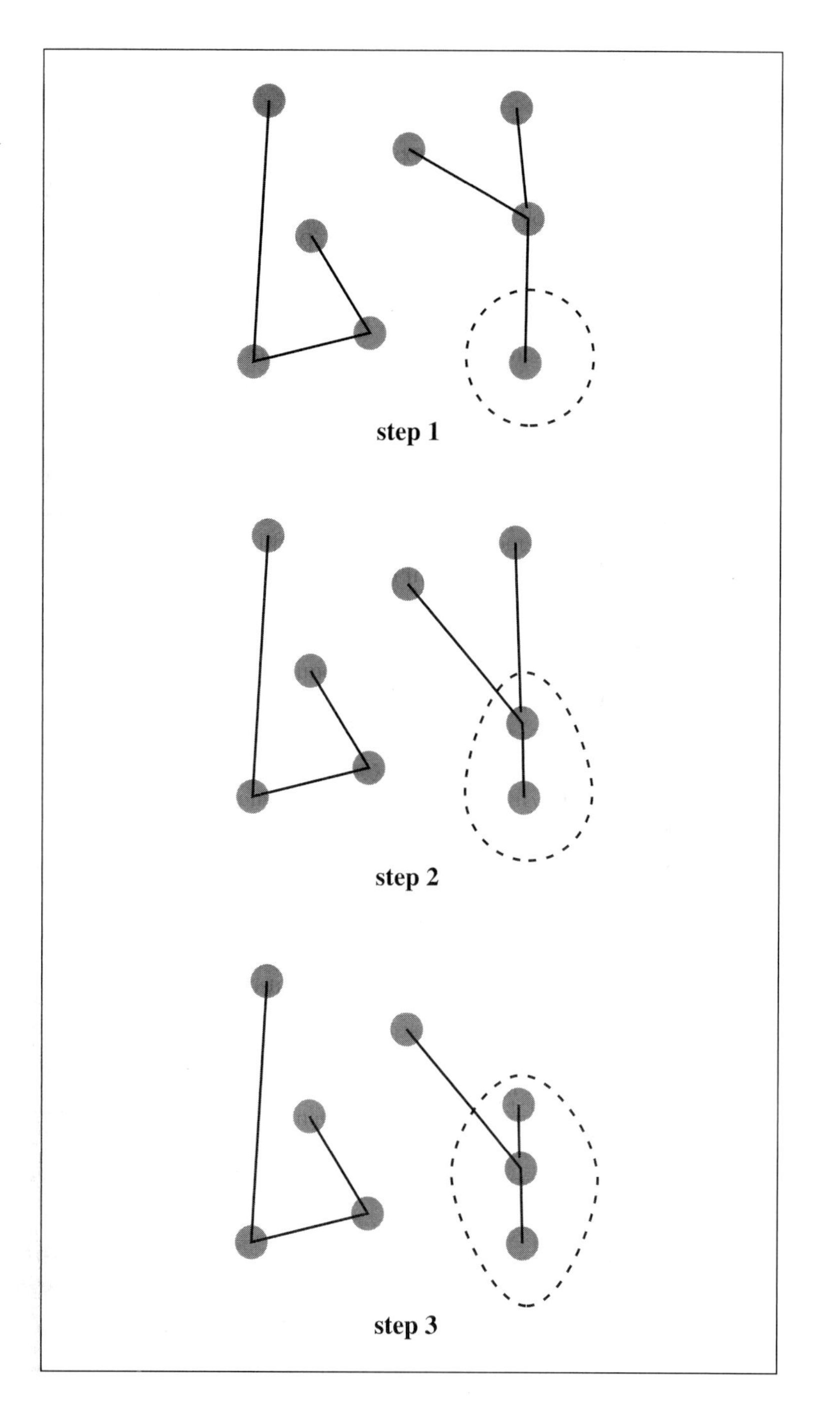

Figure 4-44 Placement by clustering.

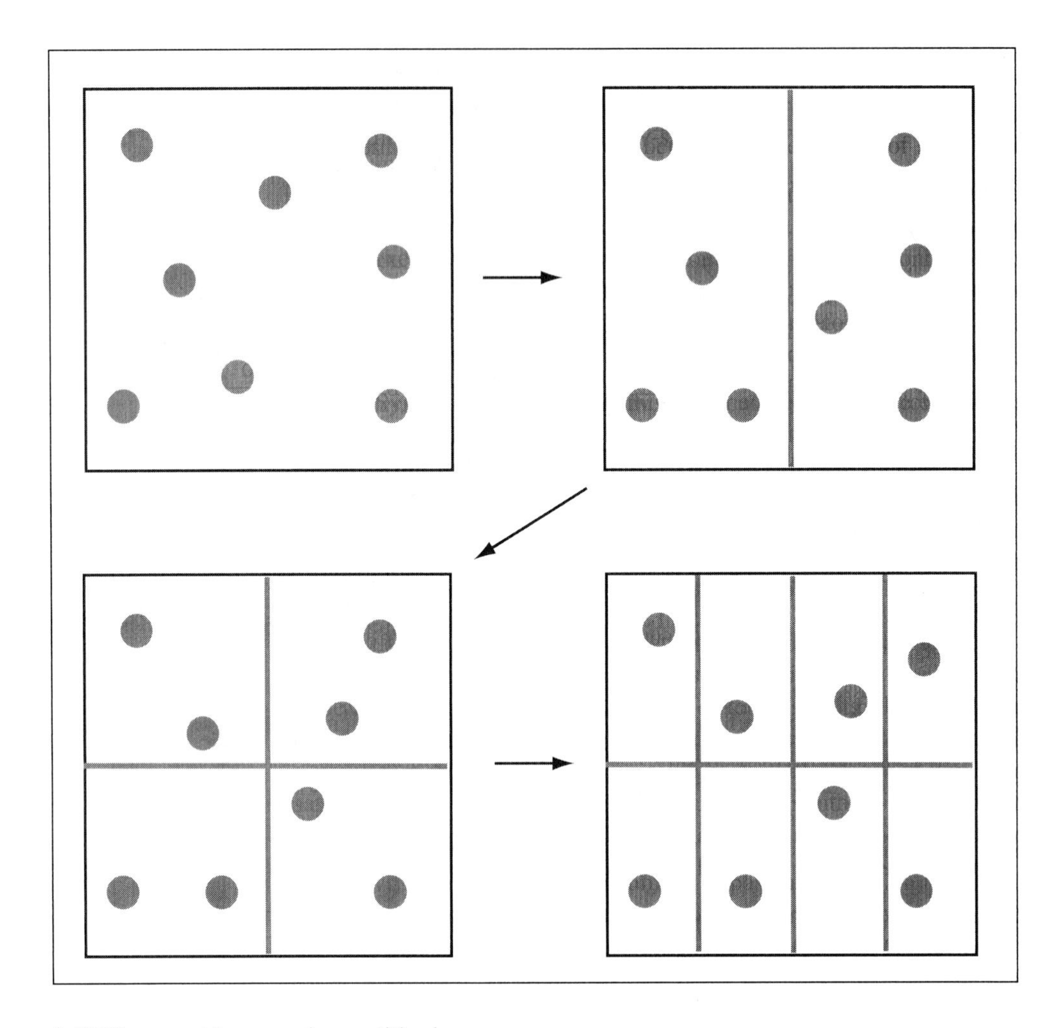

Figure 4-45 Placement by recursive partitioning.

The Kernighan-Lin algorithm [Ker70] is one well-known graph partitioning algorithm. It iteratively tries to find and move out-of-place elements in the partitions. Given an element e in partition 1 that might be moved into partition 2, we can compute some measures that indicate whether e should be moved:

- I_e is the number of nets connected to e that have all their connections in partition 1.
- E_e is the number of nets connected to e that have all their other connections in partition 2.

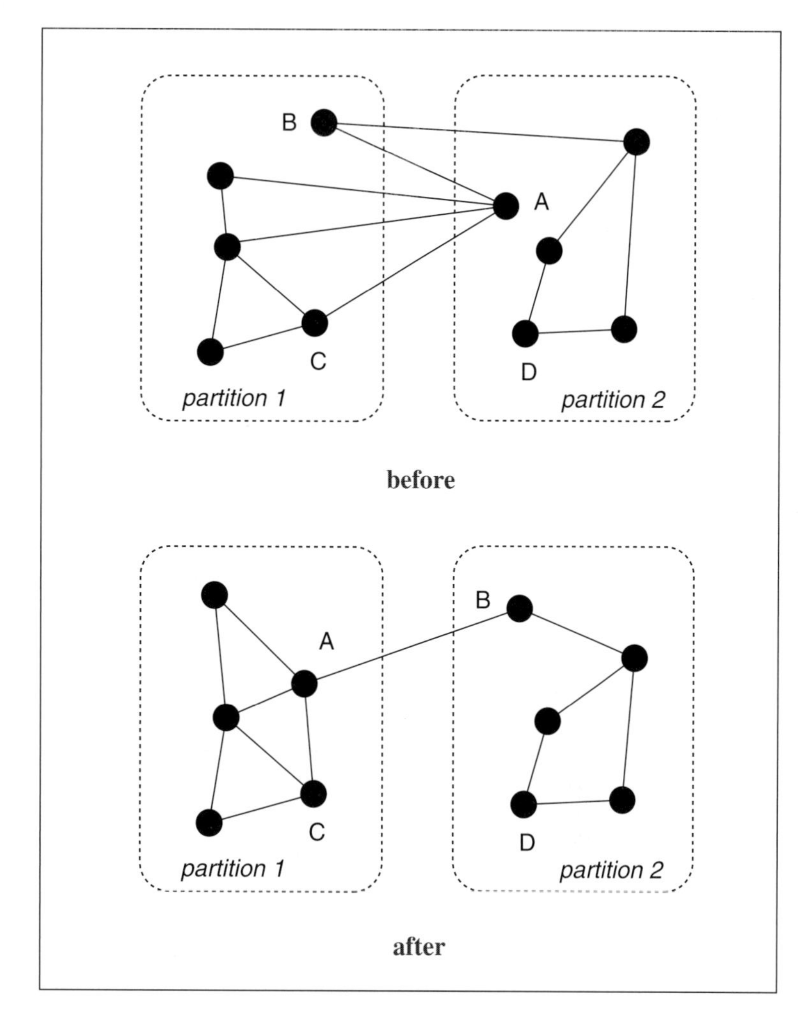

Figure 4-46 Placement by partitioning.

If e is moved from 1 to 2, all the nets counted in E_e will no longer cross the cut, while the nets counted in I_e will cross. If $D_e = E_e - I_e$, the gain obtained by exchanging e and f is given by $g = D_e + D_f - c_{ef}$, where c_{ef} is a correction factor to avoid double-counting shared nets. The Kernighan-Lin algorithm exchanges sets of nodes between partitions to reduce the number of crossing nets; by exchanging sets of nodes, rather than pairs, it can improve the partitioning even when moving any single pair of nodes will not give improvement.

Since dividing all the chip's components into two partitions doesn't give very fine direction for placement, the partitioning is usually repeated by creating subpartitions of the original partition, a process known as **hierarchical partitioning**. The levels in the hierarchy may be taken from the user's design description or by building clusters of a predetermined size.

placement by clustering

Clustering algorithms build groups of nodes from the bottom up in contrast to the top-down approach taken by partitioning. A small number of nodes are used to create initial clusters. Nodes are then added to the clusters based upon their connections with other nodes. Clustering algorithms may also use interchange steps to improve the cost of the cluster.

simulated annealing

Simulated annealing is another well-known optimization technique that has been used in placement. It is a stochastic technique that tries a large number of changes to the candidate placement, which helps the algorithm avoid being trapped in locally minimal solutions without reaching the global minimum. As the name implies, simulated annealing is analogous to annealing of metals. When a metal is heated up, then cooled down at the proper rate, the atoms naturally coalesce into a low energy structure, but the wrong cooling schedule can result in a bad structure. In simulated annealing, temperature is measured by the probability that an element will move during the current iteration. A cooling schedule is a sequence of temperatures/move probabilities over time. More elements move in an iteration when the temperature is high; fewer move at lower temperatures.

A simulated annealing placement takes place over many iterations. At each iteration, some logic elements may move, based upon probabilities parameterized by the current annealing temperature. Exactly where a selected element moves to depends on the details of the placement algorithm; it may occupy an empty spot or the algorithm may temporarily allow several logic elements to occupy the same position. The new placement is then evaluated, using a metric such as total wire length. The placer may decide to keep some moves and reject others. For the next iteration, the temperature is adjusted according to the annealing schedule and the procedure is repeated. Because moves are probabilistic, the next iteration's result may be worse than the previous iteration. But because many iterations are performed with smaller and smaller changes as time goes on, the placement tends to converge to a good result.

placement constraints

The designer may want to place constraints on the placement for any of several reasons:

- The I/O pin placement may be determined by board layout

constraints or compatibility.

- Some logic elements may be preplaced, particularly when macros are used to describe logic.

Constraints are usually satisfied by placing those elements first, then placing unconstrained elements around them. Constraints may make it difficult to find a routable layout, but those problems may not become apparent until the routing phase. We will discuss constraints in the Xilinx ISE in Example 4-15.

FPGA issues

Several example FPGA design tools illustrate methods used for FPGA placement. The Triptych FPGA [Bor95] has a fabric that emphasizes connections in the horizontal dimension. The placer for the Triptych FPGA [Ebe95] uses simulated annealing. It uses wire length and other routability measures as cost metrics during placement. Wire length tends to cluster logic elements as close together as possible, which means that wiring channels may be oversubscribed and the placement is unroutable. The Triptych placer uses two other cost measures to try to balance proximity and routability. A local routability cost function adds cost when it determines that a signal cannot be routed using global routing resources; using global routing resources would keep those resources from being used for a truly long net. Density smoothing tries to prevent local congestion. This metric looks at logic elements in groups of threes and counts the number of used LE inputs.

Sankar and Rose [San99] developed a fast placement algorithm. The first phase of their method uses bottom-up clustering. This phase randomly selects a logic element as a seed, then builds a cluster around the seed. The number of clustering levels and the size of the cluster at each level can be varied; generally larger clusters result in lower-quality results. Once the clusters have been constructed, they are placed around the chip, based on their connections to the I/O pads and the connections between the clusters. Simulated annealing is then used to improve the placement.

4.8.2 Routing

Routing selects paths for the connections that must be made between the logic elements and to the I/O pads. In an FPGA, the interconnection resources are predetermined by the architecture of the FPGA fabric. A connection must be made by finding a sequence of routing resources, all of which are otherwise unused, and which share connections such that a continuous path can be made from the source to the sink. This is a somewhat more complicated problem than the routing problem in custom VLSI because FPGA routing is more constrained.

global and detailed routing

Routing is generally divided into two phases:

- **Global routing** selects the general path through the chip but does not determine the exact wire segments to be used.
- **Detailed routing** selects the exact set of wires to be used for each connection.

As with the division into placement and routing, this is an approximation. Some global routing choices may lead to detailed routing failures. However, proper estimation of cost minimizes such failures, which in any case can often be solved by adjusting the global routing.

costs in routing

Routing has two major cost metrics: wire length and delay. Wire length approximates the utilization of routing resources; we clearly do not want to use more routing resources than are necessary for the wires, since we may want to use those routing resources later for other parts of the design. Delay may be measured by looking at the delay on paths with the largest number of levels of logic or by looking at nets whose delay is close to the maximum allowed value.

global routing

The principal job during global routing of FPGAs is to balance the requirements of various nets. Nets are routed one at a time, so the order in which nets are routed affects the final result. A net may have one of two problems: it may not be routable at all because there is no room available to make the connection; or it may take a path that incurs too much delay. These problems are harder to solve in FPGAs than in custom chip designs because the routing resources are pre-determined. Connections must be composed of pre-designed paths that may include larger wire segments than necessary, providing fairly coarse units of selection. Most FPGAs have several different categories of wiring, each with very different characteristics, and choosing the proper category for each wire without oversubscribing one of these resources only compounds the difficulty of the global routing problem.

wire ordering

Many ways have been developed to determine the order in which wires are routed. A good heuristic for an initial ordering is to route the most delay-critical nets first; one may also want to start with large-fanout nets since they consume many routing resources.

In general, a wire may be routed more than once before it finds its final route. **Ripup and reroute** is one simple strategy for choosing the order in which to route nets. When we route a net and find that we cannot find a satisfactory route for it—either because there isn't room for the connection or because the available resources would result in a net with too much delay—then we rip up the already-routed wires that are blocking a useful path and put those nets back in the queue of unrouted nets. We then route the original wire using the newly freed resources. If the design is truly unroutable, it is possible that this procedure never terminates as wires keep causing each other to be ripped up without any progress being made. Various tests can be used to detect this condition and abort the routing procedure. However, ripup and reroute is a vast improvement over a fixed ordering of nets.

routing with timing information

Frankle [Fra92] considered the problem of incorporating timing information into routing decisions. He used a method that, on each iteration of the global routing procedure, reallocated slacks. A variety of weights can be used to reallocate slacks: a constant factor; the number of fanouts of the net; delay per unit of load times the total capacitive load; fanout times delay sensitivity; and fanout times delay sensitivity times load capacitance.

maze routing for FPGAs

Rather than working with an open field, an FPGA router follows wire segments. Once a wire segment is chosen for the path, both ends of the wire segment are available for connections. We can model the FPGA interconnect as a graph whose edges are the available wire segments and the nodes are connections between wire segments. The job of the router is to traverse the graph and find the best connection between two nodes. Once a path has been used it can be marked so that those wire segments are not used again.

Nair [Nai87] developed a routing algorithm that is used in FPGAs. He assigned costs not based on Manhattan distance but on congestion. He also made sure that every net was ripped up and rerouted: the first routing pass was used to estimate congestion; the second routing pass used those estimates to redesign the wiring to minimize the effects of congestion.

The Triptych router [Ebe95] used a variation of Nair's method that applied a more gradual penalty for congestion than had Nair's original algorithm. The Triptych router calculated the cost of using a node n in a route as

$$c_n = (b_n + h_n)p_n . \quad \text{(EQ 4-21)}$$

where b_n is the base cost of using n, h_n is a cost related to the history of congestion at n during previous routing iterations, and p_n is related to the number of other signals currently using n. The p_n term is used to solve simple congestion problems. The term is gradually increased from one global routing iteration to the next in order to reduce the chance that a new signal will use a congested point. The h_n term is used to solve second-order congestion problems; it remembers the history of congested nodes and helps to push some of those signals away to other, less congested nodes.

The Triptych router adds delay to the cost of routing through node n with this formula:

$$C_n = A_{ij}d_n + (1 - A_{ij})c_n, \quad \text{(EQ 4-22)}$$

where

$$A_{ij} = D_{ij}/D_{max}. \quad \text{(EQ 4-23)}$$

A_{ij} is the ratio of the longest path containing the desired connection and the total system delay.

Swartz et al. [Swa98] developed a variation of this method that was designed to run very fast. They used depth-first search to direct the router to the desired target. They also restricted the number of routing segments added during wavefront expansion of high-fanout nets by only adding segments that are in the neighborhood of the target. This method can be viewed as an A* search algorithm.

4.9 The Logic Design Process Revisited

Now that we have studied logic design in more detail, we should review the logic design process we introduced in Section 4.2. This section uses several examples to show more detail about some steps in the logic design process using a larger example: a 16×16 array multiplier.

major phases

We can break logic synthesis for FPGAs into several major steps:

- **translation** extracts a combinational logic expression from our hardware description language;
- **logic synthesis** optimizes the extracted logic, including mapping it into the logic primitives of the FPGA;
- **placement and routing** builds a physical realization of the synthesized logic by placing the logic blocks into logic elements and choosing interconnect resources for the connections between LEs;
- **configuration generation** creates the data necessary to pro-

gram the device.

Different people and tools may use somewhat different terminology for these steps. The Xilinx ISE uses these terms in its design process:

- **synthesis** extracts logic from the hardware description language model;
- **translation** prepares for logic synthesis, including managing the design hierarchy;
- **mapping** optimizes the logic and fits it into the logic elements;
- **place and route** creates a physical realization;
- **program file generation** creates the information needed to program the device.

reports

The tools used for most of these stages generate reports that tell us what they did. It takes some practice to learn how to interpret these reports. But they can be very valuable aids in determining what went right in our logic implementation, what went wrong, and what we can do to fix the problems.

design constraints

The Verilog or VHDL model captures the function of the logic. We also need to specify non-functional requirements. Those requirements will be used by several different phases of design. Each phase tries to balance concerns to satisfy all the design constraints to the level of accuracy possible at that stage in the design process. We are generally concerned with three types of constraints:

- timing;
- area;
- pinout.

Many tools don't allow us to directly specify power consumption as a requirement. We must instead try to use timing or area constraints as substitutes or we must rewrite our hardware description to use different logic with better power consumption characteristics.

In the next example we will synthesize our 16×16 multiplier without any design constraints in order to review the process and some of the reports generated during that process. The Verilog source for this multiplier is available on the book Web site or you may construct it from the 4×4 multiplier in this chapter as an exercise. It may help you to run your own design using your tools to follow along with the example.

Example 4-12 Synthesis of a 16 x 16 multiplier

In this case we want to synthesize our multiplier without imposing any strict timing or area constraints onto it. However, we will look at some of the reports generated by the Xilinx ISE during implementation.

The synthesis phase extracts logic from our HDL model. The results of this phase are still very far from the final configuration file so the contents of the synthesis report are fairly basic. That report includes:

- a summary of the options specified for synthesis;
- a list of the source files and modules compiled;
- any syntax errors found in those files;
- synthesis results for each module, such as the adders extracted from each module;
- a summary of the low-level synthesis results;
- a final report that lists a variety of design statistics.

The final report may be the most useful part of the report. It tells us how many CLBs, IOBs, *etc.* were used, as well as how many macros of different types were used.

The translation phase prepares the design for mapping. It reads the design constraints and expands the design hierarchy. As a result, the report from this phase is short.

This phase also generates a **post-translation simulation model**. This is the first of several HDL models that can be generated during implementation; it represents the design after these initial implementation phases. The model is built from primitives that more closely reflect the structure of the logic to be used in the FPGA (though the structure of this model may not accurately reflect how the FPGA is finally configured). Here is an example component from the post-translation model:

```
X_LUT4 \p12_Madd__n0015_Mxor_Result_Xo<1>1  (
  .ADR0(x_7_IBUF),
  .ADR1(y_13_IBUF),
  .ADR2(c12[7]),
  .ADR3(row12[8]),
  .O(row13[7])
);
```

What is the difference between this model and your original Verilog model? Hopefully none so far as function goes. The two models should produce identical functional results. (This is a good place to point out that even though CAD tools do have bugs, most problems encountered by designers are due to driver error, not to the tools. If you find a discrepancy between your source model and the post-translation model's simulation results, it may be because some construct in your model does not translate in the way you thought it should.)

Mapping finishes the logic synthesis process. The mapping report gives us many statistics about the final logic design. For example, it summarizes the logic utilization:

```
Design Summary
--------------
Number of errors:      0
Number of warnings:    0
Logic Utilization:
  Number of 4 input LUTs:          501 out of   1,024   48%
Logic Distribution:
  Number of occupied Slices:        255 out of     512   49%
  Number of Slices containing only related logic:     255 out of
255  100%
  Number of Slices containing unrelated logic:        0 out of    255
0%
        *See NOTES below for an explanation of the effects of unre-
lated logic
Total Number 4 input LUTs:          501 out of   1,024   48%

  Number of bonded IOBs:             64 out of      92   69%

Total equivalent gate count for design:  3,006
Additional JTAG gate count for IOBs:  3,072
Peak Memory Usage:  64 MB
```

This summary helps us check the feasibility of our design. For example, if we compile a piece of the design and find that it uses more CLBs than we expected, we may have to rethink the logic in order to fit it into the allowable space. The report goes on to provide more detail, such as a summary of the IOB characteristics, timing information, and usage of other design resources. If we run into problems with the design we can check this report to make sure that the tools agree with us about properties of the design, such as whether a particular port should be an input or output.

This phase also generates a static timing analysis report. This report tells us about the delay along various paths. Here is a summary of results for overall timing performance:

```
Timing constraint: TS_P2P = MAXDELAY FROM TIMEGRP
"PADS" TO TIMEGRP "PADS" 99.999 uS  ;

 20135312 items analyzed, 0 timing errors detected. (0 setup
errors, 0 hold errors)
 Maximum delay is  20.916ns.
-----------------------------------------------------------------------------
```

As we will see in later examples, this part of the report will warn us if we fail to meet our timing requirements. A later part of the report gives us delays along particular paths:

```
Data Sheet report:
-----------------
All values displayed in nanoseconds (ns)

Pad to Pad
---------------+-------------------+-----------+
Source Pad  |Destination Pad|  Delay  |
---------------+-------------------+-----------+
x<0>          |p<0>              |   5.824|
x<0>          |p<10>             |  10.675|
x<0>          |p<11>             |  11.214|
x<0>          |p<12>             |  11.753|
```

These detailed timing reports help us determine how difficult it was for the tools to meet our timing requirements and how much further we can push them.

This phase also generates a new simulation model, the **post-map simulation model**. Like the post-translation model, this synthesized model is functionally equivalent to our original model but it more closely matches the structure of the logic produced thus far.

The place and route phase puts the synthesized logic into the FPGA fabric. The report generated by this phase includes summary information about the design as well as a report of the steps it took during placement and routing. For example, here is the description of the routing process:

```
Phase 1: 1975 unrouted;      REAL time: 11 secs

Phase 2: 1975 unrouted;      REAL time: 11 secs

Phase 3: 619 unrouted;      REAL time: 12 secs

Phase 4: 619 unrouted; (0)      REAL time: 12 secs
```

Phase 5: 619 unrouted; (0) REAL time: 12 secs

Phase 6: 619 unrouted; (0) REAL time: 12 secs

Phase 7: 0 unrouted; (0) REAL time: 12 secs

Routing generally requires several phases to complete all the required connections. Ripup and rerouting helps the router improve the design.

The report also tells us that there were no unrouted signals:

The NUMBER OF SIGNALS NOT COMPLETELY ROUTED for this design is: 0

This is good news. If some nets were left unrouted we would have to do something—either change our HDL description or our design constraints—to try to make the routing complete.

The place and route report gives us some basic timing, but this phase also produces a more detailed static timing report. It is similar in form to the report generated after mapping. Here is the summary line for timing after routing:

```
Timing constraint: TS_P2P = MAXDELAY FROM TIMEGRP
"PADS" TO TIMEGRP "PADS" 99.999 uS ;

 20135312 items analyzed, 0 timing errors detected. (0 setup
errors, 0 hold errors)
 Maximum delay is 38.424ns.
--------------------------------------------------------------------------------
```

Note that the delay given in this report is considerably longer than the delay given in the post-map report. That is because the system didn't have any information about interconnect delay until after the place and route phase was complete. Pre-physical design timing information gives us some information about delays but should be taken with a grain of salt.

The tools also provide a report on programming file generation that reports on the options used, etc.

The next example looks at what happens as we tighten the delay constraints.

Example 4-13 The 16 x 16 multiplier with delay constraints

Let's add a delay constraint to our multiplier to see what happens. For simplicity let's add an overall constraint that applies to all paths. We will use the timing constraints editor to add the constraint known as a pad-to-pad constraint:

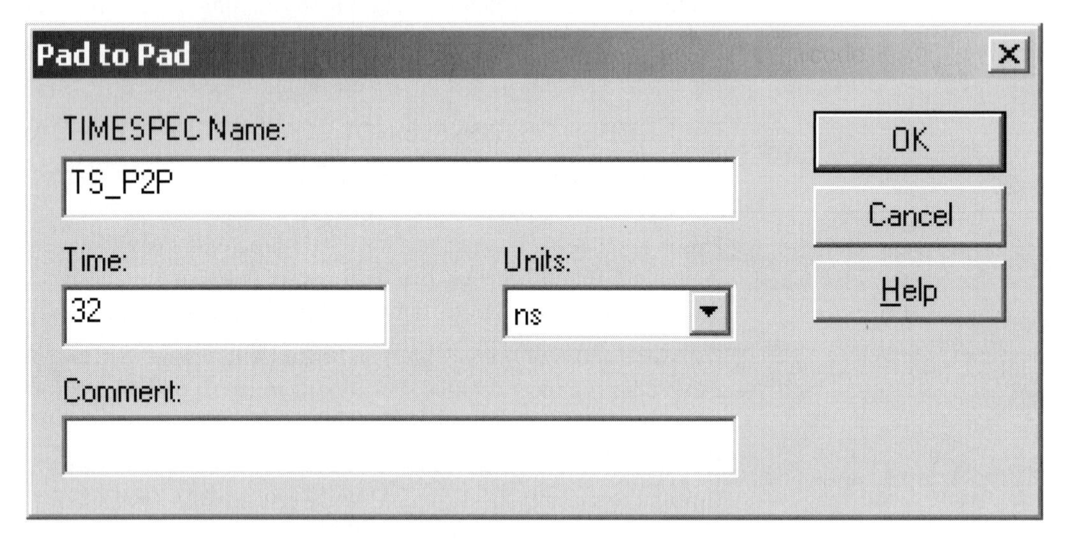

We have set the maximum delay from inputs to outputs to be 32 ns. This is less than the delay reported when we synthesized without constraints in the last example.

Here is the summary from the post-map static timing report:

```
Timing constraint: TS_P2P = MAXDELAY FROM TIMEGRP
"PADS" TO TIMEGRP "PADS" 32 nS  ;

 20135312 items analyzed, 0 timing errors detected. (0 setup
errors, 0 hold errors)
 Maximum delay is  20.916ns.
--------------------------------------------------------------------------------
```

This is the same delay as was reported for the unconstrained design. This particular design is very regular and has limited opportunities for logic synthesis to change delays by restructuring logic. In another design, such as a controller, we could see the post-map, pre-place-and-route delays change.

Here is the post-place-and-route static timing report:

```
Timing constraint: TS_P2P = MAXDELAY FROM TIMEGRP
"PADS" TO TIMEGRP "PADS" 32 nS  ;

 20135312 items analyzed, 0 timing errors detected. (0 setup
errors, 0 hold errors)
 Maximum delay is  31.984ns.
--------------------------------------------------------------------------------
```

This delay is considerably smaller than the delay we achieved in the last example. The system successfully met our timing constraint.

But how much faster will the system run? The last implementation barely met our timing constraint. The tools generally try to meet the delay goal as closely as possible to minimize area, but the closeness of this deadline suggests that we are close to the minimum delay. Let's try a tighter deadline, 25 ns, to see what happens. Here is the result of post-place-and-route static timing analysis with the new constraint:

```
Timing constraint: TS_P2P = MAXDELAY FROM TIMEGRP
"PADS" TO TIMEGRP "PADS" 25 nS  ;

 20135312 items analyzed, 11 timing errors detected. (11 setup
errors, 0 hold errors)
 Maximum delay is  31.128ns.
--------------------------------------------------------------------------------
```

The tools improved delay a little but didn't satisfy our overall timing goal. This summary states that 11 timing errors were detected. The report goes on to give details on those timing errors. Here is the timing error report for one of those paths:

```
Slack:              -6.128ns (requirement - data path)
  Source:             y<0> (PAD)
  Destination:          p<30> (PAD)
  Requirement:            25.000ns
  Data Path Delay:      31.128ns (Levels of Logic = 31)

  Data Path: y<0> to p<30>
    Location             Delay type          Delay(ns)  Physical Resource
                                                     Logical Resource(s)
    -------------------------------------------------  -------------------
    K5.I                 Tiopi                   0.825   y<0>
                                                     y<0>
                                                     y_0_IBUF
    SLICE_X2Y11.G4       net (fanout=31)         1.792   y_0_IBUF
    SLICE_X2Y11.Y        Tilo                    0.439   c2<5>
```

```
p0_Madd__n0017_Mxor_Result_Xo<1>1
   SLICE_X2Y11.F4     net (fanout=2)      0.304   row1<6>
   SLICE_X2Y11.X      Tilo                0.439   c2<5>
                                          p1_Madd__n0019_Cout1
   SLICE_X5Y16.F3     net (fanout=2)      0.784   c2<5>
   SLICE_X5Y16.X      Tilo                0.439   c3<5>
                                          p2_Madd__n0019_Cout1
   SLICE_X2Y18.G4     net (fanout=2)      0.668   c3<5>
   SLICE_X2Y18.Y      Tilo                0.439   row5<4>

p3_Madd__n0019_Mxor_Result_Xo<1>1
   SLICE_X7Y20.G4     net (fanout=2)      0.979   row4<5>
   SLICE_X7Y20.Y      Tilo                0.439   c8<4>
                                          p4_Madd__n0021_Cout1
   SLICE_X7Y12.F3     net (fanout=2)      0.647   c5<4>
   SLICE_X7Y12.X      Tilo                0.439   c6<4>
                                          p5_Madd__n0021_Cout1
   SLICE_X2Y20.F4     net (fanout=2)      1.039   c6<4>
   SLICE_X2Y20.X      Tilo                0.439   row7<4>

p6_Madd__n0021_Mxor_Result_Xo<1>1
   SLICE_X7Y22.F4     net (fanout=2)      0.622   row7<4>
   SLICE_X7Y22.X      Tilo                0.439   c8<3>
                                          p7_Madd__n0023_Cout1
   SLICE_X5Y23.F3     net (fanout=2)      0.383   c8<3>
   SLICE_X5Y23.X      Tilo                0.439   c9<3>
                                          p8_Madd__n0023_Cout1
   SLICE_X10Y25.G3    net (fanout=2)      0.939   c9<3>
   SLICE_X10Y25.Y     Tilo                0.439   c11<2>

p9_Madd__n0023_Mxor_Result_Xo<1>1
   SLICE_X10Y25.F3    net (fanout=2)      0.055   row10<3>
   SLICE_X10Y25.X     Tilo                0.439   c11<2>
                                          p10_Madd__n0025_Cout1
   SLICE_X8Y23.F4     net (fanout=2)      0.324   c11<2>
   SLICE_X8Y23.X      Tilo                0.439   c12<2>
                                          p11_Madd__n0025_Cout1
   SLICE_X8Y19.F4     net (fanout=2)      0.339   c12<2>
   SLICE_X8Y19.X      Tilo                0.439   c13<2>
                                          p12_Madd__n0025_Cout1
   SLICE_X8Y18.G3     net (fanout=2)      0.004   c13<2>
   SLICE_X8Y18.Y      Tilo                0.439   row15<2>
```

```
                                        p13_Madd__n0025_Cout1
    SLICE_X8Y18.F4      net (fanout=2)       0.024   c14<2>
    SLICE_X8Y18.X       Tilo                 0.439   row15<2>

 p14_Madd__n0025_Mxor_Result_Xo<1>1
    SLICE_X8Y15.G1      net (fanout=3)       0.889   row15<2>
    SLICE_X8Y15.Y       Tilo                 0.439   l_carry<2>
                                        Ker69511
    SLICE_X8Y15.F3      net (fanout=2)       0.040   N6953
    SLICE_X8Y15.X       Tilo                 0.439   l_carry<2>
                                        l_Madd__n0012_Cout1
    SLICE_X13Y14.G3     net (fanout=3)       0.640   l_carry<2>
    SLICE_X13Y14.Y      Tilo                 0.439   l_carry<4>
                                        Ker69811
    SLICE_X13Y14.F3     net (fanout=1)       0.035   N6983
    SLICE_X13Y14.X      Tilo                 0.439   l_carry<4>
                                        l_Madd__n0010_Cout1
    SLICE_X12Y16.G3     net (fanout=3)       0.301   l_carry<4>
    SLICE_X12Y16.Y      Tilo                 0.439   l_carry<6>
                                        Ker69761
    SLICE_X12Y16.F4     net (fanout=1)       0.002   N6978
    SLICE_X12Y16.X      Tilo                 0.439   l_carry<6>
                                        l_Madd__n0008_Cout1
    SLICE_X12Y17.G4     net (fanout=3)       0.080   l_carry<6>
    SLICE_X12Y17.Y      Tilo                 0.439   l_carry<8>
                                        Ker69711
    SLICE_X12Y17.F3     net (fanout=1)       0.035   N6973
    SLICE_X12Y17.X      Tilo                 0.439   l_carry<8>
                                        l_Madd__n0006_Cout1
    SLICE_X14Y5.G3      net (fanout=3)       0.774   l_carry<8>
    SLICE_X14Y5.Y       Tilo                 0.439   l_carry<10>
                                        Ker69661
    SLICE_X14Y5.F3      net (fanout=1)       0.035   N6968
    SLICE_X14Y5.X       Tilo                 0.439   l_carry<10>
                                        l_Madd__n0004_Cout1
    SLICE_X14Y4.G3      net (fanout=3)       0.024   l_carry<10>
    SLICE_X14Y4.Y       Tilo                 0.439   l_carry<12>
                                        Ker69611
    SLICE_X14Y4.F4      net (fanout=1)       0.002   N6963
    SLICE_X14Y4.X       Tilo                 0.439   l_carry<12>
                                        l_Madd__n0002_Cout1
    SLICE_X12Y6.G4      net (fanout=3)       0.365   l_carry<12>
    SLICE_X12Y6.Y       Tilo                 0.439   p_30_OBUF
```

```
I_Madd__n0000_Mxor_Result_Xo<0>1
   SLICE_X12Y6.F4      net (fanout=1)       0.002
I_Madd__n0000_Mxor_Result_Xo<0>
   SLICE_X12Y6.X       Tilo                 0.439   p_30_OBUF

I_Madd__n0000_Mxor_Result_Xo<1>1
   K9.O1             net (fanout=1)      1.085  p_30_OBUF
   K9.PAD            Tioop             4.360  p<30>
                                        p_30_OBUF
                                        p<30>
 ----------------------------------------------  ---------------------------
   Total                        31.128ns (17.916ns logic, 13.212ns
route)
                                     (57.6% logic, 42.4% route)
```

This report traces through the entire path. It shows us all the logic on the path and tells us how much of the delay along the path was due to logic and how much was due to interconnect. This information can help us modify the logic or the physical design to improve the delay.

The next example looks at the power report.

Example 4-14 Power analysis of the 16 x 16 multiplier

The power report is generated after place and route. Here is the report for our multiplier:

```
Power summary:                        I(mA)   P(mW)
----------------------------------------------------------------
Total estimated power consumption:              333
                   ---
            Vccint 1.50V:      0       0
            Vccaux 3.30V:    100     330
            Vcco33 3.30V:      1       3
                   ---
              Inputs:      0       0
               Logic:      0       0
             Outputs:
               Vcco33      0       0
             Signals:      0       0
                   ---
   Quiescent Vccaux  3.30V:      100     330
   Quiescent Vcco33  3.30V:        1       3
```

```
Thermal summary:
-------------------------------------------------------------
  Estimated junction temperature:              36C
            Ambient temp:  25C
              Case temp:  35C
             Theta J-A:  34C/W

Decoupling Network Summary:     Cap Range (uF)      #
-------------------------------------------------------------
Capacitor Recommendations:
Total for     Vccint :                        4
                        470.0  - 1000.0 :  1
                       0.0100  - 0.0470 :  1
                       0.0010  - 0.0047 :  2
                           ---
Total for     Vccaux :                        4
                        470.0  - 1000.0 :  1
                       0.0100  - 0.0470 :  1
                       0.0010  - 0.0047 :  2
                           ---
Total for     Vcco33 :                       10
                        470.0  - 1000.0 :  1
                        0.470  -  2.200 :  1
                       0.0470  - 0.2200 :  2
                       0.0100  - 0.0470 :  3
                       0.0010  - 0.0047 :  3
```

The first entries tell us the estimated current and power consumption for the design. They also provide the estimated temperature at which the part will operate; this helps us determine whether we need additional cooling. The last part of the report tells us how much decoupling capacitance to use across the chip's power supply pins. A decoupling capacitor goes between the power supply terminal and helps filter out power supply noise.

The next example describes how we can improve the area of our design.

Example 4-15 Area optimization of the 16 x 16 multiplier

The Xilinx ISE offers editors for the placed and routed design. We can view the placed design by double-clicking on *View/Edit Placed Design (Floorplanner)* in the Processes window pane:

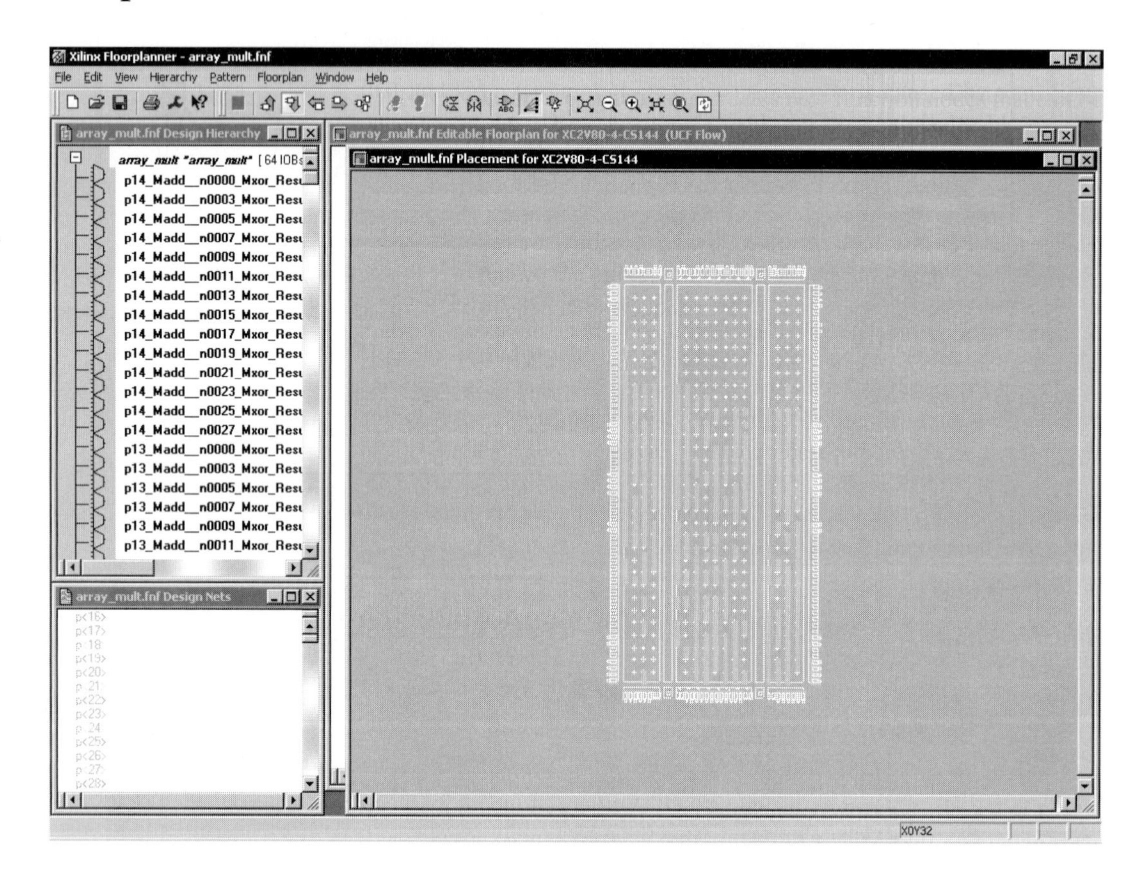

The *Design Hierarchy* pane in the floorplanner tool shows us the various components in the design. These are mapped CLB-level components. They are represented by the green rectangles in the Placement window. That window also shows a view of the positions of the CLBs and IOBs in the FPGA that we selected as a target.

One way to get a sense for the quality of the placement is to look at the positions of some of the components. If we click on one of the components in the *Design Hierarchy*, the floorplanner will show us a rat's nest view of the wires connected to that component:

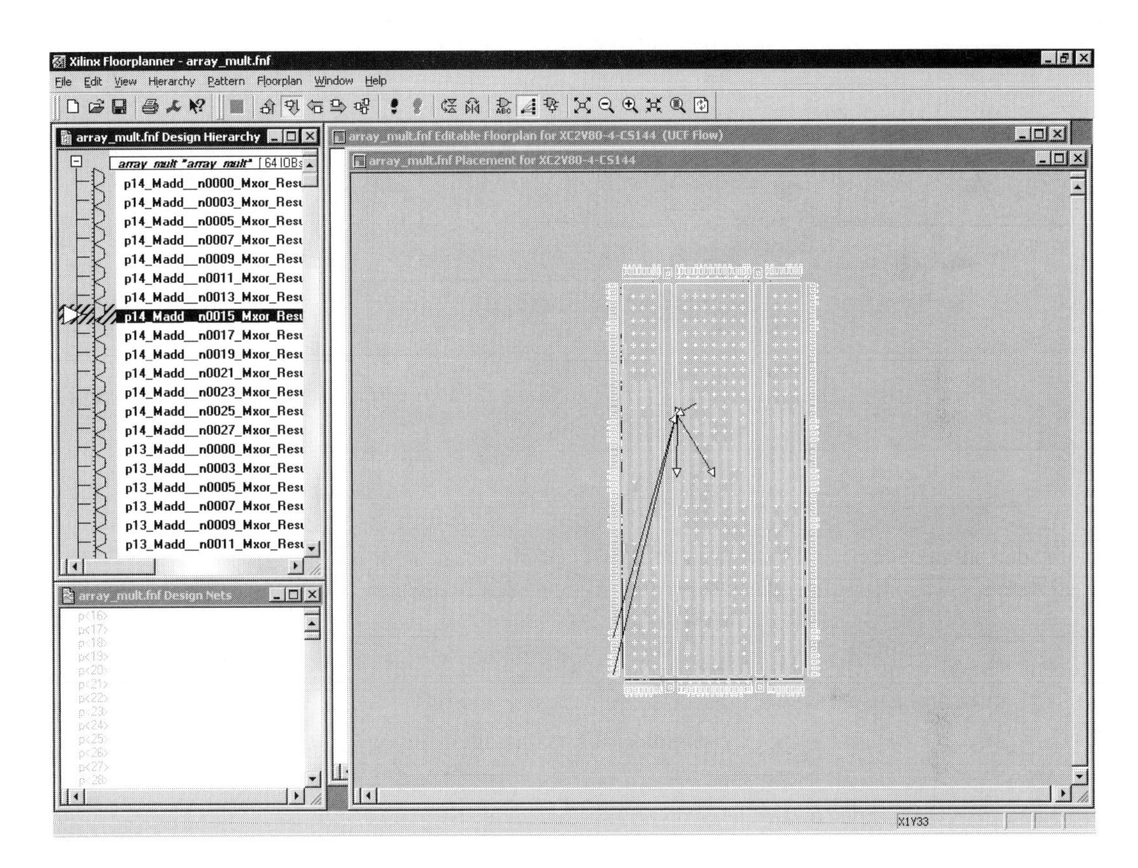

By checking a few components we can see that they are placed fairly randomly around the chip. The placement algorithm did not find a regular placement to match the regular logical structure of the array multiplier.

We can also check the routing using the FPGA editor, which we start by double-clicking on View/Edit Routed Design (FPGA Editor) in the

Project Navigator Processes window. Here is the FPGA editor window showing the routing of the design:

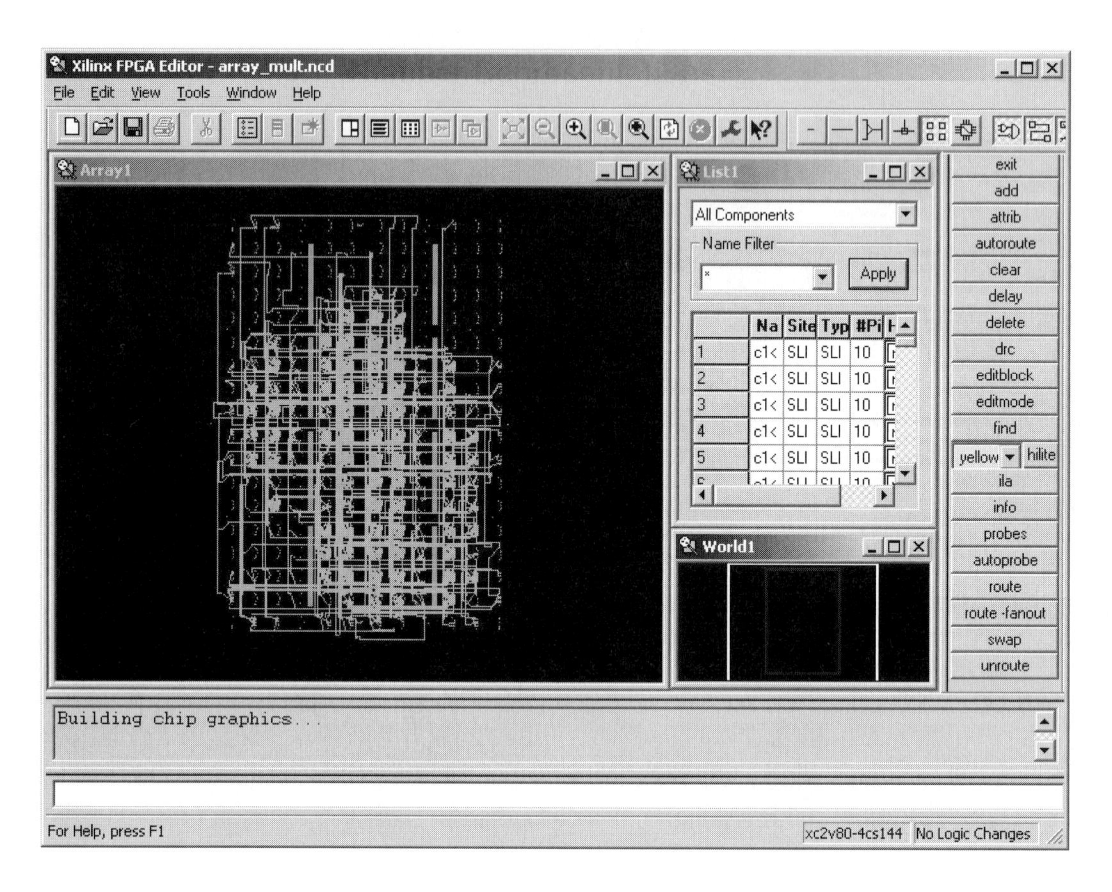

We can help the tools find a better placement by providing some constraints. We will first enter some constraint identifiers into our Verilog source. The constraint information takes the form of Verilog comments:

```
// synthesis attribute rloc of p0 is X0Y0
multrow p0(row0,x,y[1],y[0],c0,row1,c1);
```

In this case, *rloc* stands for relative location. This statement tells the system that the component *p0* needs a relative location attribute with a relative position of X=0, Y=0. We need to add one of these synthesis attribute lines for each component whose placement we want to control. The attribute takes the form of a comment because Verilog does not provide a way to describe these attributes in the language. If you write a comment that happens to start with *synthesis attribute* the tools will interpret them as a command and complain.

We will then use PACE, the constraints editor tool, to tell the system how these constraints should be arranged around the chip. We will start it by double-clicking on *Create Area Constraints* in the *Processes* pane. The PACE tool allows us to specify several types of physical constraints, including constraints on the logic itself and on the I/O pins. Here is the window for PACE:

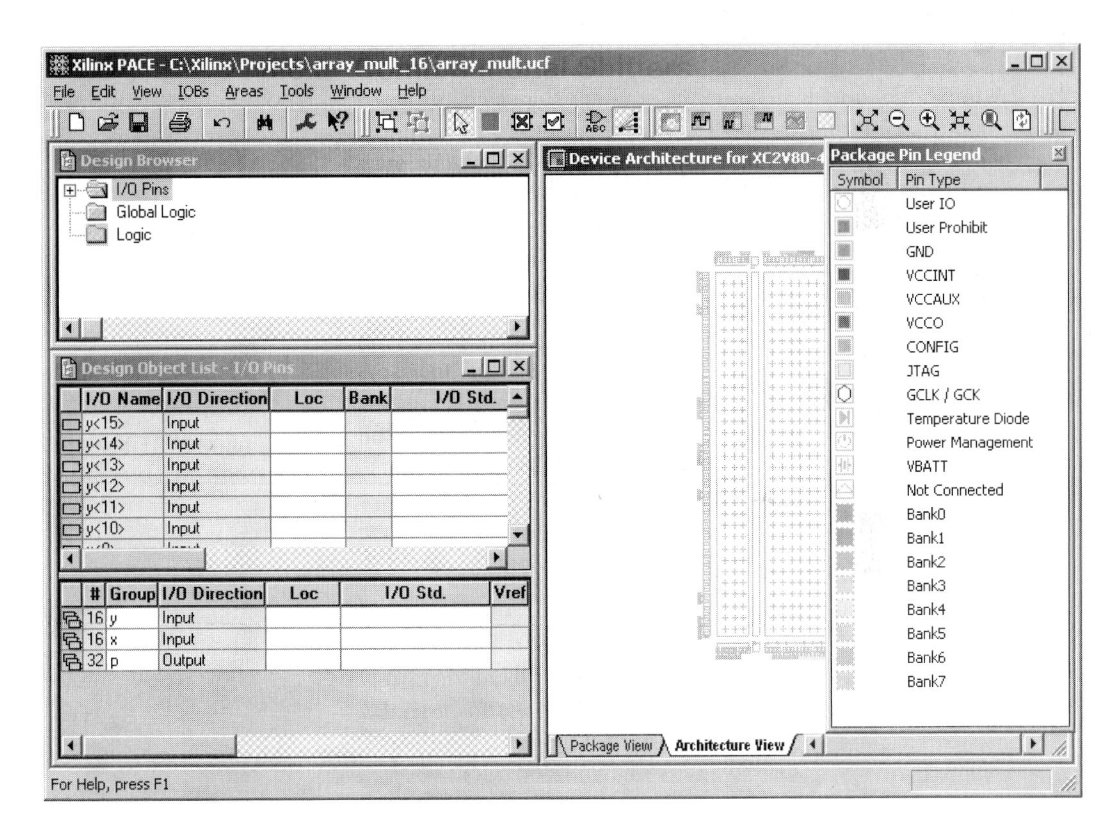

Although we will concentrate on logic placement, this tool also allows you to constrain the placement of logic as well as the assignment of chip inputs and outputs to IOBs. Pin placement is often important to the design of the printed circuit board on which the FPGA will be used.

To view and manipulate the constraints we defined in our Verilog source, we go the the *Design Browser* pane in PACE and selecting *Logic*:

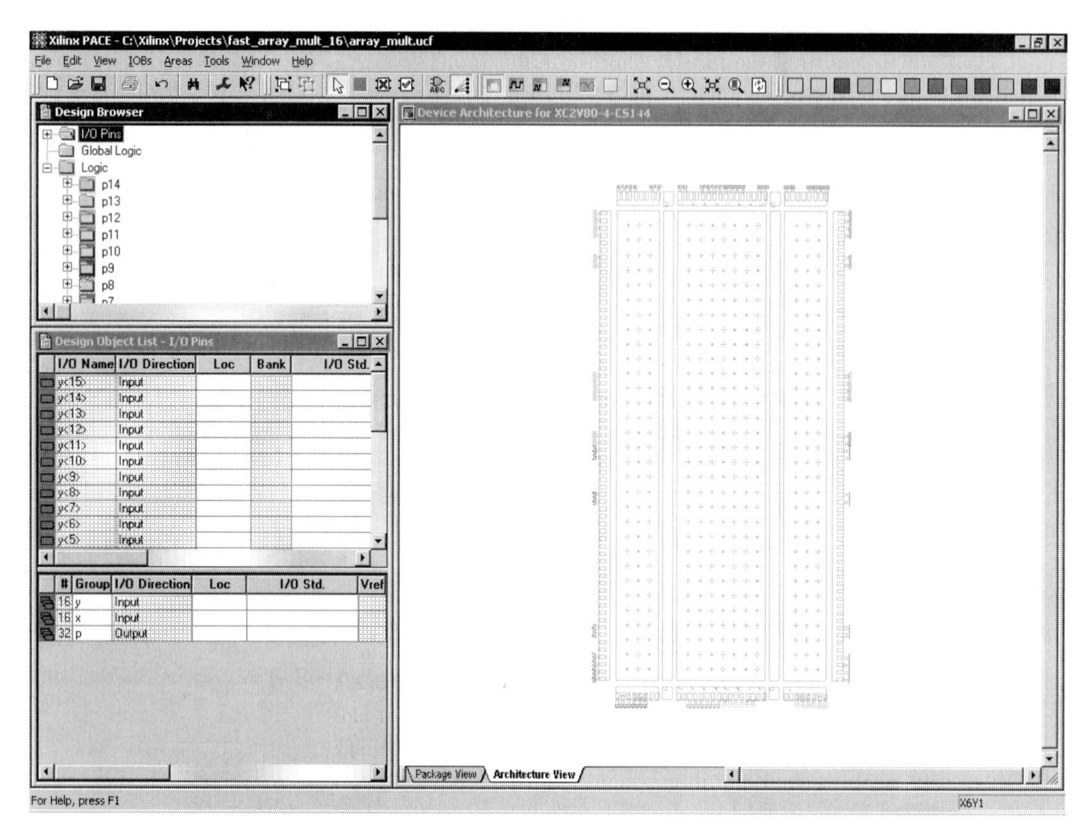

The *Logic* pane now shows us the components for which we defined constraints and the *Design Object List* describes their properties in more detail. We can place these components on the screen by selecting one of them from Design Browser and dragging it to the Device Architecture pane, which shows the positions of the CLBs and IOBs. (It helps to select *Areas->Allow Overlap*; without that option the tool won't let us overlap two constraints even temporarily.) Here is the screen after dragging and dropping p14:

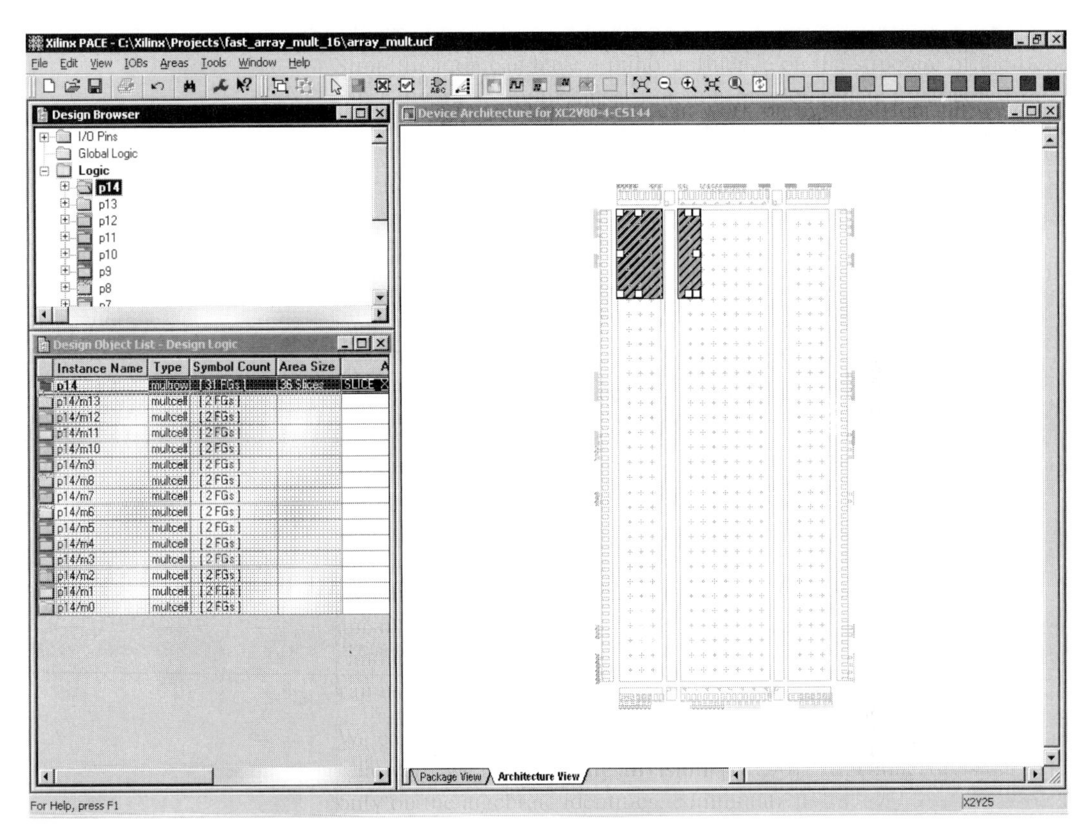

The rectangle in the device windows represents the area that will be occupied by this component's logic. That estimate was derived from the Verilog file, with some added area to make the estimate more conservative. We can use the handles on the area to change its shape. In order to arrange the rows of the multiplier in the form of the array, we can stretch it out to a wider, thinner shape:

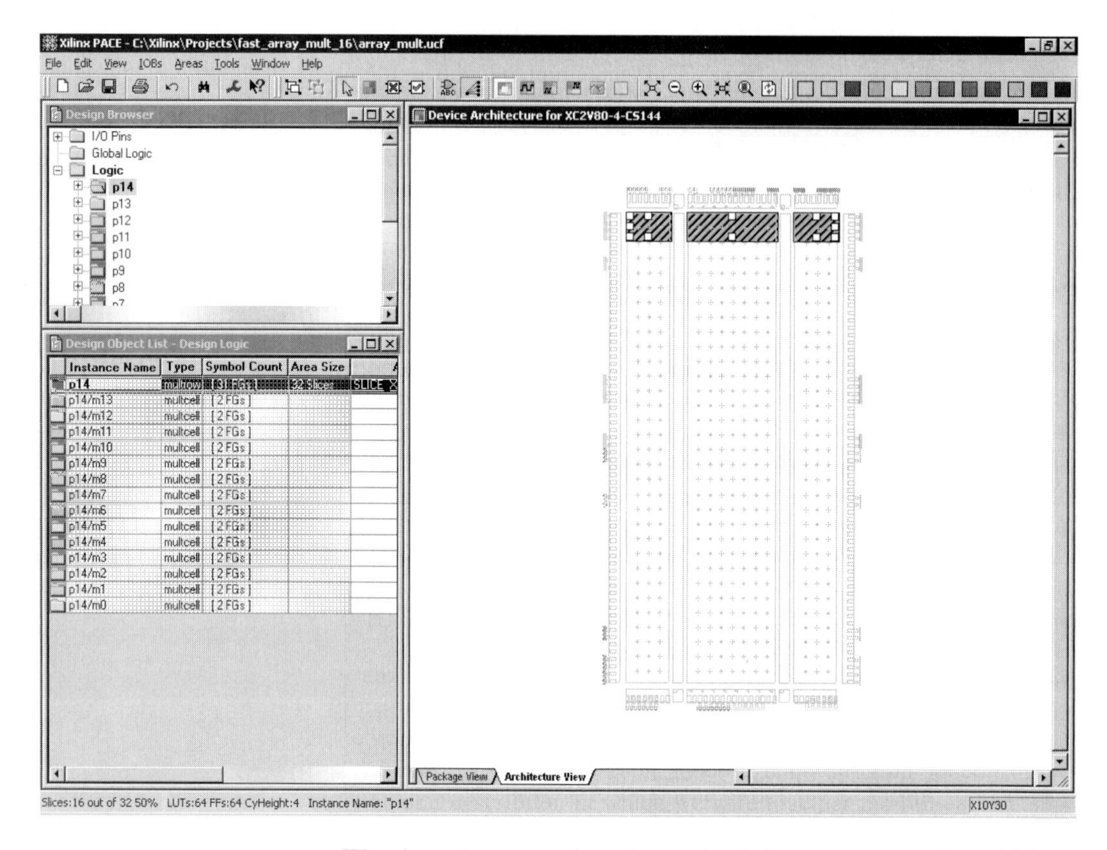

We place the constraints for each of the components for which we defined constraints. We place the rows of the multiplier one below the other in order to create the row structure in the floorplan. At the end of this process the constraints look like this:

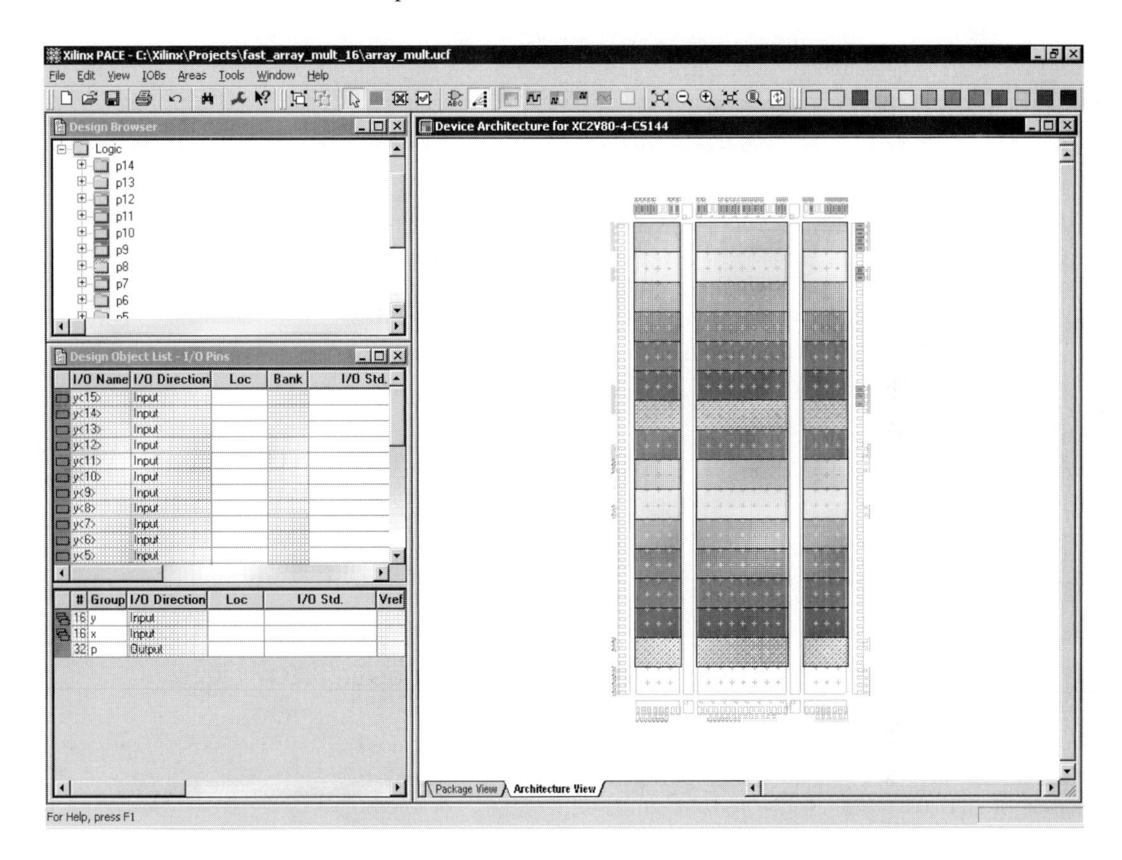

As we mentioned before, we could also add constraints on the I/O pins using the PACE editor. We do so by selecting *I/O Pins* from the *Design Browser* pane and dragging and dropping the pins.

We can now rerun the tools and see the effect of the constraints. Here is the placement built using the constraints:

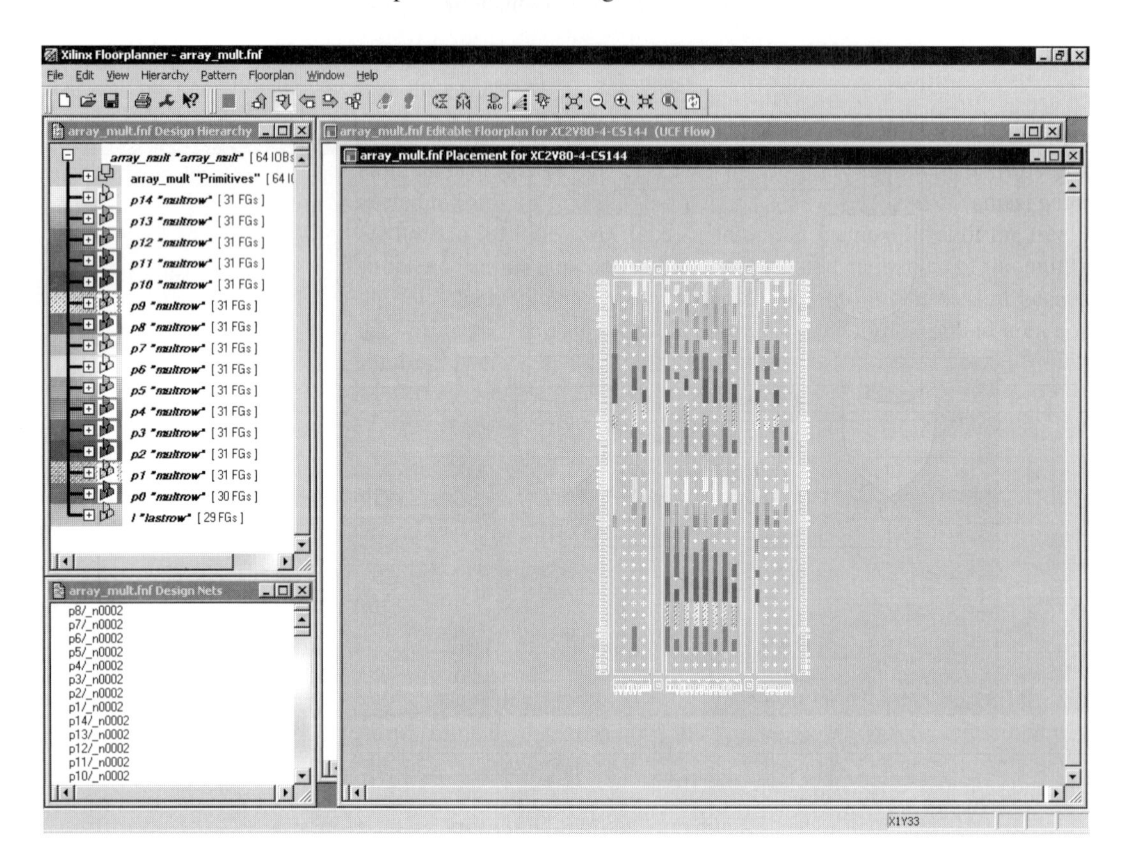

The floorplanning tool now shows us the logic in terms of the constrained components. We can see that the placement obeys the constraints and is much more regular than the unconstrained placement. It isn't, however, as regular as it could be, since the logic within the rows is not extremely regular.

The post-place-and-route timing report tells us that this new placement improved the multiplier's performance:

```
 19742142 items analyzed, 0 timing errors detected. (0 setup
errors, 0 hold errors)
 Maximum delay is  29.934ns.
--------------------------------------------------------------------------------
```

This compares to 31.1 ns for the unconstrained design.

We could further improve the placement by constraining the adders within the rows. However, we can add placement constraints only to modules, not to Verilog primitives like + operators. To add this new set of constraints we would have to rewrite *multrow.v* and *lastrow.v* to encapsulate the one-bit operators we want to constrain into components, then add synthesis attributes to those components.

The FPGA editor tool also lets you create detailed constraints to guide routing:

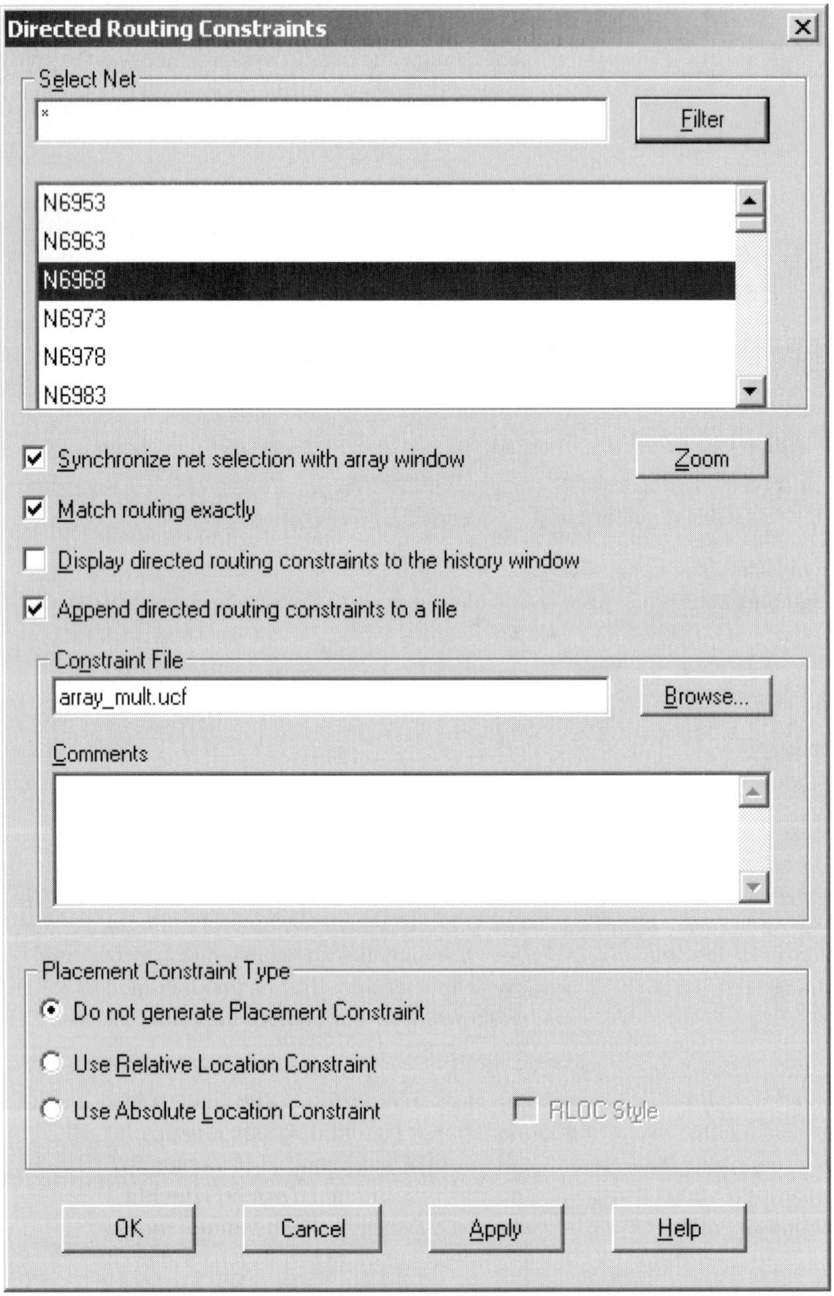

Directed Routing Constraints

Select Net

*

Filter

N6953
N6963
N6968
N6973
N6978
N6983

Synchronize net selection with array window

Zoom

Match routing exactly

Display directed routing constraints to the history window

Append directed routing constraints to a file

Constraint File

array_mult.ucf

Browse...

Comments

Placement Constraint Type

Do not generate Placement Constraint

Use Relative Location Constraint

Use Absolute Location Constraint

RLOC Style

OK Cancel Apply Help

Routing constrains require more effort than placement constraints but may be worthwhile in some performance-critical designs.

4.10 Summary

Combinational logic design is at the heart of digital system design. Given a Boolean formula, we can implement it in many different ways, resulting in different sizes of logic, delays through the logic, and energy consumption. In practice, we use tools to perform much of this work, but in order to get the most out of our tools we must understand both the basic principles of logic design and the operation of the tools themselves.

4.11 Problems

Q4-1. These logical expressions are written in C syntax. Rewrite them in Verilog syntax.

a. (a && b) || c.

b. ~a || b.

c. a ^ (b && ~c).

d. (a || ~b) && (~c || d).

e. ~a || ~b || ~c.

Q4-2. Identify the purpose of each of these Verilog syntactic elements:

a. ‘define

b. 2‘b

c. reg

d. monitor

e. $start

Q4-3. Draw schematics (using AND, OR, and inverters) that represent the results of syntax-directed translation of these statements:

a. a or (b and c)

b. not a or not b

c. a xor b

d.

```
if e then
    x <= a or b;
else
    x <= a and not b;
end if;
```

e.

```
if e then
    x <= a;
else
    y <= b;
end if;
```

Q4-4. For each example, rewrite the functions in multi-level form, introducing wherever possible additional functions for common factors.

a. $f = ab + ac + bc$; $g = ad + c' + bd$.

b. $f = abd + c'e$; $g = de' + abc$.

c. $f = ab + de'$; $g = ab + abc$; $h = a + bc + d'e$

Q4-5. Write three-input LUT specifications for these functions:

a. sum (a + b).

b. even parity.

c. difference (a - b).

Q4-6. This logic network below is simulated in an event-driven simulator:

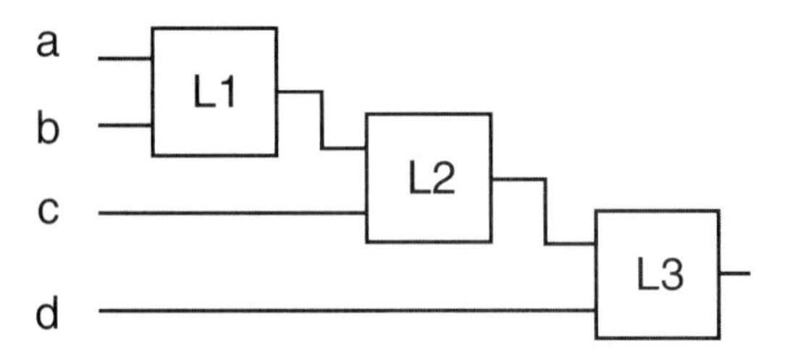

List the evaluations of logic elements in the order in which they happen, assuming that the inputs are changed in this order:

a. a; d.

b. a; c; d.

c. a; b; c; d.

Q4-7. The logic network below is simulated in an event-driven simulator:

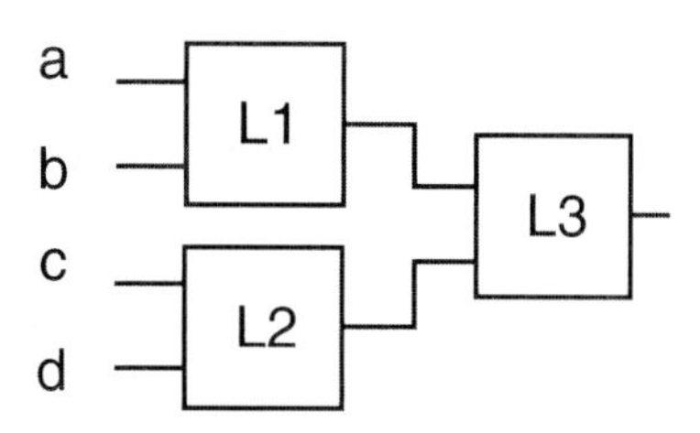

List the evaluations of logic elements in the order in which they happen, assuming that the inputs are changed in this order:

a. a; d.

b. a; b; d.

c. a; b; c; d.

Q4-8. You are given a two-bit ripple-carry counter. How many events are generated when the count changes from 01 to 10? How many when the count changes from 11 to 00?

Q4-9. Show the critical delay path(s) through each of these logic networks assuming that every logic block has equal delay:

a.

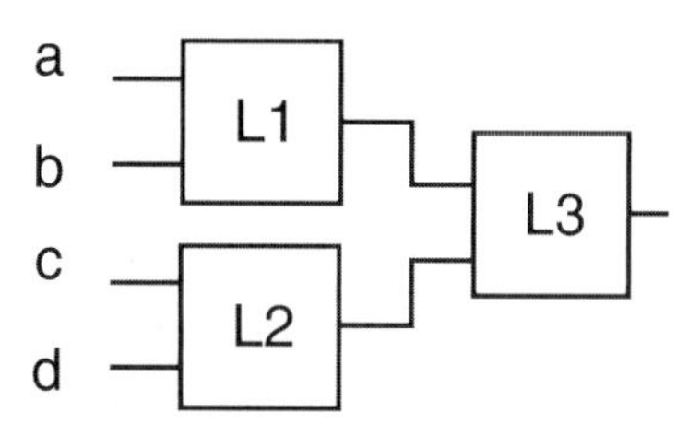

b.

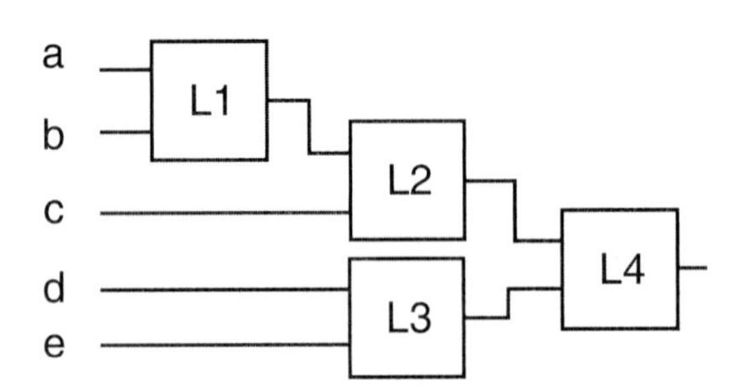

c.

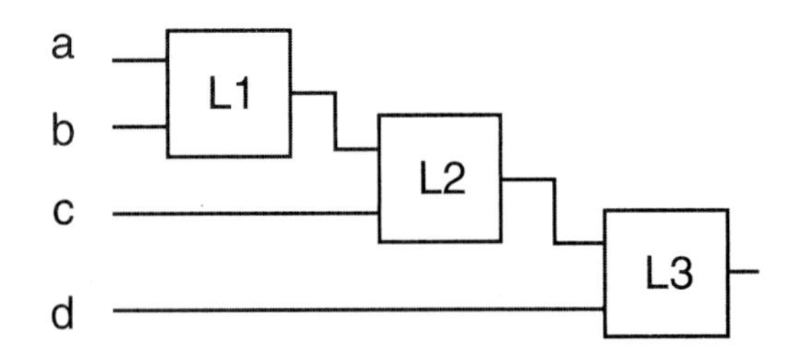

Q4-10. Write Verilog/VHDL for a four-bit barrel shifter with two four-bit data inputs.

Q4-11. Write Verilog/VHDL for an ALU that performs these functions: +, -, <, >, $\leq$, $\geq$.

Q4-12. What is the set of input values to a ripple-carry adder that gives the maximum delay?

Q4-13. Explain the worst-case delay for a carry-skip adder.

Q4-14. How many 4-input LUTs are required to build a one-bit ALU that performs all these functions: a AND b, a OR b, a XOR b, a + 1 (with carry), a + b (with carry), a - b (with borrow). Give the programming for each LUT and the connections between the LUTs.

Q4-15. How many four-input LUTs are required to build a 4×4 array multiplier?

Q4-16. How many four-input LUTs are required to build a 4×4 Booth multiplier?

Q4-17. Partition this graph into four groups of roughly equal size such that the number of edges between partitions is minimal.

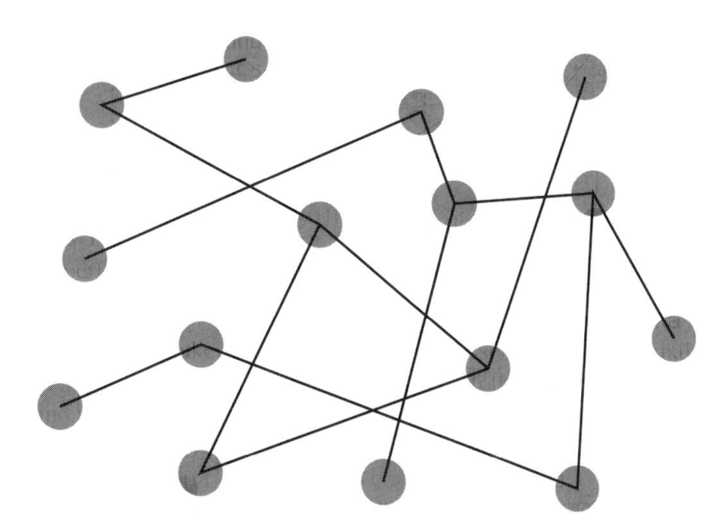

5 Sequential Machines

Descriptions of sequential machines.

Rules for legal clocking.

Performance analysis.

Power consumption.

5.1 Introduction

This chapter builds upon our experience with combinational logic to design sequential machines. Sequential machines (also known as **finite-state machines** or **FSMs**) are the basic building blocks of digital logic—the state stored in a sequential machine makes it possible to compute much more complex functions than is practical with combinational logic alone. Sequential logic introduces new constraints into the design process that we must add to the basic principles we have learned for combinational logic design.

We will study sequential machine design from the abstract to the concrete, starting with models for sequential machines and ending with performance analysis and optimization. The next section looks at the overall sequential machine process. Section 5.3 studies state transition graphs and register-transfer models, the two basic ways of specifying sequential machines. Section 5.4 looks at the timing constraints required to build sequential machines with clocks. Section 5.5 studies performance analysis of sequential machines, while Section 5.6 looks at energy/power optimization.

5.2 The Sequential Machine Design Process

A sequential machine has two basic parts:

- registers (flip-flops or latches) to hold the state of the machine;
- logic to compute some function of the values of the inputs and the registers.

Sequential machine design therefore relies heavily on combinational logic design. The techniques we have learned for optimizing combinational logic for size, performance, and power are all critical to the implementation of sequential machines. However, sequential system design imposes additional constraints that must be satisfied to ensure that the registers and logic work together properly. Those constraints must be satisfied for the sequential machine to function.

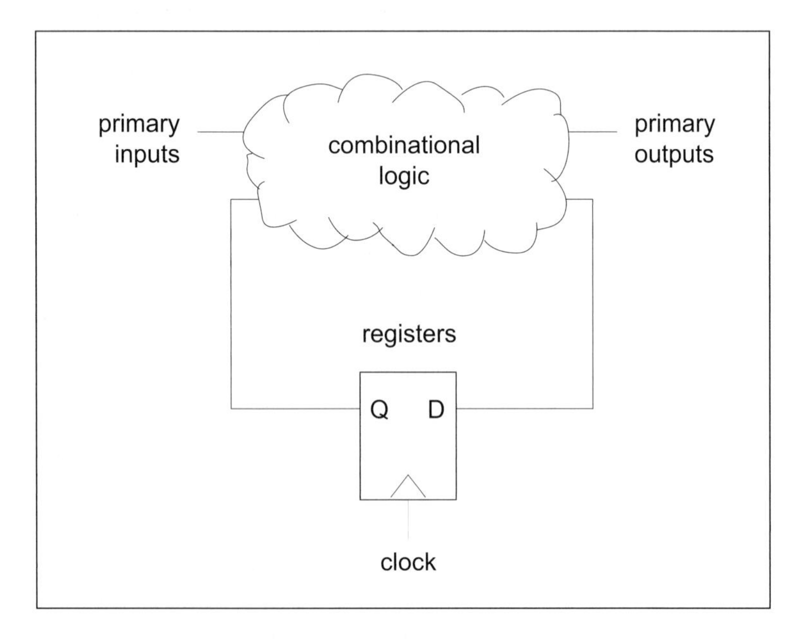

Figure 5-1 Structure of a sequential machine.

synchronous design

We will concentrate in this chapter on **synchronous** sequential systems. (In Chapter 7 we will study asynchronous machines of the kind frequently used in busses.) We use synchronous machines in order to make the design process tractable and to improve our changes of ending up with a working design. As shown in Figure 5-1, a synchronous machine uses **clock signals** to determine when events occur in the system by controlling when the registers read values computed by the combinational logic. The results of each clock cycle can be saved as **state** in the registers or as values on the primary outputs; those results are computed from combinations of the register values and the primary inputs.

Synchronous design lets us separate the overall design problem into two separable tasks:

- designing the logic to perform the function we want;
- ensuring that the logic runs fast enough to meet the system clock rate.

Separating these two tasks makes design much easier. In some asynchronous design styles, glitches on signals can cause the machine to go into an incorrect state. Even if the signal settles to the proper value, the glitch can cause a register to remember the wrong value or to forget the correct value. Determining when glitches occur requires understanding not just the logical function of the gates but also the timing properties of those gates (and of the wires as well). Such a complicated analysis is error-prone at several points. Synchronous design helps us to design reliable systems.

specification and optimization

As with combinational logic design, we want to first *specify* our system and then *optimize* it to meet our performance, power, and cost goals. Synchronous design helps us decouple the two steps—because we can cleanly specify how the logic delays relate to the clock period, we can perform much more aggressive optimizations.

There are two major ways used to specify sequential machines:

- **register-transfer models** describe the behavior as a transformation of register values;
- **state transition graphs/tables** describe the machine's behavior in terms of states and outputs.

We will describe these methods in more detail in the next section. We can describe both in hardware description languages, but one style may be easier to use for some examples and less suited to other types of examples.

Once the system has been specified, we can optimize it into an acceptable implementation. We can, of course, use all the combinational logic optimizations. We can also use other optimizations that are special to sequential machines; state assignment is an example of an optimization that is unique to sequential machines and can make a large difference in the size, speed, and power consumption of the machine.

goals of optimization

Our performance goal is generally described in terms of the desired clock rate. Although in some cases we want our machine to run as fast as possible, we often need it to run at the specified clock rate and no faster. Making it run faster than necessary can waste both area and power. One of the tasks in sequential design is turning a basic clock rate specification into constraints on the combinational logic that take into account both the timing properties of the registers and the delays inherent in distributing the clock itself.

Area is a common concern; in FPGAs area constraints are expressed in logic element and interconnect limitations. Area and performance are often at odds.

Low-energy operation often comes from high-performance design—fast logic is more efficient and uses less energy. However, there are some specific techniques we can use to reduce energy consumption in sequential machines.

5.3 Sequential Design Styles

This section takes on several core topics in the design of sequential machines. Section 5.3.1 introduces complementary ways of specifying a sequential machine. Section 5.3.2 reviews some basic facts about the properties of sequential machines. Section 5.3.3 looks at state assignment. Finally, Section 5.3.4 introduces Verilog styles for sequential machines.

5.3.1 State Transition and Register-Transfer Models

When we specify a sequential machine we don't want to tie ourselves to a particular collection of logic gates. We therefore need a way to functionally specify the machine's actions so that we can later optimize the combinational logic that implements those functions.

two ways to specify machines

The two major ways to specify sequential machines are **state transition graphs** (or *tables*) and **register-transfer** (**RT**) **models**. These are both functional models that describe what the machine does without describing all the details of how it does those things. However, because these two models describe the machine in different ways, each is better for some types of systems and not as good for others. Before we directly compare the two, let's briefly review them.

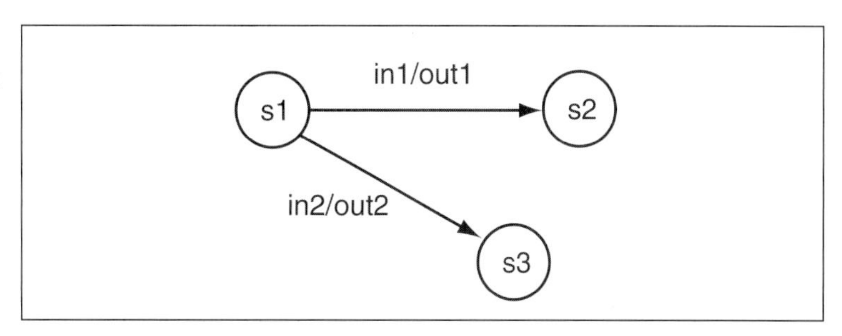

Figure 5-2 A state transition graph.

Figure 5-2 shows a fragment of a state transition graph. The nodes of the graph, such as *s1*, *s2*, and *s3*, define the states of the machine. The directed edges of the graph show the possible transitions between the states. The transitions between the states are taken on clock cycles. The transition taken out of a state depends on the inputs to the machine at that time; in this example the machine will go from s1 to s2 depending on the values of the inputs. In general, a transition may also include the outputs values to be generated. (We will develop a more formal definition of the state transition graph in Section 5.3.2).

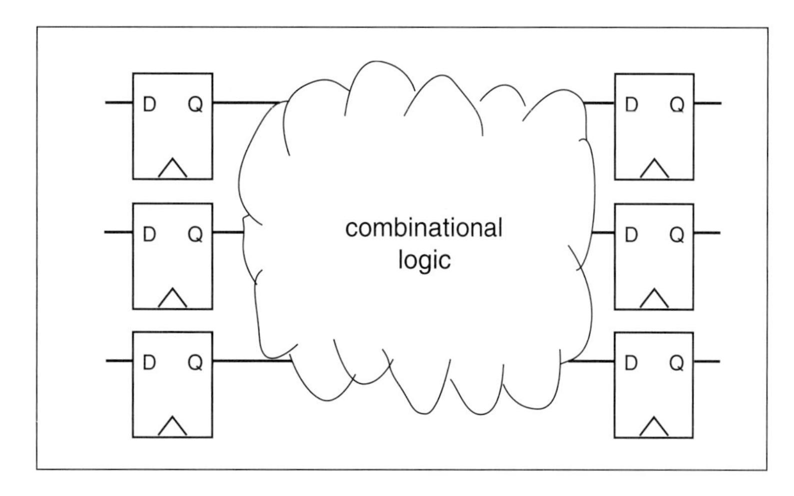

Figure 5-3 A register-transfer structure.

A register-transfer system is a pure sequential machine. It is specified as a set of registers and combinational logic functions between those elements, as shown in Figure 5-3. We don't know the structure of the logic inside the combinational functions—if we specify an adder, we don't know how the logic to compute the addition is implemented. As a result, we don't know how large or how fast the system will be.

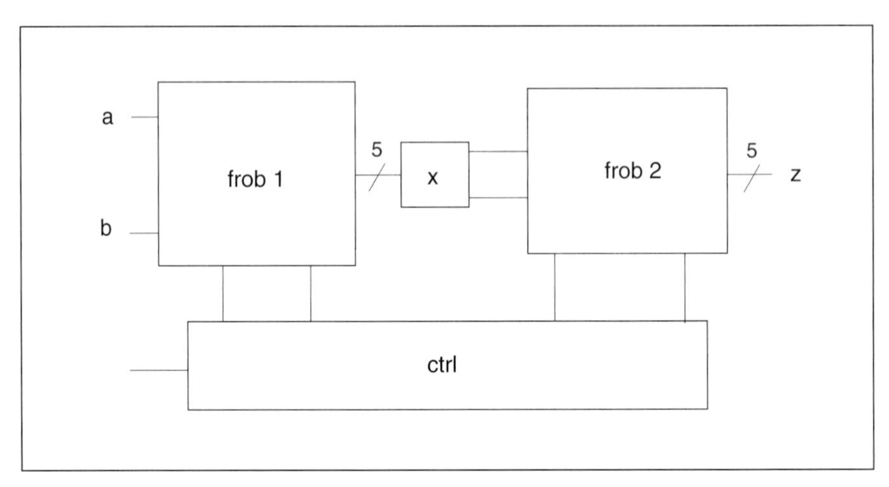

Figure 5-4 Atypical block diagram.

block diagrams

We can connect several RT machines together to form a larger machine. For example, the right-hand bank of registers in Figure 5-3, which hold the output of one block of logic, can form the input to the next block of logic. Combinations of register-transfer machines are often described as **block diagrams**. A typical block diagram is shown in Figure 5-4. A block diagram is a purely structural description—it shows the connections between boxes. (The slash and 5 identify each of those wires as a bundle of five wires, such as the five bits in a data word.)

Register-transfer designs are often described in terms of familiar components: multiplexers, ALUs, etc. Standard components can help you organize your specification and they also give good implementation hints. But don't spend too much time doing logic design when sketching your block diagram. The purpose of register-transfer design is to correctly specify sequential behavior.

symbolic signals and encoding

Sometimes we want to make a state transition graph or register-transfer machine more abstract. We can do so by specifying the inputs and outputs over symbolic, rather than binary values—for example, specifying the output of a function to range over $\{\alpha,\beta,\chi,\delta\}$. Finding binary encodings for these symbolic inputs and outputs is part of the implementation process. Sometimes these encodings for these symbols are given by the problem specification—the symbolic versions are simply easier to deal with than binary values. Sometimes we have freedom in how to encode those values. Finding input/output encodings is closely related to state assignment, which we will discuss in more detail in Section 5.3.3.

In the next two examples, we will compare the expressive power of register-transfer machines and state transition graphs.

Example 5-1 A counter

A one-bit counter consists of two components: a specialized form of adder, stripped of unnecessary logic so that it can only add 1; and a register to hold the value. We want to build an n-bit binary counter from one-bit counters.

What logical function must the one-bit counter execute? The truth table for the one-bit counter in terms of the present count stored in the latch and the carry-in is shown below. The table reveals that the next value of the count is the exclusive-or (XOR) of the current count and C_{in}, while the carry-out is the AND of those two values.

count	C_{in}	next count	C_{out}
0	0	0	0
0	1	1	0
1	0	1	0
1	1	0	1

This truth table is a specification of the combinational logic in the counter. We can also write the function as an expression:

$$c_{\text{out}} = c_{\text{in}} \cdot \text{count},$$

$$\text{next count} = c_{\text{in}} + \text{count}.$$

Here is a register-transfer schematic for a one-bit counter:

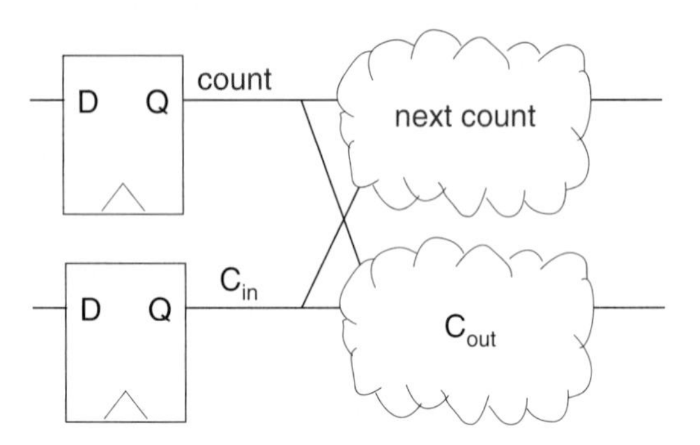

Each output is one bit wide and is computed by combinational logic that uses both the input values.

An n-bit counter's structure looks like this:

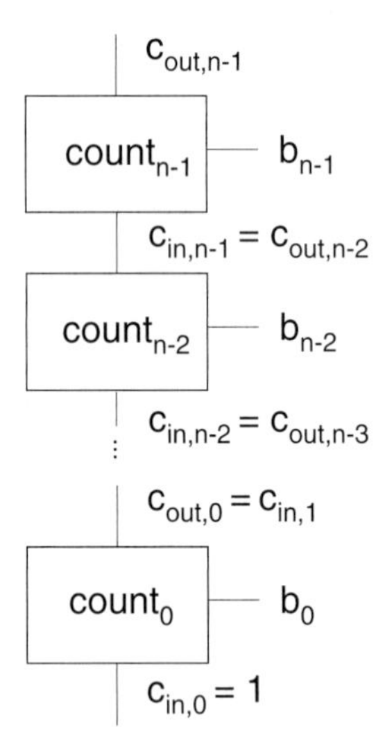

Each bit has one input and two outputs: the input $C_{in,i}$ is the carry into the ith bit; the output b_i is the current value of the count for that bit; and $C_{out,i}$ is the carry out of the bit. The carry-in value for the 0th bit is 1; on each clock cycle this carry value causes the counter to increment itself. (The counter, to be useful, should also have a reset input which forces all bits in the counter to 0; we have omitted it here for simplicity.)

What would the state transition graph for a counter look like? Here is the state transition graph for a four-bit counter:

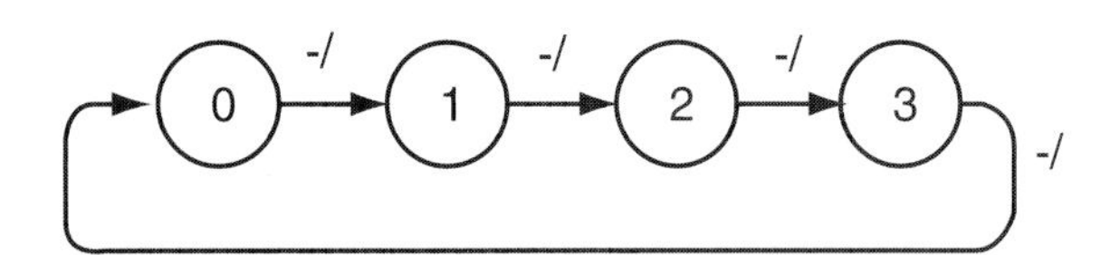

The counter unconditionally moves through the counted states. (The - input value is an input don't-care since the counter counts independent of an external input.) While this state transition graph is relatively easy to draw, the number of states in the graph is exponentially related to the number of bits in the counter: an *n*-bit counter has 2^n states in the state transition graph.

In this case, the register-transfer machine is probably the most convenient way to describe the counter. This is particularly true if we extend the counter to include resetability, loading, etc.

Example 5-2 A 01-string recognizer

Consider as an example a very simple FSM with one input and one output. If the machine's inputs are thought of as a string of 0's and 1's, the machine's job is to recognize the string "01"—the FSM's output is set to 1 for one cycle as soon as it sees "01." This table shows the behavior of the recognizer machine over time for a sample input:

time	0	1	2	3	4	5
input	0	0	1	1	0	1
present state	bit1	bit2	bit2	bit1	bit1	bit2
next state	bit2	bit2	bit1	bit1	bit2	bit1
output	0	0	1	0	0	1

We can describe the machine's behavior as either a state transition graph or a state transition table. The machine has one input, the data string, and one output, which signals recognition. It also has two states: *bit1* is looking for "0", the first bit in the string; *bit2* is looking for the trailing "1". Both representations specify, for each possible combination of input and present state, the output generated by the FSM and the next state it will assume.

Here is the state transition table:

input	present state	next state	output
0	bit1	bit2	0
1	bit1	bit1	0
0	bit2	bit2	0
1	bit2	bit1	1

And here is the equivalent state transition graph:

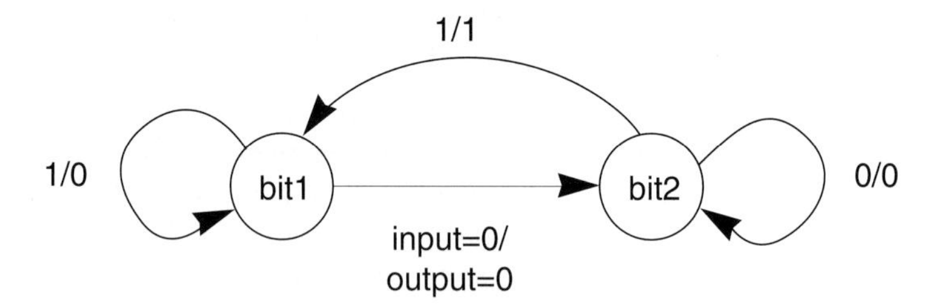

Assume that the machine starts in state *bit1* at time t=0. The machine moves from *bit1* to *bit2* when it has received a 0 and is waiting for a 1 to appear on the next cycle. If the machine receives a 0 in state *bit2*, the "01" string can still be found if the next bit is a 1, so the machine stays in *bit2*. The machine recognizes its first "01" string at t=2; it then goes back to state *bit1* to wait for a 0. The machine recognizes another "01" string at t=5.

How would we describe this function as a register-transfer machine? The output function must be true when the last bit was a 0 and the current bit is a 1. In addition to output, we will need to construct an auxiliary function:

$$\text{aux} = \text{input}\,,$$

$$\text{output} = \text{input} \cdot \text{aux}'\,.$$

The auxiliary function works because it is implemented as a register—the output function sees the old value of aux while the register accepts the new value from the input.

These functions are basically an implementation of the next-state function of the state transition graph. However, it is easier to think of the operation as transitions between states. In this case, the state transition graph or table is the easiest way to describe the function.

5.3.2 Finite-State Machine Theory

A brief review of the theory of finite-state machines will provide us with some useful tools for designing and implementing FSMs. Externally, the FSM is characterized by its primary inputs and outputs and the clock. The clock signal determines when the primary inputs are sampled and the primary outputs attain their new values; the exact times at which these occur do not matter at this level of abstraction. Internally, the machine stores a state that is updated at each tick of the clock.

Mealy and Moore machines

There are two major varieties of finite-state machines. If the primary outputs are a function of both the primary inputs and state, the machine is known as a **Mealy machine**; if the primary outputs depend only on the state, the machine is called a **Moore machine**. The Mealy machine is more general and so we will concentrate on that form.

We will refer to the primary inputs as I, the primary outputs as O, and the current state of the machine as S. Each of them is subscripted by the clock tick, so O_i refers to the output at time i. The FSM is characterized by two functions:

- the output function $O_{i+1} = \lambda(I_i, S_i)$;
- the next-state function $S_{i+1} = \Delta(I_i, S_i)$.

The current state of the FSM is determined solely by the machine's state and inputs on the last cycle. If you have recorded the total input sequence and the initial state of the machine, you can reconstruct the sequence of internal states, but without that information you cannot, in general, determine the machine's state in the past.

reachability

The state transition graph described by the next-state and output functions is a directed graph. A component of a graph is a set of nodes that can be reached from one another. A state transition graph need not describe a single connected component—some states in the machine may not be **reachable** from other states by any sequence of inputs. This condition may seem implausible in practice, but state assignments can create such machines. Consider the example of Figure 5-5, whose left-hand side shows a machine with three states. When we choose a binary

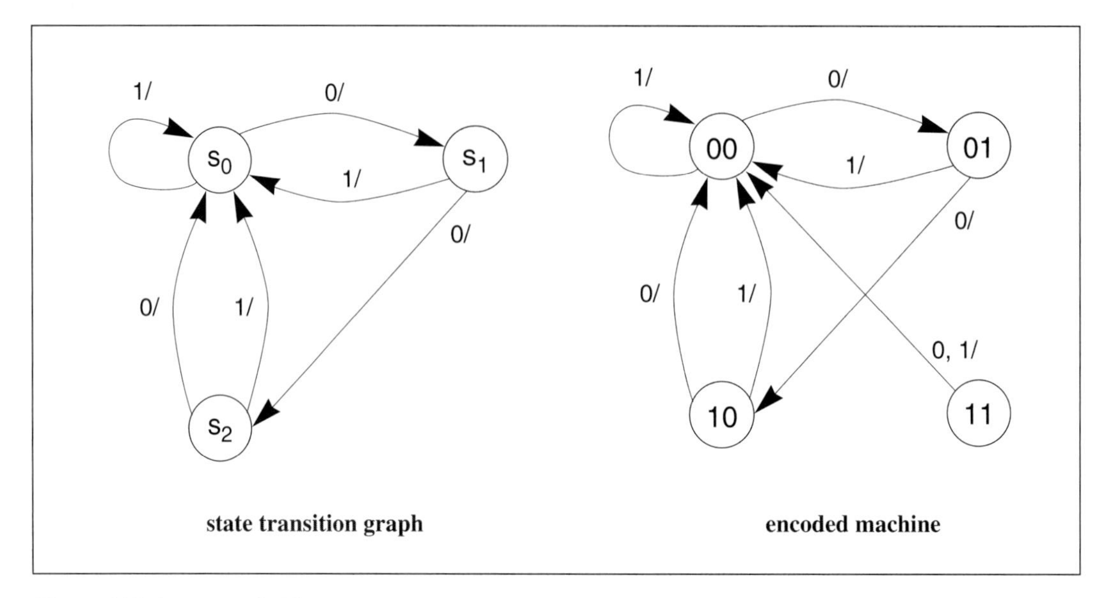

Figure 5-5 An unreachable state.

encoding for the states we add a fourth state to the machine. The transitions into and out of this state are determined by the logic used to implement the next-state and output functions for the originally defined states. Those functions ensure that, although there is a path from the 11 state to the rest of the machine, there is no sequence of inputs that will drive the machine from one of the three initial states to the 11 state. Unreachable states have a number of severe consequences: it may be possible for the machine to be stuck into a useless state; and it may not be possible to reset the machine.

homing sequences

A **homing sequence** is a sequence of inputs that will cause the machine to end up in a given state. Homing sequences are useful in testing machines as well as ensuring that the machine is in the proper state. A global reset signal is a simple form of a homing sequence.

equivalent states

An FSM may have **equivalent** states. Two states are equivalent if there is no sequence of inputs that can be used to determine which of the equivalent states the machine started in. If you cannot apply a sequence of inputs and watch the outputs to ultimately determine which state you started in, you have equivalent states. Figure 5-6 shows a simple state transition graph in which states *s2* and *s3* are equivalent—no sequence of inputs can be used to distinguish whether the machine is in *s2* or *s3*. The output and next-state functions of a machine with equivalent states

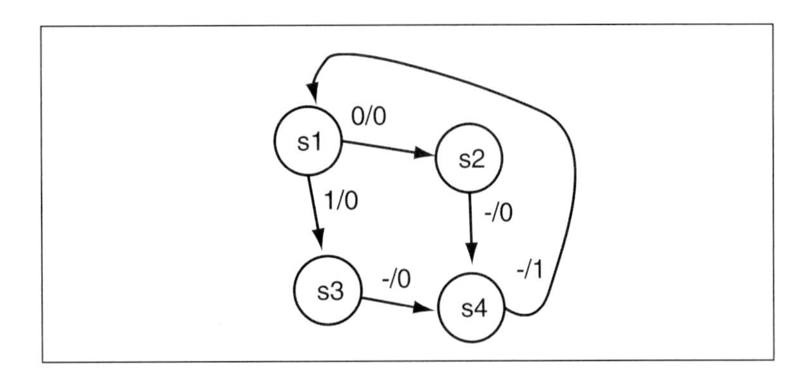

Figure 5-6 An FSM with equivalent states.

can be rewritten so that the equivalent states are combined into a single state; reducing the number of states in the machine usually (though not always) results in a better implementation.

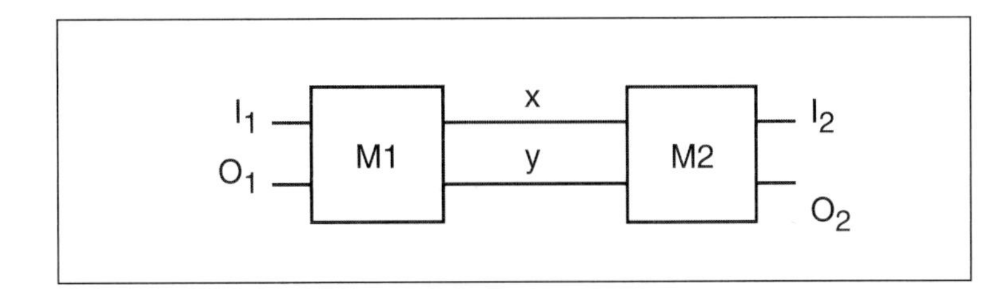

Figure 5-7 Two interconnected FSMs.

networks of FSMs

A properly interconnected set of sequential systems is also a sequential system. It is often convenient to break a large system into a network of communicating machines: if decomposed properly, the system can be much easier to understand and to design. Figure 5-7 shows two interconnected FSMs. This composition of FSMs satisfies the requirements for an FSM: the composition's inputs are the union of I_1 and I_2; its outputs are the union of O_1 and O_2; and the state of the composition is the union of the states of the two component machines.

We should note that not all compositions of Mealy machines are legal. Because a Mealy machine's output can be a function of its input, a pair of connected machines can form a combinational cycle as shown in Figure 5-8. Each machine has a combinational path from input to output, which is not illegal so long as the machines are not connected. However, the composition of the two machines forms a combinational cycle that is not broken by a register. When we compose Mealy machines we must be sure that the state transition tables of the machines don't conspire to form a combinational cycle. Moore machines don't have this problem because their output is a function only of the current state.

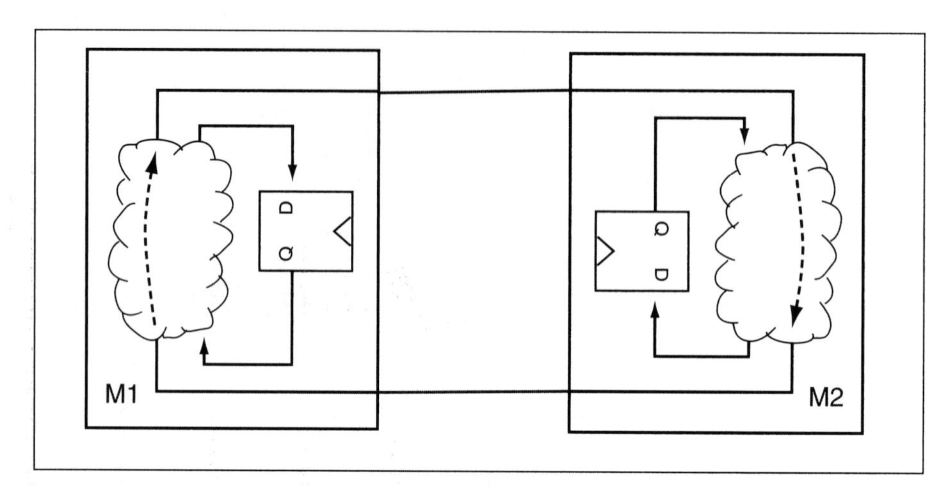

Figure 5-8 An illegal composition of Mealy machines.

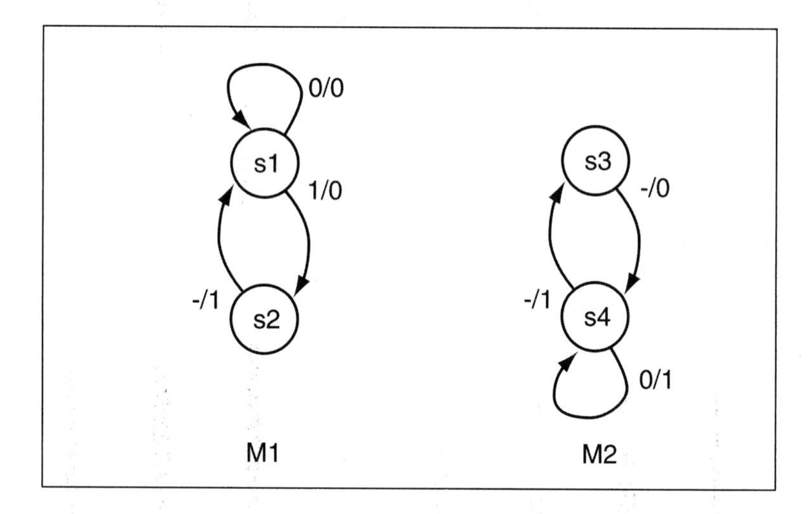

Figure 5-9 Two communicating FSMs.

The next-state and output functions of the composite FSM is given by the Cartesian product of those functions for the component FSMs; the machine is often called the **cross-product machine** after a Cartesian cross product. For a simple example of the formation of a product machine, consider the two machines in Figure 5-9. Each of these machines has one input and one output. The cross-product machine will have no inputs or outputs but will simply advance state on every clock cycle. The composite machine will have four states: s12, s23, s14, and s24. The transitions out of these composite states are determined by examining the inputs and outputs exchanged between the component machines. One way to visualize the formulation of the cross-product state transition graph is to symbolically execute the machines. If we start with *M1* in state *s1* and *M2* in state *s3*, then the machine executes the path $s13 \rightarrow s24 \rightarrow s14 \rightarrow s23$. If we start with *M1* in *s1* and *M2* in *s4*, we see that the machine repeatedly executes a cycle between states *s14* and *s23*.

factoring FSMs

A state machine can also be factored into a network of communicating FSMs. This is, in general, a difficult problem. A well-established theory exists for decomposition into serial-parallel networks of machines, though serial-parallel decomposition does not always provide interesting results in practice. Counters can also be factored out of FSMs, though this is often not necessary.

5.3.3 State Assignment

state assignment and implementation cost

State assignment is the design step most closely associated with FSMs. (Input and output signals may also be specified as symbolic values and encoded, but the state variable typically has the most coding freedom because it is not used outside the FSM.) State assignment can have a profound effect on the size of the next state logic, as shown in the next example.

Example 5-3 Encoding a shift register

Here is the state transition table for a two-bit shift register, which echoes its input bit two cycles later:

input	present state	next state	output
0	s00	s00	0
1	s00	s10	0
0	s01	s00	1
1	s01	s10	1
0	s10	s01	0
1	s10	s11	0

input	present state	next state	output
0	s11	s01	1
1	s11	s11	1

The state names are, of course, a hint at the optimal encoding. But let's first try another code: s00 = 00, s01= 01, s10 = 11, *s11* = 10. We'll name the present state bits $S_1 S_0$, the next state bits $N_1 N_0$, and the input i. The next state and output equations for this encoding are:

$$output = S_1 \overline{S_0} + \overline{S_1} S_0$$

$$N_1 = i$$

$$N_0 = i\overline{S_1} + \bar{i}S_1$$

Both the output and next state functions require logic. Now consider the shift register's natural encoding—the history of the last two input bits. The encoding is *s00* = 00, *s10* = 10, *s01* = 01, *s11* = 11. Plugging these code values into the symbolic state transition table shows that this encoding requires no next state or output logic:

$$output = S_0$$

$$N_1 = i$$

$$N_0 = S_1$$

This example may seem contrived because the shift register function is regular. But changes to the state codes can significantly change both the area and delay of sequencers with more complex state transition graphs. State codes can be chosen to produce logic which can be swept into a common factor during logic optimization—the common factors are found during logic optimization, but exist only because the proper logic was created during state assignment.

one-hot codes

A **one-hot code** is commonly used in FPGAs because they have a large number of registers. A one-hot code uses n bits to encode n states; the ith bit is 1 when the machine is in the ith state. For example, if we have three states *S1*, *S2*, and *S3*, we can assign *S1* = 001, *S2* = 010, and *S3* = 100. All the one-hot codes are equivalent: for example, *S1* = 100, *S2* =

001, and $S3$ = 010 gives the same results as our first one-hot code. We can use such a code to easily compute state subset membership by simply computing the OR of the state bits in the subset. One-hot codes require a lot of registers (64 registers for a 64-state machine, for example) but FPGAs are rich in registers.

common factors in state assignment

In some cases we may be interested in finding codes that create common factors in the FSM's combinational logic. These common factors can help us reduce the size and energy consumption of the machine's logic. State assignment creates two types of common factors: factors in logic that compute functions of the present state; and factors in the next state logic. Input encoding can best be seen as the search for common factors in the symbolic state transition table. Consider this state machine fragment:

input	present state	next state	output
0	s1	s3	1
0	s2	s3	1

If we allow combinations of the present state variable, we can simplify the state transition table as:

input	present state	next state	output
0	$s1 \vee s2$	s3	1

How can we take advantage of the OR by encoding? We want to find the smallest logic which tests for $s1 \vee s2$. For example, if we assume that the state code for the complete machine requires two bits and we encode the state variables as $s1$ = 00, $s2$ = 11, the present state logic is $\overline{S_1}\overline{S_0} + S_1 S_0$. The smallest logic is produced by putting the state codes as close together as possible—that is, minimizing the number of bits in which the two codes differ. If we choose $s1$ = 00, $s2$ = 01, the present state logic reduces to $\overline{S_1}$.

As shown in Figure 5-10, we can interpret the search for symbolic present state factors in the state transition table as a forward search for common next states in the state transition graph [Dev88]. If two states go to the same next state on the same input, the source states should be coded as close together as possible. If the transitions have similar but not identical input conditions, it may still be worthwhile to encode the source states together.

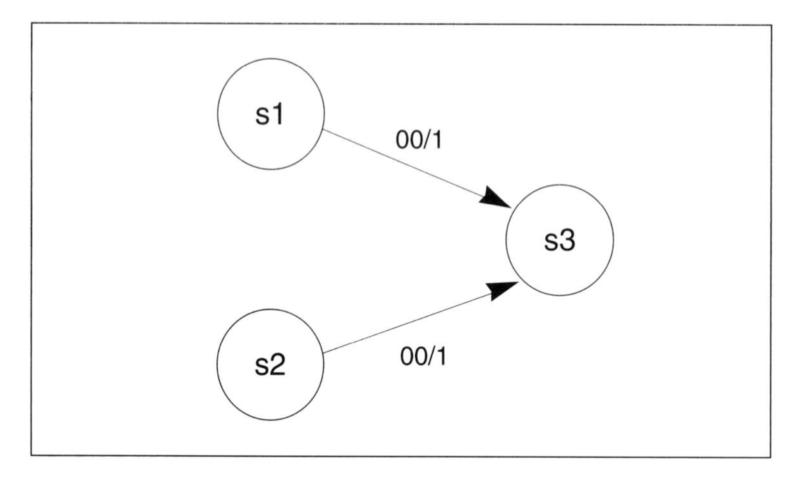

Figure 5-10 Common next states.

Figure 5-11 illustrates the relationship between bit differences and distance between codes. We can embed a three-bit code in a three-dimensional space: one axis per code bit, where each axis includes the values 0 and 1. Changing one code bit between 0 and 1 moves one unit through the space. The distance between 000 and 111 is three because we have to change three bits to move between the two codes. Putting two codes close together puts them in the same subspace: we can put two codes in the 00- subspace and four in the 1– subspace. We can generate many coding constraints by searching the complete state transition graph; the encoding problem is to determine which constraints are most important.

We can also search backward from several states to find common present states. As shown in Figure 5-12, one state may go to two different states on two different input values. In this case, we can minimize the amount of logic required to compute the next state by making the sink states' codes as close as possible to the source state's code. Consider this example:

input	present state	next state	output
0	s0	s1	1
1	s0	s2	1

We can make use of the input bit to compute the next state with the minimum amount of logic: if *s0* = 00, we can use the input bit as one bit of

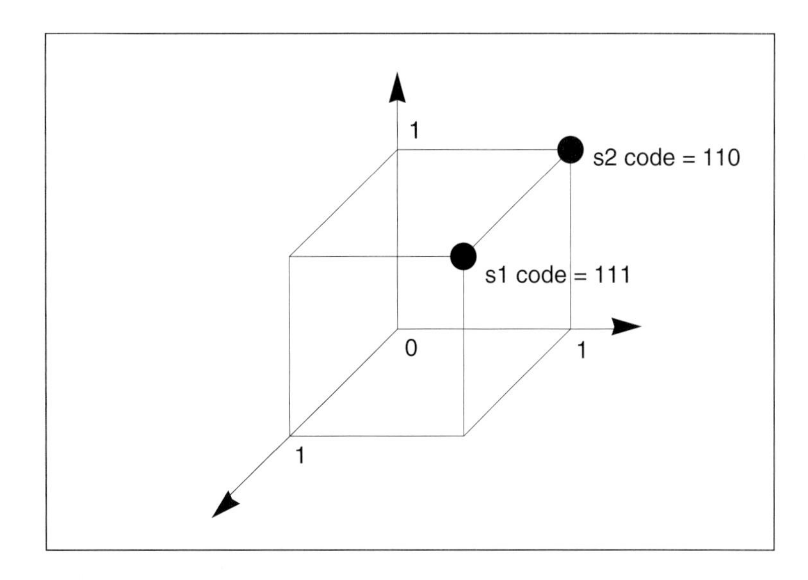

Figure 5-11 State codes embedded in a three-dimensional space.

the codes for s1 and s2: *s1* = 10, *s2* = 11. One bit of the next state can be computed independently of the input condition. Once again, we have encoded *s1* and *s2* close together so that we need the smallest amount of logic to compute which next state is our destination.

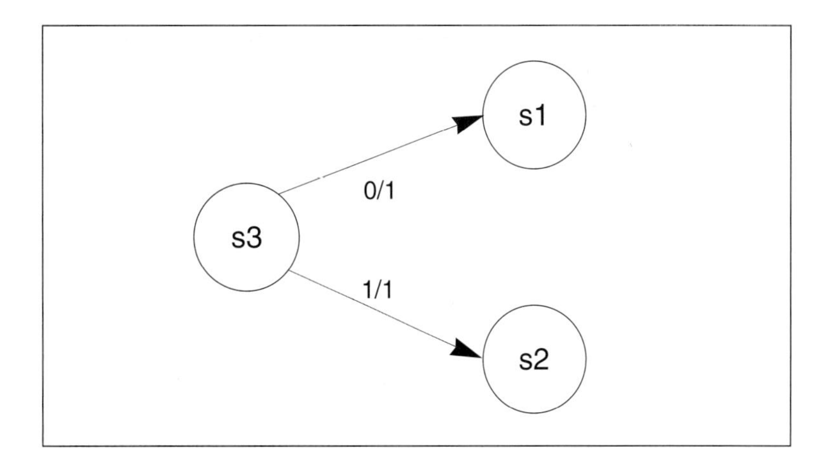

Figure 5-12 Common present states.

So far, we have looked at codes that minimize the area of the next state logic and the number of registers. State assignment can also influence the delay through the next state logic; reducing delay often requires adding state bits. Figure 5-13 shows the structure of a typical operation performed in either the next-state or the output logic. Some function *f()* of

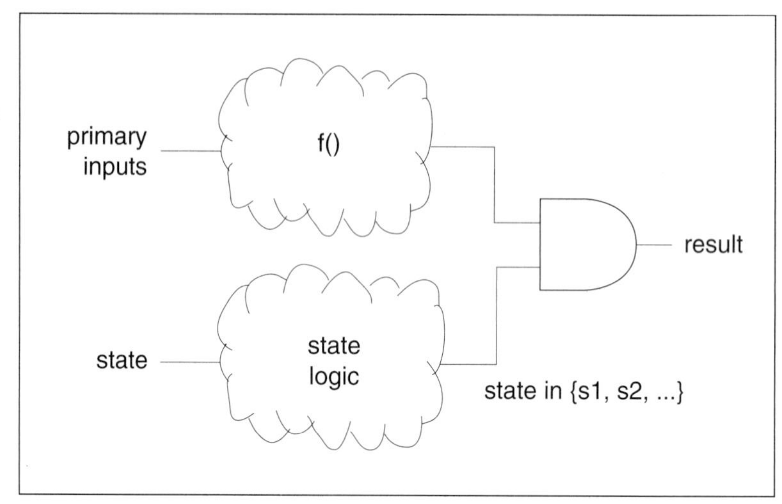

Figure 5-13 An FSM computes new values from the primary inputs and state.

the inputs is computed. This value will usually control a conditional operation: either a conditional output or a conditional change in state. Some test of the present state is made to see if it is one of several states. Then those two results are combined to determine the proper output or next state. We can't do much about the delay through *f()*, but we can choose the state codes so that the important steps on the state are easy to compute. Furthermore, the FSM probably computes several *f()*s for different operations, which in general don't have the same delay. If we can't make all computations on the state equally fast, we can choose the codes so that the fastest state computations are performed on the FSM's critical path.

As shown in Figure 5-14, state codes can add delay both on the output and next state sides. On the output logic side, the machine may need to compute whether the present state is a member of the set which enables a certain output—in the example, the output is enabled on an input condition and when the present state is either *s2* or *s4*. The delay through the logic which computes the state subset depends on whether the state codes were chosen to make that test obvious.

5.3.4 Verilog Modeling Styles

register-transfer models

We built a number of Verilog models of combinational logic in Chapter 4. We now need to learn how to write a synthesizable Verilog model of a sequential machine. The fundamental addition is of an *always* statement that lets us wait for the clock edge:

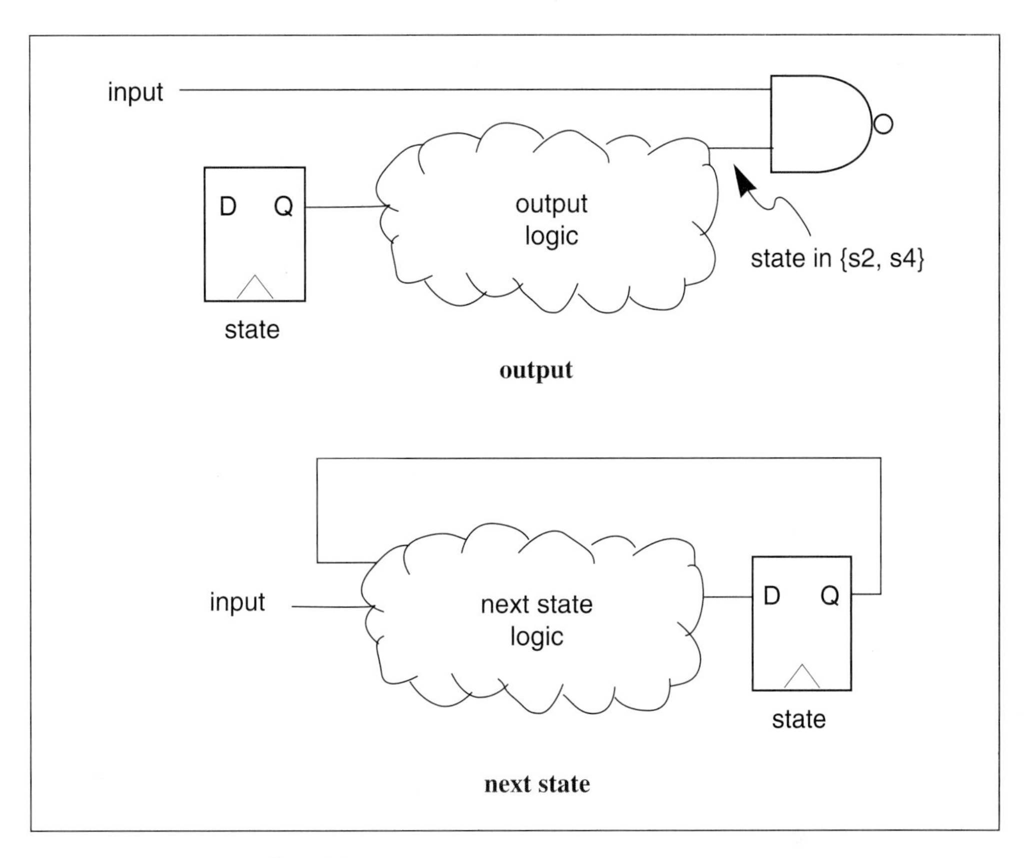

Figure 5-14 How state codes affect delay.

```
always @(posedge clock) // start execution at the clock edge
begin
    // insert combinational logic here
end
```

structural models

A structural model of a sequential machine is built exactly the same way as is a structural model of a combinational logic block. We can wire sequential machines to each other and to combinational logic to build complex machines. In doing so, we just need to obey the rules of construction for sequential machines that we will cover in Section 5.4.

state transition graph models

State transition graphs are described as conditionals within the *always* statement:

```
always @(posedge clock) // start execution at the clock edge
begin
    if (rst == 1)
```

```
        begin
            // reset code
        end
    else begin // state machine
        case (state)
            'state0: begin
                    o1 = 0;
                    state = 'state1;
                    end
            'state1: begin
                    if (i1) o1 = 1; else o1 = 0;
                    state = 'state0;
        endcase
    end // state machine
end
```

The combinational block in this module starts with a test of a reset signal and executes some reset logic if required; including a reset mode in a state machine is good practice. The state transition logic itself is included in a *case* statement. The *case* tests a register called *state* that are the outputs of the state registers. The various values of the *case* statement define what is to be done in each state. Conditional actions can be written with *if*, *case*, *etc*.

The next example is a controller for a traffic light at the intersection of two roads. This example is especially interesting in that it is constructed from several communicating finite-state machines; just as decomposing stick diagrams into cells helped organize layout design, decomposing sequential machines into communicating FSMs helps organize machine design.

Example 5-4 A traffic light controller

We want to control a road using a traffic light:

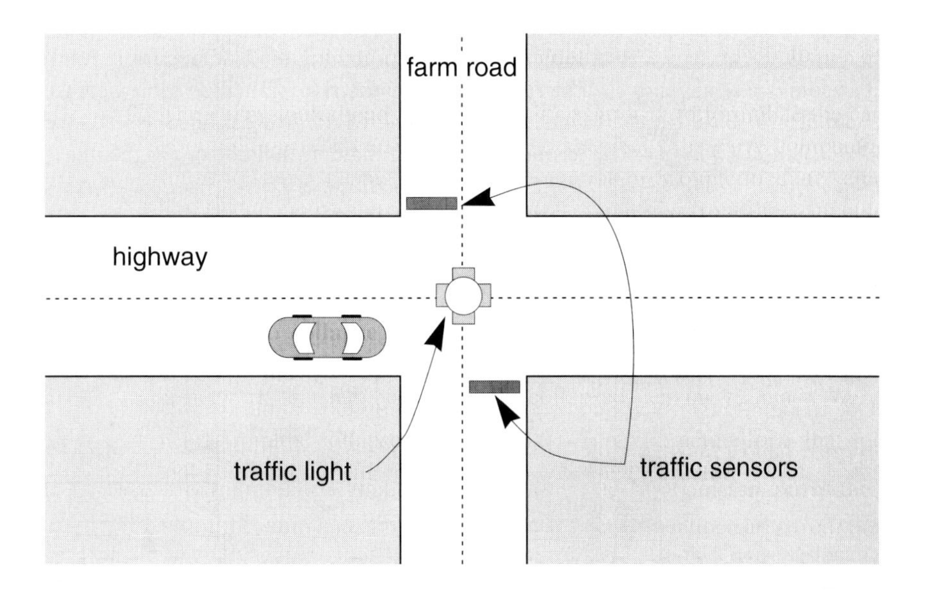

There are many possible schemes to control when the light changes. We could alternate between giving the two roads green lights at regular intervals; that scheme, however, wouldn't give us an interesting sequential machine to study. A slightly more complex and interesting system can be built by taking traffic loads into account. The highway will generally have more traffic, and we want to give it priority; however, we do not want to completely block traffic on the farm road from crossing the highway. To balance these competing concerns, we install traffic sensors on the farm road at the intersection. If there is no traffic waiting on the farm road, the highway always has a green light. When traffic stops at the farm road side of the intersection, the traffic lights are changed to give the farm road a green light as long as there is traffic. But since this simple rule allows the highway light to be green for an interval too short to be safe (consider a second farm road car pulling up just as the highway light has returned to green), we ensure that the highway light (and, for similar reasons, the farm light) will be green for some minimum time.

We must turn this vague, general description of how the light should work into an exact description of the light's behavior. This precise description takes the form of a state transition graph. How do we know that we have correctly captured the English description of the light's behavior as a state transition table? It is very difficult to be absolutely sure, since the English description is necessarily ambiguous, while the state transition table is not. However, we can check the state transition table by mentally executing the machine for several cycles and checking the result given by the state transition table against what we intuitively expect the machine to do. We can also assert several universal claims about the light's behavior: at least one light must be red at any time; lights must always follow a *green* → *yellow* → *red* sequence; and a light must remain green for the chosen minimum amount of time.

We will use a pair of communicating sequential machines to control the traffic light:

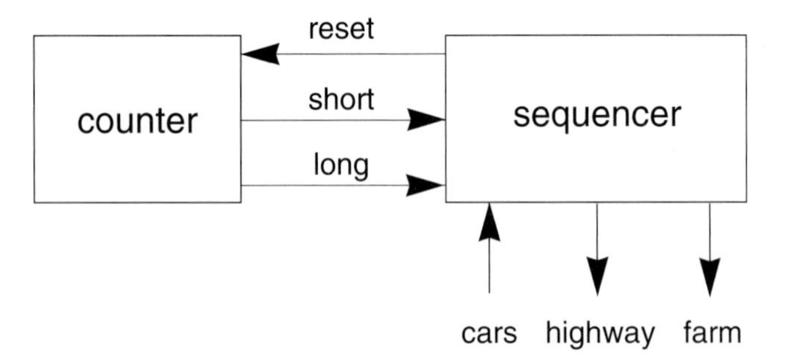

The system consists of a counter and a sequencer. Both are finite-state machines, but each serves a different purpose in the system. The counter counts clock cycles, starting when its *reset* input is set to 1, and signals two different intervals—the *short* signal controls the length of the yellow light, while the *long* signal determines the minimum time a light can be green.

The sequencer controls the behavior of the lights. It takes as inputs the car sensor value and the timer signals; its outputs are the light values, along with the timer reset signals. The sequencer's state transition graph looks like this:

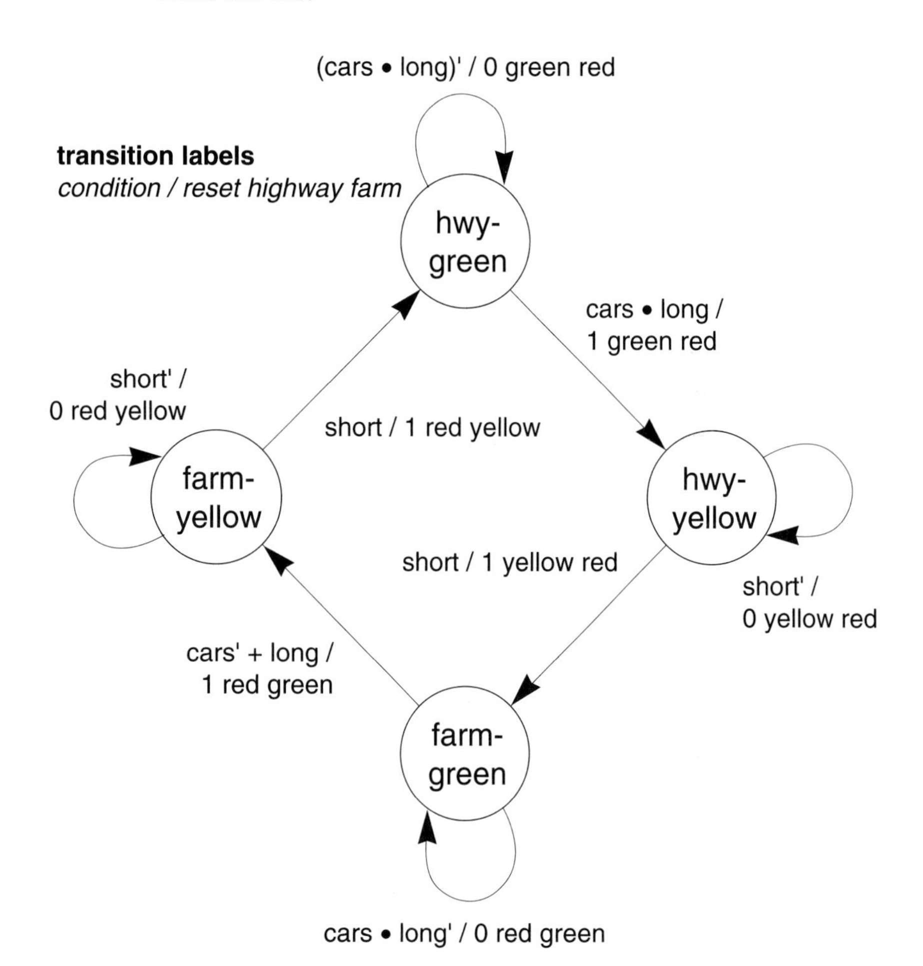

The states are named to describe the value of one of the lights; the complete set of light values, however, is presented at the machine's outputs on every cycle. Tracing through the state transition graph shows that this sequencer satisfies our English specification: the highway light remains green until cars arrive at the farm road (as indicated by the sensor) and the minimum green period (specified by the long timeout) is met. The machine then sets the highway light yellow for the proper amount of time, then sets the highway light to red and the farm light to green. The farm light remains green so long as cars pull up to the intersection, but no longer than the long timeout period. Inspection also shows that the state transition graph satisfies all our assertions: one light is always red, each light always changes colors in the *green* → *yellow* → *red* sequence; and each light, when it turns green, remains green for at least the period specified by the *long* timer.

We can also write the state transition graph as a table. Some of the transitions in the graph are labeled with OR conditions, such as *cars* + *long'*. Since each line in a state transition table can only refer to the AND of input conditions, we must write the OR conditions in multiple lines. For example, one line can specify a transition out of the *farm-green* state when *cars* = 1, while another can specify the same next state and outputs when *long* = 0.

Here is the Verilog model for the sequencer:

```
module
sequencer(rst,clk,cars,long,short,hg,hy,hr,fg,fy,fr,count_reset);
    input rst, clk; // reset and clock
    input cars; // high when a car is present at the farm road
    input long, short; // long and short timers
    output hg, hy, hr; // highway light: green, yellow, red
    output fg, fy, fr; // farm light: green, yellow, red
    reg hg, hy, hr, fg, fy, fr; // remember these outputs
    output count_reset; // reset the counter
    reg count_reset; // register this value for simplicity

// define the state codes
'define HWY_GREEN 0
'define HWY_YX 1
'define HWY_YELLOW 2
'define HWY_YY 3
'define FARM_GREEN 4
'define FARM_YX 5
'define FARM_YELLOW 6
'define FARM_YY 7

    reg [2:0] state; // state of the sequencer

    always @(posedge clk)
    begin
        if (rst == 1)
            begin
                state = 'HWY_GREEN; // default state
                count_reset = 1;
            end
    else begin // state machine
```

```
count_reset = 0;
case (state)
    'HWY_GREEN: begin
        if (~(cars & long)) state = 'HWY_GREEN;
            else begin
                state = 'HWY_YX;
                count_reset = 1;
                end
        hg = 1; hy = 0; hr = 0; fg = 0; fy = 0; fr = 1;
        end
    'HWY_YX: begin
        state = 'HWY_YELLOW;
        hg = 0; hy = 1; hr = 0; fg = 0; fy = 0; fr = 1;
        end
    'HWY_YELLOW: begin
        if (~short) state = 'HWY_YELLOW;
            else begin
                state = 'FARM_YY;
                end
        hg = 0; hy = 1; hr = 0; fg = 0; fy = 0; fr = 1;
        end
    'FARM_YY: begin
        state = 'FARM_GREEN;
        hg = 0; hy = 0; hr = 1; fg = 1; fy = 0; fr = 0;
       end
    'FARM_GREEN: begin
        if (cars & ~long) state = 'FARM_GREEN;
            else begin
                state = 'FARM_YX;
                count_reset = 1;
                end
        hg = 0; hy = 0; hr = 1; fg = 1; fy = 0; fr = 0;
        end
    'FARM_YX: begin
        state = 'FARM_YELLOW;
        hg = 0; hy = 0; hr = 1; fg = 1; fy = 0; fr = 0;
        end
    'FARM_YELLOW: begin
        if (~short) state = 'FARM_YELLOW;
            else begin
                state = 'HWY_GREEN;
                end
        hg = 0; hy = 0; hr = 1; fg = 0; fy = 1; fr = 0;
```

```
                end
            'HWY_YY: begin
                state = 'HWY_GREEN;
                hg = 1; hy = 0; hr = 0; fg = 0; fy = 0; fr = 1;
                end
            endcase
        end // state machine
    end // always
endmodule
```

The Verilog model includes more states than we saw in the specification of the state machine. We needed to add an intermediate state between each of the four main states. We will use a synchronous counter for the long and short timeouts; this timer takes a cycle between the reset and a change in the values. To compensate for this, we add a state that ignores the timer outputs for one cycle.

This model also includes some reset logic that was not included in the state transition diagram; we use *HWY_GREEN* as the default state. We define constants for the state encodings to make the Verilog code easier to read and change. The first case statement determines the next state of the machine and sets the output based upon the current state of the machine. We can check this code to be sure that at least the light is red at all times.

The sequencer and counter work in tandem to control the traffic light's operation. The counter can be viewed as a subroutine of the sequencer—the sequencer calls the counter when it needs to count out an interval of time, after which the counter returns a single value. The traffic light controller could be designed as a single machine, in which the sequencer counts down the long and short time intervals itself, but separating the counter has two advantages. First, we may be able to borrow a suitable counter from a library of pre-designed components, saving us the work of even writing the counter's state transition table. Second, even if we design our own counter, separating the machine states that count time intervals (counter states) from the machine states that make decisions about the light values (sequencer states), clarifies the sequencer design and makes it easier to verify.

Let us assume for the moment that the long counter is 4 bits and the short counter is 3 bits; These are short times for the final implementation, but they allow us to use shorter simulation sequences to simulate the system. Here is the Verilog for the counter:

```
module timer(rst,clk,long,short);
    input rst, clk; // reset and clock
```

```
    output long, short; // long and short timer outputs

    reg [3:0] tval; // current state of the timer

    always @(posedge clk) // update the timer and outputs
        if (rst == 1)
            begin
            tval = 4'b0000;
            short = 0;
            long = 0;
            end // reset
        else begin
            {long,tval} = tval + 1; // raise long at rollover
            if (tval == 4'b0100)
                short = 1'b1; // raise short after 2^2
        end // state machine
endmodule
```

And finally, here is the structural Verilog for the complete traffic light controller, composed of the sequencer and counter:

```
module tlc(rst,clk,cars,hg,hy,hr,fg,fy,fr);
    input rst, clk; // reset and clock
    input cars; // high when a car is present at the farm road
    output hg, hy, hr; // highway light: green, yellow, red
    output fg, fy, fr; // farm light: green, yellow, red

    wire long, short, count_reset; // long and short
                                    // timers + counter reset

    sequencer s1(rst,clk,cars,long,short,
                 hg,hy,hr,fg,fy,fr,count_reset);
    timer t1(count_reset,clk,long,short);

endmodule
```

5.4 Rules for Clocking

In this section we will consider the rules that must be followed to build a properly functioning sequential system. This is not the same as optimizing the system for performance. These rules ensure there is *some speed* at which the machine will properly implement the output and next-state functions. When we optimize the system for performance, we must make sure that we do not violate the rules that ensure proper operation of the machine.

We will first look at the flip-flops and latches used to store the machine's state. We will then consider the rules for clocking.

5.4.1 Flip-Flops and Latches

Building a sequential machine requires **registers** that read a value, save it for some time, and then can write that stored value somewhere else, even if the element's input value has subsequently changed. A Boolean logic gate can compute values, but its output value will change shortly after its input changes. Each alternative circuit used as a register has its own advantages and disadvantages.

register characteristics

A generic register has an internal memory and some circuitry to control access to the internal memory. Access to the internal memory is controlled by the *clock* input—the register reads its *data* input value when instructed by the clock and stores that value in its memory. The output reflects the stored value, probably after some delay. Registers differ in many key respects:

- exactly what form of clock signal causes the input data value to be read;
- how the behavior of *data* around the read signal from *clock* affects the stored value;
- when the stored value is presented to the output;
- whether there is ever a combinational path from the input to the output.

latches vs. flip-flops

Introducing a terminology for registers requires caution—many terms are used in slightly or grossly different ways by different people. We choose to follow Dietmeyer's convention [Die78] by dividing registers into two major types:

- **Latches** are transparent while the internal memory is being set from the data input—the (possibly changing) input value is transmitted to the output.
- **Flip-flops** are not transparent—reading the input value and changing the flip-flop's output are two separate events.

The main attraction of latches in VLSI design is that they require fewer transistors and are smaller than flip-flops. However, most FPGA cells include both latches and flip-flops, so area is not a consideration. The latch vs. flip-flop dichotomy is important because the decision to use latches or flip-flops dictates substantial differences in the structure of the sequential machine.

Figure 5-15 Setup and hold times.

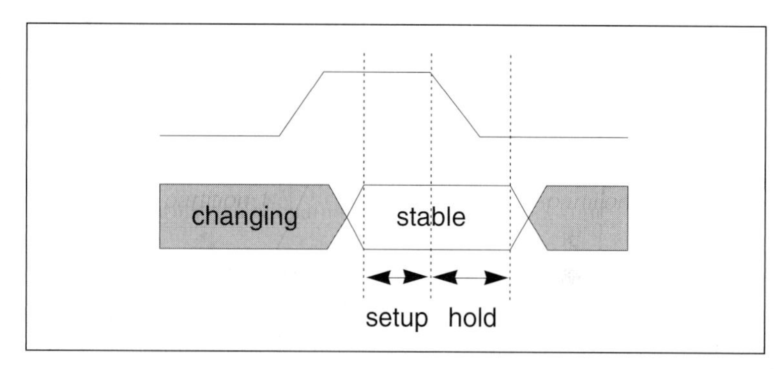

Registers can also be categorized along another dimension, namely the types of data inputs they present:

- The most common data input type in VLSI design is the **D-type** register. Think of "D" as standing for data—the Q output of the register is determined by the D input value at the clocking event.
- The **T-type** register toggles its state when the T input is set at the clocking event.
- The **SR-type** register is either set by the S input or reset by the R input (the S and R inputs are not allowed to be 1 simultaneously).
- The **JK-type** is similar but its J and K inputs can both be 1. The other register types can be built using the JK-type as a component.

register behavior

The behavior of a register can vary depending on the type of register, but all registers store a value—retaining a state based on that value—under control of a clock signal. We will refer to a change in the clock signal as a **clock event**. In a typical clock cycle there will be two clock events: the 0-to-1 transition and the 1-to-0 transition. A register usually stores its value at one of those events. The data value to be latched must remain stable around the time of the clock event to ensure that the memory element retains the proper value. In Figure 5-15, the register stores the input value around the clock's falling edge, as is typical of a latch.

timing characteristics

The two most commonly quoted parameters of a memory element are its **setup time** and **hold time**, which define the relationship between the clock and input data signals. The setup time is the minimum time the data input must be stable before the clock signal changes, while the hold time is the minimum time the data must remain stable after the clock changes.

Another important parameter of a register is its **propagation time**, which is the amount of time required for a new data value to reach the register's output. The setup and propagation times, along with the delay times through the combinational logic, determine how fast the system can run.

register features

Registers may provide additional inputs. A load input causes the register to be loaded only when the load signal is asserted and the appropriate clock event happens; if the clock event arrives but load is not asserted then the register does not change state. A clear input sets the register's state to 0. A clear input may be synchronous (clears only at the clock event) or asynchronous (clears at any time).

clock signal characteristics

The characteristics of the clock signal used to control the register are important. An error in the clock signal can cause a register to overwrite its state with a different, bad value or to incorrectly store an incoming value. The exact characteristics of the clock depend on the circuits used in the registers that the clock controls, but we generally require that the clock signal not have glitches and that the clock edge events have fast rise/fall times. A parameter of the clock signal is its **duty cycle**—the fraction of the clock period for which the clock is active.

5.4.2 Clocking Disciplines

We need reliable rules that tell us when a circuit acts as a sequential machine—we can't afford to simulate the circuit thoroughly enough to catch the many subtle problems which can occur. A **clocking discipline** is a set of rules that tell us:

- how to construct a sequential system from gates and registers;
- how to constrain the behavior of the system inputs over time.

Adherence to the clocking discipline ensures that the system will work at *some* clock frequency. Making the system work at the required clock frequency requires additional analysis and optimization.

The constraints on system inputs are known as **signal types**, which define both how signals behave over time and what signals can be combined in logic gates or registers. By following these rules, we can ensure that the system will operate properly at some rate; we can then worry about optimizing the system to run as fast as possible while still functioning correctly.

common clocking rules

Different register types require different rules, so we will end up with a family of clocking disciplines. All disciplines have two common rules, however. The first is simple:

> **Clocking rule 1**: *Combinational logic gates cannot be connected in a cycle.*

Gates connected in a cycle form a primitive register and cease to be combinational—the gates' outputs depend not only on the inputs to the cycle but the values running around the cycle. In fact, this rule is stronger than is absolutely necessary. It is possible to build a network of logic gates which has cycles but is still combinational—the values of its outputs depend only on the present input values, not past input values. However, careful analysis is required to ensure that a cyclic network is combinational, whereas cycles can be detected easily. For most practical circuits, the acyclic rule is not overly restrictive.

The second common rule is somewhat technical:

> **Clocking rule 2:** *All components must have bounded delay.*

This rule is easily satisfied by standard components, but does rule out synchronizers for asynchronous signals.

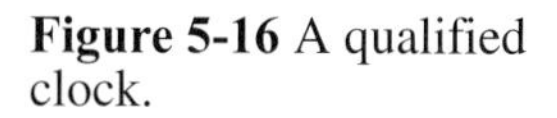

Figure 5-16 A qualified clock.

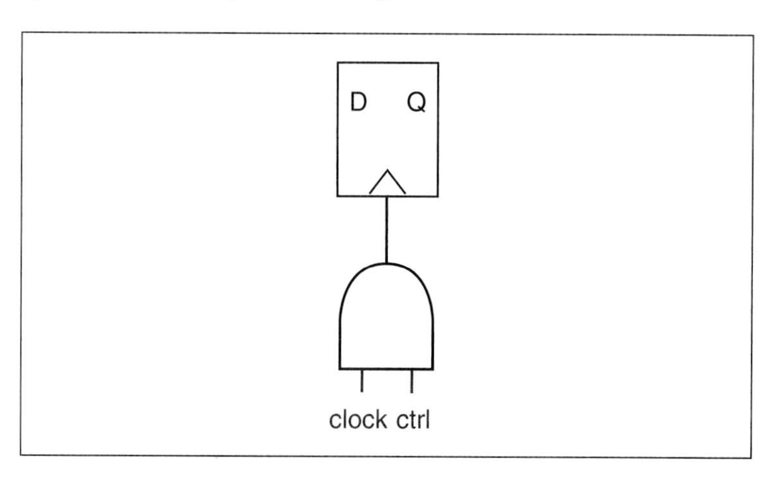

qualified clocks

In addition to clocks, we occasionally use the **qualified clock.** A qualified clock is a logical combination of a clock with another signal. The register's clock input is controlled by $\mathrm{clock} \cdot \mathrm{ctrl}$. If *clock* = 1 but *ctrl* = 0, the register will not be clocked. The *ctrl* acts to gate the clock and selectively load the register. However, qualified clock signals must be generated carefully to avoid timing violations. We need to minimize the timing differences between changes on the clock and the qualified clock. This means that we want to use a small amount of logic to generate the qualified clock signal and design the wiring to minimize the delay from the clock signal to the register's clock input.

clocking discipline for flip-flops

The clocking discipline for systems built from flip-flops is simplest, so let's consider that first. A flip-flop system looks very much like that of the generic sequential system, with a single rank of registers.

We can define conditions that the clock and data signals must satisfy which are conservative but safe. A flip-flop system has one type of clock signal, ϕ, and one type of data signal, S, as shown in Figure 5-17. The figure assumes that the flip-flops read their inputs on the positive ($0 \rightarrow 1$) clock edge. The data inputs must have reached stable values at the flip-flop inputs on the rising clock edge, which gives this requirement on the primary inputs:

> **Flip-flop clocking rule 1:** *All primary inputs can change only in an interval just after the clock edge. All primary inputs must become stable before the next clock edge.*

The length of the clock period is adjusted to allow all signals to propagate from the primary inputs to the flip-flops. If all the primary inputs satisfy these conditions, the flip-flops will latch the proper next-state values. The signals generated by the flip-flops satisfy the clocking discipline requirements.

clocking discipline for latches

A single rank of flip-flops cutting the system's combinational logic is sufficient to ensure that the proper values will be latched—a flip-flop can simultaneously send one value to its output and read a different value at its input. Sequential systems built from latches, however, are normally built from two ranks of latches. To understand why, consider the relationships of the delays through the system to the clock signal which controls the latches, as illustrated in Figure 5-18. The delay from the present state input to the next state output is very short. As long as the latch's clock is high, the latch will be transparent. If the clock signal is held high long enough, the signal can make more than one loop around the system: the next-state value can go through the latch, change the value on the present state input, and then cause the next-state output to change.

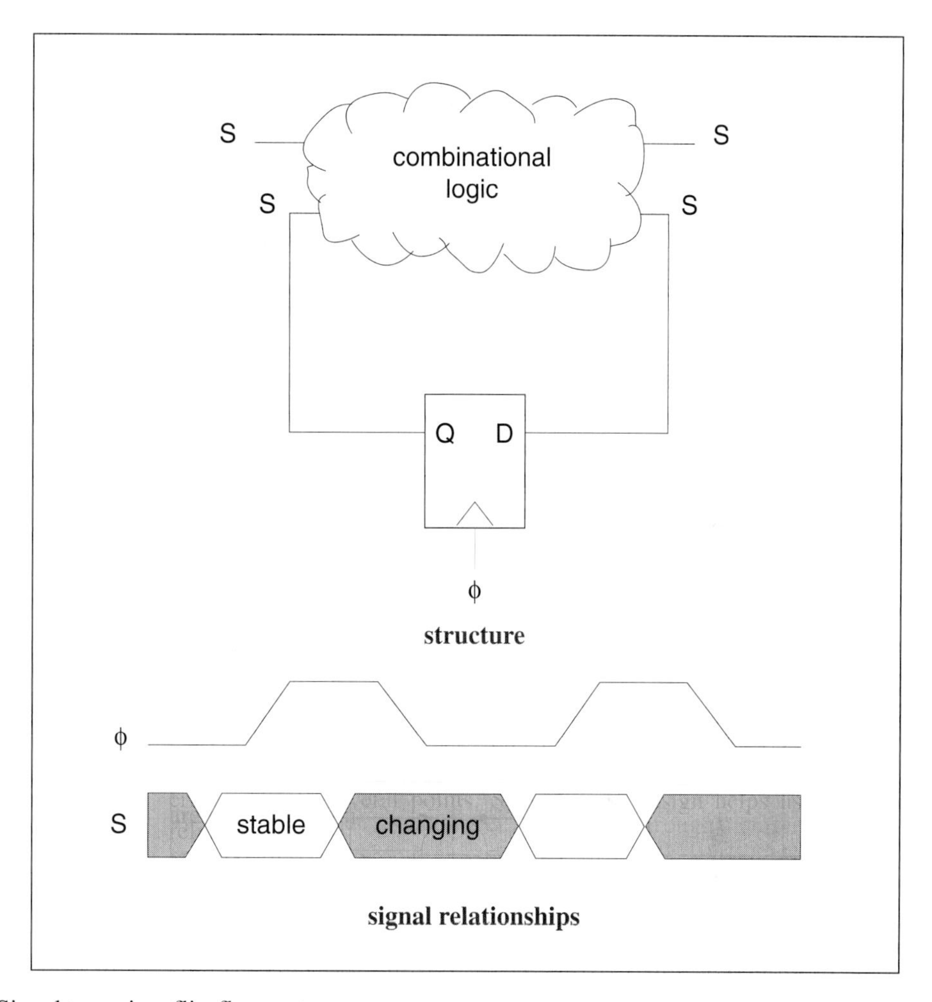

Figure 5-17 Signal types in a flip-flop system.

two-sided clocking constraint

In such a system, the clock must be high long enough to securely latch the new value, but not so long that erroneous values can be stored. That restriction can be expressed as a **two-sided** constraint on the relative lengths of the combinational logic delays and the clock period:

- the latch must be open less than the shortest combinational delay;
- the period between latching operations must be longer than the longest combinational delay.

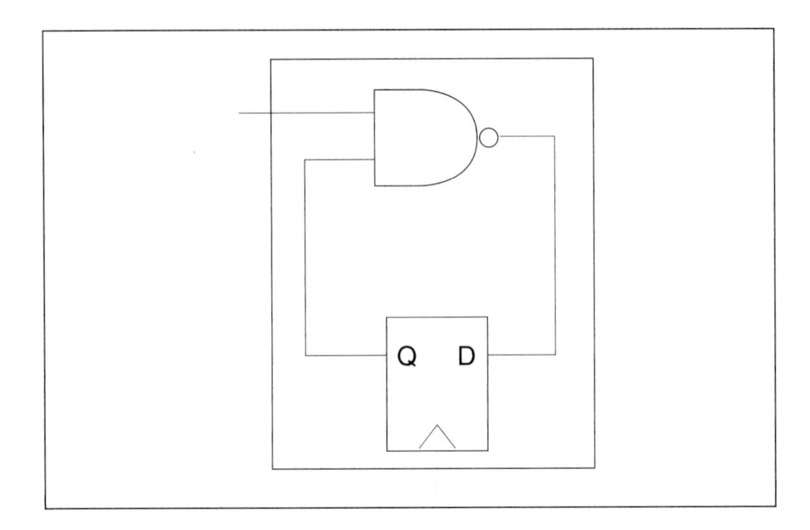

Figure 5-18 Single latches may let data shoot through.

It is possible to meet two-sided constraint, but it is very difficult to make such a circuit work properly.

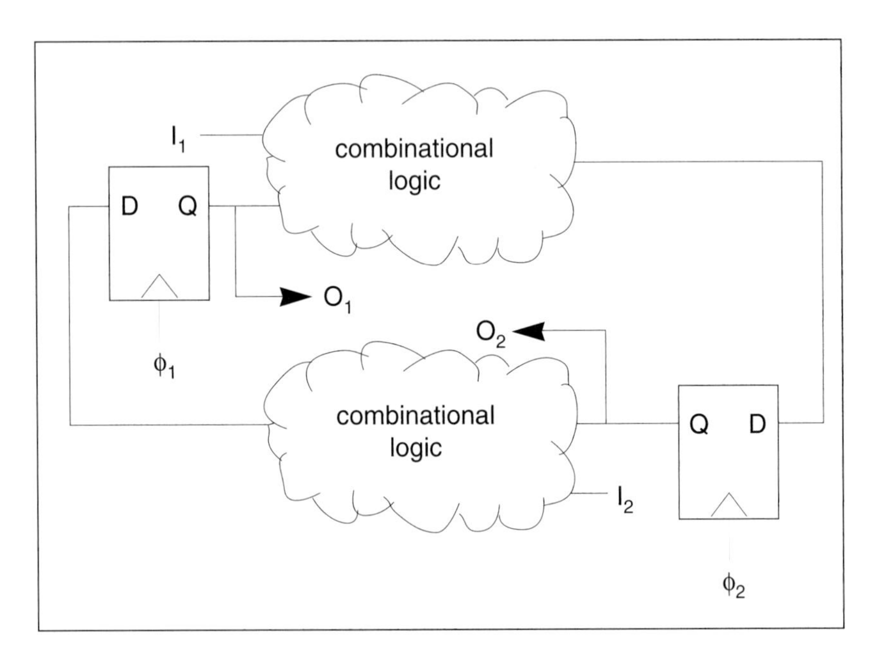

Figure 5-19 The strict two-phase system.

two-phase clocking discipline

A safer architecture—the **strict two-phase** clocking discipline system—is shown in Figure 5-19. Each loop through the system is broken by two ranks of latches:

> **Two-phase clocking rule 1:** *Every cycle through the logic must be broken by n ϕ_1 latches and n ϕ_2 latches.*

The latches are controlled by the **non-overlapping** clock phases clocks shown in Figure 5-20. A ϕ_1-high, ϕ_2-high sequence forms a complete clock cycle. The non-overlapping clocks ensure that no signal can propagate all the way from a latch's output back to its output. When ϕ_1 is high, the ϕ_2-controlled latches are disabled; when ϕ_2 is high, the ϕ_1 latches are off. As a result, the delays through combinational logic and clocks in the strict two-phase system need satisfy only a **one-sided** timing constraint: each phase must be longer than the longest combinational delay through that phase's logic. A one-sided constraint is simple to satisfy—if the clocks are run slow enough, the phases will be longer than the maximum combinational delay and the system will work properly. (A chip built from dynamic latches that is run so slowly that the stored charge leaks away won't work, of course. But a chip with combinational logic delays over a millisecond wouldn't be very useful anyway.)

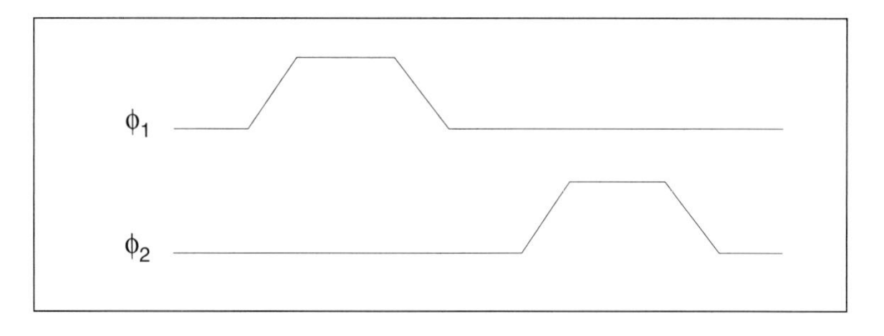

Figure 5-20 A two-phase, non-overlapping clock.

correctness of the two-phase discipline

It is easy to see that we can stretch the clock phases and inter-phase gaps to ensure that the strict two-phase system works. The inputs to the combinational logic block at the ϕ_1 latch outputs are guaranteed to have settled by the time ϕ_1 goes low; the outputs of that block must have settled by the time ϕ_2 goes low for the proper values to be stored in the ϕ_2 latches. Because the block is combinational there is an upper bound on the delay from settled inputs to settled outputs. If the time between the falling edges of ϕ_1 and ϕ_2 is made longer than that maximum delay, the correct state will always be read in time to be latched. A similar argument can be made for the ϕ_1 latches and logic attached to their outputs.

Therefore, if the clock cycle is properly designed, the system will function properly.

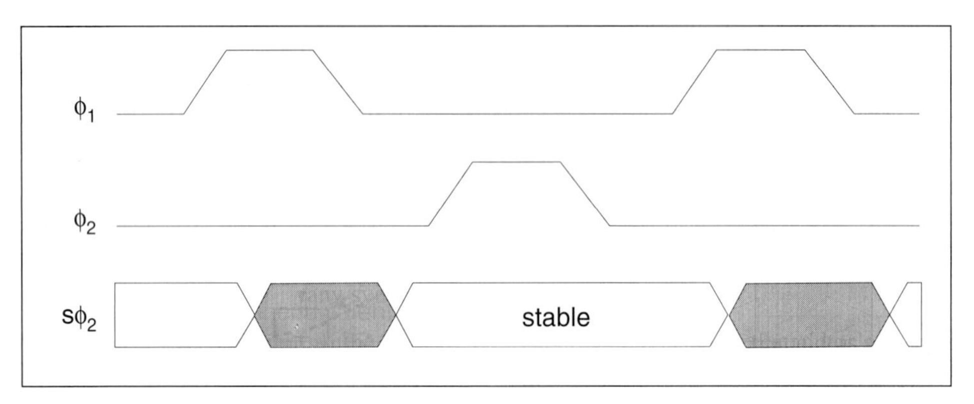

Figure 5-21 A stable ϕ_2 signal.

two-phase clocking types

The strict two-phase system has two clock types, $\phi_1$and ϕ_2. Each clock type has its own data type [Noi82], the **stable** signal, which is equivalent to a **valid** signal on the opposite clock phase. Figure 5-21 shows the two clock phases and the output of a ϕ_1-clocked latch. The latch output changes only during a portion of the ϕ_1 phase. It therefore meets the setup and hold requirements of the succeeding ϕ_2 latch once the clock phase durations are properly chosen. Because the signal's value is settled during the entire ϕ_2 portion of the clock cycle, it is called **stable** ϕ_2, abbreviated as **sϕ_2**. The output of a ϕ_2 -clocked latch is stable ϕ_1. A sϕ_2 signal is also called **valid** ϕ_1, abbreviated as **vϕ_1**, since it becomes valid around the time the ϕ_1 latch closes. Similarly, a signal that is stable during the entire ϕ_1 portion of the clock is known as **stable** ϕ_1 or sϕ_1.

combining clocking types

Figure 5-22 summarizes how clocking types combine. Combinational logic preserves signal type: if all the inputs to a gate are sϕ_1 then its output is sϕ_1. Clocking types cannot be mixed in combinational logic: a gate cannot have both sϕ_1 and sϕ_2 inputs. The input to a ϕ_1-controlled latch is sϕ_1 and its output is sϕ_2.

two-coloring

Figure 5-23 shows how signal types are used in the strict two-phase system. The system can have inputs on either phase, but all inputs must be stable at the defined times. The system can also have outputs on either phase. When two strict two-phase systems are connected, the connected inputs and outputs must have identical clocking types. Assigning clock-

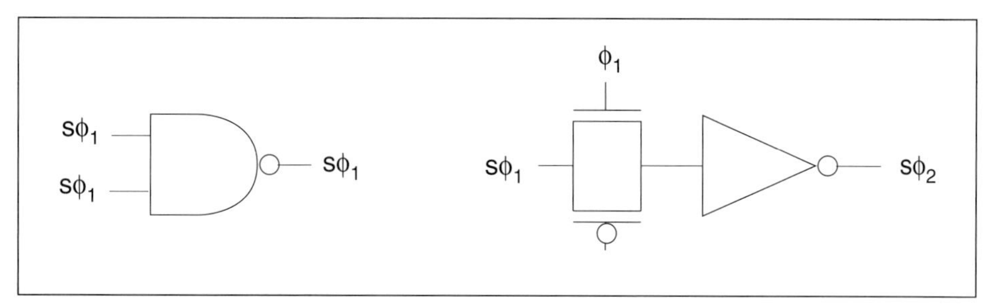

Figure 5-22 How strict two-phase clocking types combine.

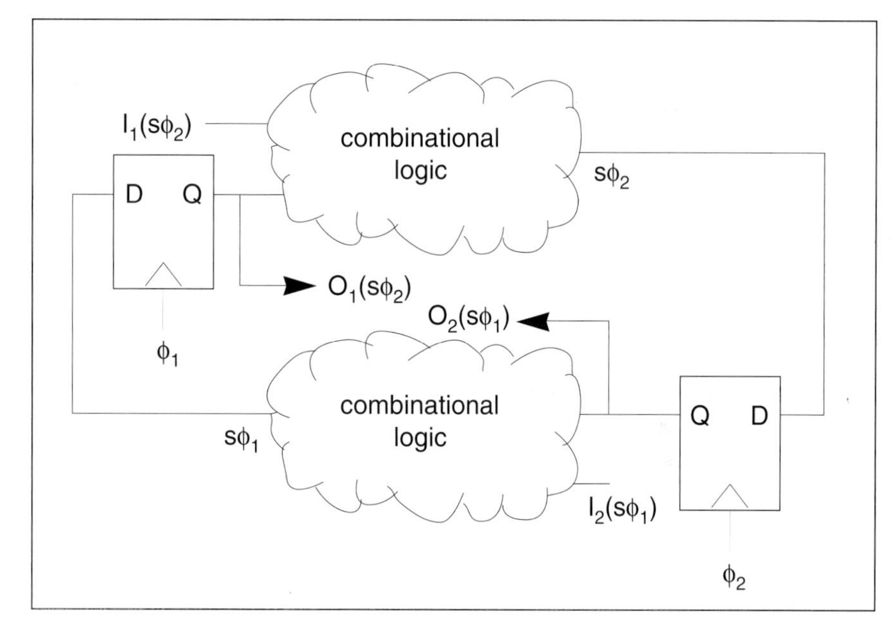

Figure 5-23 Clocking types in the strict two-phase system.

ing types to signals in a system ensures that signals are properly combined, but it will not guarantee that all loops are broken by both ϕ_1 and ϕ_2 latches. This check can be performed by **two-coloring** the block diagram. To two-color a schematic, color ϕ_1 and all signals derived from it red, and all ϕ_2-related signals green. For the system to satisfy the two-phase clocking discipline, the two-colored diagram must satisfy these rules:

- No latch may have an input and output signal of the same color.

- The latch input signal and clock signal must be of the same color.
- All signals to a combinational logic element must be of the same color.

The two-coloring check is a simple way to ensure that the rules of the clocking discipline are satisfied.

qualified clocks and latches

In the strict two-phase system there are two qualified clock types, qualified ϕ_1 ($q\phi_1$) and qualified ϕ_2 ($q\phi_2$). Qualified clocks may be substituted for clocks at latches. Since a static latch controlled by a qualified clock is no longer refreshed on every clock cycle, the designer is responsible for ensuring that the latch is reloaded often enough to refresh the storage node and to ensure that at most one transmission gate is on at a time. Qualified clocks are generated from the logical AND of a stable signal and a clock signal. For instance, a $q\phi_1$ signal is generated from a $s\phi_1$ signal and the ϕ_1 clock phase. When the clock is run slowly enough, the resulting signal will be a stable 0 or 1 through the entire ϕ_1 period.

5.5 Performance Analysis

Clocking disciplines help us construct systems that will operate at some speed. However, we usually want the system to run at some minimum clock rate. We saw in Section 4.4 how to determine the delays through combinational logic. We need additional analysis to ensure that the machine will run at the clock rate we want.

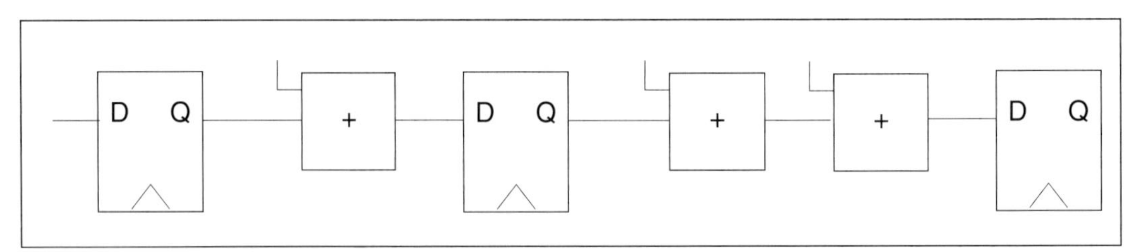

Figure 5-24 A sequential system with unbalanced delays.

logic delays

The desired clock period determines the maximum allowable delays through the combinational logic. Because most machines are built from several blocks of combinational logic and the clock period is determined by the *longest* combinational delay, the worst-case delay through the system may not be obvious when you look at the component blocks of combinational logic in isolation. Consider the system of Figure 5-24: one path from flip-flop to flip-flop has one adder while the other has two. The system's clock period will be limited by the two-adder path. In some cases, the result of a long path is not used every cycle—for example, the flip-flop may be conditionally loaded. The system's clock period is still limited by the delay through this occasionally-used path, since we can't predict the cycles on which that path will be used.

register characteristics

Our analysis must also take into account the timing characteristics of the registers used in the design. We have already seen that latch-based systems have more complicated timing relationships even when we consider only correctness; determining the clock rate of a latch-based system is also somewhat harder.

skew

Finally, we must consider the effects of **clock skew**. One of the basic assumptions of sequential machine design is that the clock is ideal—the clock arrives instantaneously at all points in the design. That assumption doesn't hold for systems of any reasonable size. Clock skew is the relative delay between the arrival of the clock signal at different physical points in the system. We can factor skew into our performance analysis and understand how it limits clock rate; this analysis helps us determine what parts of our system are most sensitive to clock skew.

FPGA design reports

Design tools report to us about the design characteristics relevant to sequential design. These reports tell us about paths relative to the registers.

5.5.1 Performance of Flip-Flop-Based Systems

To start our analysis of machines with flip-flops, let us make some semi-ideal assumptions:

- The clock signal is perfect, with no rise or fall times and no skew. The clock period is P.
- We will assume for simplicity that the flip-flops store new values on the rising clock edge.
- The flip-flops have setup time s and propagation time p.
- The worst-case delay through the combinational logic is C.

The structure of our sequential machine is shown in Figure 5-25. This is a very generic structure that describes machines with multiple flip-flops and as many combinational logic blocks as necessary.

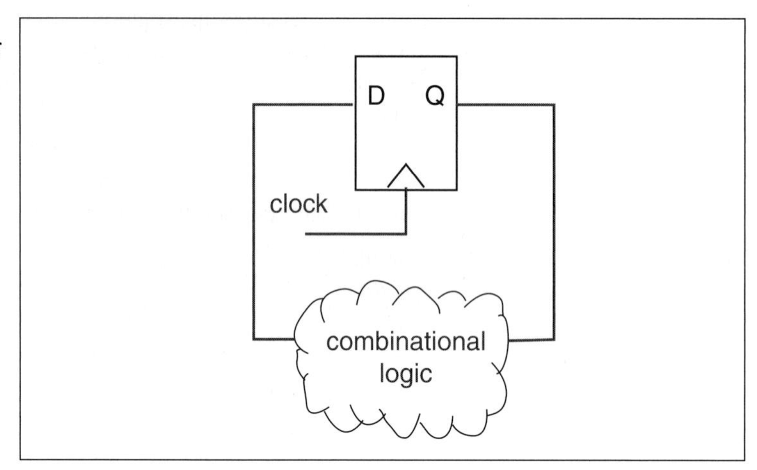

Figure 5-25 Model system for performance analysis of flip-flop-based machines.

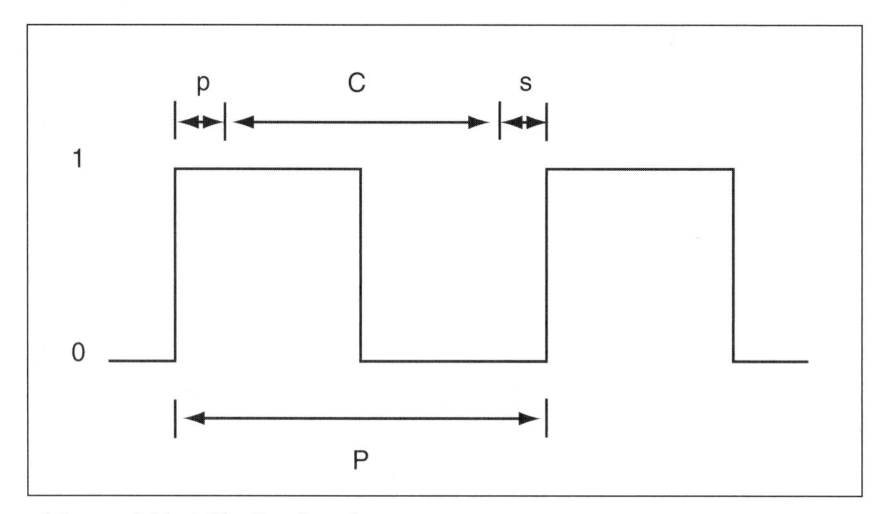

Figure 5-26 Timing of the semi-ideal flip-flop-based system.

Figure 5-26 shows how the various delays through the system fall into the clock period. The physical parameters of the combinational logic and flip-flops contribute to overall delay:

- The flip-flop's propagation time (p) determines how long it takes for the new value to move from the flip-flop's input to its output.
- Once the new value reaches the flip-flop's data output, the new values start to propagate through the combinational logic. They require C time to compute the results.

- The results must arrive at the flip-flop one setup time (s) before the next rising clock edge.

clock period

We can therefore write our constraint on the clock period as

$$P \geq C + s + p\,. \quad \text{(EQ 5-1)}$$

Longer clock periods (lower clock frequencies) will also work. Notice that this discussion does not rely on the **duty cycle** of the clock (the duty cycle is the percentage of the clock period for which the clock is high). The duty cycle does not matter here because the flip-flop is edge-triggered.

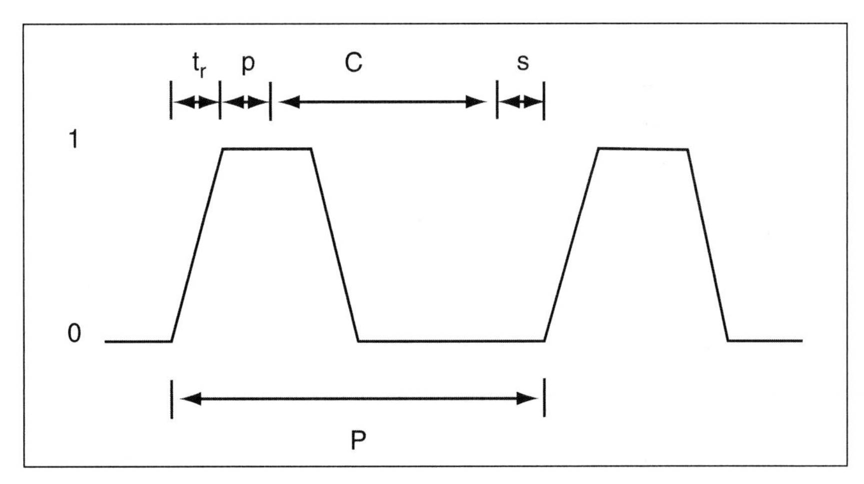

Figure 5-27 Constraints with rise and fall times.

rise and fall times

If we relax one of our assumptions about the clock, we end up with a clock signal with non-zero rise and fall times, as shown in Figure 5-27. In practice, the clock's rise and fall times are noticeable when compared to the other system delays because clock nets are large and hard to drive at high speeds. The rise (t_r) and fall time (t_f) add to the overall clock period:

$$P \geq C + s + p + t_r\,. \quad \text{(EQ 5-2)}$$

min/max delays

One additional non-ideality that you may occasionally see mentioned is **minimum** and **maximum delays**. We have assumed so far that delays are known—we can provide a single number that accurately reflects the delay through the component. However, delay may vary for several reasons:

Figure 5-28 Min/max delays.

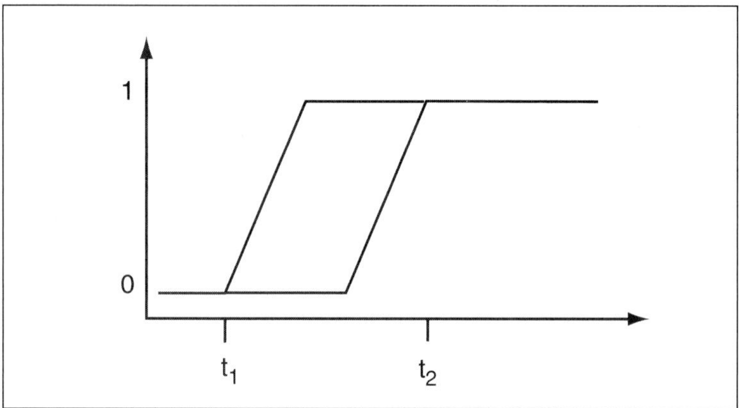

- Manufacturing variations may cause different parts to exhibit different delays.
- Delay may vary with temperature.

Min/max delays provide bounds on the delay—if the delay is given as $[t_1, t_2]$ then the delay is at least t_1 and at most t_2 but can vary anywhere in between. Figure 5-28 shows a timing diagram with min-max delays on a clock signal that goes between 0 and 1, but they can also be applied to logic stable/changing values in general logic signals.

In the worst case, min/max delays can cause substantial problems. If each component exhibits min/max delays and we do not know anything about the relationships between the components' delays, we must assume that delay bounds add—after going through two components, the delay bounds would be $[2t_1, 2t_2]$ and so on. Since we must assume that the signal is not valid during any part of the min/max delay window, we must run the system more slowly unless we can more accurately determine the actual delays through the logic.

However, on a single chip the delays through the components are in fact correlated. The physical parameters that determine delay through the logic and wires vary only slowly over the chip. This means that it is unlikely that one gate on the chip would have the best-case delay while the other would have the worst-case delay. The best-case/worst-case range characterizes variations over all the chips. Therefore, within a chip, min/max bounds tend to be considerably smaller than the min/max bounds across chips. When designing multi-chip systems, we need to make much more pessimistic assumptions.

5.5.2 Performance of Latch-Based Systems

The analysis of latch-based systems follows the same general principles. But the paths through the combinational logic are more complex because latches are transparent.

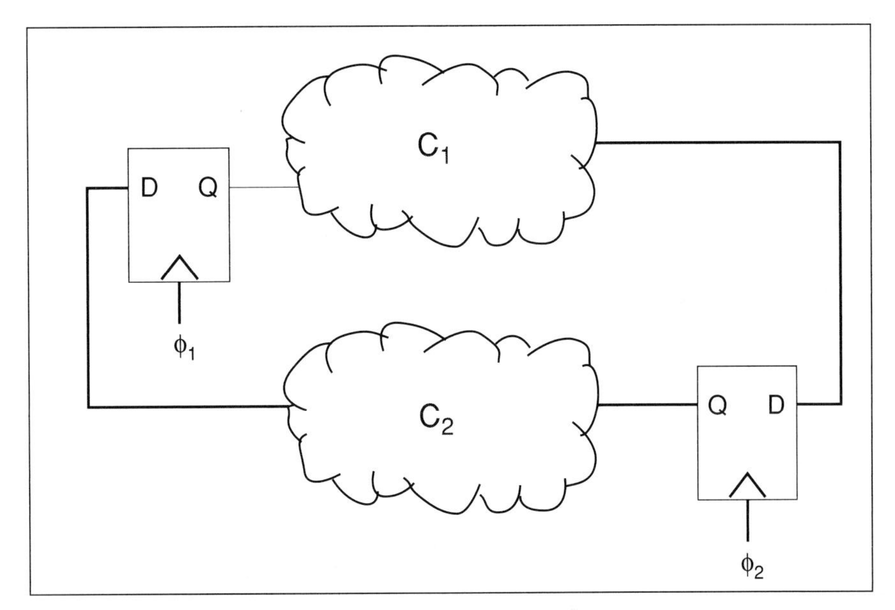

Figure 5-29 Model system for performance analysis of latch-based machines.

Figure 5-29 shows the model system we will use to analyze a two-phase latch-based machine. The delays through the two blocks of combinational logic are C_1 and C_2; we will assume that all the latches have the same setup and propagation times of s and p. We will also assume for convenience that all latches close at the downward edge of their clock.

Figure 5-30 shows the timing chart for the two-phased system. First consider the upper block of logic C_1, which is fed by the ϕ_1-controlled latch. The inputs to that block are not stable until h time units after the downward transition of ϕ_1. They then propagate through the block with C_1 delay and must arrive s time units before the downgrade transition of ϕ_2. A similar analysis can be made for propagation through the C_2 block. This gives a constraint on the clock period of

$$P \geq C_1 + C_2 + 2s + 2p \,. \qquad \text{(EQ 5-3)}$$

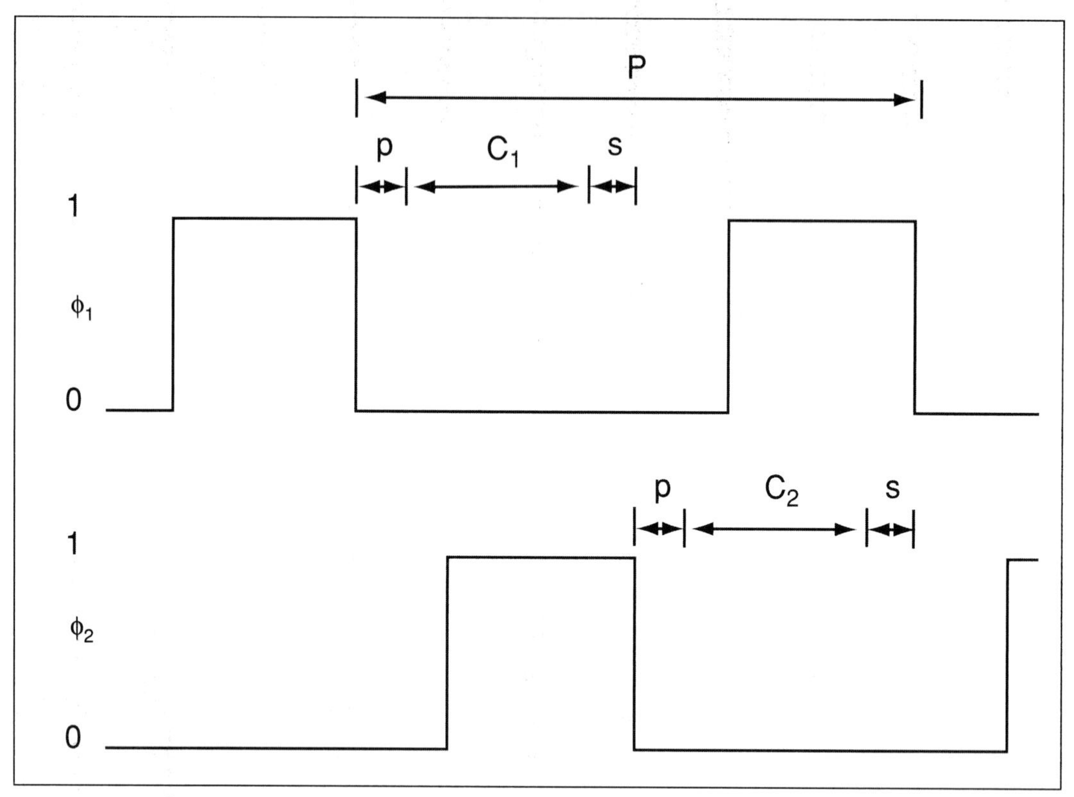

Figure 5-30 Timing chart for a two-phased system.

However, we can improve these results (though implementing this scheme is tricky and not recommended). If the signal from C_1 arrives early, it can shoot through the ϕ_2 latch and start to propagate through the C_2 block. This head start can reduce the total period if a short path in C_1 feeds a long path in C_2. In a latch-based system we can equalize the length of each phase to 50 ns, as shown in Figure 5-31, by taking advantage of the transparency of latches. Ignore for a moment the setup and hold times of the latches to simplify the explanation. Signals that become valid at the end of ϕ_2 propagate through the short-delay combinational logic. If the clock phases are arranged so that $\phi_1 = 1$ when they arrive, those signals can shoot through the ϕ_1 latch and start the computation in the long-delay combinational block. When the ϕ_1 latch closes, the signals are kept stable by the ϕ_1 latch, leaving the ϕ_2 latch free to open and receive the signals at the end of the next 50 *ns* interval. How-

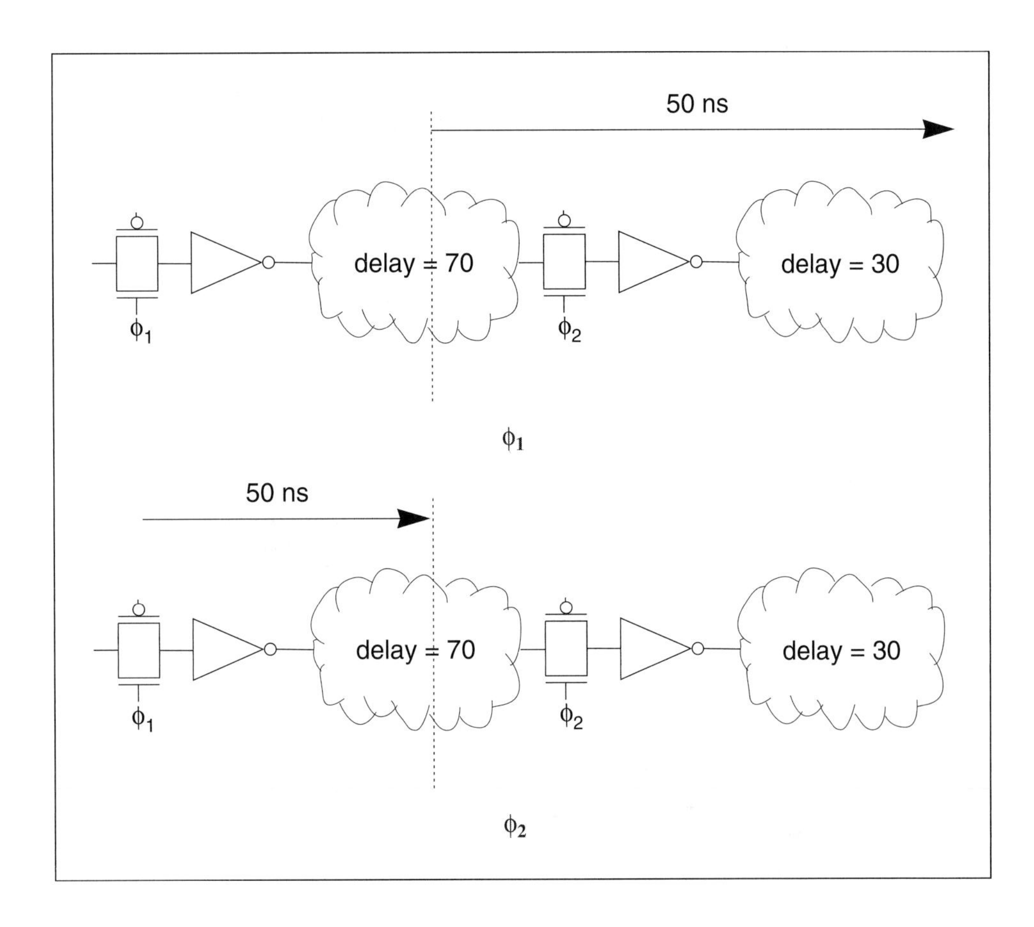

Figure 5-31 Spreading a computation across two phases in a latch-based system.

ever, this scheme violates the strict two-phase clocking discipline. Making sure that you have properly measured the delays and have not violated a timing constraint can be tricky.

5.5.3 Clock Skew

skew problems

Skew is a relative delay between two signals. Skew causes problems when we think we are combining two sets of values but are in fact combining a different set of values. We may see skew between two data signals, a data signal and a clock, or between clock signals in a multi-clock system. In Figure 5-32, the registers that provide inputs a and b produce

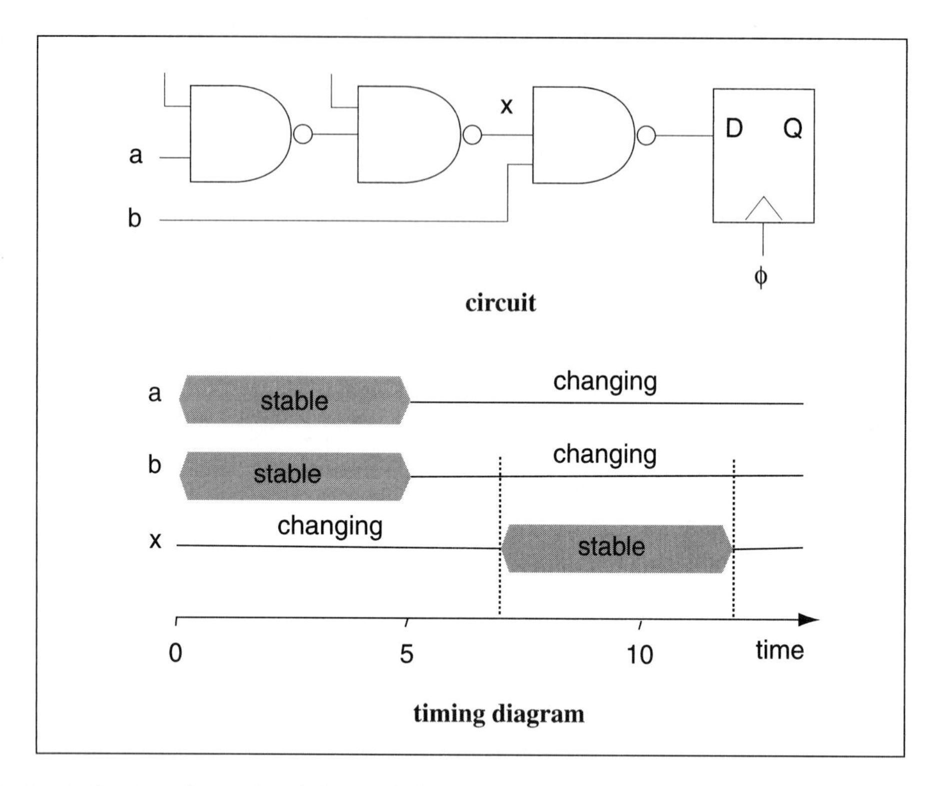

Figure 5-32 A circuit that introduces signal skew relative to a clock.

Figure 5-33 A circuit that can suffer from clock skew.

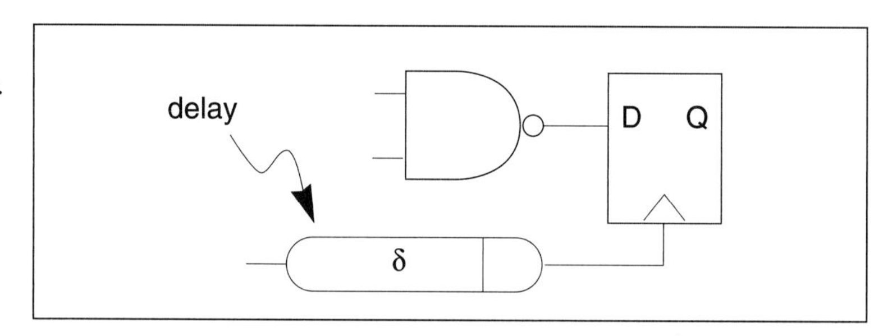

valid signals over the range [0,5 ns]. At those a and b inputs, the two signals are aligned in time to be simultaneously valid. By the time a's signal has propagated to point x, however, the combinational logic has delayed that signal so that it does not become valid until after b has ceased to be valid. As a result, the gate that combines b and x produces garbage during the time window marked by the dotted lines. However, this sort of problem shouldn't occur in a system that satisfies a clocking discipline, since a and b should remain stable until the end of the clock period.

clock skew

In synchronous design, skew relative to one of the clocks is the most important variety. Figure 5-33 illustrates clock skew: the signal provided to the latch is valid from 0 to 5 ns, but the clock does not close the latch until 6 ns; as a result, the latch stores a garbage value. The difficulty of solving this problem depends on the source of the delay on the clock line.

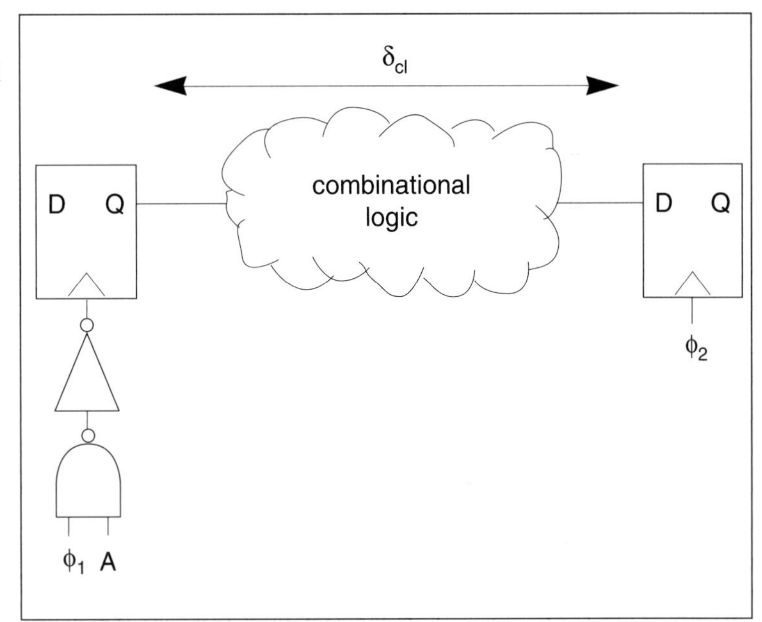

Figure 5-34 Clock skew and qualified clocks.

Clocks can be skewed for several reasons:

- If they are from external sources, they may arrive skewed.
- The delays through the clock network may vary with position on the chip.
- Qualified clocks may add delay to some clock destinations and not others.

Qualified clocks should be used carefully because they introduce clock skew that can invalidate the assumptions made about the relationships between combinational logic delays and clock values. Consider the circuit of Figure 5-34: the ϕ_1 latch is run by a qualified clock while the ϕ_2 latch is not. When the ϕ_1 signal falls at the system input, the clock input to the latch falls δ_{clk} time later. In the worst case, if δ_{clk} is large enough, the ϕ_1 and ϕ_2 phases may both be 1 simultaneously. If that occurs, signals can propagate completely through latches and improper values may be stored.

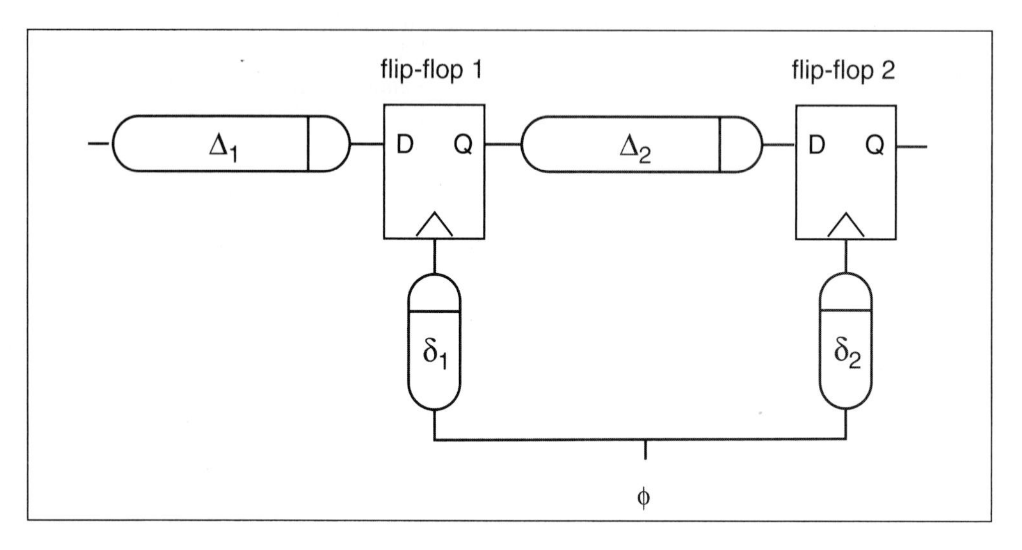

Figure 5-35 Model system for clock skew analysis in flip-flop-based machines.

skew in flip-flop systems

We can build a simple model to understand how clock skew causes problems in flip-flop systems. The model is shown in Figure 5-35. The clock ϕ is distributed to two flip-flops; each connection has its own skew. The combinational logic blocks feeding each flip-flop each have their own delays δ_1 and δ_2. Clock skew is measured from one point to another: the skew from the clock input of flip-flop 1 to the clock input of flip-flop 2 is $s_{12} = \delta_1 - \delta_2$ and the skew from flip-flop 2's clock input to flip-flop 1's clock input is $s_{21} = \delta_2 - \delta_1$. The clock controls when the D input to each flip-flop is read and when the Q output changes. Clock skew gives the signal at the D input more time to arrive but it simultaneously gives the signal produced at the Q output less time to reach the next flip-flop. If we assume that each flip-flop instaneously reads its D input and changes its Q output, then we can write this constraint on the minimum clock period T:

$$T \geq \Delta_2 + \delta_1 - \delta_2 = \Delta_2 + s_{12}. \qquad \text{(EQ 5-4)}$$

This formula tells us that the clock period must be adjusted to take into account the late arrival of the clock at flip-flop 1. This formula also makes it clear that if the clock arrives later at flip-flop 2 than at flip-flop 1, we actually have more time for the signal to propagate.

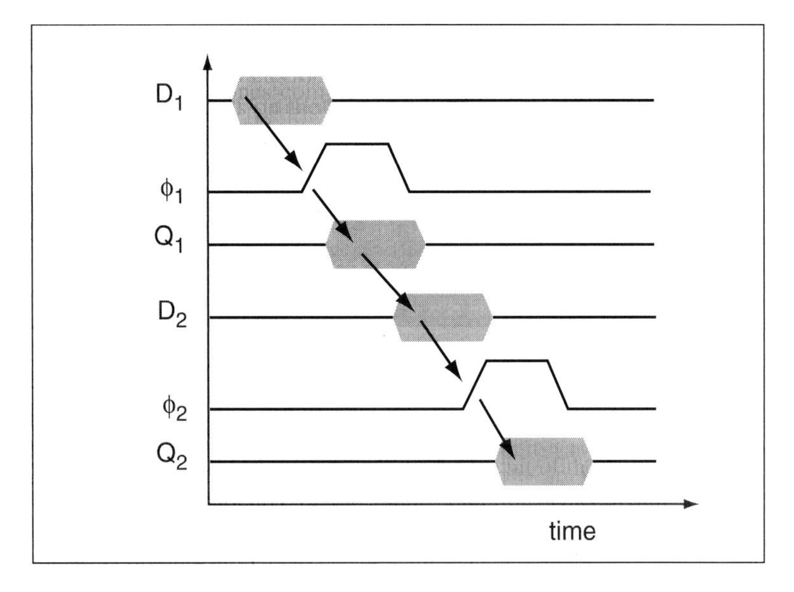

Figure 5-36 Timing in the skew model.

Figure 5-36 shows that as the clock skew $\delta_1 - \delta_2$ increases, there is less time for the signal to propagate through the combinational logic. As the clock edge ϕ_1 moves forward in time, the output of flip-flop 1 Q_1 is delayed. This in turn delays the input to flip-flop 2 D_2, pushing it closer to missing the clock edge ϕ_2.

Equation 5-4 is easy to use in the case when we can choose the clock period after we know the combinational logic delay and the skew. However, the more common situation is that we are given a target clock period, then design the logic. It is often useful to know how much skew we can tolerate at a given flip-flop. For this case, we can rewrite the relation as

$$s_{12} \geq T + \Delta_2 . \qquad \text{(EQ 5-5)}$$

taming clock skew

What can we do about clock skew? Ideally, we can distribute the clock signal without skew. The clock distribution networks of FPGAs are designed to deliver the clock with minimum skew. In custom chips, a great deal of effort goes into designing a low-skew clock distribution network.

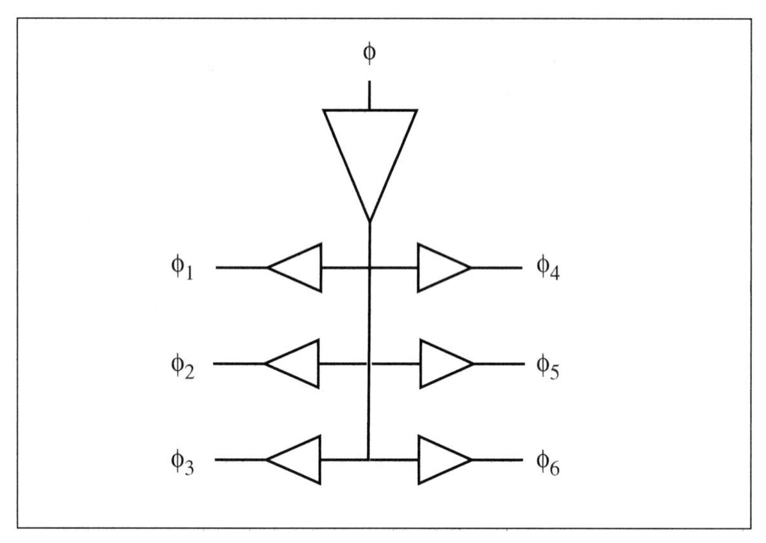

Figure 5-37 Skew in a clock distribution tree.

In practice, we cannot always eliminate skew. In these cases, we can exploit physical design to minimize the effects of skew. Consider the clock distribution tree of Figure 5-37. Each output of the tree has its own delay so there are many possible skew values between pairs of taps on the clock tree. The skew between taps that are physically close is often less than the skew between taps that are further apart. By proper placement of the combinational logic we can minimize the skew between adjacent ranks of flip-flops. The next example looks at how to deal with clock skew.

Example 5-5 Dealing with clock skew

The delays in our clock distribution network are distributed like this:

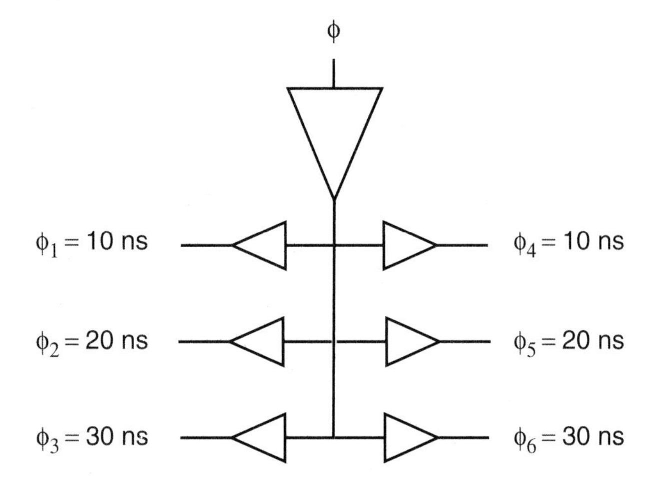

The placement of logic around the clock tree will affect both the combinational logic delays and the clock skew. Let's consider several examples.

Here is a very bad case:

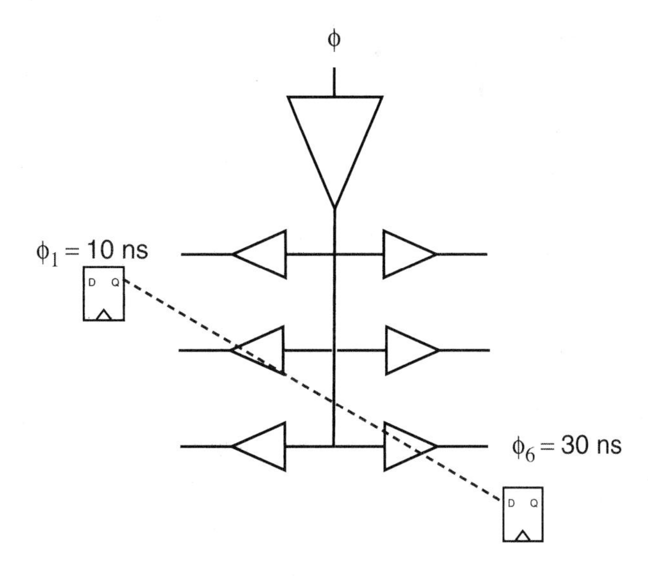

This design has both large clock skew and long wires that lead to large combinational delays.

This case is better:

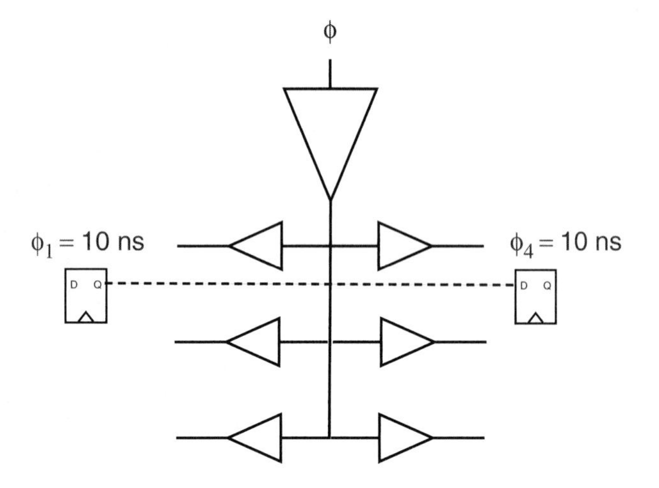

We have no clock skew here but we still have long wires.

This case is even better:

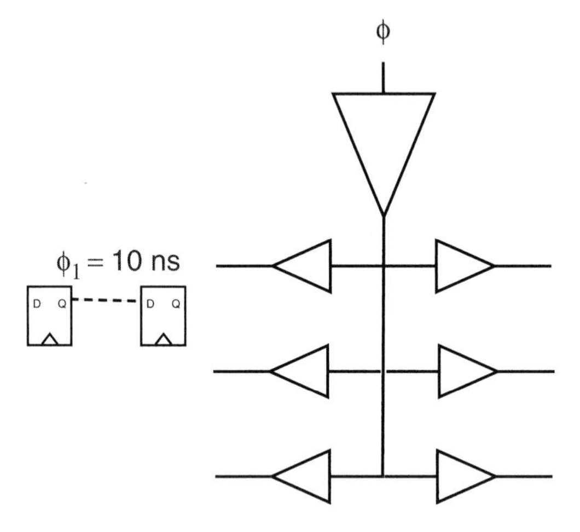

This design reduces the combinational delay and keeps the flip-flops within one island of clock skew.

---

delay-locked loops

Many FPGAs provide circuitry to minimize the delay from the off-chip clock input to the internal clock signals; this can be particularly important for asynchronous designs. The **delay-locked loop** shown in Figure

Figure 5-38 A delay-locked loop.

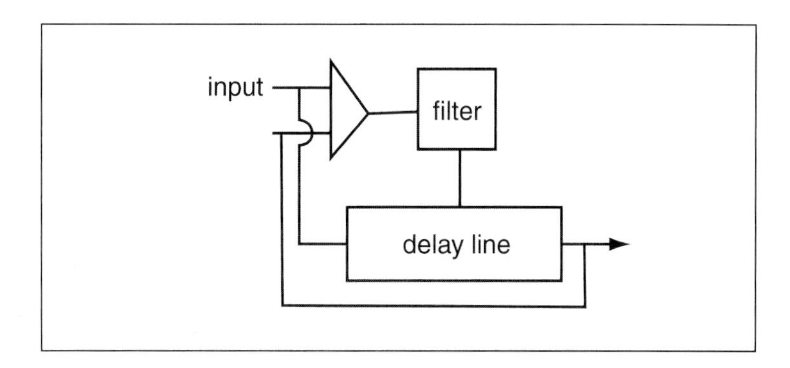

5-38 is one circuit commonly used for this purpose. It compares the input clock to the internal clock using a filter. The output of the filter controls the delay in a variable delay line so as to align the clock edges on the input and internal clock lines. (A phase-locked loop is another circuit used for this purpose; it uses a delay element as an oscillator.)

flip-flop skew optimization

An algorithm for optimizing the constraints on the clock for flip-flop-based systems was provided by Fishburn [Fis90]. This formulation places no restrictions on the system's interconnections but is more difficult to solve. Fishburn formulates two linear programs. The first minimizes the clock period P subject to clocking correctness constraints. In these formulas, the ith flip flop receives the clock after a delay of length x_i; the arrival time of the clock may vary within the range $[\alpha x_i, \beta x_i]$. The minimum and maximum logic delays between a the flip flops i and j are $MIN(i,j)$ and $MAX(i,j)$, respectively. The minimum delay that can be introduced by logic is *MINDELAY.* The linear program to be solved can be written as:

$$\begin{array}{lcl} \text{minimize } P \text{ subject to} & & \\ \alpha x_i - \beta x_j & \geq & HOLD\text{-}MIN(i,j) \\ \alpha x_i - \beta x_j + P & \geq & SETUP + MAX(i,j) \\ x_i & \geq & MINDELAY \end{array} \quad \text{(EQ 5-6)}$$

The second problem maximizes the minimum margin for error in clocking constraints—this formulation distributes the slack in clock arrival times more evenly. This problem can be written as a linear program by introducing a variable M which is added to each of the main inequalities; when M is maximized, it will be the minimum slack over all the inequalities. In this formulation, P is given and therefore constant. The problem can be written as:

$$\begin{aligned} &\text{maximize } M \text{ subject to} \\ &\alpha x_i - \beta x_j - M \geq HOLD\text{-}MIN(i,j) \\ &\alpha x_i - \beta x_j - M \geq SETUP + MAX(i,j) - P \\ &x_i \geq MINDELAY \end{aligned} \quad \text{(EQ 5-7)}$$

Either of these problems can be solved by a standard linear programming package.

skew in latch-based systems

Sakallah, Mudge, and Olukotun developed a set of constraints which must be obeyed by a latch-controlled synchronous system [Sak92]. Their formulation allows an arbitrary number of phases and takes into account propagation of signals through the latches. While the constraints must be solved by an algorithm for problems of reasonable size, studying the form of the constraints helps us understand the constraints which must be obeyed by a latch-controlled system.

The system clock period is T_c. The clock is divided into k phases $\phi_1, \ldots, \phi_k$, each of which is specified by two values: the start time s_i, relative to the beginning of the system clock period, of the ith phase; and T_i, the duration of the active interval of the ith phase. Connectivity is defined by two $k \times k$ matrices. $C_{ij} = 1$ if $i \geq j$ and 0 otherwise; it defines whether a system clock cycle boundary must be crossed when going from phase i to phase j. $K_{ij} = 1$ if any latch in the system takes as its input a signal from phase ϕ_i and emits as its output a signal of phase ϕ_j and is 0 otherwise. We can easily write basic constraints on the composition of the clock phases:

- periodicity requires that $T_i \leq T_c, i = 1, \ldots, k$ and $s_i \leq T_c, i = 1, \ldots, k$;
- phase ordering requires that $s_i \leq s_{i+1}, i = 1, \ldots, k\text{-}1$;
- the requirement that phases not overlap produces the constraints $s_i \geq s_j + T_j - C_{ji} T_c, \forall (i,j) \ni K_{ij} = 1$;
- clock non-negativity requires that $T_c \geq 0, T_i \geq 0, i = 1, \ldots, k$, and $s_i \geq 0, i = 1, \ldots, k$.

We now need constraints imposed by the behavior of the latches. The latches are numbered from 1 to l for purposes of subscripting variables which refer to the latches. The constraints require these new constraints and parameters:

- p_i is the clock phase used to control latch i; we need this mapping from latches to phases since we will in general have several latches assigned to a single phase.

- A_i is the **arrival time**, relative to the beginning of phase p_i, of a valid signal at the input of latch i.
- D_i is the **departure time** of a signal at latch i, which is the time, relative to the beginning of phase p_i, when the signal at the latch's data input starts to propagate through the latch.
- Q_i is the earliest time, relative to the beginning of phase p_i, when latch i's data output starts to propagate through the combinational logic at i's output.
- Δ_{DCi} is the setup time for latch i.
- Δ_{DQi} is the propagation delay of latch i from the data input to the data output of the latch while the latch's clock is active.
- Δ_{ij} is the propagation delay from an input latch i through combinational logic to an output latch j. If there is no direct, latch-free combinational path from i to j, then $\Delta_{ij} = -\infty$.The Δ array gives all the combinational logic delays in the system.

The latches impose setup and propagation constraints:

- Setup requires that $D_i + \Delta_{DCi} \leq T_{pi}, i = 1, \ldots, l$. These constraints ensure that a valid datum is setup at the latch long enough to let the latch store it.
- Propagation constraints ensure that the phases are long enough to allow signals to propagate through the necessary combinational logic. We can use a time-zone-shift equation to move a latch variable from one clock phase to another: $S_{ij} \equiv s_i\text{-}(s_j + C_{ij}T_c)$. A signal moving from latch j to latch i propagates in time $Q_j + \Delta_{ij}$, relative to the beginning of phase p_j. We can use the time-zone-shift formula to compute the arrival time of the signal at latch i measured in the time zone p_i, which is $Q_j + \Delta_{ji} + S_{pipj}$. The signal at the input of latch i is not valid until the latest signal has arrived at that latch: the time $A_i = max_i(Q_j + \Delta_{ij} + S_{pipj})$. To make sure that propagation delays are non-negative, we can write the constraints as $D_i = max(0, A_i), i = 1, \ldots, l$.
- Solving the constraints also requires that we constrain all the D_i's to be non-negative.

Optimizing the system cycle time requires minimizing T_c subject to these constraints.

5.5.4 Retiming

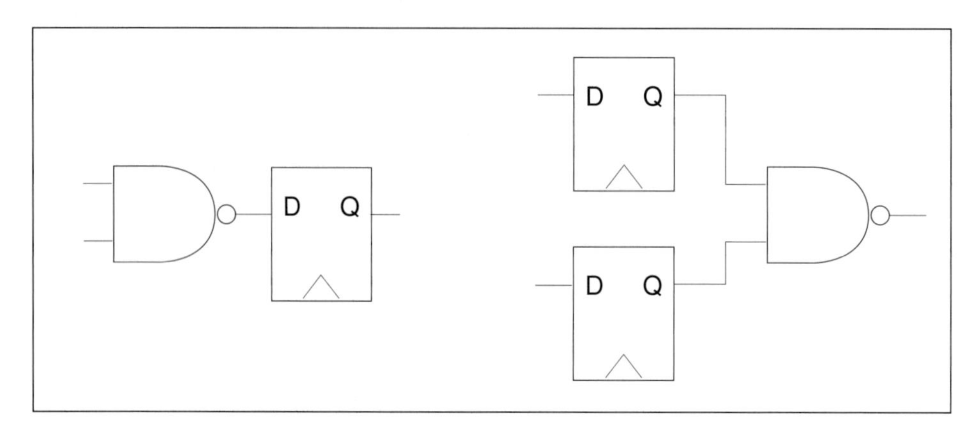

Figure 5-39 Retiming preserves combinational function.

In many cases, we can move registers to balance combinational delays. A simple example of **retiming** [Lei83] is shown in Figure 5-39. Moving the register from the output of the NAND to its inputs doesn't change the combinational function computed, only the time at which the result is available. We can often move registers within the system to balance delays without changing the times of the signals at the primary inputs and outputs. In the example of Figure 5-24, we could move the middle flip-flop to split the middle addition in two. CAD tools can retime logic by using an optimization algorithm.

5.6 Power Optimization

Eliminating glitching is one of the most important techniques for power reduction in CMOS logic. Glitch reduction can often be applied more effectively in sequential systems than is possible in combinational logic. Sequential machines can use registers to stop the propagation of glitches, independent of the logic function being implemented.

Many sequential timing optimizations can be thought of as retiming [Mon93]. Figure 5-40 illustrates how flip-flops can be used to reduce power consumption by blocking glitches from propagating to high capacitance nodes. (The flip-flop and its clock connection do, of course, consume some power of their own.) A well-placed flip-flop will be posi-

Figure 5-40 Flip-flops stop glitch propagation.

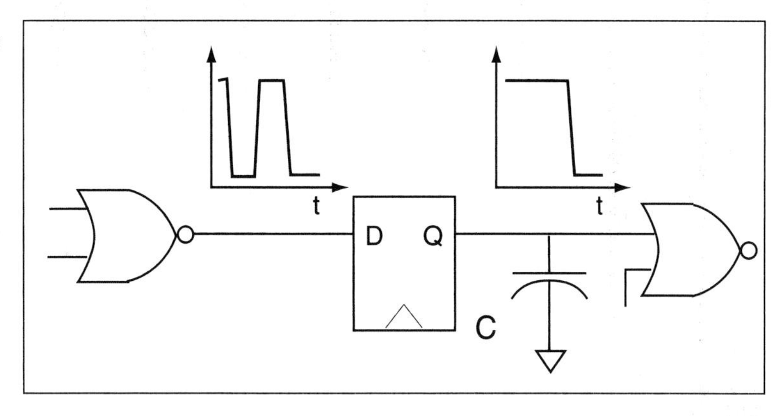

tioned after the logic with high signal transition probabilities and before high capacitance nodes on the same path.

Beyond retiming, we can also add extra levels of registers to keep glitches from propagating. Adding registers can be useful when there are more glitch-producing segments of logic than there are ranks of flip-flops to catch the glitches. Such changes, however, will change the number of cycles required to compute the machine's outputs and must be compatible with the rest of the system.

Proper state assignment may help reduce power consumption. For example, a one-hot encoding requires only two signal transitions per cycle—on the old state and new state signals. However, one-hot encoding requires a large number of registers. The power consumption of the logic that computes the required current-state and next-state functions must also be taken into account.

5.7 Summary

Sequential logic design builds upon combinational logic design. Sequential machines, when properly designed, provide powerful capabilities. However, sloppy design can lead to unreliable machines. Following a clocking discipline is an important step in ensuring that your sequential machine will operate properly.

5.8 Problems

Q5-1. Use the register-transfer diagram of the one-bit counter of Example 5-1 to draw an RT diagram of a four-bit counter. The counter should operate in parallel—the four-bit result should be computed in one clock cycle. The diagram should show the register for every bit in the counter, clouds for combinational logic, and all the input values used by each combinational logic cloud.

Q5-2. Use the register-transfer diagram of the one-bit counter of Example 5-1 to draw an RT diagram of a four-bit counter. The counter should operate in serial—the four-bit result should be computed in four clock cycles. The diagram should show the register for every bit in the counter, clouds for combinational logic, and all the input values used by each combinational logic cloud.

Q5-3. Design a resetable counter. It has a *reset* input that causes the counter's next state to be 0 when *reset* = 1.

a. Draw the state transition graph for a four-bit resetable counter.

b. Draw the register-transfer machine for a four-bit resettable counter.

c. Write a Verilog/VHDL description of the four-bit resettable counter.

Q5-4. Design a loadable counter. It has a *load* input that causes the counter's next state to be equal to an input value when *load* = 1.

a. Draw the state transition graph for a four-bit loadable counter.

b. Draw the register-transfer machine for a four-bit loadable counter.

c. Write a Verilog/VHDL description of the four-bit loadable counter.

Q5-5. Modify the output values on the transitions of the state transition graph of Figure 5-6 so that *s2* and *s3* are not equivalent.

Q5-6. Animate the transistor-by-transistor operation of the D-latch: show which transistors are active and how data flows through the circuit first for *CK* = 1, then for *CK* = 0.

Q5-7. Do the two phases of a two-phase, non-overlapping clock have to be of equal length?

Q5-8. Design the system block diagram and one-bit cell for a conditional counter, which counts only when its input *count* = 1.

Q5-9. Write the state transition table for an eight-bit conditional counter. It has two inputs, *count* and *reset* (which returns the counter to the 0 count), and a three-bit binary output of the current count.

Q5-10. Determine the present state, next state, and output of the "01"-string recognizer machine for the input 10010101110. Assume the machine starts in state *bit1*.

Q5-11. Write a state transition table for the counter of the traffic light controller, assuming that the short timeout occurs four clock cycles after the counter is reset and that the long timeout occurs several clock cycles after reset.

Q5-12. Draw the state transition graph for the product machine of Figure 5-9.

Q5-13. Consider a two-phase sequential system in which all the combinational logic is connected between the outputs of the ϕ_1 latches and the inputs of the ϕ_2 latches.

a. Draw a two-colored block diagram of such a machine.

Draw two such systems, connected so that the outputs of one feed the inputs of the next. Does this system satisfy the two-phase clocking discipline requirements? Explain.

Q5-14. Write Verilog/VHDL code to introduce glitch control to the inputs of an adder. The adder receives an additional signal *chg* that is 1 when the adder's inputs are changing.

Q5-15. A ripple carry counter has a total clock skew of x time units. Show the distribution of that clock skew to the bits in the counter that results in:

a. the slowest maximum speed for the counter;

b. the fastest maximum speed for the counter.

Q5-16. In the model of Figure 5-35, let $\Delta_1 = 20ns$ and $\Delta_2 = 25ns$. Determine the minimum allowable clock period for these skew values:

a. $\delta_1 = 0$, $\delta_2 = 5ns$.

b. $\delta_1 = 5ns$, $\delta_2 = 0$.

c. $\delta_1 = 8ns$, $\delta_2 = 7ns$.

Q5-17. In the model of Figure 5-35, let $\Delta_1 = 12ns$ and $\Delta_2 = 12ns$. How much skew from flip-flop 1 to flip-flop 2 can we allow if the clock period is:

a. $T = 12ns$.

b. $T = 15ns$.

c. $T = 19ns$.

6 Architecture

Register-transfer and behavioral design.

Pipelining.

Design methodologies.

Design verification.

Design example.

6.1 Introduction

In this section we move to design with components larger than logic elements. We first look at register-transfer design and behavioral optimizations such as scheduling and allocation. We then look at design methodologies and in particular design verification. We conclude with some example designs to illustrate what we have learned so far.

6.2 Behavioral Design

In this section we will study design abstractions above register-transfer. **Behavioral design** moves from a more abstract description of the operations to be performed to a register-transfer-level model of when and how those operations will be performed. Behavioral design is often performed manually in order to create the register-transfer model to be used for synthesis. Capturing the constraints on the system's behavior and understanding how to efficiently map the behavior into hardware is key to designing high-performance and compact machines.

We will first review the data path-controller style of architecture. We will then consider scheduling and allocation and how to create data path-controller systems. Finally, we will consider pipeline design.

6.2.1 Data Path-Controller Architectures

Figure 6-1 Data and control are equivalent.

```
if (i1 = 0)
    o1 = a;
else
    o1 = b;
end;
```

control

```
o1 = ((i1 == 0) & a) | ((i1 == 1 & b);
```

data

data path and controller

One very common style of register-transfer machine is the **data path-controller** architecture. We typically break architectures into data and control sections for convenience—the data section includes loadable registers and regular arithmetic and logical functions, while the control section includes random logic and state machines. Since few machines are either all data or all control, we often find it easiest to think about the system in this style.

data and control are equivalent

The distinction between data and control is useful—it helps organize our thinking about the machine's execution. But that distinction is not rigid; data and control are equivalent. The two Verilog statements in Figure 6-1 correspond to the same combinational logic: the *if* statement in the control version corresponds to an *or* in the data version that determines which value is assigned to the *o1* signal. We can use Boolean data operations to compute the control flow conditions, then add those conditions to any assignments to eliminate all traces of the control statement. The process can be reversed to turn data operations into control.

Operators such as adders are easily identifiable in the architectural description. As shown in Figure 6-2, some hardware is implicit. The *if* statement defines conditions under which a register is loaded and the source of its new value. Those conditions imply, along with control logic to determine when the register is loaded, a multiplexer to route the desired value to the register. We generally think of such multiplexers as data path elements in block diagrams, but there is no explicit mux operator in the architectural description.

Very few architectures are either all control (simple communications protocols, for example) or all data (digital filters). Most architectures

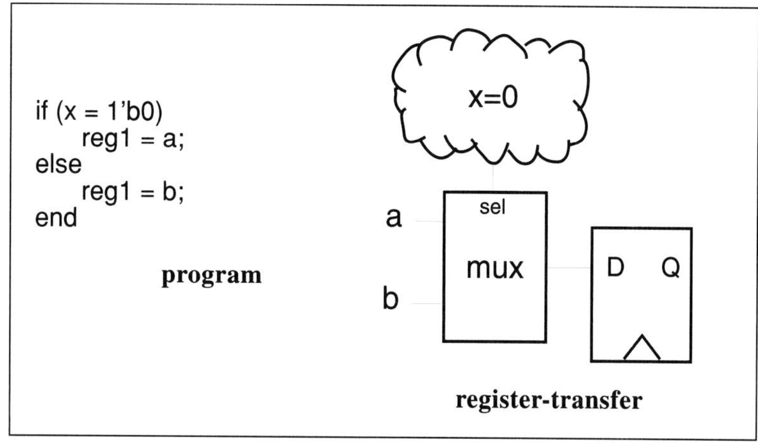

Figure 6-2 Multiplexers are hidden in architectural models.

require some control and some data. Separating the design into a controller and a data path helps us think about the system's operation. Separating control and data is also important in many cases to producing a good implementation—data operators have specialized implementations and control structures require very different optimization methods from those used for data. Figure 6-3 shows how we can build a complex system from a single controller for a single data path, or by dividing the necessary functions between communicating data path-controller systems.

6.2.2 Scheduling and Allocation

behavior vs. register-transfer

A register-transfer isn't the most abstract, general description of your system. The register-transfer assigns each operation to a clock cycle, and those choices have a profound influence on the size, speed, and testability of your design. If you think directly in terms of register-transfers, without thinking first of a more abstract **behavior** of your system, you will miss important opportunities. Consider this simple sequence of operations:

```
x = a + b;
y = c + d;
if (z > 0)
    w = e+ f;
end
```

How many clock cycles must it take to execute these operations? The assignments to x and y and the test of z are all unrelated, so they could be performed in the same clock cycle; though we must test z before we

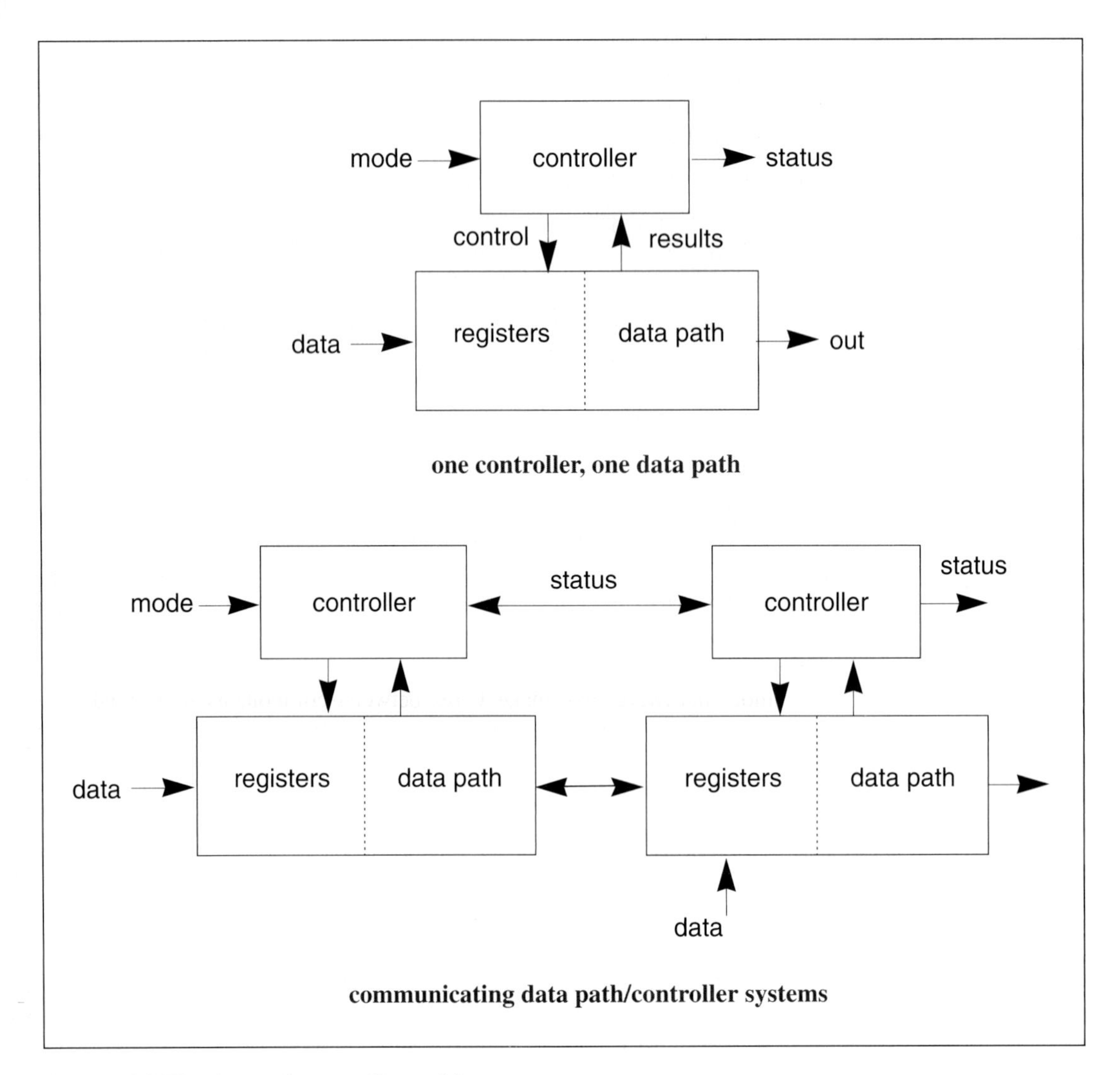

Figure 6-3 The data path-controller architecture.

perform the conditional assignment to *w*, we could design logic to perform both the test and the assignment on the same cycle. However, performing all those operations simultaneously costs considerably more hardware than doing them in successive clock cycles.

High-level synthesis (also known as **behavioral synthesis**) constructs a register-transfer from a behavior in which the times of operations are not fully specified. The external world often imposes constraints on the times at which our chip must execute actions—the specification may, for example, require that an output be produced within two cycles of receiving a given input. But the behavior model includes only necessary constraints on the system's temporal behavior.

Even if we don't use a behavioral synthesis tool, we can use some of the techniques of high-level synthesis to help us design our system. High-level synthesis methods help us understand the design space and come up with a design that meets all our requirements.

scheduling and allocation

The primary jobs in translating a behavior specification into an architecture are **scheduling** and **allocation** (sometimes called **binding**). The specification program describes a number of operations which must be performed, but not the exact clock cycle on which each is to be done. Scheduling assigns operations to clock cycles. Several different schedules may be feasible, so we choose the schedule which minimizes our costs: delay and area. The more hardware we allocate to the architecture, the more operations we can do in parallel (up to the maximum parallelism in the hardware), but the more area we burn. As a result, we want to allocate our computational resources to get maximum performance at minimal hardware cost. Of course, exact costs are hard to measure because architecture is a long way from the final layout: adding more hardware may make wires between components longer, adding delay which actually slows down the chip. However, in many cases we can make reasonable cost estimates from the register-transfer design and check their validity later, when we have a more complete implementation.

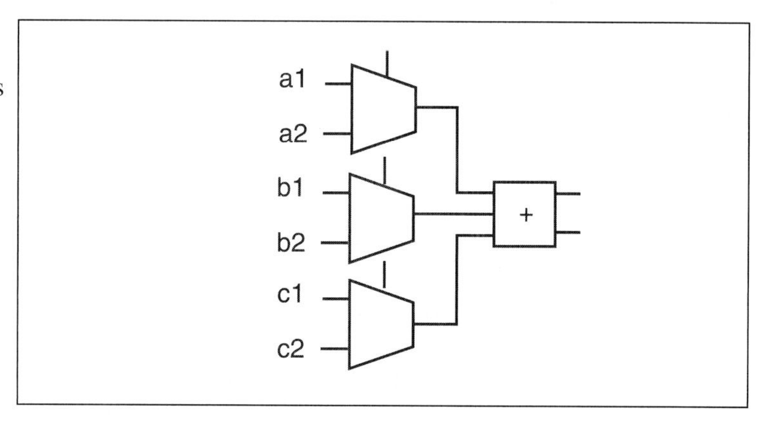

Figure 6-4 Sharing in FPGAs can be expensive.

sharing in FPGAs

In particular, sharing logic using multiplexers can be problematic in FPGAs. Figure 6-4 shows the logic required to share a one-bit full adder. We must use a multiplexer for each of the bits input to the adder.

We also need a control signal to each multiplexer. This gives a total of 7 inputs, more than will fit into a typical logic element. If we split this into more than one FPGA, we would need two or three logic elements just to implement the multiplexers in a typical FPGA fabric. As a result, we would be better off using two adders rather than trying to share one with multiplexers. However, if the multiplexers are needed for another purpose or if the element to be shared is larger, sharing might make sense.

behavioral models

A program that models a chip's desired function is given a variety of names: *functional model, behavioral model, architectural simulator*, to name a few. A specification program mimics the behavior of the chip at its pins. The internals of the specification need have nothing to do with how the chip works, but the input/output behavior of the behavior model should be the same as that of the chip.

Figure 6-5 Fragment of a Verilog behavioral model.

```
o1 = i1 | i2;
if (i3 = 1'b0)
    begin
    o1 = 1'b1;
    o2 = a + b;
    end
else
    o1 = 1'b0;
end
```

behavior in Verilog

Figure 6-5 shows a fragment of a simple Verilog behavioral model. This code describes the values to be computed and the decisions to be made based on inputs. What distinguishes it from a register-transfer description is that the cycles on which these operations are to occur are not specified. We could, for example, execute $o1 = 1'b1$ *and* $o2 = a + b$ on the same cycle or on different cycles.

behavioral HDL dialects

We should note that behavioral synthesis subsets of HDLs are not as standardized as are register-transfer synthesis subsets. We will use a fairly simple, generic form of behavioral Verilog, but these examples do not necessarily illustrate a subset that is accepted by any particular tool.

data dependencies and DFGs

Reading inputs and producing outputs for a functional model requires more thought than for a register-transfer model. Since the register-transfer's operations are fully scheduled, we always know when to ask for an input. The functional model's inputs and outputs aren't assigned particular clock cycles yet. Since a general-purpose programming language is executed sequentially, we must assign the input and output statements a particular order of execution in the simulator. Matching up the results of behavioral and register-transfer simulations can be frustrating, too. The most important information given by the functional model is the constraints on the order of execution: *e.g.*, $y = x + c$ must be executed after $x = a + b$. A **data dependency** exists between the two statements because x is written by the first statement and used by the second; if we use x's value before it is written, we get the wrong answer. Data flow constraints are critical pieces of information for scheduling and allocation.

The most natural model for computation expressed entirely as data operations is the **data flow graph** (**DFG**). The data flow graph captures all data dependencies in a behavior which is a **basic block**: only assignments, with no control statements such as *if*. The following example introduces the data flow graph by building one from a language description.

Example 6-1 Program code into data flow graph

We will use a simple example to illustrate the construction of data flow graphs and the scheduling and allocation process. For the moment we will ignore the possibility of using algebraic transformations to simplify the code before implementing it.

The first step in using a data flow graph to analyze our basic block is to convert it to single-assignment form:

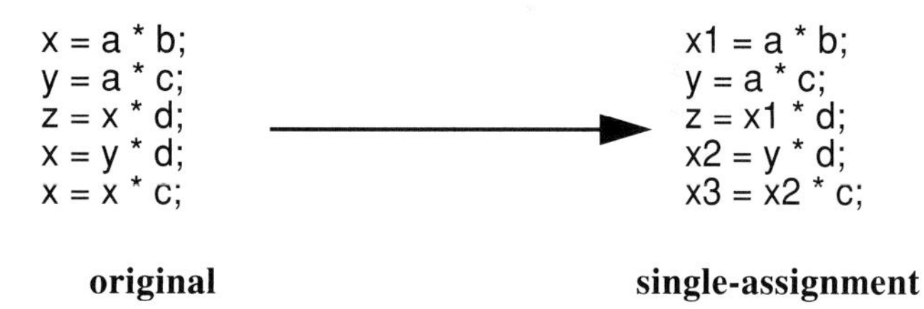

Now construct a graph with one node for each data operator and directed edges for the variables (each variable may have several sinks but only one source):

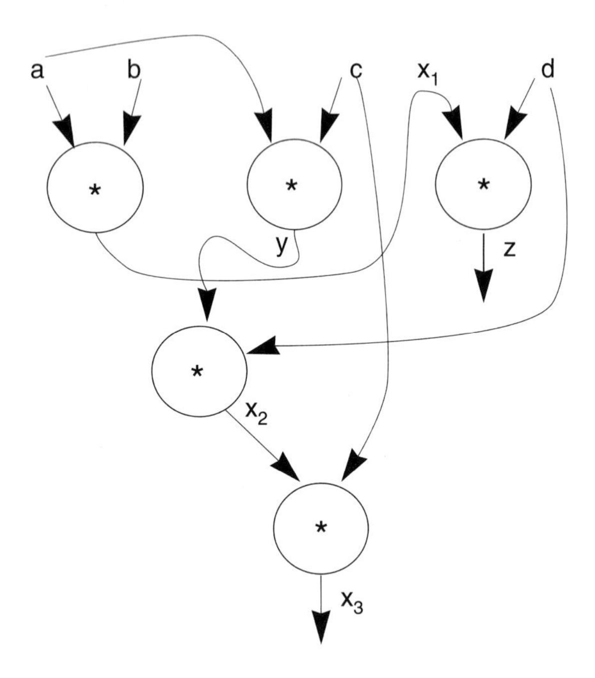

The data flow graph is a **directed acyclic graph** (DAG), in which all edges are directed and there is no cycle of edges that form a path from a node back to that node. A data flow graph has primary inputs and primary outputs like those in a logic network. (We may want to save the value of an intermediate variable for use outside the basic block while still using it to compute another variable in the block.) We can execute this data flow graph by placing values for the source variables on their corresponding DAG edges. A node *fires* when all its incoming edges have defined values; upon firing, a node computes the required value and places it on its outgoing edge. Data flows from the top of the DAG to its bottom during computation.

hardware implementation

How do we build hardware to execute a data flow graph? The simplest—and far from best—method is shown in Figure 6-6. Each node in the data flow graph of the example has been implemented by a separate hardware unit which performs the required function; each variable carrier has been implemented by a wire. This design works but it wastes a lot of hardware. Our execution model for data flow graphs tells us that not all of the hardware units will be working at the same time—an operator fires only when all its inputs become available, then it goes idle. This direct implementation of the data flow graph can waste a lot of area—the deeper the data flow DAG, the higher the percentage of idle hardware at any moment.

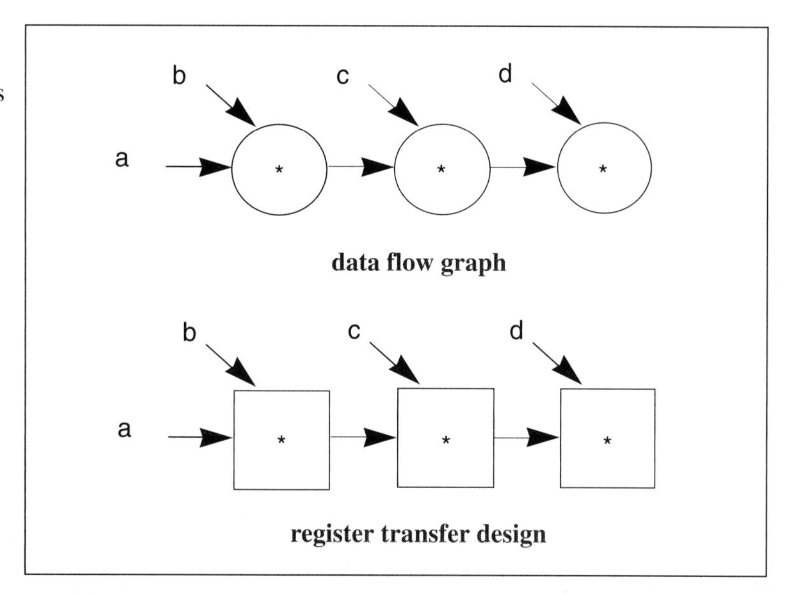

Figure 6-6 An overgenerous implementation of a data flow graph.

sharing data path hardware by adding a controller

We can save hardware for the data operators at the cost of adding hardware for memory, sequencing, and multiplexing. The result is our canonical data path-plus-controller design. The data path includes registers, function units, and multiplexers which select the inputs for those registers and function units. The controller sends control signals to the data path on each cycle to select multiplexer inputs, set operations for multi-function units, and to tell registers when to load.

The next example shows how to schedule and bind a data flow graph to construct a data path/controller machine.

Example 6-2 From data flow to data path/ controller

We will use the data flow graph of Example 6-1. Assume that we have enough chip area to put two multipliers, in the data path. We have been vague so far about where primary inputs come from and where primary output values go. The simplest assumption for purposes of this example is that primary inputs and outputs are on pins and that their values are present at those pins whenever necessary. In practice, we often need to temporarily store input and output values in registers, but we can decide how to add that hardware after completing the basic data path-controller design.

We can design a schedule of operations for the operations specified in the data flow graph by drawing cut lines through the data flow—each line cuts a set of edges which, when removed from the data flow graph, completely separate the primary inputs and primary outputs. For the schedule to be executable on our data path, no more than two multiplications can be performed per clock cycle. Here is one schedule that satisfies those criteria:

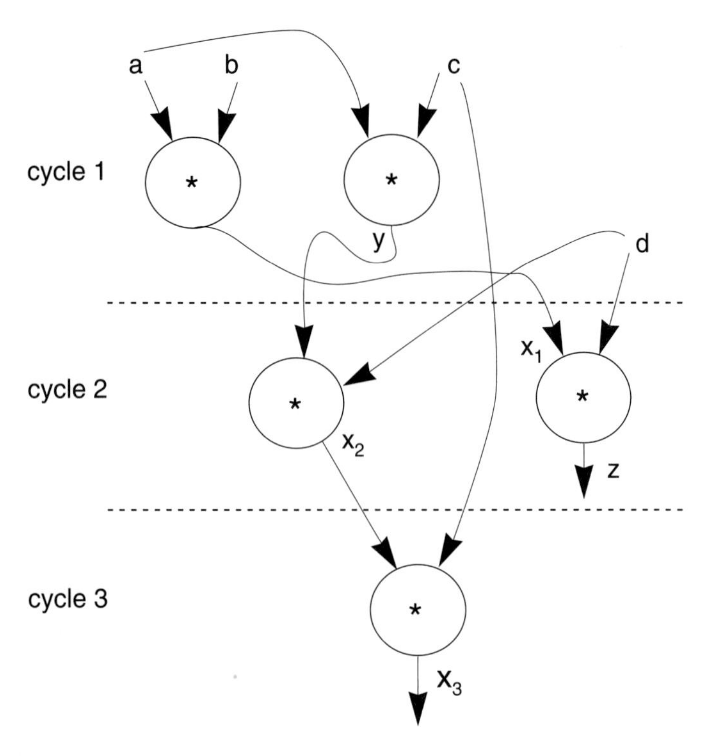

All the operations between two cut lines are performed on the same clock cycle. The next step is to bind variables to registers. Values must be stored in registers between clock cycles; we must add a register to store each value whose data flow edge crosses a cut. For the moment we will show one register per cut edge:

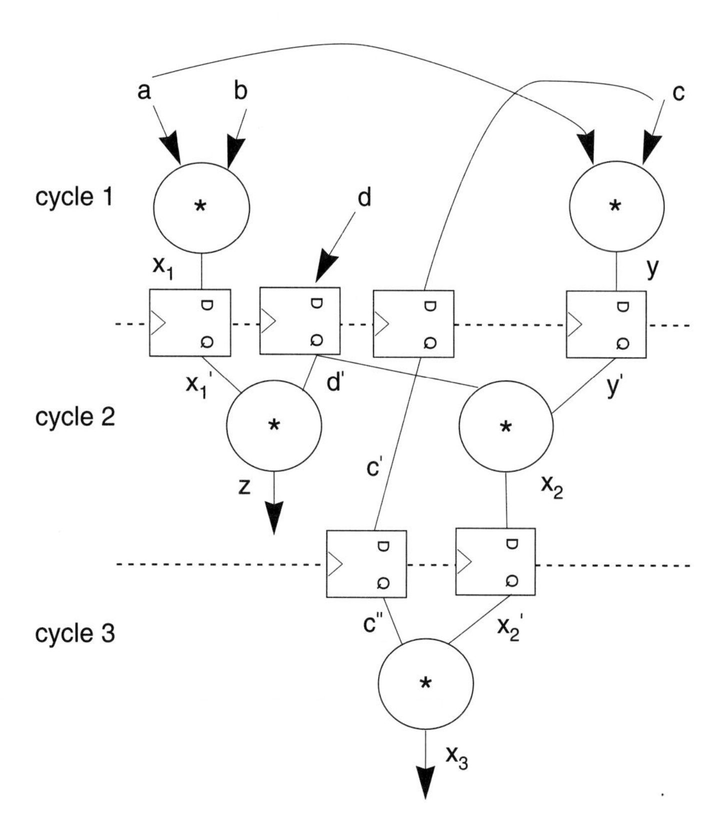

In this figure, *c*' is the *c* signal delayed by one clock cycle and *c*'' is *c* delayed by two clock cycles. But we can make the design more efficient by allocating the data values that cross edges to share registers. As with data path operators, one-to-one register allocation wastes area because not all values must be stored at the same time. We need at most four registers since that is the maximum number of values cut by a cycle boundary.

Here is one feasible sharing of registers using the input value to identify the value or values in the register:

register	values
R1	x_1
R2	y, x_2
R3	c, c'
R4	d

We can see fairly clearly from the last figure how to share the two available multipliers: we group together the left-hand multipliers in the first two cycles along with the single multiply in cycle 3, then put the remaining multiplications in the other multiplier.

Now that we have scheduled operations, bound data operations to data function units, and allocated values to registers, we can deduce the multiplexers required and complete the data path design. For each input to a shared function unit or register, we enumerate all the signals which feed the corresponding input on the operator; all of those signals go into a multiplexer for that input.

The final data path looks like this:

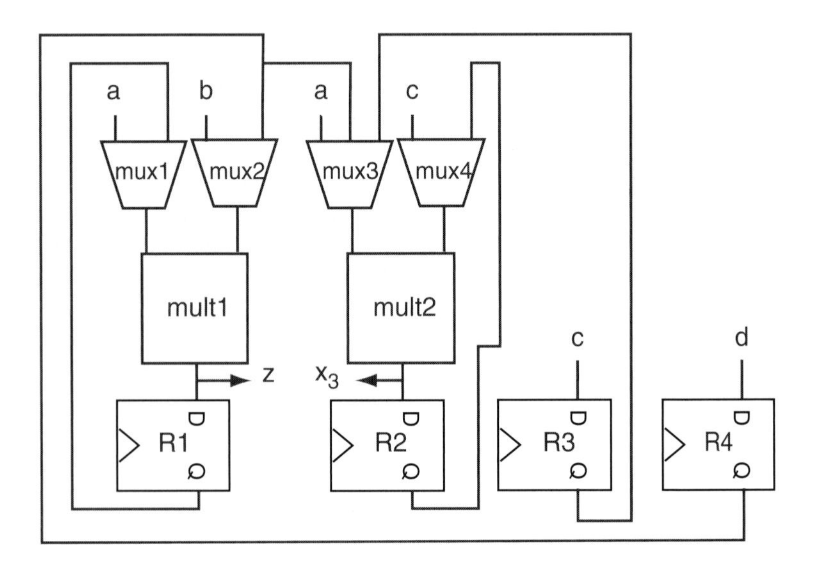

If we assume that we have combinational blocks for the multiplier, the Verilog description of the data path looks like this:

```
module dp(reset,clock,a,b,c,d,muxctrl1,muxctrl2,muxctrl3,
      muxctrl4,loadr1,loadr2,loadr3,loadr4,x3,z);
  parameter n=7;
  input reset;
  input clock;
  input [n:0] a, b, c, d; // data primary inputs
  input muxctrl1, muxctrl2, muxctrl4; // mux control
  input [1:0] muxctrl3; // 2-bit mux control
  input loadr1, loadr2, loadr3, loadr4; // register control
  output [n:0] x3, z;

   reg [n:0] r1, r2, r3, r4; // registers
   wire [n:0] mux1out, mux2out, mux3out, mux3bout, mux4out,
         mult1out, mult2out;

   assign mux1out = (muxctrl1 == 0) ? a : r1;
   assign mux2out = (muxctrl2 == 0) ? b : r4;
   assign mux3out = (muxctrl3 == 0) ? a :
                    (muxctrl3 == 1 ? r4 : r3);
   assign mux4out = (muxctrl4 == 0) ? c : r2;
   assign mult1out = mux1out * mux2out;
   assign mult2out = mux3out * mux4out;
   assign x3 = mult2out;
   assign z = mult1out;
   always @(posedge clock)
      begin
   if (reset)
         begin
         r1 = 0; r2 = 0; r3 = 0; r4 = 0;
         end
      if (loadr1) r1 = mult1out;
      if (loadr2) r2 = mult2out;
      if (loadr3) r3 = c;
      if (loadr4) r4 = d;
      end
endmodule
```

Now that we have the data path, we can build a controller that repeatedly executes the basic block. The state transition graph consists of a single loop, with each transition in the loop executing one cycle's operation. The controller requires no inputs, since it makes no data-dependent branches. Its outputs provide the proper control values to the data path's multiplexers and function units at each step. The controller looks like this:

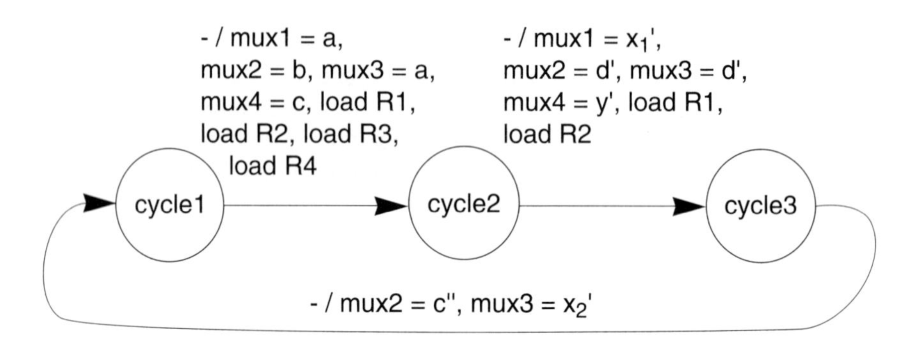

If we choose a state assignment of *cycle1=0*, *cycle2=1*, *cycle3=2*, the Verilog for the controller looks like this:

```
module ctrl(reset,clock,muxctrl1,muxctrl2,muxctrl3,muxctrl4,
        loadr1,loadr2,loadr3,loadr4);
   input reset, clock;
   output muxctrl1, muxctrl2, muxctrl4;
   output [1:0] muxctrl3;
   output loadr1, loadr2, loadr3, loadr4;
   reg muxctrl1, muxctrl2, muxctrl4,
      loadr1, loadr2, loadr3, loadr4;
   reg [1:0] muxctrl3;

     reg [1:0] state;

     always @(posedge clock)
          begin
          if (reset == 1)
               begin
               muxctrl1 = 0; muxctrl2 = 0;
               muxctrl3 = 0; muxctrl4 = 0;
               loadr1 = 0; loadr2 = 0;
               loadr3 = 0; loadr4 = 0;
               state = 0;
               end
        else begin
               case (state)
                    0:
                    begin
```

```
                muxctrl1 = 0; muxctrl2 = 0;
                muxctrl3 = 0; muxctrl4 = 0;
                loadr1 = 1; loadr2 = 1;
                loadr3 = 1; loadr4 = 1;
                state = 1;
                end
                1:
                begin
                muxctrl1 = 1; muxctrl2 = 1;
                muxctrl3 = 1; muxctrl4 = 1;
                loadr1 = 1; loadr2 = 1;
                loadr3 = 0; loadr4 = 0;
                state = 2;
                end
                2:
                begin
                muxctrl1 = 0; muxctrl2 = 0;
                muxctrl3 = 2; muxctrl4 = 1;
                loadr1 = 0; loadr2 = 0;
                loadr3 = 0; loadr4 = 0;
                state = 0;
                end
            endcase
        end // else
    end // always

endmodule
```

Once we wire together the data path and controller, the implementation is complete.

ALUs and allocation

We are not required to map each operator in the behavior to a separate function unit in the hardware. We can combine several operators into a single module like an ALU. The controller must also generate the function code to control the ALU. The next example illustrates allocating functions into ALUs.

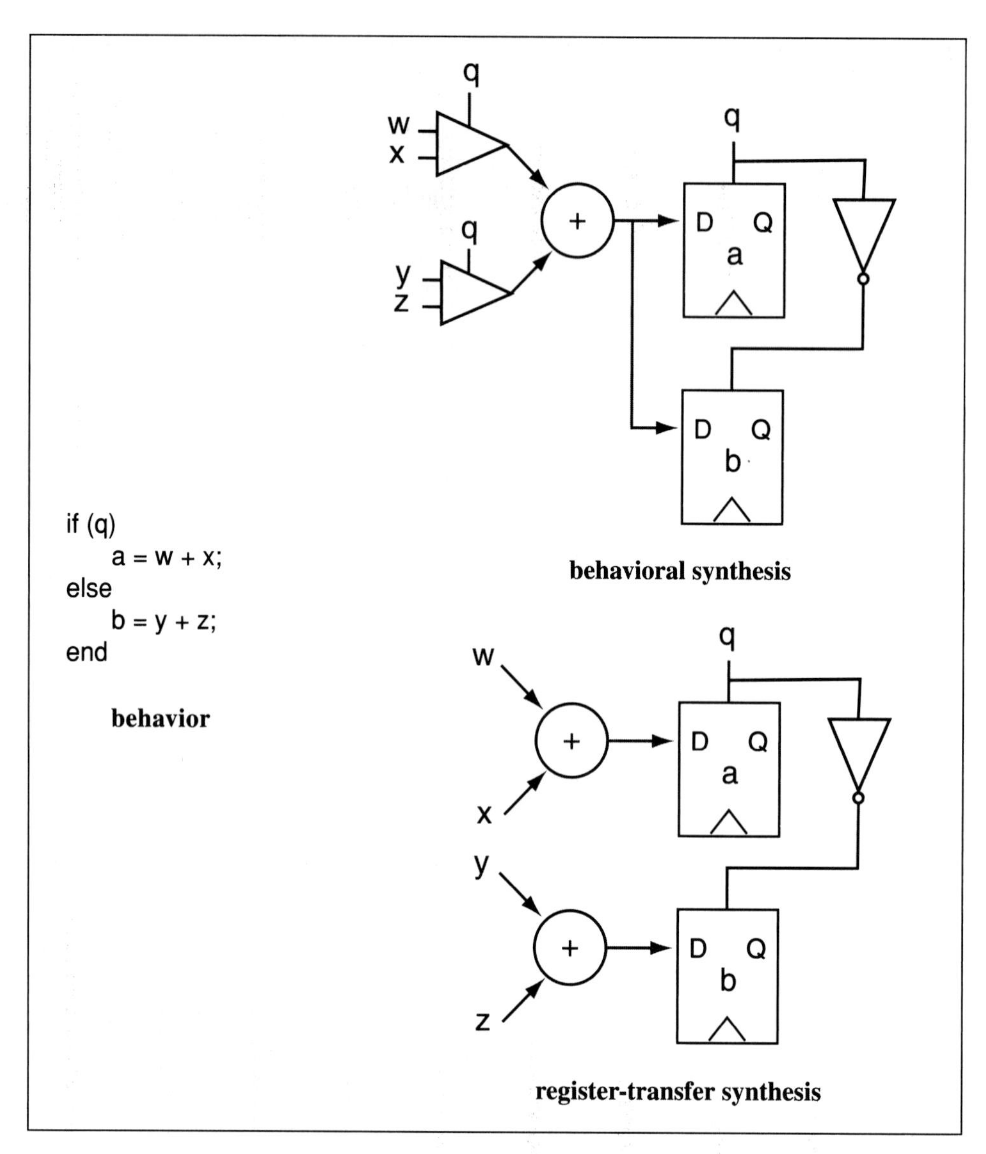

Figure 6-7 Behavioral *vs.* register-transfer synthesis.

Example 6-3 Allocating ALUs

Let us use this scheduled data flow diagram as an example:

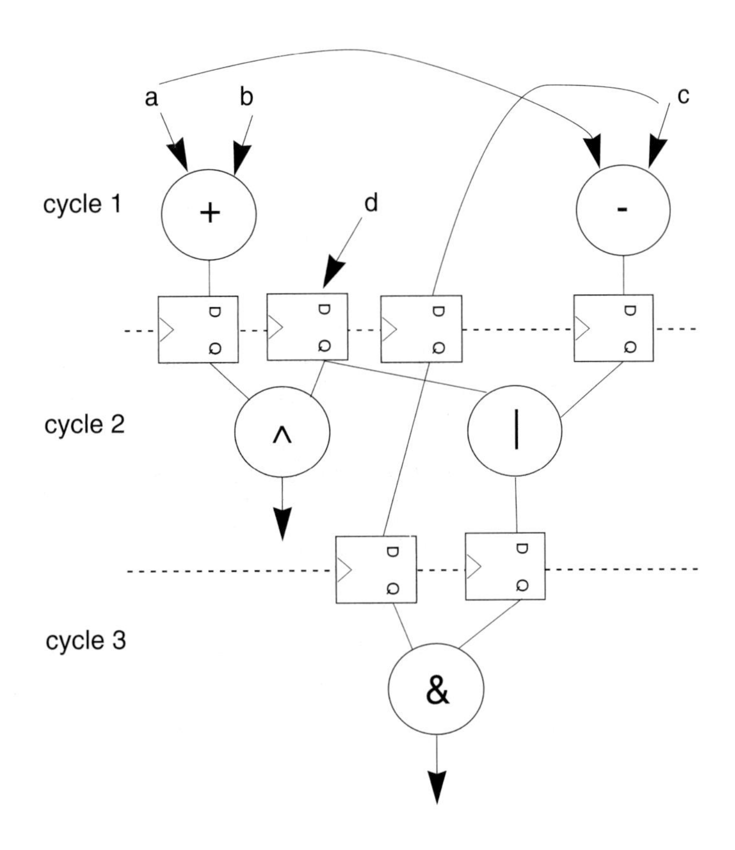

If we are limited to using two ALUs, we must allocate the functions in the data flow graph to the two ALUs. There are several ways to do this; one of them is:

- ALU1: +, ^, &
- ALU2: -, |

Note that we could end up with less logic if we could allocate the ALUs such that one performed addition and subtraction and the other performed the logical operations. However, in the current schedule, we must perform the addition and subtraction in the same cycle.

If we are limited to one ALU, then we must reschedule the design to perform one operation per cycle. The single ALU would implement all the functions in the data flow graph.

---

limitations of RT synthesis

It is important to remember that register-transfer synthesis cannot figure out how to share adders, registers, and the like, nor can it change the schedule of operations. Consider the Verilog fragment in Figure 6-7. The two additions in this code are mutually exclusive—the machine will not perform $w+x$ and $y+z$ at the same time. If we were to use behavioral synthesis techniques to design this logic and share hardware, we would end up with the single adder design shown in the figure. This design uses multiplexers to guide the proper inputs to the single adder. But register-transfer synthesis can only generate the two adder design shown. It cannot determine how to share the adder so it instantiates one piece of hardware for each operation in the behavioral description. The multiplexers in the behavioral synthesis result cause it to run a little slower, but behavioral synthesis can actually generate either of these hardware solutions, depending on whether we choose delay or area as the primary constraint. However, register-transfer synthesis by its nature does not determine how to share hardware.

exploring the design space

In Example 6-2, we made a number of arbitrary choices about when operations would occur and how much hardware was available. The example was designed to show only how to construct a machine that implements a data flow graph, but in fact, the choices for scheduling and allocation are the critical steps in the design process. Now that we understand the relationship between a data flow graph and a data path-controller machine, we need to study what makes one data path-controller implementation better than another. This understanding will help us explore the design space and choose the best design, not just a feasible design.

Obviously, scheduling and allocation decisions depend on each other. The choice of a schedule limits our allocation options; but we can determine which schedule requires the least hardware only after allocation. We need to separate the two decisions as much as possible to make the design task manageable, but we must keep in mind that scheduling and allocation depend on each other.

hardware costs

To a first approximation, scheduling determines time cost, while allocation determines area cost. Of course, the picture is more complex than that: allocation helps determine cycle time, while scheduling adds area for multiplexers, registers, etc. But we always evaluate the quality of a schedule by its ultimate hardware costs:

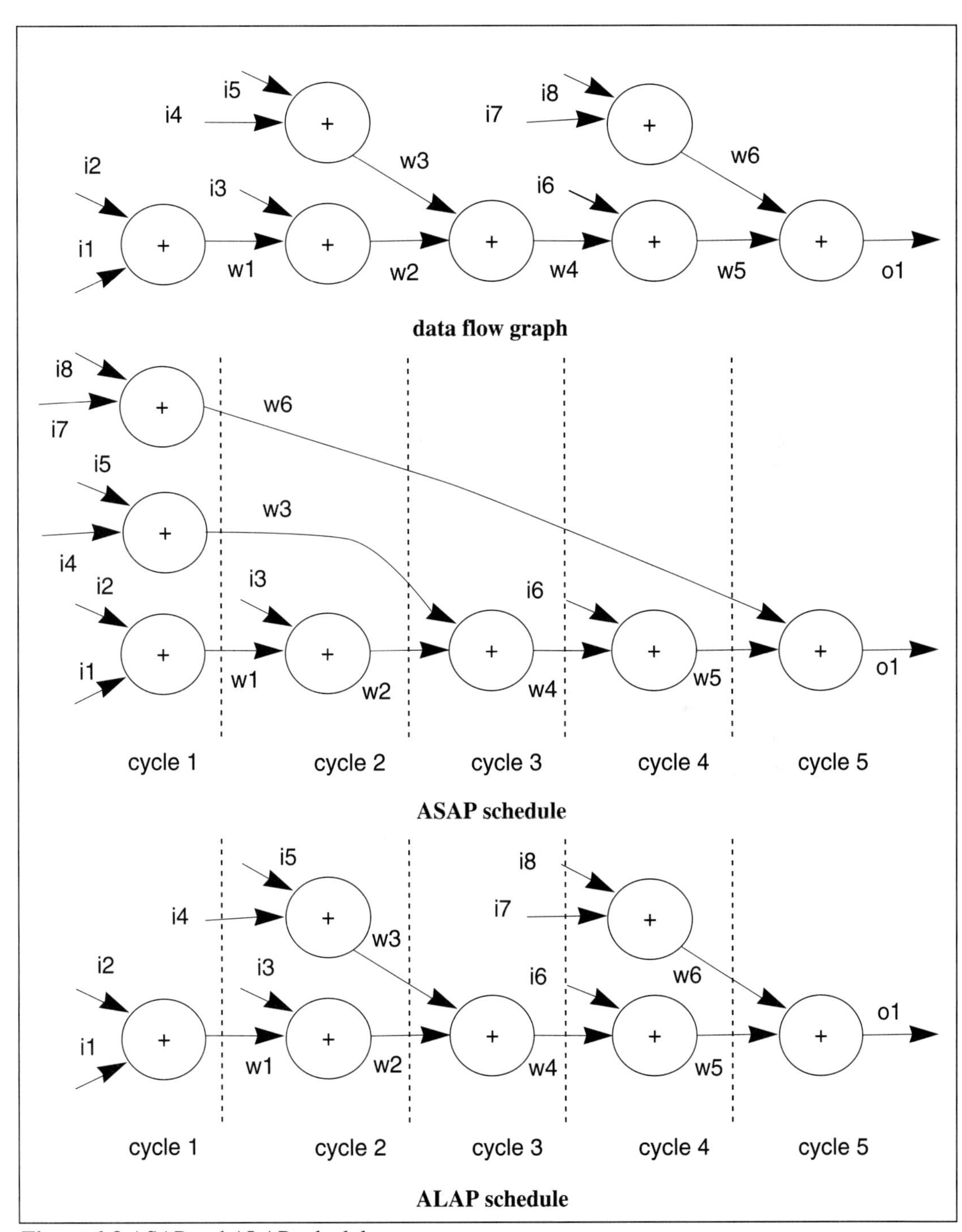

Figure 6-8 ASAP and ALAP schedules.

- **Area**. Area of the data path-controller machine depends on the amount of data operators saved by sharing vs. the hardware required for multiplexing, storage, and control.
- **Delay**. The time required to compute the basic block's functions depends on the cycle time and the number of cycles required. After the easy victories are won by obvious data hardware sharing, we can generally reduce area only by increasing delay—performing data operations sequentially on fewer function units.
- **Power**. The power consumption of the system can be affected by scheduling and allocation—a more parallel schedule consumes more power.

ASAP and ALAP schedules

There are many possible schedules which satisfy the constraints in a data flow graph. Figure 6-8 shows how to find two simple schedules. In this example we assume that we can perform as many additions as possible in parallel but no more than one addition in series per cycle—**chained** additions stretch the clock period. The **as-soon-as-possible** (ASAP) schedule is generated by a breadth-first search from the data flow sources to the sinks: assign the source nodes time 0; follow each edge out to the next rank of nodes, assigning each node's time as one greater than the previous rank's; if there is more than one path to a node, assign its time as the latest time along any path. The simplest way to generate the **as-late-as-possible** (ALAP) schedule is to work backward from the sinks, assigning negative times (so that the nodes just before the sinks have time -1, etc.), then after all nodes have been scheduled, adjust the times of all nodes to be positive by subtracting the most negative time for any node to the value of each node. The ASAP and ALAP schedules often do not give the minimum hardware cost, but they do show the extreme behaviors of the system.

critical path in schedule

The ASAP and ALAP schedules help us find the critical paths through the data flow graph. Figure 6-9 shows the critical path through our data flow graph—the long chain of additions determines the total time required for the computation, independent of the number of clock cycles used for the computation. As in logic timing, the critical path identifies the operations which determine the minimum amount of time required for the computation. In this case, time is measured in clock cycles.

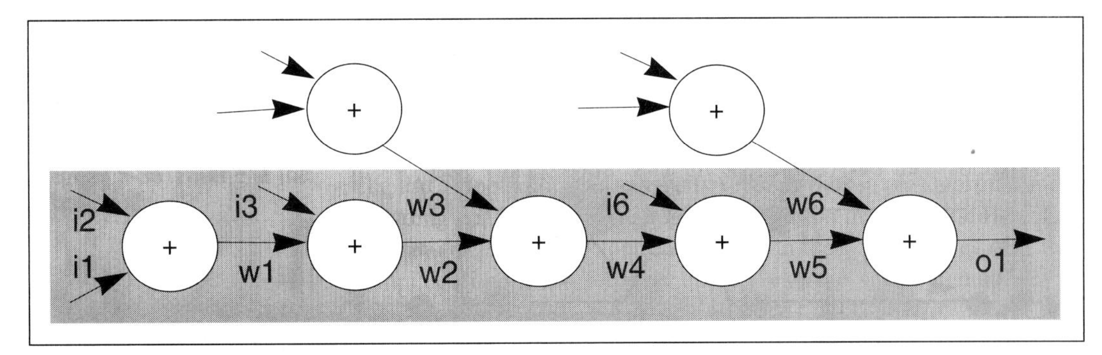

Figure 6-9 Critical path of a data flow graph.

Example 6-4 ASAP and ALAP schedules in Verilog

We will use a structural description style to see how scheduling changes the Verilog description. In Figure 6-8 we did not show the registers that need to be inserted at the clock boundaries. We will name each register after the wire it cuts: *w1reg*, for example. Some signals will need more than one register, such as *w6reg1*, *w6reg2*, *w6reg3*, and *w6reg4* for the four registers needed for *w1* in the ASAP schedule. The ASAP and ALAP Verilog descriptions will differ in where those registers are placed.

Here is the Verilog for the ASAP schedule:

```
reg [n-1:0] w1reg, w2reg, w6reg1, w6reg2, w6reg3,
    w6reg4, w3reg1, w3reg2, w4reg, w5reg;

always @(posedge clock)
    begin
    // cycle 1
    w1reg = i1 + i2;
    w3reg1 = i4 + i5;
    w6reg1 = i7 + i8;
    // cycle 2
    w2reg = w1reg + i3;
    w3reg2 = w3reg1;
    w6reg2 = w6reg1;
    // cycle 3
    w4reg = w3reg2 + w2reg;
    w6reg3 = w6reg2;
    // cycle 4
    w5reg = i6 + w4reg;
```

```
    w6reg4 = w6reg3;
    // cycle 5
    o1 = w6reg4 + w5reg;
    end
```

And here is the Verilog for the ALAP schedule:

```
reg [n-1:0] w1reg, w2reg, w6reg, w6reg2,
            w6reg3, w3reg, w4reg, w5reg;

always @(posedge clock)
    begin
    // cycle 1
    w1reg = i1 + i2;
    // cycle 2
    w2reg = w1reg + i3;
    w3reg = i4 + i5;
    // cycle 3
    w4reg = w3reg + w2reg;
    w6reg3 = w6reg2;
    // cycle 4
    w5reg = i6 + w4reg;
    w6reg = i7 + i8;
    // cycle 5
    o1 = w6reg + w5reg;
    end
```

The Verilog code makes it clear that the ALAP schedule has fewer registers.

---

scheduling costs

Before we consider more sophisticated scheduling methods, we should reflect on what costs we will use to judge the quality of a schedule. We are fundamentally concerned with area and delay; can we estimate area and delay from the data path-controller machine implied by a schedule without fleshing out the design to layout?

Consider area costs first. An allocation of data path operators to function units lets us estimate the costs of the data operations themselves. After assigning values to registers we can also estimate the area cost of data storage. We can also compute the amount of logic required for multiplexers. Estimating the controller's area cost is a little harder because area cannot be accurately estimated from a state transition graph. But we can roughly estimate the controller's cost from the state transitions, and if we need a more accurate estimate, we can synthesize the controller to logic or, for a final measure, to layout.

Now consider delay costs: both the number of clock cycles required to completely evaluate the data flow graph and the maximum clock rate. We have seen how to measure the number of clock cycles directly from the data flow graph. Estimating cycle time is harder because some of the data path components are not directly represented in the data flow graph.

Figure 6-10 Delay through chained adders is not additive.

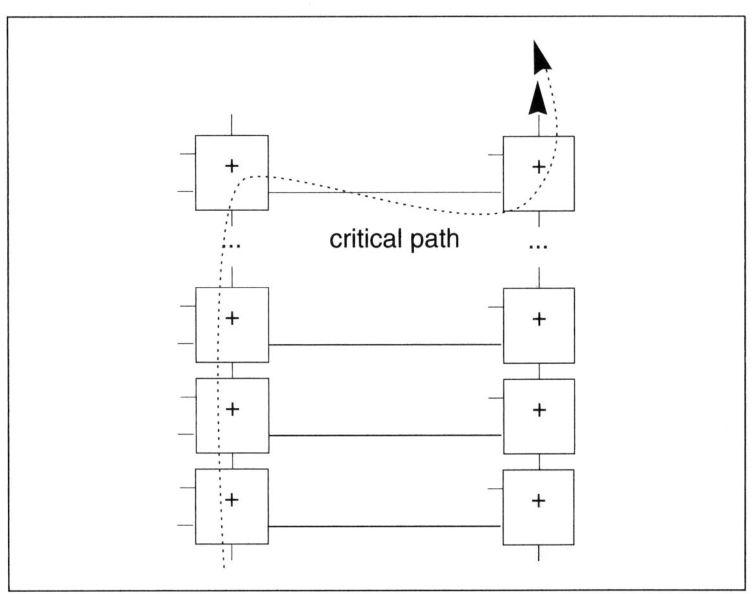

One subtle but important problem is illustrated by Figure 6-10: the delay through a chain of adders (or other arithmetic components) is not additive. The simplest delay estimate from the data flow graph is to assign a delay to each operator and sum all the delays along a path in each clock cycle. But, as the figure shows, the critical path through a chain of two adders does not flow through the complete carry chain of both adders—it goes through all of the first adder but only the most significant bit of the second adder. The simple additive model for delay in data flow graphs is wildly pessimistic for adders of reasonable size. For accurate estimates, we need to trace delays through the data path bit by bit.

If you are worried about delay, multiplexers added for resource sharing should concern you. The delay through a multiplexer can be significant, especially if the multiplexer has a large number of data inputs.

arithmetic vs. control

One important reason to separate control from data is that arithmetic-rich and control-rich machines must be optimized using very different techniques to get good results—while optimization of arithmetic machines concentrates on the carry chain, the optimization of control requires identifying Boolean simplification opportunities within and between transitions. We typically specify the controller as a state transition graph, though we may use specialized machines, such as counters, to implement the control.

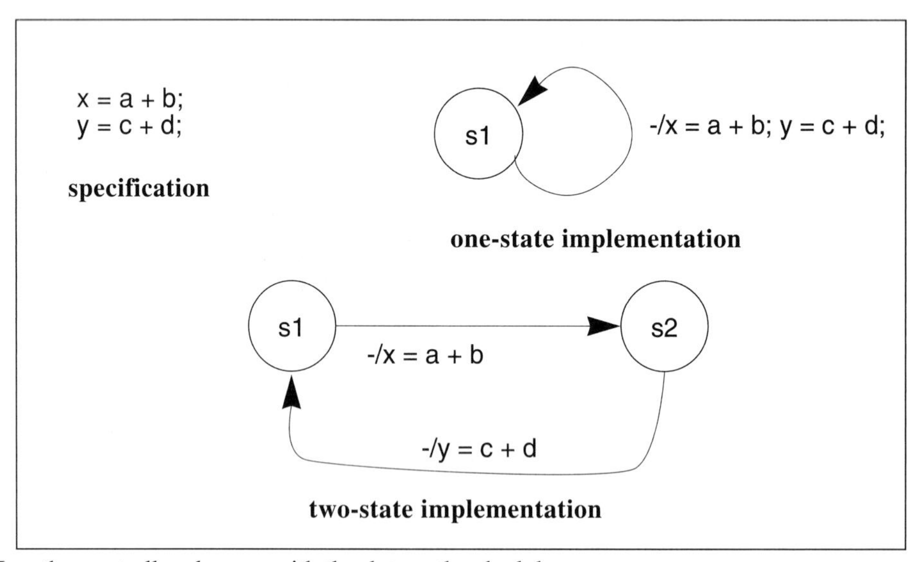

Figure 6-11 How the controller changes with the data path schedule.

creating the controller

The high-level synthesis problem for control is one step more abstract than creating the truth table that implements a state transition graph—we must design the state transition graph which executes the desired algorithm. Consider the simple example of Figure 6-11. The two controllers are clearly not equivalent in the automata-theoretic sense: we can easily find one input sequence which gives different output sequences on the two machines, since the two machines don't even use the same number of cycles to compute the two additions. But even though the two controllers are not sequentially equivalent, they both satisfy the behavior specification.

When do we want to schedule control? Primarily when we have two controllers talking to each other. If one of the other controllers has already been designed, it probably fixes the times at which events must occur. But if we are designing both controllers simultaneously, we may be able to change the times of control operations either to optimize the other controller or to optimize the associated data paths.

controller costs

How do we judge the quality of a controller which implements the control section of a program? That, of course, depends on our requirements. As usual, we are concerned with the area and delay of the FSM. The behavior specification may give us additional constraints on the number of cycles between actions in the program. We may have to satisfy strict sequencing requirements—when reading a random RAM, for example, we supply the address on one clock cycle and read the data at that location exactly one clock cycle later. We often want to minimize the number of cycles required to perform a sequence of operations—the number of cycles between reading a value and writing the computed result, for instance. To compute a result in the minimum number of cycles, we must perform as many operations as possible on each clock cycle. That requires both scheduling operations to take best advantage of the data path, as we saw in the last section; it also requires finding parallelism within the control operations themselves.

controller design

For now we will assume that the data path is given and consider how to design and implement the controller. The construction of a controller to execute a behavior specification proceeds as follows:

- Each statement in the behavior model is annotated with data path signals: arithmetic operations may require operation codes; multiplexers require selection signals; registers require load signals.
- Data dependencies are identified within each basic block.
- In addition, **control dependencies** are identified across basic blocks—a statement that is executed in only one branch of a control statement must be executed between the first and last states of that conditionally executed basic block. If the same statement appears in every branch, it is not dependent on the control signal and can be moved outside the control statement.
- External scheduling constraints, which reflect the requirements of the other machines to which this one will be connected, are added. External scheduling constraints are those which cannot be determined by looking at the behavior specification itself but which are required when the machine is connected to its intended working environment.
- Each program statement is scheduled—assigned an execution clock cycle which satisfies all the data and control dependencies.

- The controller's state transition graph can be constructed once the schedule is known.

Figure 6-12 Constructing a controller from a program.

```
o1 = 1'b1;
if (i1 = 1'b1)
    begin
    o1 = 1'b0;
    o2 = 1'b1;
    o2 = 1'b0;
    end
else
    begin
    o3 = 1'b1;
    o2 = 1'b1;
    end
end
o5 = 1'b0;
```

code

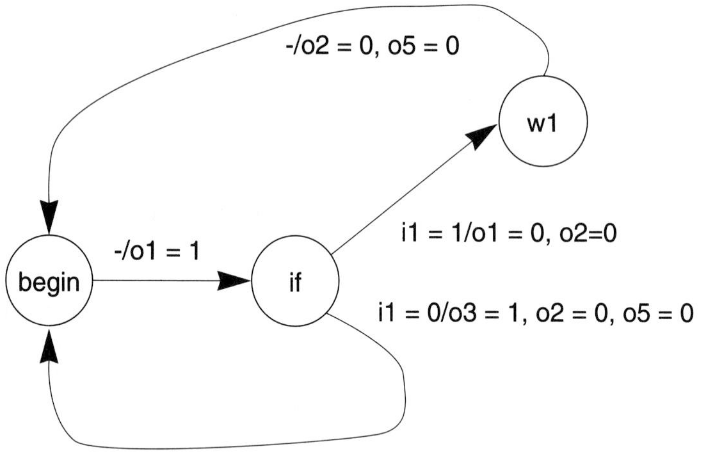

controller

exposing parallelism

Figure 6-12 shows how some opportunities for parallelism may be hidden by the way the program is written. The statements *o1* = *'1'* and *o5* = *'0'* are executed outside the *if* statement and, since they do not have any data dependencies, can be executed in any order. (If, however, one of the *if* branches assigned to *o5*, the *o5* = *'0'* assignment could not be performed until after the *if* was completed.) The assignment *o2* = *'0'* occurs within *both* branches of the *if* statement and data dependencies do not tie it down relative to other statements in the branches. We can therefore pull out the assignment and execute a single *o2 = 1'b0* before or after the *if*. If a statement must be executed within a given branch to maintain correct behavior, we say that statement is control-dependent on the branch.

```
x = a - b;
if (x < y)
     o1 = 1'b0;
end;
```

behavior specification

```
source_1 = a_source; source2 = b_source; op = subtract; load_x =1'b0;
source_1 = x_source; source_2 = y-source; op = gt;
if (gt_result)
     o1_mux = zero_value;
end;
```

controller operations

Figure 6-13 Rewriting a behavior in terms of controller operations.

matching controller and data path

If we want to design a controller for a particular data path, two complications are introduced. First, we must massage the behavior specification to partition actions between the controller and data path. A statement in the behavior may contain both data and control operations; it can be rewritten in terms of controller inputs and outputs which imply the required operations. Figure 6-13 gives a simple example. The first assignment statement is replaced by all the signals required to perform the operation in the data path: selecting the sources for the ALU's operands, setting the operation code, and directing the result to the proper register. The condition check in the *if* statement is implemented by an ALU operation without a store. We must also add constraints to ensure that these sets of operations are all executed in the same cycle. (Unfortunately, such constraints are hard to write in Verilog or VHDL and are usually captured outside the behavior model.) Those constraints are external because they are imposed by the data path—the data path cannot, for example, perform an ALU operation on one cycle and store the result in a temporary register for permanent storage on a later cycle. We also need constraints to ensure that the ALU operation for the assignment is performed on the same cycle as the test of the result, or the comparison result will be lost.

The second complication is ensuring that the controller properly uses the data path's resources. If we have one ALU at our disposal, the controller can't perform two ALU operations in one cycle. The resource constraints are reflected in the controller's pins—a one-ALU data path will have only one set of ALU control signals. We may, however, have to try different sequences of data path operations to find a legal implementation with both a good controller and the desired data path.

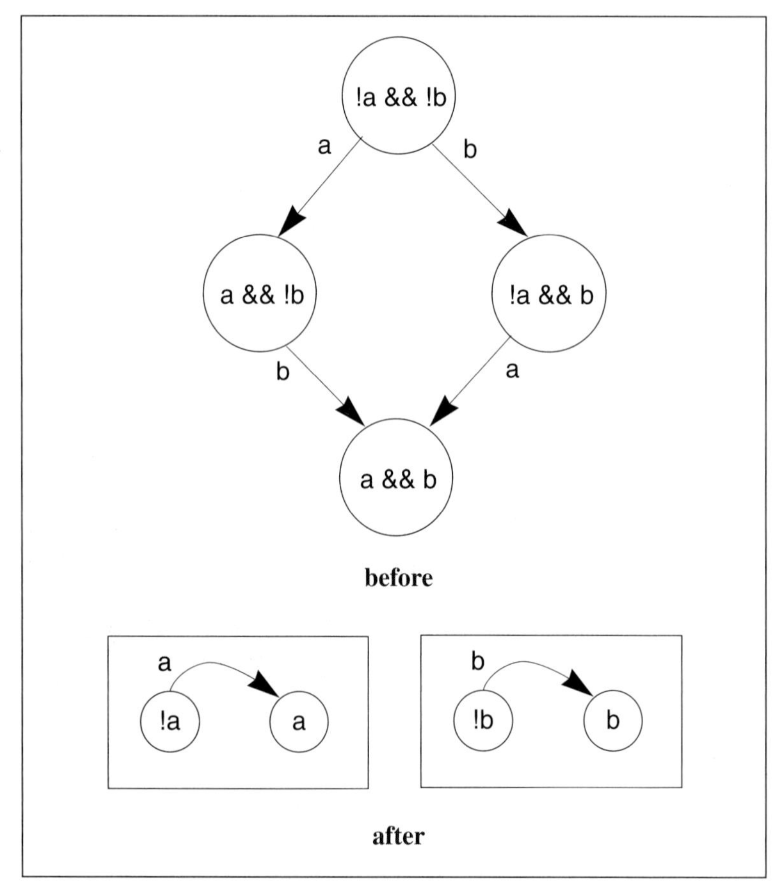

Figure 6-14 Breaking a pair of tests into distributed control.

controller styles

Finally, a word about controller implementation styles. You may have learned to implement a controller as either a **hardwired machine** or a **microcoded machine**. As shown in Figure 6-15, a hardwired controller is specified directly as a state transition graph, while a microcoded controller is designed as a microcode memory with a microprogram counter. (The microcoded controller also requires control logic to load the μPC for branches.) It is important to remember that these are implementation styles, not different schedules. The hardwired and microcoded controllers for a given design are equivalent in the automata-theoretic sense—we can't tell which is used to implement our system by watching only its I/O behavior. While one may be faster, smaller, or easier to modify than another for a given application, changing from one style to another doesn't change the scheduling of control operations in the controller. You should first use control scheduling methods to design the controller's I/O behavior, then choose an implementation style for the machine.

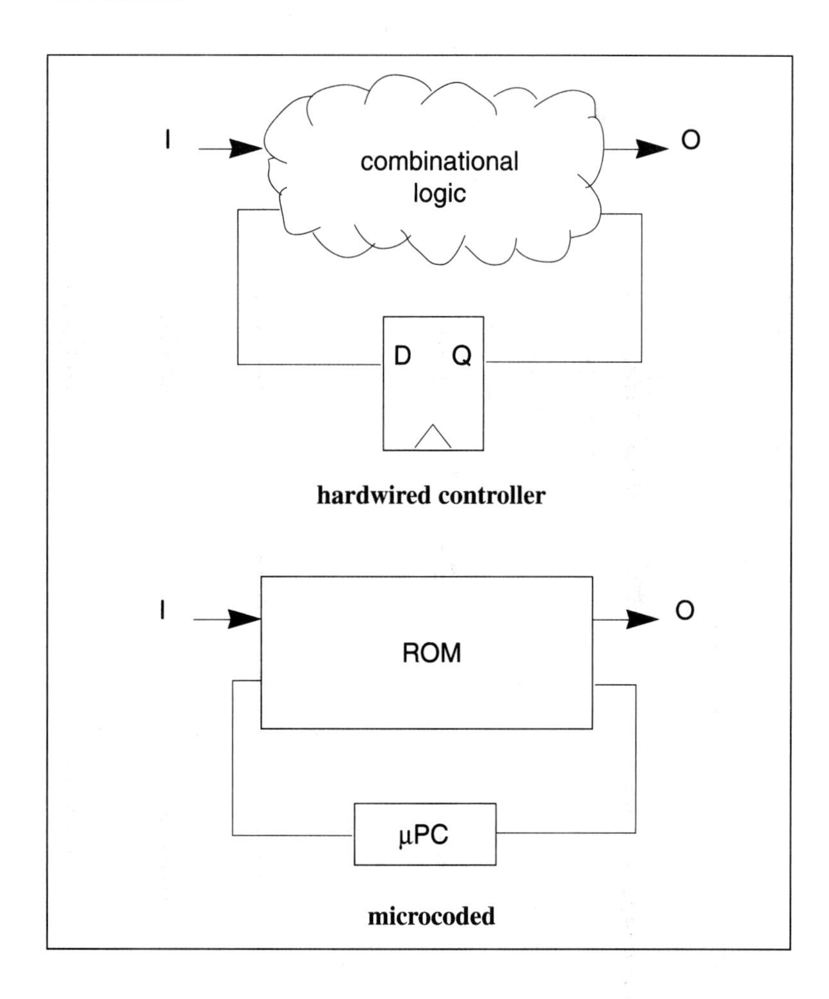

Figure 6-15 Controller implementation styles.

data path/controller interactions

So far, we have designed the data path and controller separately. Dividing architecture design into sub-problems makes some issues clearer, but it doesn't always give the best designs. We must consider interactions between the two to catch problems that can't be seen in either alone. Once we have completed an initial design of the data path and controller individually, we need to plug them together and optimize the complete design.

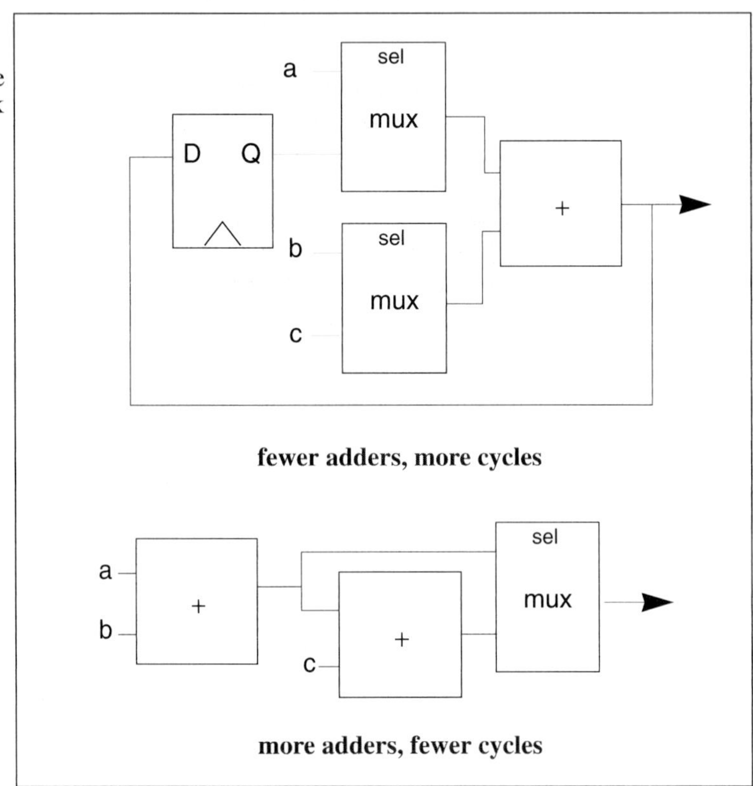

Figure 6-16 Adding hardware to reduce the number of clock cycles required for an operation.

The first, obvious step is to eliminate superfluous hardware from the data path. A schedule may have been found for a controller that doesn't require all the hardware supplied by the data path. A more sophisticated step is to add hardware to the data path to reduce the number of cycles required by the controller. In the example of Figure 6-16, the data path has been designed with one adder. The *true* branch of the *if* can be executed in one cycle if another adder is put into the data path. Of course, the second adder is unused when the false branch is executed. The second adder also increases the system's clock period; that delay penalty must be paid on every cycle, even when the second adder is not used.

Whether the second adder should be used depends on the relative importance of speeding up the true branch and the cost in both area and delay of the second adder.

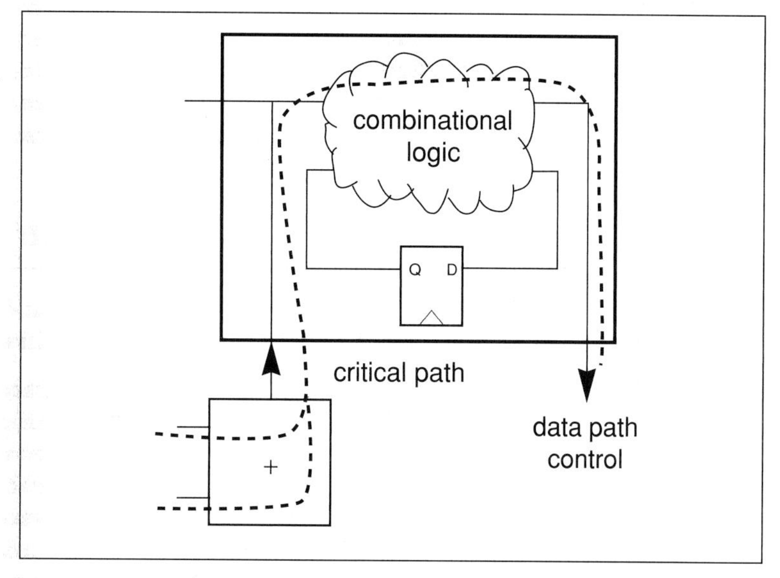

Figure 6-17 Delay through a data path controller system.

Another important optimization is adjusting the cycle time through the combined system. Even though the delay through each subsystem is acceptable, the critical path through the combination may make the cycle time too long. Overly long critical paths are usually caused by computations that use the result of a data path operation to make a control decision, or by control decisions that activate a long data path operation. In the example of Figure 6-17, the critical path goes through the carry chain of the ALU and into the next-state logic of the controller. We can speed up the clock by distributing this computation over two cycles: one cycle to compute the data path value, at which point the result is stored in a memory element; and a second cycle to make the control decision and execute it in the data path. One way to view the effect of this pipelining is that it moves the control decision ahead one cycle, increasing the number of cycles required to compute the behavior. However, it may not always be necessary to add cycles—if the adder is free, rather than move the control decision forward, we can move the addition back one cycle, so that the result is ready when required by the controller.

6.2.3 Power

In this section, we will discuss two important architectural methods for reducing power consumption: **architecture-driven voltage scaling** and **power-down modes**. The first method increases parallelism, while the second method tweaks the sequential behavior to eliminate unnecessary activity.

architecture-driven voltage scaling

As was noted in Section 2.4.3, the power consumption of static CMOS gates varies with the square of the power supply voltage. The delay of a gate does not decrease as quickly as power consumption. Architecture-driven voltage scaling [Cha92] takes advantage of this fact by adding parallelism to the architecture to make up for the slower gates produced by voltage scaling. Even though the parallel logic adds power, the transformation still results in net power savings.

This effect can be understood using the generic register-transfer design of Figure 6-18. A basic architecture would evaluate its inputs (clocked into registers in this case) every clock cycle using its function unit. If we slow down the operating frequency of the function unit by half, we can still generate outputs at the same rate by introducing a second function unit in parallel. Each unit gets alternate inputs and is responsible for generating alternate outputs. Note that the effective operation rate of the system is different in different components: the outputs are still generated at the original clock rate while the individual function units are running at half that rate. Parallelism does incur overhead, namely the extra capacitance caused by the routing to/from the function units and the multiplexer. This overhead is, however, usually small compared to the savings accrued by voltage scaling.

Parallelism can also be introduced by pipelining. If the logic has relatively little feedback and therefore is amenable to pipelining, this technique will generally reduce in less overhead capacitance than parallel-multiplexed function units.

The power improvement over a reference power supply voltage V_{ref} can be written as [Cha92]:

$$P_n(n) = \left[1 + \frac{C_i(n)}{nC_{\text{ref}}} + \frac{C_x(n)}{C_{\text{ref}}}\right]\left(\frac{V}{V_{\text{ref}}}\right), \quad \text{(EQ 6-1)}$$

where n is the number of parallel function units, V is the new power supply voltage, C_{ref} is the reference capacitance of the original function unit, C_i is the capacitance due to interprocessor communication logic,

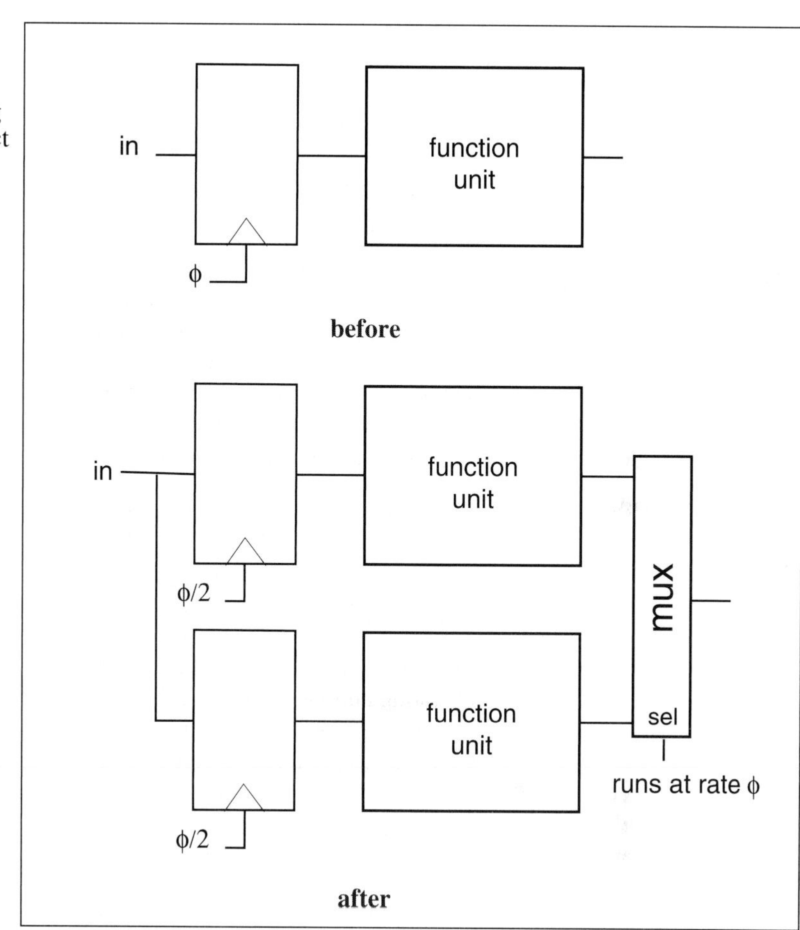

Figure 6-18 Increasing parallelism to counteract scaled power supply voltage.

and C_x is the capacitance due to the input/output multiplexing system. Both C_i and C_x are functions of the number of parallel function units.

power-down modes

Glitch reduction can reduce power consumption by eliminating unnecessary circuit activity. A power-down mode is a technique for eliminating activity in a large section of the circuit by turning off that section. Power-down modes are common in modern microprocessors and are used in many ASICs as well.

A power-down mode is more control-oriented than is architecture-driven scaling. Implementing a power-down mode requires implementing three major changes to the system architecture:

- conditioning the clock in the powered-down section by a power-down control signal;
- adding a state to the affected section's control which corresponds to the power-down mode;
- further modifying the control logic to ensure that the power-down and power-up operations do not corrupt the state of the powered-down section or the state of any other section of the machine.

The conditional clock for the power-down mode must be designed with all the caveats applied to any conditional clock—the conditioning must meet skew and edge slope requirements for the clocking system. Static or quasi-static registers must be used in the powered-down section for any state that must be preserved during power-down (it may be possible to regenerate some state after power-up in some situations). The power-down and power-up control operations must be devised with particular care. Not only must they put the powered-down section in the proper state, they must not generate any signals which cause the improper operation of other sections of the chip, for example, by erroneously sending a *clear* signal to another unit. Power-down and power-up sequences must also be designed to keep transient current requirements to acceptable levels—in many cases, the system state must be modified over several cycles to avoid generating a large current spike.

6.2.4 Pipelining

Pipelining is a well-known method for improving the performance of digital systems. Pipelining exploits concurrency in combinational logic in order to improve system throughput.

Figure 6-19 illustrates the fundamentals of pipelining. We start with a block of combinational logic that computes some function *f(a)*. That logic has some intrinsic delay. If we introduce a rank of registers into the logic properly, we can split it into two separate blocks of combinational logic. (We must be sure that the registers cut all the paths between the two logic blocks.) Each resulting block has a smaller delay; if we have done a good job, each block has about half the delay of the original block. Because each block has its own registers, we can operate the two blocks independently, working on two values at the same time. The left-hand block would start computing *f(a)* for a new input while the right-hand block would complete the function for the value started at the last cycle. Furthermore, we have reduced the cycle time of the machine

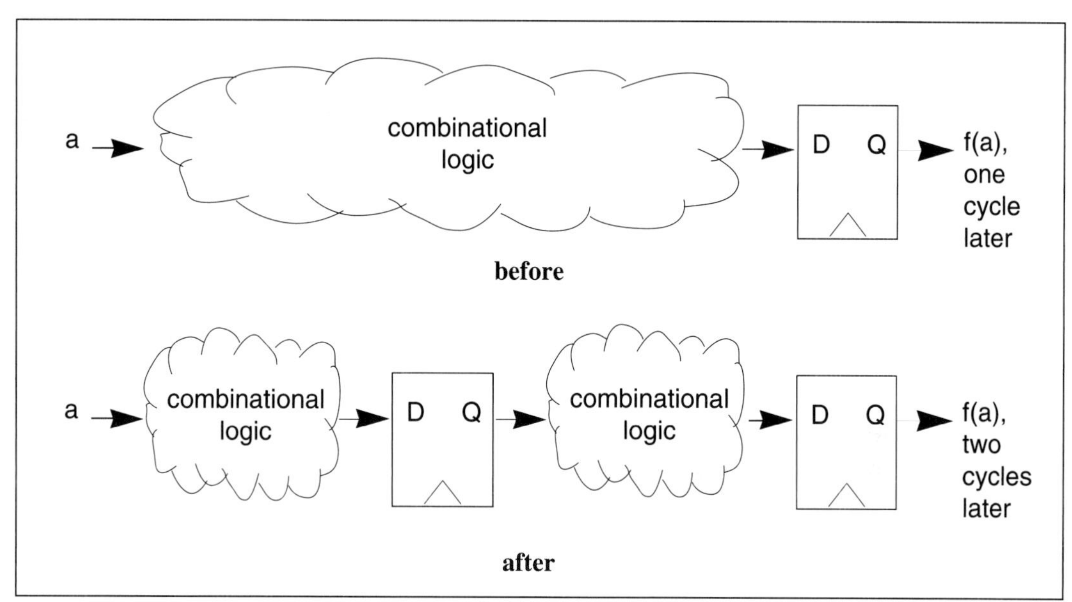

Figure 6-19 Adding pipelining.

because we have cut the maximum delay through the combinational logic.

throughput and latency

It is important to keep in mind that pipelining does not reduce the amount of time it takes to compute $f(a)$. In fact, because the register in the middle of the machine has its own setup and propagation times, the time it takes for a value to go through the system goes up slightly with pipelining. This time is known as **latency**. But we have increased the machine's **throughput**—the number of results that we can compute in a given amount of time. Assume that the delay through a single combinational logic block that computes $f(a)$ is D. Call latency L and throughput T. If we ignore register setup and hold times for the moment, we can write these definitions for the unpipelined system:

$$L = D, \quad \text{(EQ 6-2)}$$

$$T = 1/D. \quad \text{(EQ 6-3)}$$

If we pipeline the logic perfectly so that it is divided into two blocks with equal delay, then L remains the same but T becomes $2/D$. If we perfectly pipeline the logic into n blocks, we have

$$L = D, \quad \text{(EQ 6-4)}$$

$$T = n/D. \quad \text{(EQ 6-5)}$$

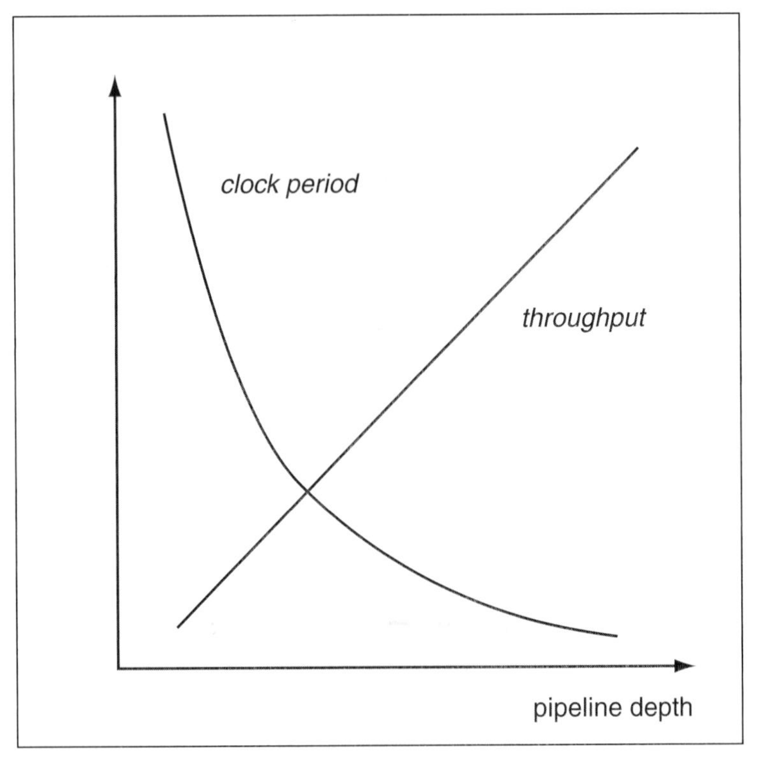

Figure 6-20 Clock period and throughput as a function of pipeline depth.

The clock period is determined by the delay D and the number of pipeline stages:

$$P = \frac{D}{n} \quad \text{(EQ 6-6)}$$

Figure 6-20 shows how clock period and throughput change with the number of pipeline stages. Throughput increases linearly with pipeline depth. Clock period decreases dramatically with the first few pipeline stages. But as we add more pipeline stages, we are subdividing a smaller and smaller clock period and we obtain smaller gains from pipelining. We could ultimately pipeline a machine so that there is a register between every pair of gates, but this would add considerable cost in registers.

adding pipeline registers

If we want to pipeline a combinational system, we need to add registers at the appropriate points. Luckily, FPGAs have lots of registers so an FPGA-based logic design can usually be pipelined inexpensively. The

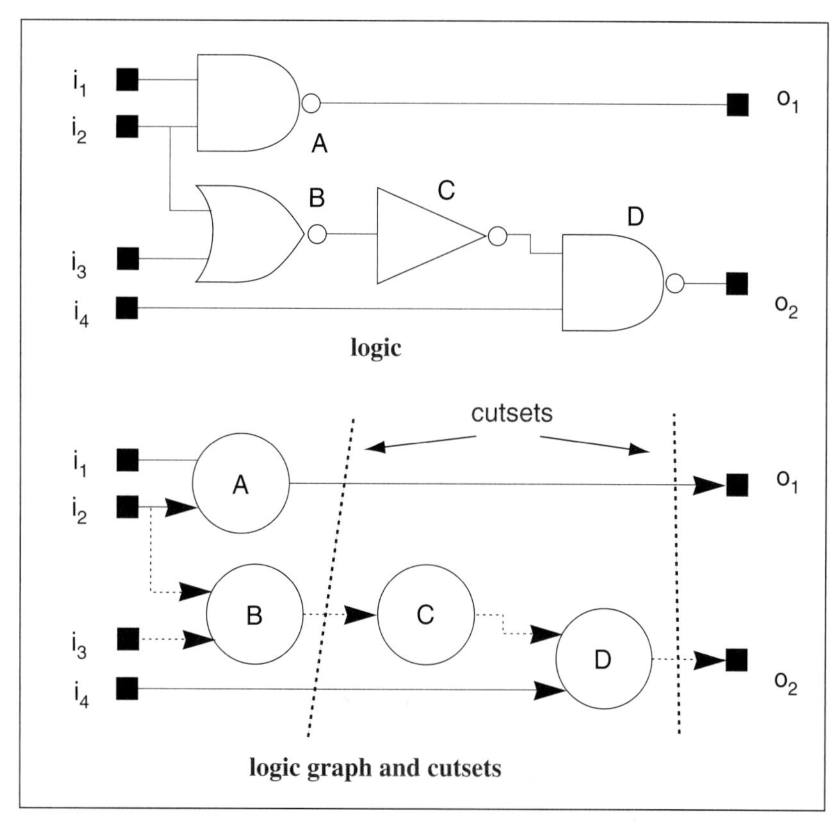

Figure 6-21 Logic and cutsets for pipelining.

register addition process resembles the timing analysis we performed in Section 4.4.4. We can use a graph to represent the structure of the logic, with one node per LE or gate and an edge that represent electrical connections. In timing optimization we must improve timing across a cutset of the timing graph. Similarly, we must add a **rank** of registers across a cutset of the logic graph. (Remember, a cutset of a graph is a set of edges that breaks all paths between two sets of nodes. In this case, we want to add registers between all possible paths from inputs to outputs.) Figure 6-21 shows the logic of Figure 6-14 and two of the many possible cutsets through the logic graph. Adding registers across any cutset is sufficient to ensure that the logic's functionality is not destroyed. However, the best cutsets divide the logic into blocks of roughly equal delay. We can use retiming, introduced in Section 5.5.4, to determine the best places to put the pipelining registers. Retiming will duplicate and merge registers as necessary to preserve functionality.

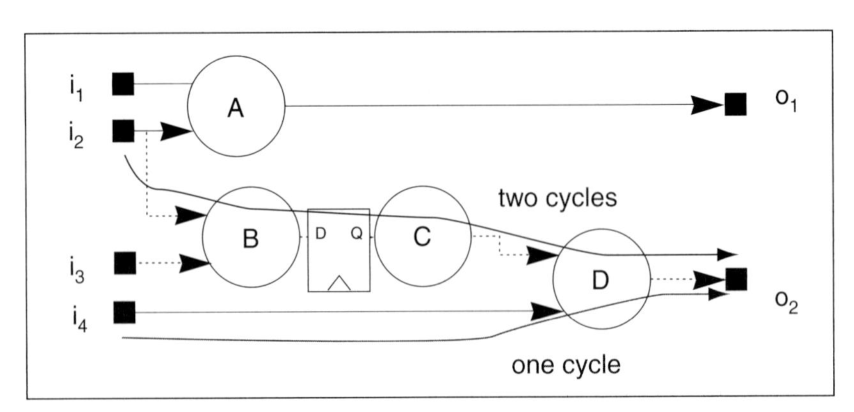

Figure 6-22 A bad cutset for pipelining.

What happens if we do not add registers across an entire cutset? Consider Figure 6-22, in which a cut through the logic is made that does not place registers along all the paths from inputs to outputs. There is no register along the path from i_4 to o_2, but there is one register along the i_2- o_2 path. As a result, when the inputs are presented to the logic, the D gate will receive its two inputs from two different clock cycles. The resulting output will be garbage.

A pipelined unit produces its outputs on a later clock cycle than does a non-pipelined unit. This is fine in some cases, while other logic designs are intolerant to the latency introduced by pipelining. We need to be careful when we are adding pipelines to a more abstract design than the logic gates since the abstraction may obscure some of the important details.

In steady state, an ideal pipeline produces an output on every cycle. However, when you start the pipe, it takes n cycles for an n-stage pipeline to produce a result. Similarly, when you stop putting data in the pipe, it takes n stages for the last value to come out of the pipe. The initialization time reduces the average pipeline utilization when we have short bursts of data. If we have D clock cycles' worth of data in an n-stage pipeline, then the utilization is

$$\frac{D}{D+n}. \qquad \text{(EQ 6-7)}$$

As D approaches infinity—that is, as we move toward operating the pipeline continuously—this utilization approaches 1. When D and n are near equal, then the utilization goes down.

pipelines with control

More complex pipelines, such as those in CPUs, have controllers that cause the pipeline to do different things on different cycles. In the case of the CPU, the pipeline must perform different operations for different instructions. When designing non-ideal pipelines with control, we must be sure that the pipeline always operates properly.

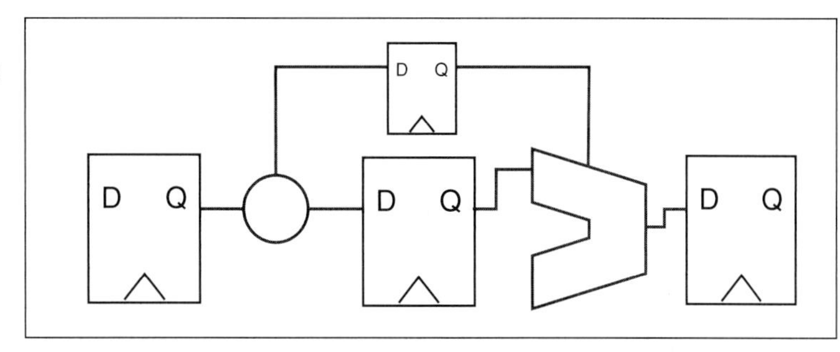

Figure 6-23 A pipeline with a feedforward constraint.

Some control operations are straightforward. Figure 6-23 shows a pipeline in which one stage computes the opcode for the ALU to be used in the next stage. So long as we pipeline the opcode so that it arrives at the right time, this works fine.

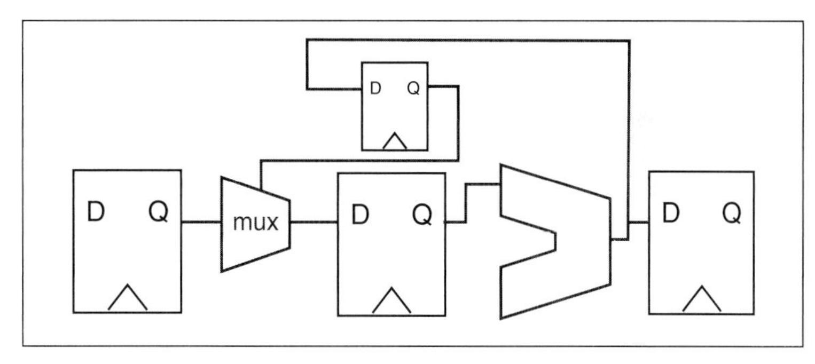

Figure 6-24 A pipeline with a backward constraint.

Figure 6-24 shows a more complex case. Here, the result of the ALU operation causes us to change what we do in a previous stage. A good example is a conditional jump in a CPU, where the result of one instruction causes us to change which instructions are to be executed next in time (corresponding to earlier in the pipe). Here we must make sure that the pipeline does not produce any erroneous results as a result of the change in condition.

Figure 6-25 shows a still more complicated situation. In this case, a single ALU is shared by two different stages. Small CPUs may share an ALU or other logic in order to reduce the size of the logic. Here we must

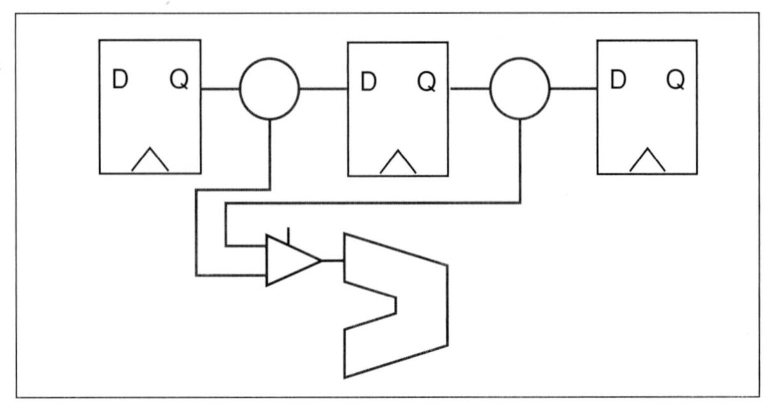

Figure 6-25 A pipeline with hardware shared across several stages.

be very careful that the two stages do not try to use the shared hardware at the same time. The losing stage of the pipeline would probably not be able to detect the error and would take a bad result.

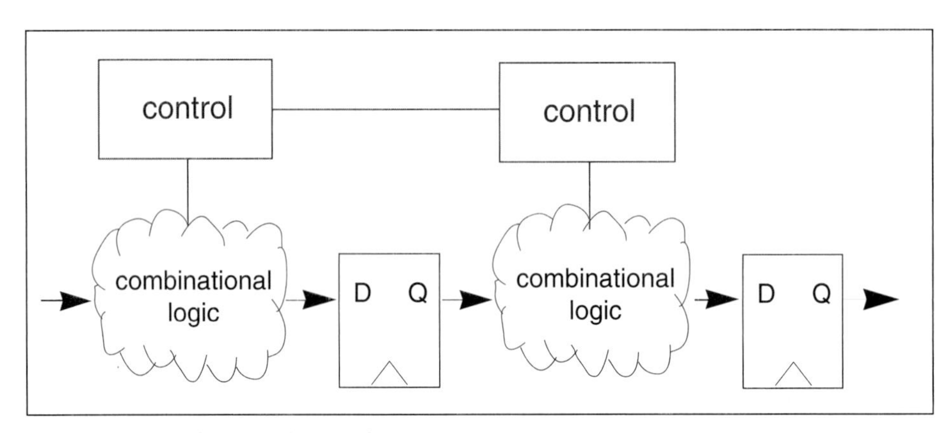

Figure 6-26 A pipeline with distributed control.

pipeline control

How do we design controllers for pipelines? The most common form of pipeline control is **distributed control**, illustrated in Figure 6-26. Rather than design one large state machine that simultaneously controls all the stages of the pipe, we typically design one controller for each stage. As shown in the figure, the controllers communicate in order to coordinate the actions across stages. Distributed control has some advantages in both specification and implementation. It is often easier to write the description of what a stage does than it is to write the entire pipeline operation for a single cycle. But the main reason for distributed control is that distributing the logic to be near the stage it controls helps to reduce delays.

When we design a pipeline with control, we must also worry about verifying that the pipeline operates correctly. Because they have a large amount of state, pipelines can be hard to verify. Distributed control makes some aspects of design and verification easier and some other aspects harder. Distributed control is harder simply because we do not have all the control information in one place. When we distribute the control formulas, it is harder for a person to look at the control actions and see that they always do what was intended.

When we write the control for a pipeline stage, we often use **symbolic FSM** notation. The transitions in a symbolic FSM may contain not just constants but also the names of registers in the pipeline data path. If we look at the state of the entire machine, we would have to write states for every different data path register value. But because we are interested in the operations on these states, symbolic FSMs allow us to factor out the control from the data path.

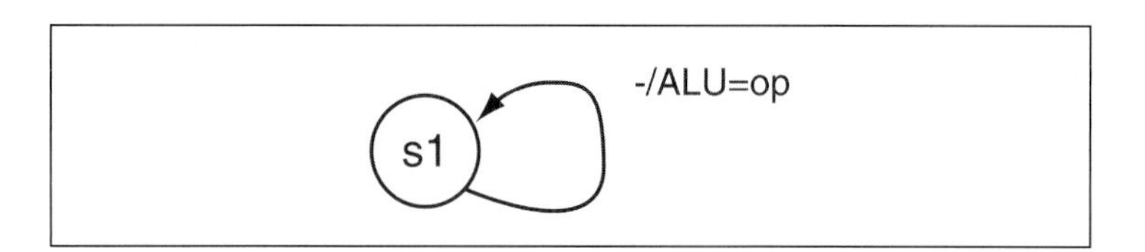

Figure 6-27 A simple symbolic FSM for pipeline control.

A simple case of a pipeline control machine is shown in Figure 6-27. This machine describes the control for the ALU of the pipeline in Figure 6-23.The controller makes no decisions but needs a register to hold the ALU opcode. On every clock cycle, the ALU opcode register may receive a new value from the previous pipeline stage, causing the ALU to perform a different operation, but this does not require an explicit control decision.

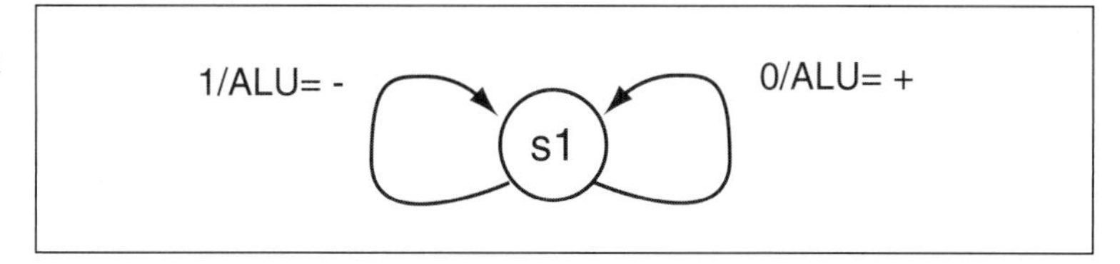

Figure 6-28 A condition in pipeline control.

Figure 6-28 shows a symbolic FSM with a condition. Here, some operation, probably from the previous stage, may cause the state machine to issue an ALU opcode of either + or -. Because the pipeline stage is only one cycle long, the conditions all return to the initial state.

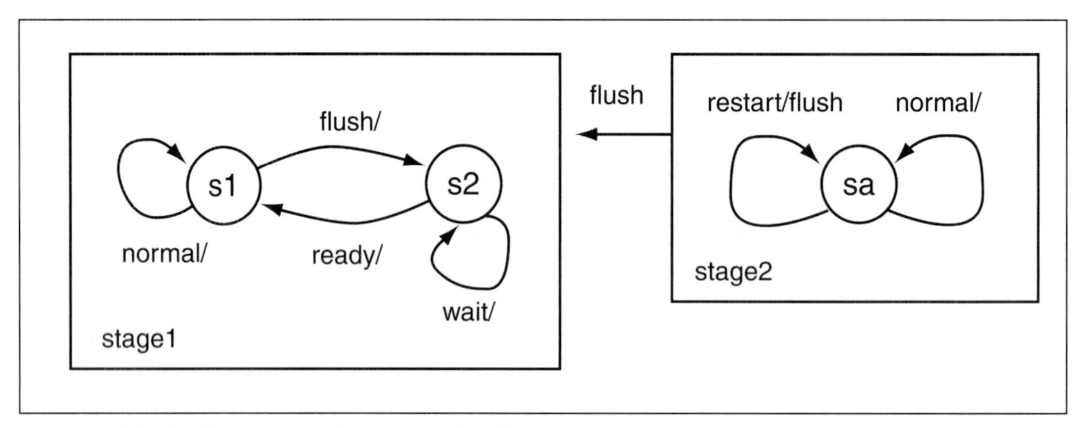

Figure 6-29 Distributed control for a pipeline flush.

Figure 6-29 shows one possible description of distributed control for a pipeline flush like that illustrated in Figure 6-24. Stage 2 receives a restart signal from its own logic. This causes it to issue a flush command to stage 1. Stage1 then enters a separate state in which it waits for the flush to end; this may be controlled by a timer or some other external event. The stage then returns to normal operation.

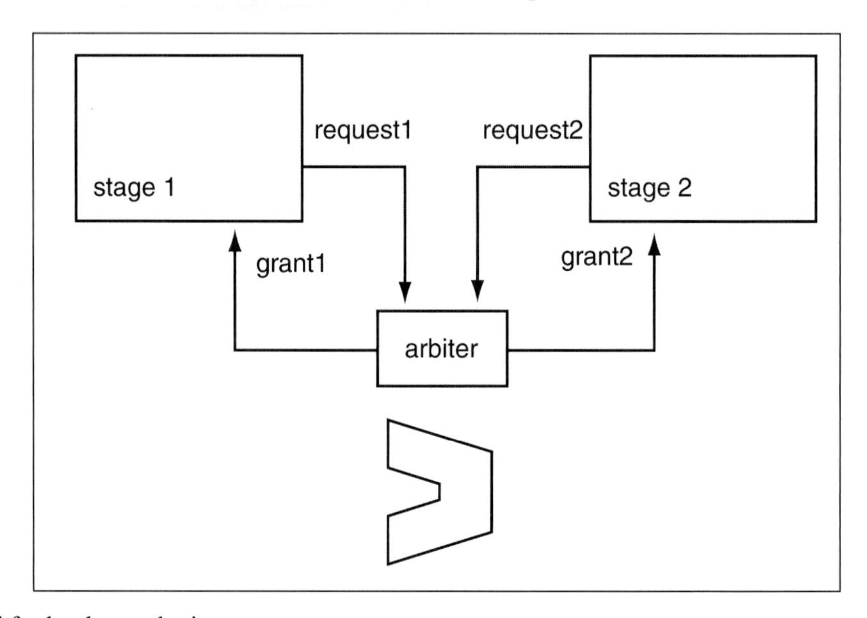

Figure 6-30 Control for hardware sharing.

Figure 6-30 shows control logic for shared hardware like that of Figure 6-25. This scheme is similar to arbitration on a bus. Arbitration logic determines which stage gets the shared hardware (the ALU) on each cycle. Each stage uses a request line to signal the arbiter; each stage watches its own grant line to see when it has the shared logic. In this setup, if each pipe stage is only one cycle long, the arbiter would have to be combinational to ensure that the answer came within the same cycle. The stages would have to hold request for the remainder of the cycle to ensure that the arbiter output didn't change.

Figure 6-31 Product machine formed from distributed control.

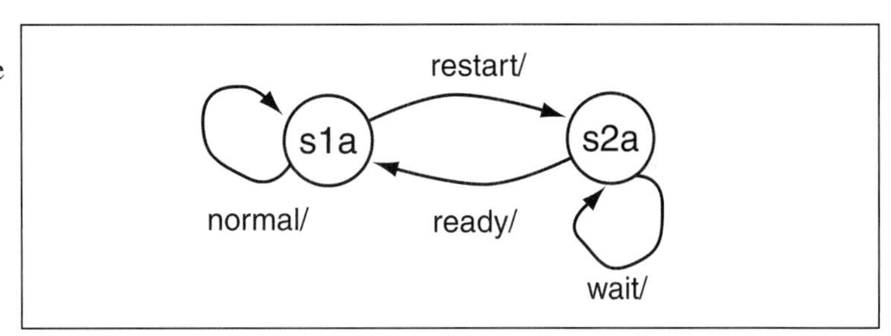

pipeline verification

Distributed control can be hard to verify because we do not have a single description that tells us what all of the pipeline does in every stage. However, we can form the product machine of the distributed control stages to help us see the global results of control. Figure 6-31 shows the product machine for the distributed pipeline flush of Figure 6-29. Forming the product machine of a symbolic FSM takes some more work than forming the product of a pure FSM, but it is not difficult. The product machine makes it easier to verify that:

- the control enters all the required states and performs all the required operations in those states;
- the control does not enter any undesired states.

One technique used for pipeline verification is **symbolic simulation**. If we tried to simulate the pipeline with all known data and control register values, we would quickly find that the simulation was much too long. Pipelines have large numbers of states that blow up simulation times. However, by simulating the data register values symbolically, we can collapse many of those states. A relatively small number of simulations can then cover the various cases in the control logic.

6.3 Design Methodologies

In this section we will look at the overall design process. Designing a complex chip requires care. A large design takes time and money to design. The cost of mistakes is particularly high if they are found after the mistakes are found only after the design is shipped, but errors can be costly even if they are found too late in the design process. Methodologies are especially important because most designs are completed by more than one person and poor communication can cause all sorts of problems. The goals of a **design methodology** are to design the chip as quickly as possible and with as few errors as possible.

Of course, the exact sequence of steps you follow to design a chip will vary with your circumstances: what type of chip you are designing; size and performance constraints; the design time allowed; the CAD tools available to you; and many other factors. In this section we will look at

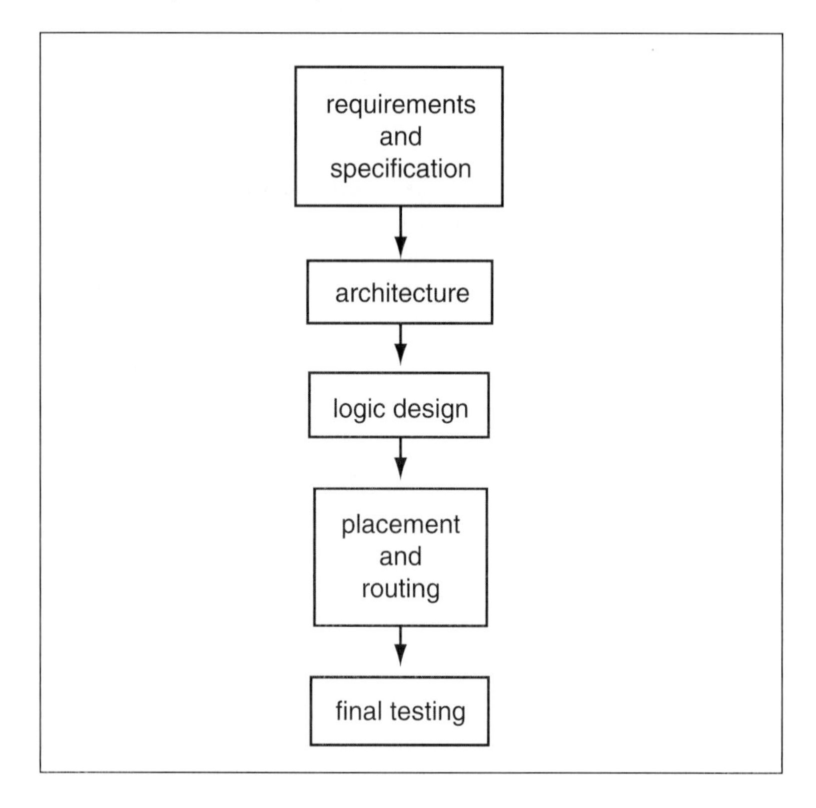

Figure 6-32 A generic integrated circuit design flow.

the basic characteristics of most design methodologies without prescribing a particular detailed methodology.

6.3.1 Design Processes

design flows

A design methodology is often called a **design flow** since the flow of data through the steps in the methodology may be represented in a block diagram. Figure 6-32 shows a generic design flow for VLSI systems. While all design methodologies will vary from this in practice, this flow shows some basic steps that must be considered in almost any design. Let's consider this process in more detail:

- **Requirements and specification.** Before diving into the details of the design, we need to understand what we are supposed to design. A requirements document generally refers to a written description of the basic needs of the system design, while a specification is a more detailed technical description of what the chip does. Requirements typically come from customers and marketing, while a specification is used to translate the requirements into specifics useful to the design team. Getting the requirements and specification right is essential—delivering the chip design to the customer and finding out that it isn't what they wanted is an expensive way to debug the specification.
- **Architecture.** If the chip being designed is a rework of an existing design—a design shrink, a few added features, etc.—then the architectural design is simple. But when designing something new, a great deal of work is required to transform the requirements into a detailed microarchitecture ready for logic design. Architectural design requires construction of a microarchitectural simulator that is sufficiently detailed to describe the number of clock cycles required for various operations yet fast enough to run a large number of test vectors. A test suite must also be constructed that adequately covers the design space; if the design is a rework of a previous architecture, then the vectors for that system are a starting point which can be augmented with new tests. Architectural design requires extensive debugging for both functionality and performance; errors that are allowed to slip through this phase are much more expensive to fix later in the design process.
- **Logic design.** Logic design will usually be performed with logic synthesis tools for FPGA designs, though it is possible to design the logic manually. In either case, the design will probably go through several refinement steps before completion. Initial design verification steps will concentrate on logical correctness and basic timing properties. Once the basic struc-

ture of the logic has taken shape, scan registers can be inserted and power consumption can be analyzed. A more detailed set of timing checks can also be performed, including delay, clock skew, and setup/hold times. In extreme cases it may be necessary to make more drastic changes to the logic to correct problems found late in the logic design process.

- **Placement and routing.** The physical design steps of placement and routing are usually performed primarily by tools, though some hand tweaking may be necessary to get the performance and size you require. This is the stage at which performance, size, and power become apparent. Careful physical design may be necessary to meet those goals or, in extreme cases, you may need to go back and redesign the logic. If redesign is necessary, the placement and routing tools should help you identify the problem areas that need to be fixed.
- **System integration and final testing**. The design will often be done in pieces by several designers. Though test runs should be made before the end of the design process, the final stages necessarily include putting together all of the pieces and making sure that they work together. Testing should include both functionality and non-functional requirements like performance and power.

design iterations

Ideally, we would like to start from a basic description of the chip to be designed and work steadily to a final configuration file for the FPGA. However, most projects are usually not that simple. We always start the design with incomplete information. Early design stages rely on estimates which may be supplied by tools. As we progress through the design process, we gather new information that may tell us some of our earlier assumptions were wrong. We may have to go back and rework some of the earlier design work. Hopefully, those changes will be small and will change only relatively recent design decisions. Generally, the further back in time we go to change a design decision, the harder and more expensive the change will be. Experience with digital design in general and the type of system being designed in particular is the best guard against excessive design iterations.

project management

Most engineers take a dim view of managers, but some sort of management is essential in any multi-person project. Most interesting designs are large enough to require several people, so project management is usually unavoidable. Even if no one is in charge, using processes to make sure that everyone on the project has the information they need is essential.

6.3.2 Design Standards

Design methodologies embody **best practices** that have been learned over the years. Those practices may take several forms: tools, features, documentation, management methods, etc. In this section we will look at a few types of methods that we can carry over from project to project. Design verification is of course critical; we will devote the next section to that topic.

design features

We can identify some features that are useful in almost any design. A simple example is a reset signal that puts the chip in a known state. Resets make it possible to test the design; a surprising number of chips have been designed without reset signals. Boundary scan is another example of a generally useful feature for design verification and debugging.

source code control

Source code control is normally associated with software projects, but since so much hardware design is based on HDLs, source code control is very important in hardware projects as well. A source code control system allows you to keep track of the versions of various files and helps manage access. Any file to be managed by the control system is checked in at the start. A checked-in file may be checked-out as a read-only file for general use. If a designer wants to modify a file, it can be checked out for writing, but the source code control system ensures that only one person at a time has a modifiable version of the file. The modification to the file is often called a **delta**. When the file is checked back into the system, the source code control keeps the old version as well as the new one. (These systems generally store the changes to the file at each delta to save disk space.) The system can provide either the newest version of the file as well as older versions. Keeping around older versions helps you track down where errors were introduced and simplifies recovery from mistakes.

documentation

One constant through all circumstances is the importance of good design documentation. You should write down your intent, your process, and the result of that process at each step. Documentation is important for both you and the others with whom you work.

Documentation may be for internal or external use. The specification is a good example of an external document. Although the spec is used by the design team, it is generally given to the customer as a guide to how the chip works. Internal documentation is produced by the design team for their own use. Written descriptions and pictures help you remember what you have done and understand complex relationships in the design. A paper trail also makes the design understandable by others. If you are hit by a truck while designing a complex chip, leaving only a few scribbled notes on the backs of envelopes and napkins, the grief of your loved ones will be matched by the anguish of your employer who must figure out how to salvage your incomplete design.

Even after the chip is done, documentation helps fix problems, answer questions about its behavior, and serves as a starting point for the next-generation design. Although you may feel that you understand your own design while you are working on it, a break of only a few months is more than enough time for most people to forget quite a few details. Good documentation helps you remember what you did. You will find that a little time spent on documentation as you go more than pays for itself in the long run.

Typical external documents include:

- **Overviews**. A short description of the major features of the chip. This document may include a block diagram, performance figures, etc.
- **Data sheet**. A detailed description of the chip's operation. This description is for the user of the chip, but it should include all the information required for a designer using the chip in a larger design: operating modes, registers, performance, etc.
- **Application notes**. Describes a system designed using the chip. Talks about the important considerations in designing with the chip, the advantages of the chip, and suggests some methods for how to use the chip to handle a particular application.

Typical internal documents include:

- **Requirements**. The requirements are gathered from customers, but the requirements document is typically widely circulated since it could be used by competitors.
- **Architecture specification**. The architecture document is a complete description of what the chip needs to do and how it will do it. The external description of the chip's operation from this document may wind up in the final data sheet but the internal architectural description is kept private.
- **Module descriptions**. The major modules in the design should have their own documents describing their inputs and outputs, timing requirements, and internal architecture.
- **Test plan**. The test plan ensures that the design is thoroughly

tested. It describes what interfaces will be tested, what tests will be applied, and what results are expected.

- **Design schedule**. The schedule lays out milestones for the project and the number of people (and possibly equipment) needed for each stage.

design reviews

Design reviews are very valuable tools for finding bugs early. A design review is a meeting in which the designer presents the design to another group of designers who comment on it. Design reviews were developed in the 1970's for software projects [Fag76] but are equally useful for hardware projects, particularly when HDLs are used.

The key players in the early stages of the design review process are the designer and the **review leader.** The designer is responsible for preparing a presentation for the meeting, while the review leader organizes the review logistics. In preparation for review, the designer prepares documentation that describes the component or system being designed:

- purpose of the unit (not all of the reviewers may not be familiar with this piece of the design);
- high-level design descriptions, including specifications and architecture;
- details of the designs, including the HDL code, schematics, etc.;
- procedures used to test the design.

The review meeting itself includes the designer, the leader, the audience, and a **review scribe**. During the design review, the audience listens to the designer describe the design and comment on it (politely, of course). This description is more than a brief overview. The review should start with the introductory material to familiarize the audience with the design and then move to the implementation itself. The designer should, for example, explain the HDL code line by line to relate the code back to the unit's required function. The designer should also describe the test procedures in detail.

During the meeting, the scribe takes notes of suggestions for improvement. After the meeting, the designer uses these notes to improve the design. The designer should use the notes as a checklist to ensure that all the bugs found are fixed. The review leader should work with the designer to fix the problems. If the problems found were extensive enough, a second review may be useful to ensure that the new design is satisfactory.

Many bugs will simply be found by the designer during the course of preparing for the meeting; many others will be identified by the audience. Design reviews also help the various members of a team synchronize—at more than one design review, two members of the same design team have realized that they had very different understandings of the interface between their components. Design reviews won't catch all the bugs in the design but they can catch a surprisingly large fraction of the total bugs in the system and do so early enough to minimize the cost of fixing those bugs.

6.3.3 Design Verification

verification goals

Chip designs require extensive design verification efforts. Not only are most chip designs complex, requiring sophisticated tests, but the cost of shipping bad designs can be high. Verification has several goals:

- functional verification ensures that the chip performs the desired functions;
- performance verification tests the speed at which the chip will run;
- in some cases, we may run power tests to verify the power/energy consumption of the design.

verification methods

You may have access to **formal verification** tools, which can automatically compare a combinational logic or sequential machine description against the implementation, using tautology or FSM equivalence algorithms. Some formal verification tools can also be used for higher-level tasks like bus verification. You are more likely to use simulation to validate your design. You can simulate a single description of your machine, such as the register-transfer description, to be sure you designed what you wanted; you can also compare the results of two different simulations, such as the logic and register-transfer designs, to ensure that the two are equivalent.

verification by simulation

Verilog and VHDL provide generic simulation engines. By writing hardware descriptions at different levels of abstraction, you can control the detail and speed of your simulation. Earlier in the design process, architectural or register-transfer designs can be used to verify functionality. Later in the process, gate-level structural descriptions can be used to verify functionality as well as performance or power.

performance verification

Performance verification—making sure the system runs fast enough—can be separated from functionality if the system is properly designed. If we have obeyed a clocking methodology, we know that the system will work if values arrive at the registers within prescribed times.

We can use both static timing analysis and simulation to verify performance. Static timing analysis is relatively fast and provides valuable debugging information for paths that are too long. However, it can be pessimistic—it may identify long paths that cannot be exercised. (Sensible designers, however, assume that paths identified by static timing analysis are valid unless compelling proof shows otherwise. At least one major chip design had to be redone because a path identified by static timing analysis was ignored.)

Simulation can be used to complement static timing analysis. Information from placement and routing can be used to **back annotate** the schematic with information on loading and wire delay. That information allows us to verify the performance as well as the functionality of the design.

simulation vectors

Simulation uses input vectors and their associated expected outputs. Choosing the right input vectors is possibly the most important aspect of simulation. It takes far too many input sequences to exhaustively test a digital system, but a reasonable number of simulation vectors is generally sufficient to adequately exercise the system and disclose a large number of the bugs in the system. However, simulating with not enough or the wrong vectors can instill false confidence while leaving too many bugs latent in the system.

There are two major types of simulation vectors:

- **black-box vectors** are generated solely from the specification without knowledge of the implementation;
- **clear-box vectors** (sometimes called white-box vectors) are derived from an analysis of the implementation.

Each has its uses. Black-box vectors, because they are derived from the specification, are unbiased by errors in the implementation. However, certain types of errors are less likely to be obvious from an analysis of the specification. Clear-box vectors can be used to expose internal contradictions in the design, such as paths that are never exercised. Clear-box tests can often be generated automatically by tools. However, they tend to concentrate on low-level details of the design.

Black-box tests can be generated in several ways. The specification can be analyzed and used to generate tests, usually manually. For example, a set of vectors for a multipliers is relatively easy to generate from the specifications for the multiplier. **Random tests** may be generated. Using random tests requires some other system description so that the expected output can be determined. Random tests are not a panacea but they are a useful tool in the arsenal of tests. **Regression tests** are tests generated for previous versions of a design. Using regression tests helps ensure that old bugs do not reappear (for example, by reusing old code) as well as helping to uncover new bugs.

White-box tests are created by analyzing the design and generating tests to ensure that a local property of the design is globally obeyed. For example, just because a certain value is generated during processing does not mean that the value makes it to the primary outputs as it should. White-box testing methods generally emphasize either control or data. An example of a control-oriented test is ensuring that each transition in an FSM is exercised at least once. An example of a data-oriented test is ensuring that the variables in an expression are given values that test certain ranges of the expression's value.

6.4 Design Example

In this section we will exercise our design skills by designing a simple programmable processor.

6.4.1 Digital Signal Processor

In this section we will design a simple programmable **digital signal processor** (**DSP**). This DSP is not pipelined and should not be considered an exercise in high-performance CPU design. However, the DSP does allow us to design a data path-controller system.

Figure 6-33 DSP connected to dual memories.

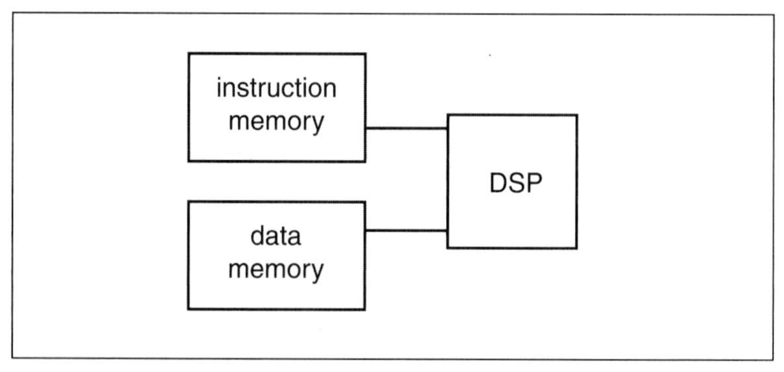

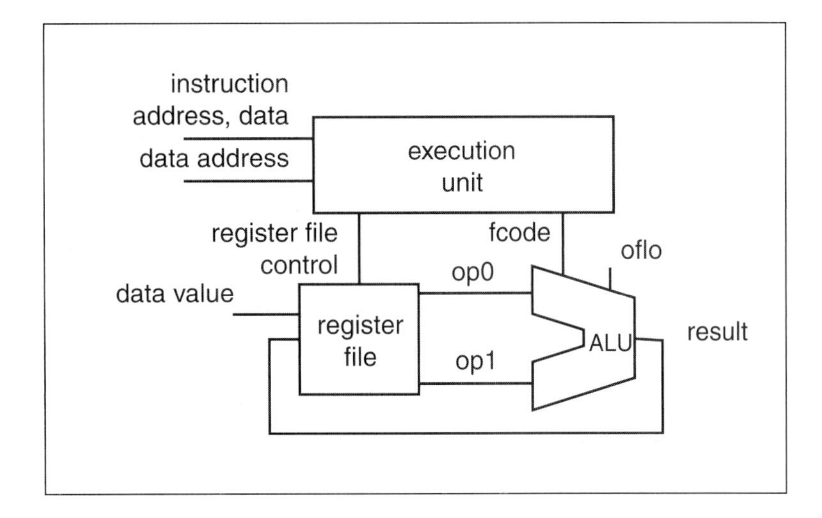

Figure 6-34 Microarchitecture of the DSP.

The term DSP is somewhat overused at this point, but one common meaning is a computer with separate instruction and data memories as shown in Figure 6-33. This **Harvard architecture** provides higher memory bandwidth. Figure 6-34 shows the microarchitecture of the DSP. It consists of three major units: the register file, the ALU, and the execution unit. The register file and the ALU together form the data path. The execution unit is responsible for fetching and decoding instructions as well as guiding the data path through the execution of the instruction.

Figure 6-36 DSP instruction encoding.

opcode	**encoding**
ADD	0
SUB	1
LOAD	2
STORE	3
JMP	4
JZ	5
NOP	6

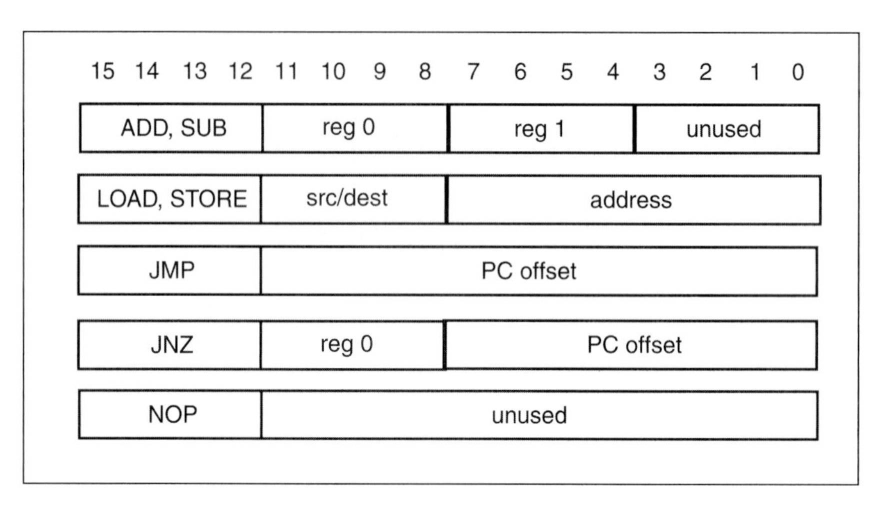

Figure 6-35 DSP instruction formats.

The DSP uses 16-bit words for both instructions and data. Figure 6-35 shows the formats of the DSP's instructions:

- ADD and SUB perform *reg0* = *reg0* + *reg1* and *reg0* = *reg0* - *reg1*, respectively.
- LOAD and STORE load or store the respective address from/to the src/dest register. The address is padded with zeroes.
- JMP jumps to the specified address (padded with zeroes).
- JNZ tests the specified register and, if it is zero, adds the specified PC offset to the PC.
- NOP has an opcode but the rest of the instruction is unused.

The encodings of the opcodes are shown in Figure 6-36.

Now let's design the three major units in the DSP. The ALU must be able to perform several different types of functions as required by the instruction set. For the ALU, we will use a modified version of the one we designed for Example 4-10. That ALU implements five operations: addition, subtraction, increment, decrement, and negate. Unlike our original ALU, this one provides registers to hold the output value (including overflow), which will simplify the connection between the ALU and the register file. The registers allow us to use a synchronous design style in the Verilog model:

```
// ALU module
```

```
'define PLUS 0
'define MINUS 1
'define PLUS1 2
'define MINUS1 3
'define NEG 4

module alu(fcode,op0,op1,result,oflo,clk);
    parameter n=16, flen=3;
    input clk;
    input [flen-1:0] fcode; // operation code
    input [n-1:0] op0, op1; // operands
    output [n-1:0] result; // operation result
    output oflo; // overflow
     reg oflo;
    reg [n-1:0] result;

always @(fcode)
begin
    result = 16'bxxxxxxxxxxxxxxxx;
    oflo = 1'bx;

    case (fcode)
        'PLUS: begin {oflo, result} = op0 + op1; end
        'MINUS: begin {oflo, result} = op0 - op1; end
        'PLUS1: begin {oflo, result} = op0 + 1; end
        'MINUS1: begin {oflo, result} = op0 - 1; end
        'NEG: begin {oflo, result} = -op0; end
    endcase // case (fcode)
end
endmodule
```

This model can be modified to provide more ALU operations.

The register file is a block of memory that can be read and written. We will use a register file with two ports—it can support two simultaneous reads or writes. Each port has its own read input and write output. A bulk memory would use the same data port for both reading and writing but our design is simplified by separating read and write data. Here is the Verilog for the register file:

```
// Register file module

//clk: clock signal
//reg(a/b)val: reg file out port 0 for register a/b
```

```
//reg(a/b)adrs: reg file index add  for register a/b
//reg(a/b)rw: read/write control sig  for register a/b
//reg(a/b)write: reg file input port  for register a/b

module reg_file(clk,regaval,regaadrs,regarw,regawrite,regb-
val,regbadrs,regbrw,regbwrite);

    parameter n = 16; // instruction word width
    parameter nreg = 4; // number of registers

    input clk;
    input [nreg-1:0] regaadrs, regbadrs;  // addresses for a, b
    output[n-1:0] regaval, regbval; // a, b data values
    input regarw, regbrw; // read/write for a, b
    input [n-1:0] regawrite, regbwrite;

    reg[n-1:0] RegisterFile [0:nreg-1];  // the registers

    // read values
    assign regaval = RegisterFile[regaadrs];
    assign regbval = RegisterFile[regbadrs];

    always @(posedge clk)
     begin
       if (regarw) RegisterFile[regaadrs] = regawrite;
       if (regbrw) RegisterFile[regbadrs] = regbwrite;
      end
endmodule
```

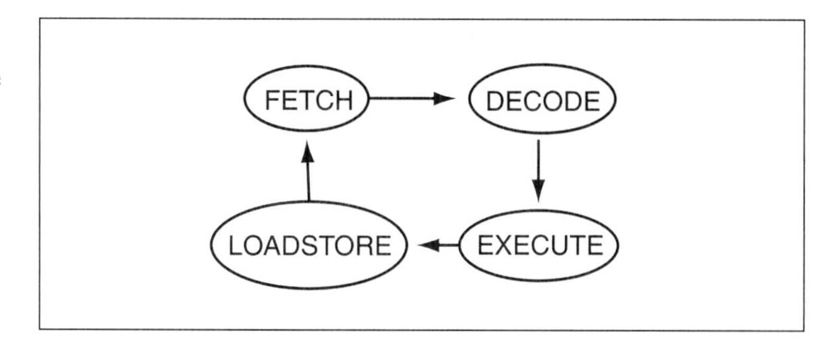

Figure 6-37 State machine for the execution unit.

The execution unit fetches, decodes, and executes instructions. Figure 6-37 shows the state machine for the execution unit. Because this is a simple DSP, it always follows sequence of four states: fetch, decode, execute, loadstore. The fetch state requests the next instruction from the instruction memory. The decode state extracts arguments from the instruction based upon the opcode. The execute state performs the actual operation. The loadstore state is an additional state required to complete a load to or store from data memory; this state is unnecessary if the instruction is not a *LOAD* or *STORE* but we simplify the machine somewhat by always executing it.

Here is the Verilog for the execution unit:

```
// Execution unit

module ex(rst,clk,iadres,inst,dadres,data,aluop,carryin,carry-
out,reg0,reg1,regarw,regbrw,drw,regaval);

    parameter n = 16; // instruction word width
    parameter nreg = 4; // number of bits in register field
    parameter nalu = 3; // width of alu opcode

    input clk, rst;
    input [n-1:0] inst,data,regaval;
        // instruction and data from memory
    input carryout; // carryout from ALU
    output carryin; // carryin to ALU
    output [n-1:0] iadres; // address to instruction memory
    output [n-1:0] dadres; // address to data memory
    output [nalu-1:0] aluop; // opcode to ALU
    output [nreg-1:0] reg0, reg1; // addresses to register file
    output regarw, regbrw, drw;
        // read or write (same for both) 0 = read, 1 = write
    reg [n-1:0] ir; // instruction register
    reg [2:0] exstate; // state of the execution unit
    reg [n-1:0] iadres, dadres;
    reg regarw, regbrw, drw, carryin;
    reg [nreg-1:0] reg0;
    reg [nreg-1:0] reg1;
    reg [nalu-1:0] aluop;

// state codes for EX box
'define FETCH 0
'define DECODE 1
'define EXECUTE 2
'define LOADSTORE 3

// instruction format
```

```
// 15 14 13 12 11 10 9  8  7  6  5  4  3  2  1  0
// ADD,SUB      | reg0       | reg1       | xxxx
// LOAD,STORE   | src/dest   | address
// JMP          | PC offset
// JZ           | reg        | PC offset
//
// instruction fields
'define OPA n-1
'define OPZ n-4
'define ARG0A n-5
'define ARG0Z n-8
'define ARG1A n-9
'define ARG1Z n-12

// opcodes
'define ADD 0
'define SUB 1
'define LOAD 2
'define STORE 3
'define JMP 4
'define JZ 5
'define NOP 6

always @(rst or posedge clk) // execute state machine
begin
  if(rst == 1)
   begin
    exstate = 'FETCH;
    regarw = 0;
    regbrw = 0;
    drw = 0;
    aluop = 'NOP;
    iadres = 16'b0000000000000000;
   end
  else
   begin
       case (exstate)
           'FETCH :
             begin
               ir = inst; regarw = 0; regbrw = 0; drw = 0;
               exstate = 'DECODE; aluop = 'NOP;
             end
           'DECODE :
```

```
begin
  exstate = 'EXECUTE;
  if (ir['OPA:'OPZ] == 'JZ) // set up the test
      begin
      reg0 = ir['ARG0A:'ARG0Z];
      end
  else if(ir['OPA:'OPZ] == 'JMP)
      begin
      end
  else if (ir['OPA:'OPZ] == 'LOAD)
      begin
      // set up dest
      reg1 = ir['ARG0A:'ARG0Z];
      // fetch me
      dadres = {8'b00000000, ir['ARG1A:0]};
      end

  else if (ir['OPA:'OPZ] == 'STORE)
      begin
      reg1 = ir['ARG0A:'ARG0Z];
      // store to me
      dadres = {8'b00000000, ir['ARG1A:0]};
      end

  else if (ir['OPA:'OPZ] == 'ADD)
      begin
      reg0 = ir['ARG0A:'ARG0Z];
      reg1 = ir['ARG1A:'ARG1Z];
      end

  else if (ir['OPA:'OPZ] == 'SUB)
      begin
      reg0 = ir['ARG0A:'ARG0Z];
      reg1 = ir['ARG1A:'ARG1Z];
      end
  end

'EXECUTE :
 begin
    exstate = 'LOADSTORE;
    case (ir['OPA:'OPZ]) // opcode
       'ADD :
       begin //ADD
```

```
                iadres = iadres + 1;
                aluop = 'ADD;
                carryin = 0;
                regarw = 1;
                end //ADD

                'SUB:
                begin
                iadres = iadres + 1;
                aluop = 'SUB;
                carryin = 0;
                regarw = 1;
                end //SUB

                'LOAD :
                begin //LOAD
                iadres = iadres + 1;
                regbrw = 1;
                end //LOAD

                'STORE :
                begin //STORE
                iadres = iadres + 1;
                drw = 1;
                end

                'JMP :
                begin //JMP
                iadres = { 4'b0000, ir['ARG0A:0] };
                end //JMP

                'JZ :  // use result from decode cycle
                begin //JZ
                if (regaval == 0)
                    iadres = { 8'b00000000, ir['ARG1A:0] };
                end //JZ

            endcase //(opcode)
          end

        'LOADSTORE:
          begin
          exstate = 'FETCH;
```

```
                end
            endcase //(exstate)
        end //(clk)
end // always

endmodule
```

Now we can put together the units to form the complete DSP. Here is the structural Verilog for the DSP:

```
// The complete DSP (without data or instruction memory).

module dsp(clk,rst,ival,iadres,dval,dadres,drw,regbval);

parameter n = 16; // instruction word width
parameter flen=3; // length of ALU opcode
parameter nreg=4; // number of registers

input clk, rst;
input [n-1:0] ival, dval; // value for instruction and data
inout [n-1:0] iadres, dadres; // address for instruction and data
output drw; // read/write for data memory
output [n-1:0] regbval;

wire [nreg-1:0] rega, regb; // selected a, b registers
wire [n-1:0] regaval, regbval, regawrite;
    // data values from register file, data value to reg file
wire [flen-1:0] aluop;
wire regarw, regbrw;
wire carryin, overflow;

// don't connect execute output carryin
ex execute(rst,clk,iadres,ival,dadres,dval,aluop,,over-
flow,rega,regb,regarw,regbrw,drw,regaval);
alu alu1(aluop,regaval,regbval,regawrite,overflow,clk);
reg_file regfile(clk,regaval,rega,regarw,regawrite,regb-
val,regb,regbrw,dval);

endmodule
```

6.5 Summary

Our experience in combinational and sequential logic design is the foundation for building large-scale systems. However, some design decisions must be made at higher levels of abstraction above the logic gate level. Register-transfer design and behavioral optimizations like scheduling and allocation provide us with models for major design decisions that determine how much logic we will need and how fast it will run.

6.6 Problems

Q6-1. Find an assignment of variables to multiplexers for the example of Example 6-2 that gives the minimum number of multiplexer inputs.

Q6-2. Design data path and controller to repeatedly receive characters in 7-bit, odd parity format. You can assume that the bits are received synchronously, with the most significant data bit first and the parity bit last. The start of a character is signaled by two consecutive zero bits (a 1 bit is always attached after the last bit to assure a transition at the start of the next character). At the end of a character, the machine should write the seven-bit character in bit-parallel form to the *data* output and set the output *error* to 1 if a parity error was detected.

Q6-3. Design a data path and controller for a programmable serial receiver, extending the design of Question Q6-2. The receiver should be able to take seven-bit or eight-bit characters, with zero, one, or no parity. The character length is given in the *length* register—assume the register is magically set to seven or eight. The parity setting is given in the *parity* register: 00 for zero parity, 01 for zero parity, and 10 for one parity.

Q6-4. Design a data path and controller for this code:

```
x = a + b;
if (x = 3'b101)
    x = a;
else
    y = b;
end
```

Assume that *i1* is a primary inputs, *o1* is an external output, and *a*, *b*, *x*, and *y* are registers. Show:

a. a block diagram showing the signals between the data path and the controller and the connections of each to the primary inputs and outputs;

b. the structure of the data path;

c. the state transition graph of the controller, including tests of primary inputs and data path outputs and assignments to primary outputs and data path inputs.

Q6-5. Draw data flow graphs for the following basic blocks:

a. c = a + b; d = a + x; e = c + d + x;

b. w = a - b; x = a + b; y = w - x; z = a + y;

c. w = a - b; x = c + w; y = x - d; z = y + e;

Q6-6. For the basic block below:

```
t1 = a + b;
t2 = c * t1;
t3 = d + t2;
t4 = e * t3;
```

a. Draw the data flow graph.

b. What is the longest path(s) from any input to any output?

c. Use the laws of arithmetic to redesign the data flow graph to reduce the length of the longest path.

Q6-7. Can the data flow graph of Example 6-3 be rescheduled to allow the addition and subtraction operations to be allocated to the same ALU in a two ALU design? If not, explain why. If so, give the schedule and an allocation of the functions to two ALUs such that the addition and subtraction are performed in the same ALU.

Q6-8. This code fragment is repeatedly executed:

```
if (i1) then
    c = a - b;
    d = a + b;
else
    c = a + e;
end;
```

i1 is a one-bit primary input to the system, *e* is an *n*-bit primary input, and *a*, *b*, *c*, and *d* are all stored in *n*-bit registers. Assume that *e* is magically available when required by the hardware.

a. Design a data path with one ALU that executes this code.

b. Design a controller that executes the code on your data path.

Q6-9. Design data path/controller systems to perform the function (a + b) + (c + d) + (e + f) + (g + h). Show the data path block diagram and the controller state transition graph. Identify all control signals. The system should run in non-pipelined mode—it should accept no new inputs until the current computaiton has finished. Allocate function units and registers, showing what values are operated on or stored in each function unit/register. Construct two designs:

a. One ALU.

b. Two ALUs.

Q6-10. Construct a pipelined version of the data path/controller systems from Question Q6-9. The system will acccept a new set of inputs and produce a new output on every 7th clock cycle. The data path should contain two ALUs. Show the data path block diagram, controller state transition graph, and all control signals.

Q6-11. Construct a low-power version of the pipelined system of Question Q6-10. The system should use two units, each running at half the clock rate and achieve the same performance of the original system. Show the block diagram and any necessary control.

Q6-12. Assume that all arithmetic operations take one clock cycle and that no arithmetic simplifications are allowed. For each expression show the ASAP schedule and the ALAP schedule.

a. ((a + b) + c) + (d + e)

b. (a + b) * (c + d)

c. ((((a + b) * c) * d) + (e + f) * (g + h))

Q6-13. Schedule the data flow graph of Figure 6-8 assuming that two additions can be chained in one clock cycle:

a. Show the ASAP schedule.

b. Show the ALAP schedule.

Q6-14. Schedule the data flow graph of Figure 6-8 assuming that no more than one addition can be performed per clock cycle.

Q6-15. The carry out of an adder is used as an input to a controller. How might you encode the controller's states to minimize the delay required to choose the next state based on the value of the carry out?

Q6-16. Extend Equation 6-5 to take into account nonzero setup and hold times in the registers. You may model setup and hold time of the register as a single parameter r.

Q6-17. Construct a logic example similar to Figure 6-22 in which an improperly formed pipelining cutset causes a logic gate to receive inputs from two different clock periods.

Q6-18. Add a single pipeline stage to a 4×4 array multiplier that results in the maximum performance.

7 Large-Scale Systems

Busses.

Platform FPGAs.

Multi-FPGA systems.

Novel FPGA architectures.

7.1 Introduction

Modern FPGAs contain so many transistors that we can build complex systems on a single FPGA. Even so, large system designers often use multiple FPGAs to build multi-chip systems. In this chapter we will consider four topics related to the design of complex systems. First, we will look at busses used to connect CPUs and other components in large systems. Next, we will consider platform FPGAs that provide several different types of implementation media on a single chip. We will then move onto multi-FPGA systems and how to partition a large design into several communicating FPGAs. Finally, we will look at new architectures that use FPGAs and new architectures for FPGAs themselves.

7.2 Busses

In this section we will examine the design of **bus** interfaces. Buses are widely used to connect to microprocessors and as a common interconnect. Buses also allow us to explore a new style of logic design, asynchronous logic.

busses implement protocols

A bus is, most fundamentally, a common connection. It is often used to refer to a physical connection that carries a **protocol** for communication between processing elements. The physical and electrical characteristics of the bus are closely matched to the protocol in order to maximize the cost-effectiveness of the bus system.

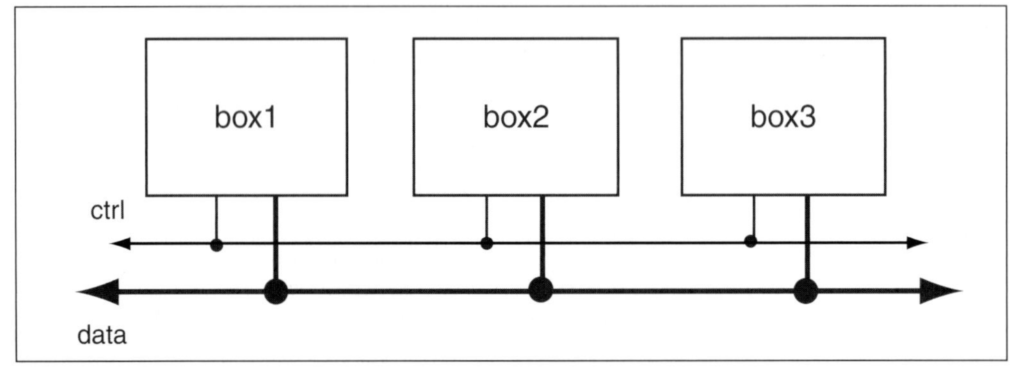

Figure 7-1 A bus-based system.

A simple bus-based system is shown in Figure 7-1. The bus allows us to construct a system out of communicating components. The components in the system communicate via a series of wires. Some of these wires carry data while others carry the control information for the protocol. The bus specification includes electrical characteristics of these components: voltages, currents, maximum capacitive loads, etc. The bus specification also includes the protocol for how to use the control and data signals to communicate between the components connected to the bus.

In the next section we will look at bus protocols and the specification of busses. We will then look at logic design for busses. Finally, we will consider an example bus.

7.2.1 Protocols and Specifications

A protocol is an agreed-upon means for communication. While the protocol ultimately describes the complete system, the best way to understand a protocol is often by looking at the behavior of the components that are communicating. Once we understand what each component expects from the other component and what it does in return, then it is easier to understand the flow of communication embodied in the protocol.

events on signals

We often talk about events on signals as **assertions** or **deassertions** rather than as 1 or 0. An assertion event may be logically true but at a zero voltage because that can be more reliably signaled over the particular physical medium in use. Assert/deassert terminology gives us some independence from the physical representation of signals that is often useful.

protocols and state transition graphs

Protocols for digital systems are often described as state transition graphs. Each component in the system has a state; inputs from other components cause it to move to different states and to emit outputs. Those new states may be good if the component gets the protocol signal that it expects; the states may also represent bad conditions if the component doesn't see the behavior it expects from other components.

The state machines used to describe protocols are not the synchronous state machines that we have used for logic design. We use **event-driven** state machines to describe protocols. These state machines are similar in behavior to event-driven simulators—the machine changes state only when it observes an input event. An event is, in general, a change in a signal. One might implement a protocol using a synchronous state machine that polls the input signal, but at the interface the user can only tell that the machine responds to these events.

timing diagrams and protocols

We introduced timing diagrams in Section 4.4.1. The timing diagram specifies a part of the protocol used by the bus. It describes one scenario of the bus operation; several such scenarios are generally needed to fully specify the protocol. Even in the simple case of Figure 4-11, the timing diagram specifies two events that are part of one step of the protocol—the data values is presented on the *d* line and then it is removed from *d*.

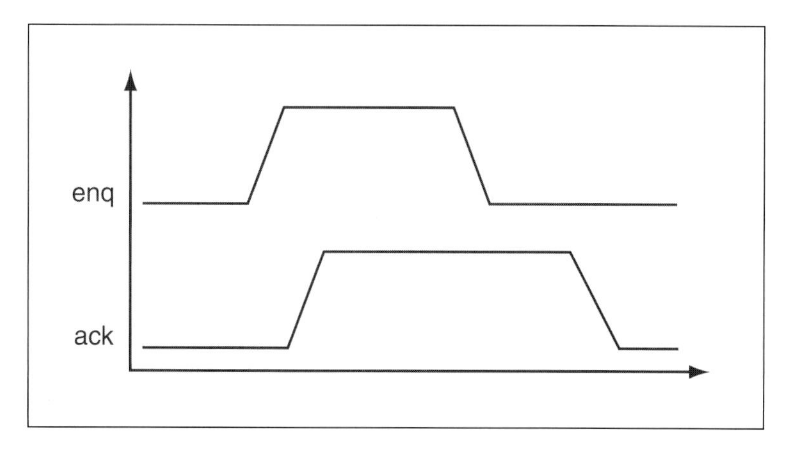

Figure 7-2 Events in a four-cycle handshake.

four-cycle handshake

Let us use a simple example to show how protocols can be described. Figure 7-2 shows the activity during a **four-cycle handshake**, a protocol that is the basic building block for many more complex protocols. A four-cycle handshake is used to reliably communicate between two systems. At the end of the protocol, not only is some information transformed (either implicitly by performing the handshake or by passing data in the middle of the handshake) but both sides know that the communication was properly completed.

The four-cycle handshake uses two signals, **enq** (*enquiry*) and **ack** (*acknowledge*). Each signal is output by one component and received by the other. The handshake allows the two components to reliably exchange information.

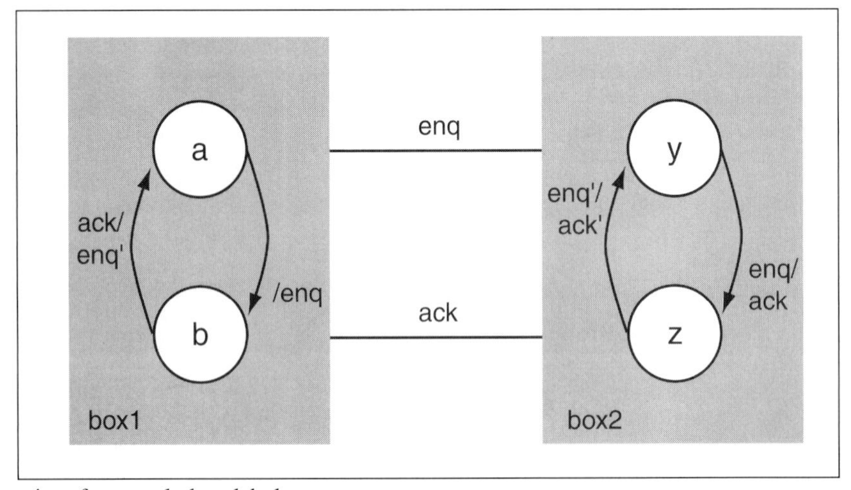

Figure 7-3 Components in a four-cycle handshake.

Figure 7-3 shows the two components in more detail. The *enq* and *ack* signals move between the components. The figure shows the state machine describing the protocol in each component machine. The protocol allows *box1* to signal *box2*; that signal could be a simple completion signal or it could be used to tell *box2* that *box1* has some data ready.

Let us first consider *box1*:

1. *Box1* raises *enq* to tell *box2* that it is ready. (This action is instigated by some other activity within box1.)
2. When *box1* sees an *ack*, it lowers *enq* to tell *box2* that it saw the acknowledgement.

Once *box1* has responded to the *ack* signal from *box2*, it returns to its original state and is ready for another round.

Let us consider the same transaction from the side of *box2*:

1. When *box2* sees *enq* go high, it goes into a new state and sends an *ack* to *box1*.
2. When *box2* sees *enq* go low, it sets *ack* low and returns to its original state.

Just as *box1* returns to its original state once the handshake is complete, *box2* also returns to its original state and is ready to perform the handshake protocol once again. If we want to use the handshake simply to allow *box1* to tell *box2* that it is ready, then the handshake itself is enough. If we want to pass some additional data, we would pass it after *box2* has raised *ack* and before *box1* lowers *enq*.

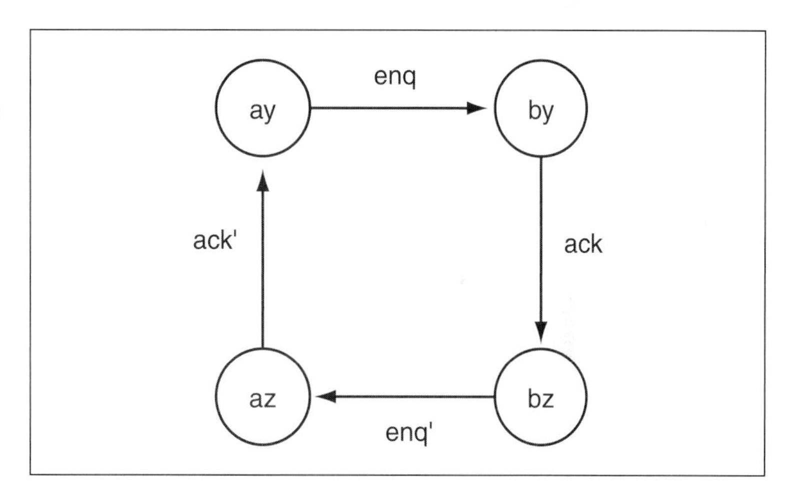

Figure 7-4 The combined state transition graph for the four-cycle handshake.

If we want to see the overall action of the protocol, we can form the Cartesian product of the two state machines that describes the two components. The product machine for the four-cycle handshake is shown in Figure 7-4. Forming the Cartesian product actually causes the *enq* and *ack* signals to disappear but we have shown them as events on the transition between states. The names of each Cartesian product state is the combination of the names of the two component states that combined to make it. We can now see that the combination of *box1* and *box2* go through four states during the four-cycle handshake:

1. *ay* is the initial state in the protocol. The system leaves that state when *enq* is asserted.

2. *by* is the state in which *enq* is active but has not yet been acknowledged.
3. *bz* has both *enq* and *ack* asserted. Data can be passed between *box1* and *box2* in this state.
4. *az* has *enq* deasserted by *ack* still asserted. The system leaves this state for *ay* when *ack* is deasserted.

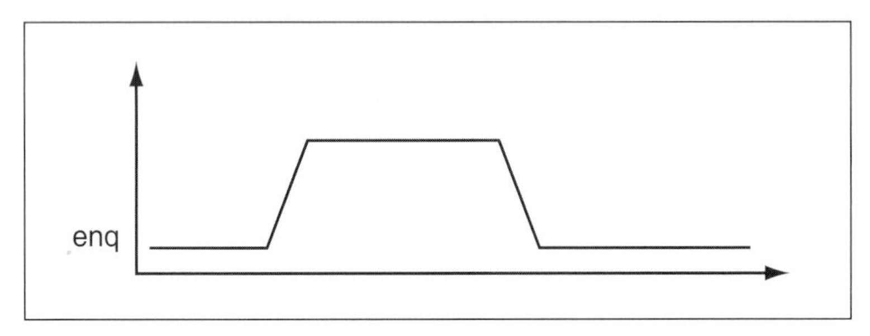

Figure 7-5 Events in a two-cycle handshake.

two-cycle handshake

An even simpler protocol is the two-cycle handshake. This protocol is less reliable but is sometimes good enough for basic signaling. As shown in Figure 7-5, the two-cycle handshake uses *enq* but not *ack*. The enquiry is simply asserted and then deasserted.

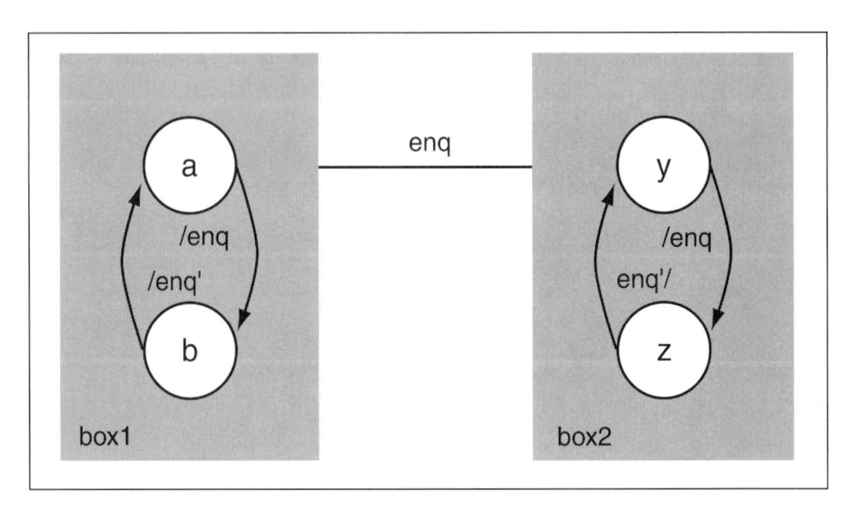

Figure 7-6 Components in a two-cycle handshake.

Figure 7-6 shows the component state machines for the two-cycle handshake. Because *box1* does not receive an *ack*, it must guess as to when to deassert *enq*. This makes the protocol less reliable but it does provide some basic signaling. We generally use some sort of timer to determine how long to wait before deasserting *enq*. Either a counter or a logic path with a known delay can provide us with the necessary delay.

7.2.2 Logic Design for Busses

asynchronous logic in busses

Why use timing diagrams to describe protocols? Why not use state transition graphs? Why design systems that depend on particular delay values? Because busses often connect components that do not (and cannot) share a common, synchronized clock. As a result, the bus must use asynchronous logic for communication. Because the types of asynchronous logic used in busses often depends upon timing values, we use timing diagrams to show the necessary timing constraints. In this section we will study the design of busses using asynchronous logic.

Asynchronous busses represent a compromise between performance and cost. Busses are generally used to connect physically distributed components that are far enough apart that significant propagation delays are incurred when sending signals from one component to another. If all communications were run to a common clock, that clock would run very slowly. It certainly doesn't make sense to force the components to run their internal clocks at the same rate as the external bus. So the bus must be designed to hide timing problems from the components.Many modern busses do in fact use clock signals distributed on the bus with all bus signals synchronized to that clock. However, the bus clock runs much more slowly than and independent of the components' internal clocks.

Figure 7-7 An asynchronous element.

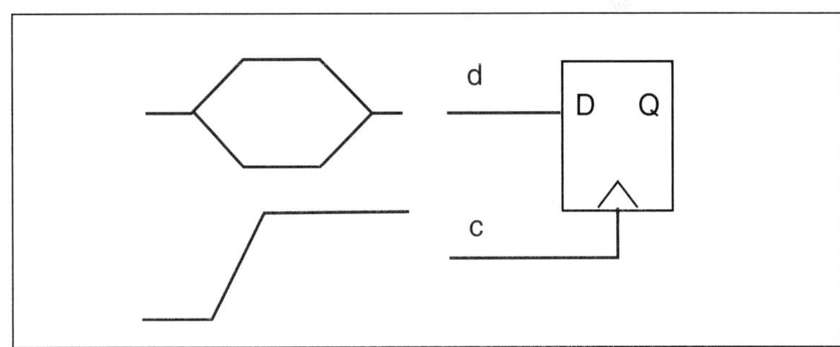

control signals as clocks

Consider the simple circuit of Figure 7-7. A flip-flop is used to capture a data value from the outside world. The data comes into the flip-flop on the *d* signal; *d* is guaranteed to be stable for a certain minimum period. But when does it arrive? The flip-flop requires a clock signal that obeys the setup and hold times of the flip-flop. We use the *c* signal to tell the flip-flop when the data on *d* is ready. This costs an extra signal but it provides a good deal of timing flexibility. Even if we cannot guarantee absolute delays between the components on the bus, we can design the bus so that the relative delays between signals on the bus are closely matched. (This requires carefully controlling crosstalk, capacitive load, *etc.*, but it can be done.) Because the relative delays of d and c are known, generating them with the proper timing at the source ensures that they will arrive at the destination with the same timing relationship. Thus, we can generate the timing information required by the flip-flop and send it along with the data itself.

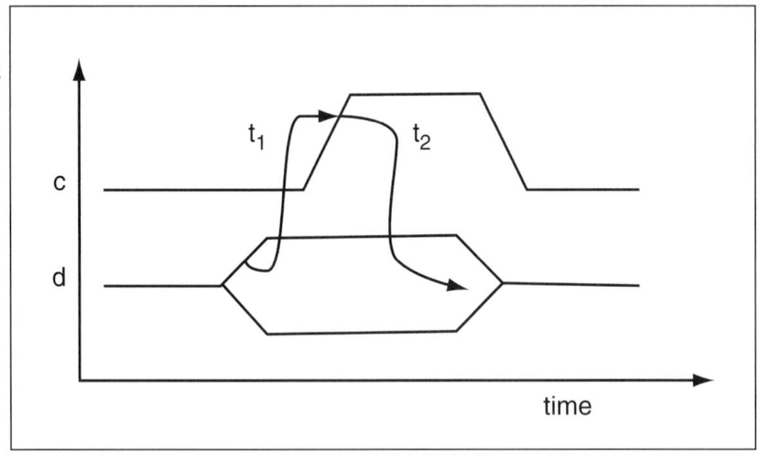

Figure 7-8 Timing constraints on a flip-flop.

capturing events

We will use flip-flops (or latches) to capture signals sent over the bus. We must be more careful about timing constraints when we design logic for a bus. In a fully synchronous system, we tend to separate combinational logic delays and clock skew, controlling the clock skew so that we only have to check a straightforward delay requirement for all the logic. Every signal on a bus, in contrast, may have its own timing. The fundamental requirements for a flip-flop are its setup and hold times. These constraints then become constraints on the incoming data and control signals. In Figure 7-8, the t_1 constraint comes from the flip-flop's setup time while the t_2 constraint comes from its hold time. We can draw a similar timing diagram for a latch-based bus receiver.

timing constraints

These two constraints are defined in terms of the events on the c and d lines. We can name the events as follows:

- t_{d1} = time at which d becomes stable;
- t_{d2} = time at which d stops being stable;
- t_{c1} = time at which c rises.

If the setup time of a flip-flop is t_s and its hold time is t_h then we can write t_1 and t_2 as

$$t_1 = t_{c1} - t_{d1} \geq t_s, \quad \textbf{(EQ 7-1)}$$

$$t_2 = t_{d2} - t_{c1} \geq t_h. \quad \textbf{(EQ 7-2)}$$

The equations for t_1 and t_2 define them in terms of events on the bus while the inequalities constrain the minimum time between the events. We don't in general know the exact times at which t_{d1} and t_{d2} happen. But the constraint is written in terms of the difference of the two times, which we can directly relate to the flip-flop setup and hold times. Given a particular technology, we determine its setup and hold times and substitute those values into the inequalities.

communication and timing

All the timing constraints on the bus ultimately come from the components used to build the bus. How we view them depends on our point of view in the bus. Our own component imposes **timing constraints** that must be satisfied from the inside. The component with which we want to communicate imposes **timing requirements** from the outside.

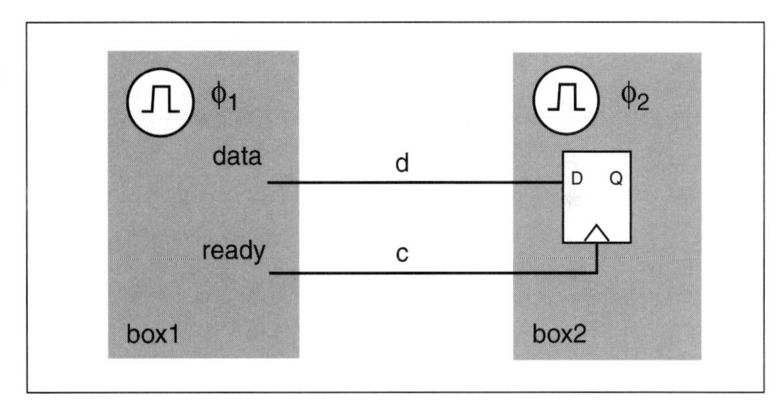

Figure 7-9 Two components communicating over a bus.

Figure 7-9 shows the flip-flop-based bus receiver in the context of the bus. *Box1* wants to send a value to *box2*. Each component has its own clock: ϕ_1 for *box1* and ϕ_2 for *box2*. For the moment, let us assume that *box1* somehow generates the data signal and a ready signal that has the proper timing relationship to the data. Those two signals are sent along the bus's *d* and *c* wires. The *c* control signal causes the flip-flop to remember the data on *d* as it arrives.

metastability

One problem we must consider when transmitting signals asynchronously is **metastability** [Cha72]. A flip-flop or latch typically remem-

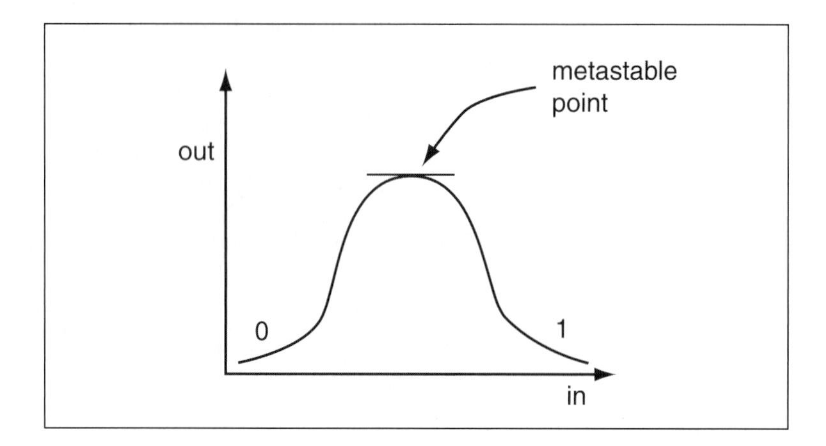

Figure 7-10 A metastable state in a register.

bers a 1 or 0 reliably. As shown in Figure 7-10, the register has two stable states representing 0 and 1. However, if the data does not satisfy the setup and hold times and is changing around the critical interval for the clock, then a bad value may be stored. If the captured value is near the stable 0 or 1 state, then the memory element will quickly roll down the hill to the stable value. However, if the value is in the metastable region, then the memory element captures a value that is finely balanced between the stable 0 and 1 states. This metastable point is not totally stable because the value will eventually move toward either 0 or 1. However, the amount of time it takes the memory element to move to a 0 or 1 is unbounded. Either the receiver must be able to detect a metastable state and wait for it to resolve or the receiver will get a corrupt value.

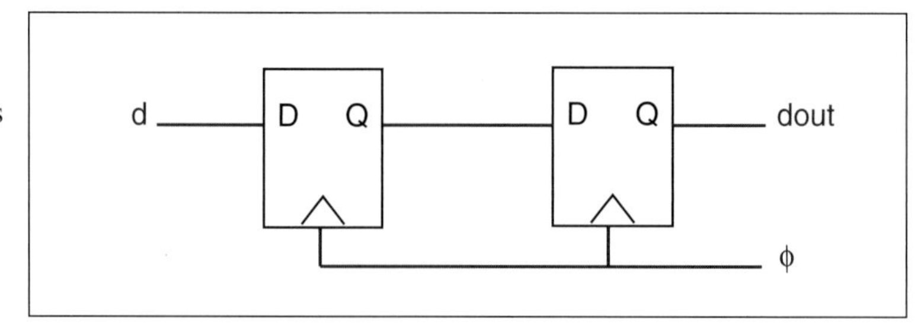

Figure 7-11 A multi-stage synchronizer that minimizes metastability problems.

We can minimize the chance of metastability with the **multi-stage synchronizer** shown in Figure 7-11. It has two flip-flops, both under control of the clock that controls the receiving logic. The key part of the design is that the data is captured twice: the signal is sampled by the first flip-flop and that result is resampled by the second flip-flop. Because the synchronizer is basically a shift register, it takes two clock cycles to see the received value. But even if the first register goes metastable, it is less likely that the second register will also go metastable. Metastability cannot be eliminated, however, and there is always a small chance that the output of the multistage synchronizer will be metastable. It is possible to build a multistage synchronizer with more stages, but the additional stages will only slightly improve the probability of eliminating metastability and will definitely add latency.

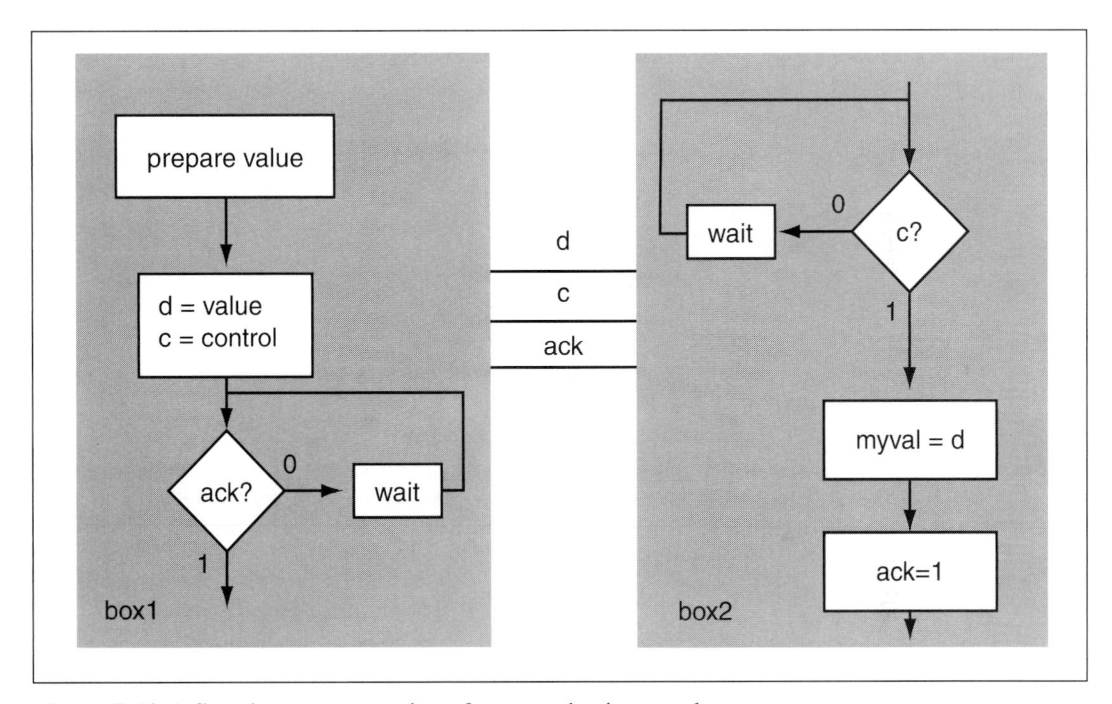

Figure 7-12 A flowchart representation of communication on a bus.

bus protocols and events

Now that we understand how to receive asynchronous signals, we can start to consider the complete bus system. Before delving back into the components and their timing requirements, let us step back to remember the functionality of a bus transaction. Figure 7-12 shows a flowchart that functionally approximates the behavior of the components on the bus. We say *approximates* because flowcharts are not designed for asynchronous behavior, but this chart gives you a basic idea of the flow of control between the two components. This chart shows a handshake so that we can consider how to generate an acknowledgment from the receiver. Once *box1* has prepared the value it wants to send, it then transmits both the data and an associated control signal. Meanwhile, *box2* is waiting for that control signal. When it sees the control, it saves the value and sends an acknowledgment. Once *box1* sees the *ack* event, it goes on to further processing. We have already seen the circuitry used for *box2* to receive the event; we also need circuitry for *box1* to generate the event and for *box2* to generate an acknowledgment.

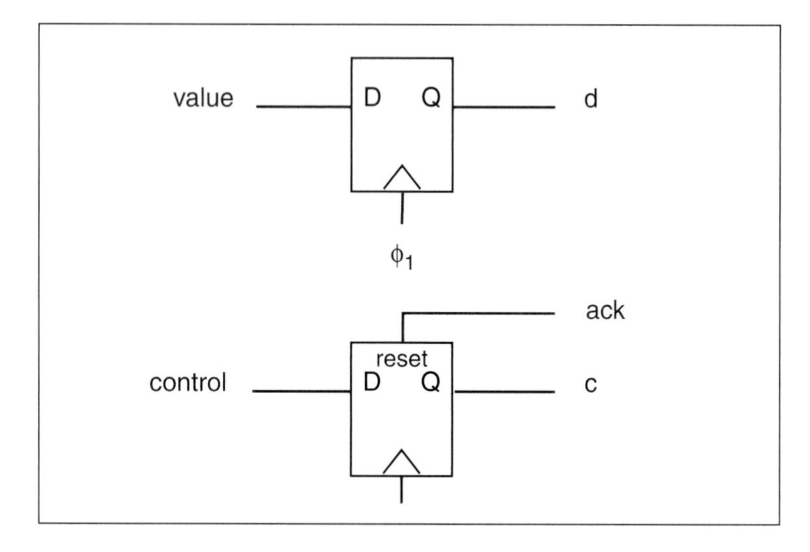

Figure 7-13 Bus logic for *box1*.

logic implementations

Figure 7-13 shows the bus logic in *box1*. The data value is stored in a flip-flop to be held on the bus. The control signal for the bus is generated by another flip-flop. The *ack* signal causes the flip-flop that holds the control value to reset itself. Because both flip-flops are clocked by the internal clock ϕ_1, we know that they will acquire their values at the same time. If we need to change the delay of *c* relative to *d* in order to meet setup and hold constraints on the other side, we can add delay to one of the signals using a delay element.

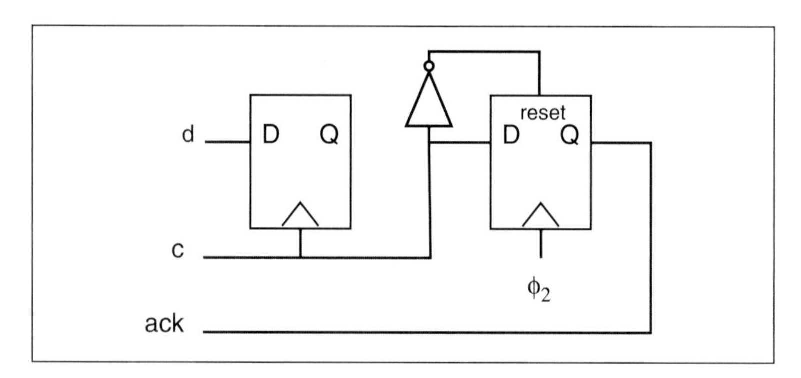

Figure 7-14 Bus logic for *box2*.

Figure 7-14 shows the logic on the *box2* side of the bus. The first flip-flop captures the data signal using *c* as a clock for activation. The next flip-flop samples *c* to tell when the data has arrived. After one tick of the internal clock ϕ_2 that flip-flop sends out an acknowledge signal. When *c* goes low, it resets the second flip-flop to drop the *ack* signal.

Figure 7-15 A timing diagram for the bus.

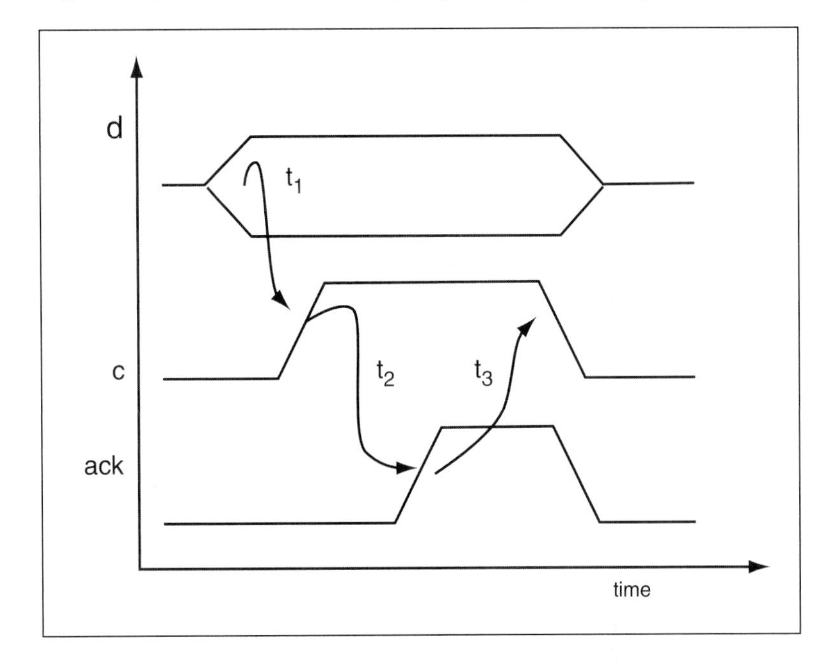

bus timing

Figure 7-15 shows a timing diagram for the bus. We can write some timing relations for this bus, using these names for the events:

- t_{d1} = time at which *d* becomes stable;
- t_{d2} = time at which *d* stops being stable;
- t_{c1} = time at which *c* rises;
- t_{c2} = time at which *c* falls;
- t_{ack1} = time at which *ack* rises.

If the setup and hold times of all the flip-flops are t_s and t_h respectively then we can write the constraints as

$$t_1 = t_{c1} - t_{d1} \geq t_s, \quad \textbf{(EQ 7-3)}$$

$$t_2 = t_{\text{ack1}} - t_{c1} \geq t_h. \quad \textbf{(EQ 7-4)}$$

$$t_3 = t_{c2} - t_{\text{ack1}} \geq t_h .$$ **(EQ 7-5)**

We could also constrain the fall time of the *ack* against the next bus cycle to be sure that enough time is left to properly capture the value.

post-synthesis timing simulation

Many tool sets, such as the Xilinx ISE, will generate post-synthesis simulation models with timing annotations. These models will provide more accurate timing information when simulated. An accurate timing simulation is very important in asynchronous logic design because the logic design depends on assumptions about delays. Simulation helps verify the proper operation of the logic.

7.2.3 Microprocessor and System Busses

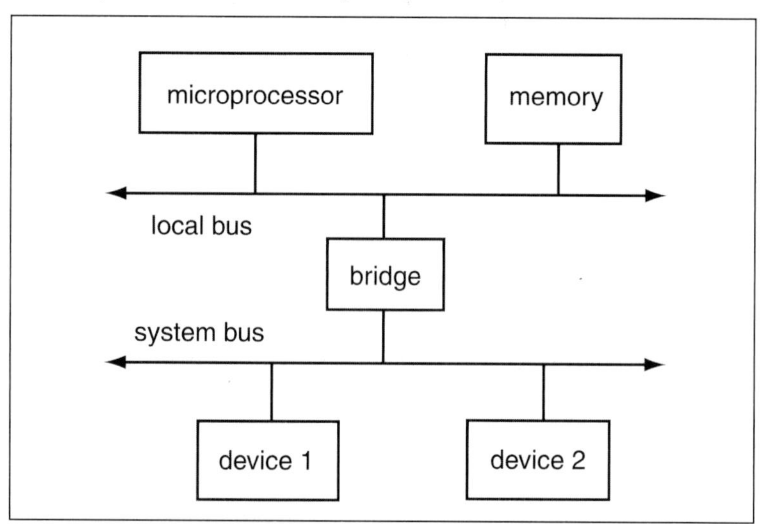

Figure 7-16 Busses connected by a bridge.

types of busses

Busses are often used to connect microprocessors to memories and peripherals. Microprocessor busses have come to influence the design of other busses as well. As shown in **Figure 7-16**, a **local bus** is used to connect the microprocessor to high-speed memory while a **system bus** is used to connect the local bus to peripherals. The component that connects two busses is called a **bridge**. A local bus must provide very high performance to avoid slowing down the microprocessor during the fetch-execute cycle; system busses come in a wide variety of cost/performance points. However, these two different types of busses share some common characteristics due to their common heritage.

Busses usually talk to devices using addresses. Each device on the bus is assigned an address or a range of addresses. The device responds when it sees its address on the address lines. Separate lines are used to send and receive data. Further lines are used to provide the control signals required to show when data and addresses are valid, etc.

master-slave operation

Busses are generally **master-slave** systems: one component controls the operation of the bus and the other devices follow along. A slave device may send data on the bus but only when the master tells it to. The master, in contrast, initiates all the various types of operations on the bus. In a microprocessor system the microprocessor typically acts as the bus master. The other components in the system are assigned addresses to identify them.

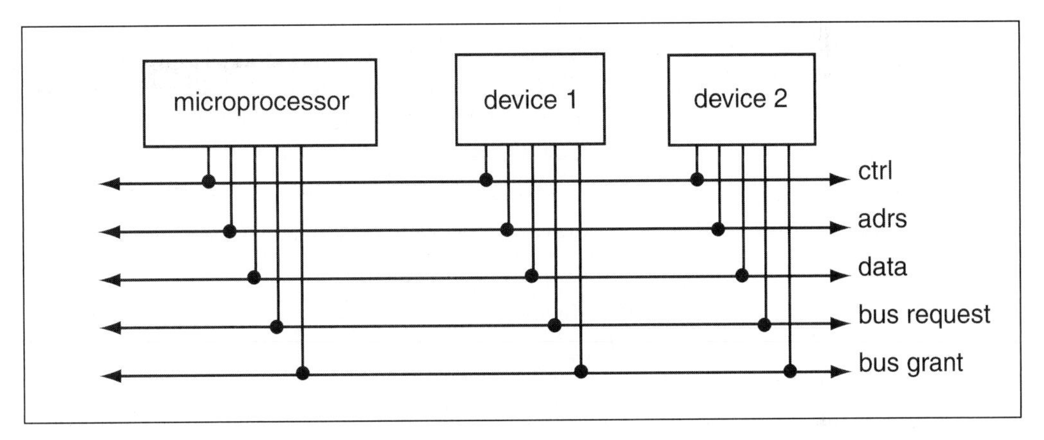

Figure 7-17 Basic signals in a bus.

signals on busses

Figure 7-17 shows some of the basic signals in the bus. The *ctrl*, *adrs*, and *data* lines are used for basic bus transactions, such as reads and writes. When another device wants to become the bus master for some time, it uses the *bus request* and *bus grant* signals. This protocol is typically a four-cycle handshake, with the handshake completed only when the new bus master has finished and returns control to the default bus master. The process of choosing a master for the bus is known as **bus arbitration**. Bus arbitration logic must be carefully designed: not only must it be fast, it also cannot afford to grant two devices mastership of the bus under any circumstance.

bus characteristics and standards

Busses differ greatly in their details:

- **physical** The bus standard often specifies size of the connector and cards.

- **electrical** Different voltages and currents may be used for signaling. The bus may also define the maximum capacitive load allowed.
- **protocols** Busses may use different protocols that trade off speed and flexibility.

The Peripheral Component Interconnect (PCI) bus [Min95] is a widely used standard originally developed for PCs. The PCI standard is very complex. The next example touches upon some aspects of PCI to illustrate some bus concepts.

Example 7-1 The PCI bus

PCI was developed as a high-speed system bus to replace ISA and Micro Channel in PCs. The standard has been extended several times and PCI variants are now used in a wide variety of digital systems.

The standard describes both 33 MHz and 66 MHz implementations. PCI uses a non-terminated bus and its electrical design takes advantage of reflections to reduce the switching time on the bus. The bus includes a clock signal, CLK, that has a 30 ns period at 33 MHz and a 15 ns period at 66 MHz.

The bus has several types of signals (the PCI spec uses # to mean negation):

- System signals:
    - **CLK** The system clock is an input to all PCI devices.
    - **RST#** The reset signal initializes all PCI configuration registers, state machines, and drivers.
- Address/data bus:
    - **AD[31:0]** The combined address and data bus, normally 32 bits but can be extended to 64 bits.
    - **C/BE#[3:0]** Command or byte enable defines the type of transaction.
    - **PAR** Set by the sender to ensure even parity on the address lines and C/BE#.
- Transaction control signals:
    - **FRAME#** Indicates the start and duration of the transaction.
    - **TRDY#** Target ready is driven by the currently-addressed target and is asserted when the target is ready to complete the current data transfer.

- **IRDY#** Initiator ready is driven by the bus master and indicates that the initiator is driving valid data onto the bus.
- **STOP#** Allows the target to tell the initiator to stop the transaction.
- **IDSEL** Initiation device select is used as a chip select while accessing device configuration registers.
- **LOCK#** Locks the currently-addressed memory target.
- **DEVSEL#** Device select is asserted by the target when it has decoded its address.

- Arbitration signals:
  - **REQ#, GNT#** Request and grant lines from each device are connected to the bus arbiter. The arbiter determines which device will be the next master.
- Interrupt request signals:
  - **INTA#, INTB#, INTC#, INTD#** Allow devices to request interrupts.
- Error reporting signals:
  - **PERR#, SERR#** Used to report parity and system errors, respectively.

PCI handshakes to transfer data. The source of the data must assert its ready signal when it drives data onto the bus. The receiver does not have to respond immediately and only raises its ready line when it is ready to receive. If the receiver takes more than one clock cycle to respond, the intervening clock periods are known as **wait states**. Once a sender or receiver indicates that it is ready to complete a data phase, it cannot change its control lines until that phase is done.

Here is a timing diagram for a sample read transaction, in which an initiator reads several locations from a device:

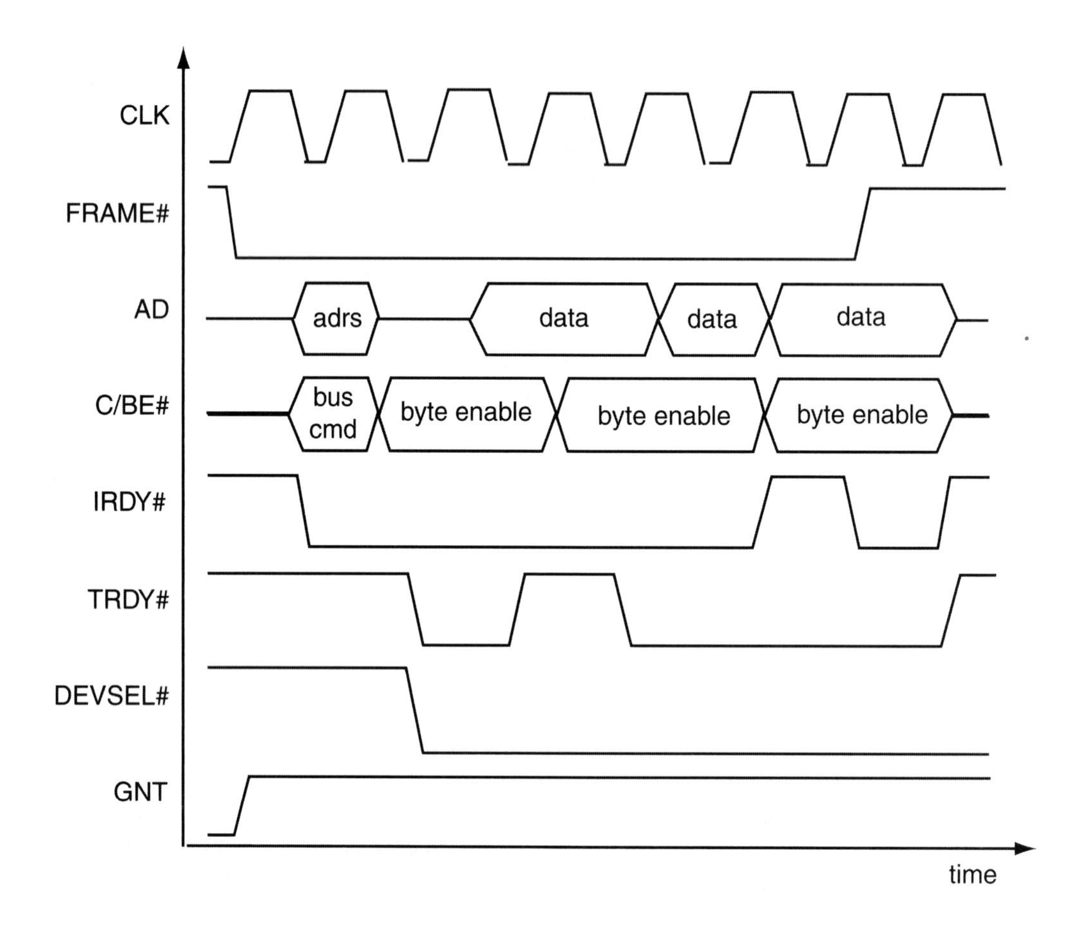

The read transaction starts with the initiator asserting a valid start address and command on the bus and raising FRAME#. The address is written onto AD and the command asserted onto C/BE.

On the next cycle, the initiator stops driving AD. This cycle is used to turn around the AD lines, ensuring that two devices don't try to drive them at the same time. The initiator uses the C/BE lines on this cycle to signal which byte lanes will be used. It also asserts IRDY# to show that it is ready to receive the first datum from the device.

On the third cycle, the target device asserts DEVSEL# to show that it has recognized its address. It also starts to drive the data onto the AD bus and asserts TRDY# to show that the data is on AD. Since the second cycle was used to turn around AD, the target device can now safely drive the data onto AD without electrical conflicts.

PCI supports multi-word transfers. When the initiator continues to assert IRDY# but does not deassert FRAME#, the transaction continues to the next data item. The address is sent at the start of the transaction but not with every datum. The target must remember the starting address and increment it as necessary to keep track with the current address.

If the target needs a wait state, it can deassert TRDY# to tell the initiator that it needs more time. A total of three data elements are transferred on this bus transaction. Two of them require wait states—each datum can take a different number of wait states.

At the end of the transfer, the initiator deasserts IRDY# and the target deasserts TRDY# and DEVSEL#. The PCI bus is now ready for the next transaction.

The PCI bus is arbitrated by a central arbiter. Each potential master is connected to the arbiter by its own REQ# and GNT# signals. A device can remain master as long as it wants after it gains bus mastership. The arbiter's main task is to choose the next bus master when several devices simultaneously request the bus. The PCI standard does not define the arbitration algorithm but it does require that a fairness algorithm be used to avoid deadlocks on the bus. *Round-robin* is a common example of a fair arbitration scheme.

PCI allows the bus arbitration process to occur while a transfer is taking place. The new master gains control of the bus at the end of the current transfer. *Hidden arbitration* improves bus performance by overlapping arbitration time with other activities.

FPGAs may provide support for common bus standards. Many FPGAs, for example, can be programmed to provide PCI-compatible electrical signals at their pins.

7.3 Platform FPGAs

The **platform FPGA** is a relatively recent category of chip that combines several different types of programmable components. A platform FPGA has all the components necessary to build a complete system and should require few, if any, additional chips. Of course, exactly what constitutes a system depends on your point of view. As a result, there are some significant differences in what they include. Platform FPGAs clearly include FPGA fabrics. They may also include CPUs, embedded memory, memory interfaces, high-speed serial interfaces, and bus interfaces.

advantages of platform FPGAs

The most obvious advantage of platform FPGAs is high levels of integration. Moving more functions onto a single chip generally provides several advantages:

- Smaller physical size.
- Lower power consumption.
- Higher reliability.

Integrating several functions is also important in realizing complex functions. Some subsystem-to-subsystem interconnections require large number of connections that may not be feasible in chip-to-chip connections but quite possible on-chip. Moving chip-to-chip connections onto the chip also makes them run much faster.

The other advantage of platform FPGAs is that they can more efficiently implement many system-level functions than is possible with pure FPGA fabrics. We have seen how even basic FPGA fabrics add additional logic for arithmetic functions. Platform FPGAs take this approach to the subsystem level. By adding specialized logic, platform FPGAs make it possible to squeeze all the necessary logic for a complete system onto a single chip.

Specialized high-speed I/O can also be seen as a generalization of programmable FPGA I/O pins. High-speed serial protocols are widely used for disks and networks. While the various standards for these protocols work on similar principles, the details are quite different. Programmable I/O subsystems allow the system designer to choose which I/O features to implement using a collection of basic circuits.

7.3.1 Platform FPGA Architectures

Let us look at two platform FPGAs in more detail. The next two examples show that these platform FPGAs, while both supporting system design, provide quite different building blocks.

Example 7-2 Xilinx Virtex-II Pro Platform FPGA

The Virtex-II Pro family [Xil02] has several major features:

- One or more IBM PowerPC RISC CPUs.
- Multi-gigabit I/O circuitry.
- Embedded memory.
- An FPGA fabric based on the Virtex-II architecture.

Here is a photograph of the chip showing some of its features:

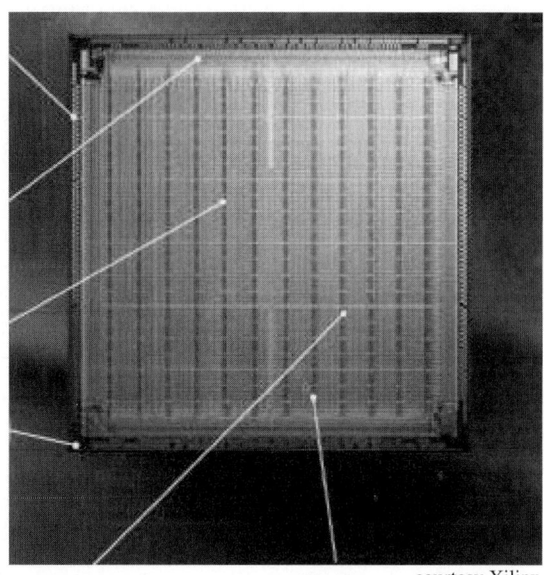

courtesy Xilinx

The smallest Virtex-II Pro does not include a PowerPC. The smallest one with a CPU also has 3,168 logic cells, 500 Kbits of RAM, 28 multipliers, 4 clock management blocks, 4 gigabit transceivers, and 348 user I/O pins. The largest Virtex-II Pro has four PowerPC CPUs, 125,136 logic cells, 10 Mbits of block ram, 556 multipliers, 12 clock management blocks, 24 gigabit I/O transceivers, and 1200 user pins.

The gigabit transceiver units, known as Rocket I/O, operates in a range of 622 Mb/s to 3.125 Gb/s. It can be used to implement a variety of standard protocols, such as Fibre Channel, Gigabit Ethernet, Infiniband, and Aurora. Each transceiver has a Physical Media Attachment that serializes and deserializes the data. It also has a Physical Coding Layer that contains CRC, elastic buffers, and 8-to-10 bit encoding/decoding.

The CPU is a PowerPC 405DC core that can operate at over 300 MHz on chip. It is a 32-bit Harvard architecture machine. It supports both 32-bit and 64-bit fixed-point arithmetic; floating-point operations are trapped and can be performed in software. The CPU includes separate instruction and data caches. The CPU can address a 4 GB address space. Its memory management unit (MMU) provides address translation, memory protection, and storage attribute control. The MMU supports demand-paged virtual memory. The CPU also has a complete interrupt system and a set of integrated timers.

The IBM CoreConnect bus is used to connect the PowerPC to other parts of the FPGA. The CoreConnect architecture provides three busses:

- The processor local bus is used for high-speed transactions. It supports 32-, 64- and 128-bit data paths and provides separate read and write paths. It is a pipelined bus that supports multiple masters, burst transactions, and split transactions.
- The on-chip peripheral bus supports lower-speed devices. It supports multiple masters and has separate read and write paths.
- The device control register bus is a low-speed, daisy-chained bus for configuration and status information.

The on-chip peripheral bus connects to the CPU using a bridge to the processor local bus:

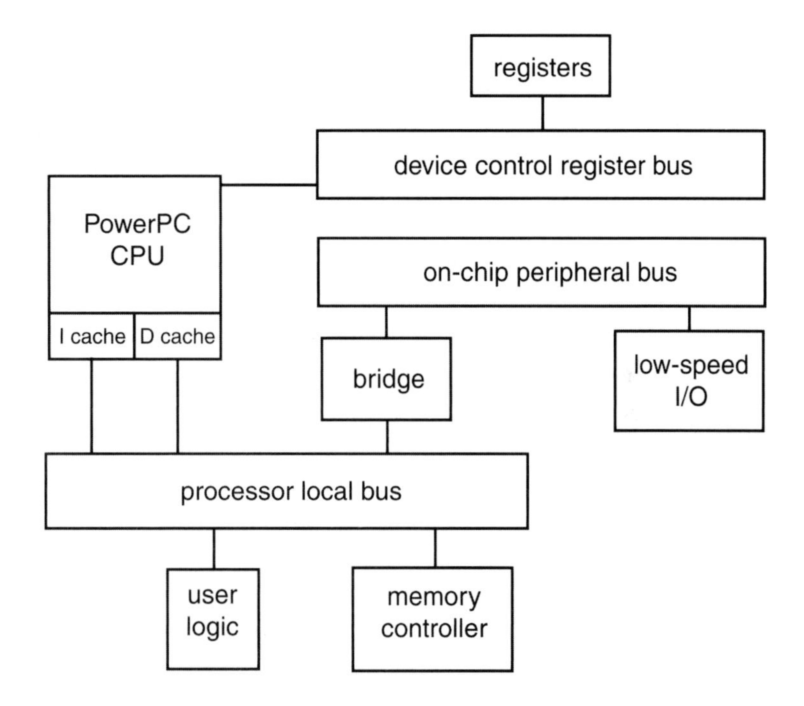

Much of the CoreConnect system is implemented as soft IP on the Virtex-II Pro.

The configurable logic fabric is built from an array of CLBs. Each CLB has four slices in it. Each slice contains:

- carry and arithmetic logic;
- wide function multiplexers;
- two registers configurable as flip-flops or latches;
- two four-input function generators (lookup tables) that can be configured as lookup tables, 12-bit shift registers, or 16-bit shift registers.

Each slice connects to both local interconnect and to the general routing matrix. The function generators can be configured as either single-port or dual-port memories and in a variety of sizes; the memories can be used as either RAM or ROM. The chip also contains larger blocks of memory known as SelectRAM. Each SelectRAM block is 18 Kb and is a dual-port memory; it can be configured into several single-port and dual-port configurations.

The Virtex-II Pro provides 18 x 18 multipliers that are designed to be faster than multipliers implemented in the CLBs. A multiplier may be associated with a SelectRAM block or may be used independently.

The digital clock manager (DCM) provides logic for manipulating and managing clock signals. The DCM can de-skew clock signals, generate new clock frequencies, and shift clock phases. The global clock multiplexer buffers are the interface between the clock pins and the clock distribution network. The global clock multiplexers can be driven by the clock pins or by the DCM.

Example 7-3 Altera Stratix Platform FPGA

The Altera Stratix family of parts [Alt03] combines an FPGA fabric with memory blocks and DSPs. The smallest of the Stratix devices contains 10,570 logic elements, 920K bits of RAM, 48 embedded multipliers, 6 PLLs, and 426 pins; the largest includes 79,040 logic elements, 7.4 megabits of RAM, 176 embedded multipliers, 12 PLLs, and 1,238 pins.

The chip is organized like this:

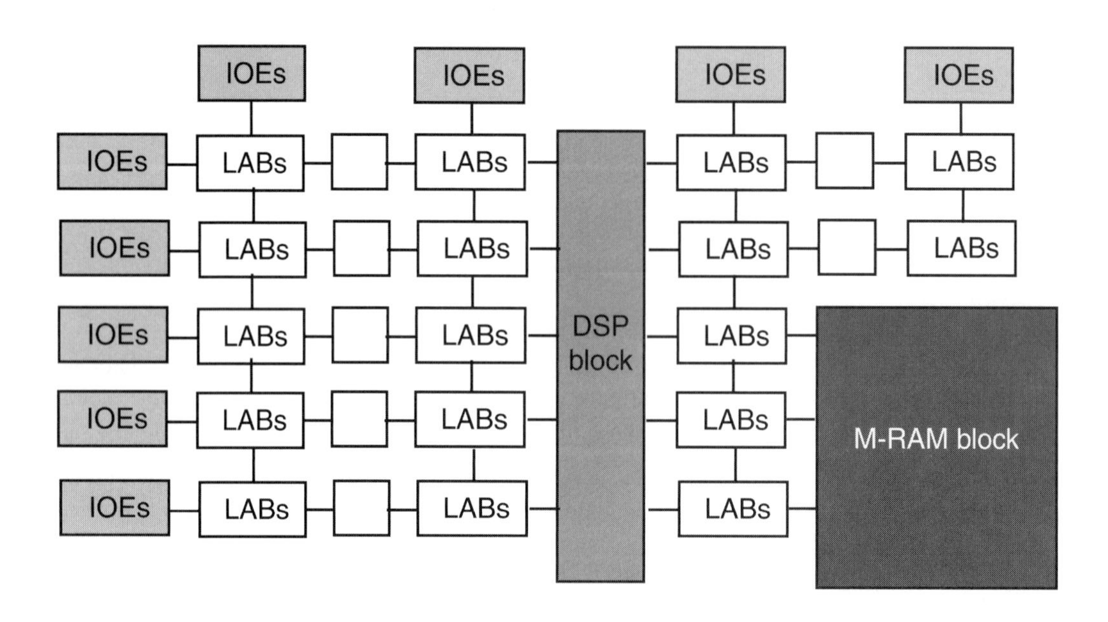

The logic array blocks (LABs) and the associated interconnect form the FPGA core. The DSP blocks are interspersed between the LABs. The M-RAM blocks are also interspersed within the FPGA array; this memory is separate from the configuration/logic element array. The I/O elements (IOEs) are arranged at the edges of this heterogeneous array.

The LABs in the Stratix are reminiscent of the LABs for other Altera devices. Each LAB includes 10 logic elements, a carry chain, local interconnect, LUT chains, and register chains. An LE can operate in either normal or dynamic arithmetic mode; the dynamic arithmetic mode makes use of the carry chain.

The global interconnection system is called the MultiTrack interconnect structure. It consists of fixed-length row and column interconnections. Along rows, wires may span 1, 4, 8, or 24 blocks. A terminal block and its neighbor can drive a length-4 wire. A length-24 interconnect can be driven by every fourth block. The column routing resources include length 4, 8, and 16 interconnections as well as the LUT and register chains. The MultiTrack interconnect can be used to connect to LABs, DSPs, or memory blocks.

The units in the DSP block are not general-purpose computers. They are designed to support multiplication-intensive arithmetic functions. They can be chained together in various combinations to perform 9-bit, 18-bit, or 36-bit multiplications. Each DSP block has enough units to support one 36-bit X 36-bit multiplication. The smallest Stratix device has six DSP blocks while the largest has 22.

The DSP block, when configured to perform 18 X 18 multiplications, looks like this:

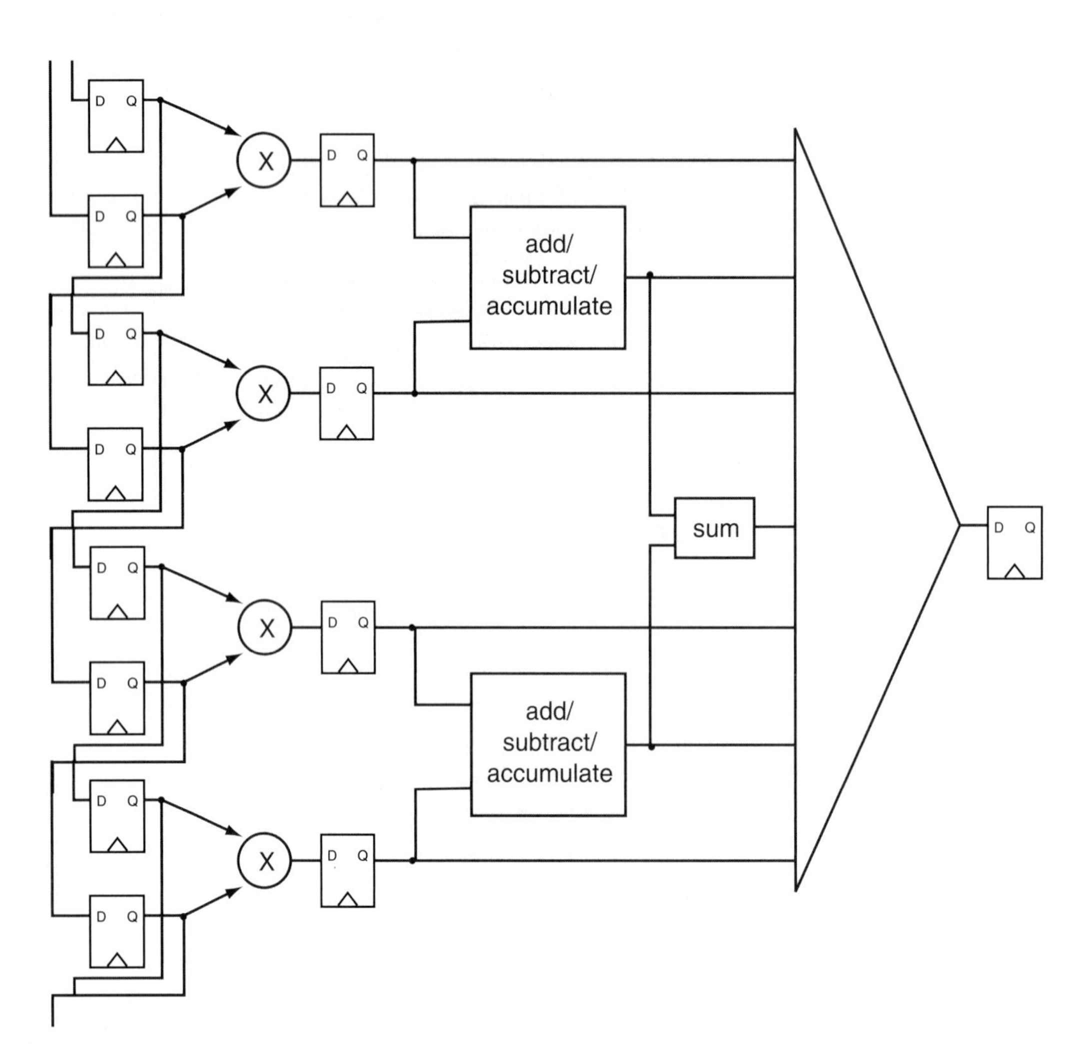

The input registers can be organized into a shift register or can feed the multipliers directly. The multiplier supports either signed or unsigned arithmetic. The add/subtract/accumulate unit either passes the output through, performs a sum or difference, or accumulates the sum. The final sum unit adds together the four partial results. The output multiplexer selects between the various partial results as final outputs. The register on the multiplexer's output is optional. Various signals control which of these options are in use.

These units can be programmed to operate in several different ways. For example, they can be configured in two-multiplier mode to perform the real and imaginary components of a complex operation. The shift registers can be used to feed in data and coefficients for FIR filters.

The Stratix architecture provides three different type of RAM:

- M512 RAM (32 X 18 bits);
- M4K RAM (128 X 36 bits);
- M-RAM (4K X 144 bits).

Besides the differences in size, these provide somewhat different features: The M4K and M-RAM can be used as true dual-ported memory while the M512 has a more limited dual-port mode; the M512 is also not byte enabled; M-RAM cannot be configured as ROM. The blocks support a parity bit for each byte. They can be configured as FIFOs as well as for random access. Memory blocks provide input and output registers for a synchronous RAM interface. The memory blocks can be configured as shift registers for applications such as pseudo-random number generation.

The Stratix devices include a number of clocking resources:

- 16 global clock networks;
- 16 regional clock networks (four per quadrant);
- 8 dedicated regional fast clock networks.

This organization provides for up to 48 different clock domains on the chip. The chip also includes multiple PLLs that can be used for clock generation.

The pins are highly programmable. The Stratix provides support for a variety of differential signaling standards that are used in high-speed chip-to-chip communication.

In order to better understand how to use platform FPGAs, we will use the next few sections to consider several of the categories of subsystems in platform FPGAs.

7.3.2 Serial I/O

serial I/O requirements

Serial I/O needs specialized circuitry for two reasons: it runs at very high speed; and it requires non-digital circuit interfaces. Serial I/O is commonly used to connect large systems together, including disk drives and networks. High-speed serial I/O provides high-performance at a reasonable cost.

There are several different high-speed serial I/O standards in use. While the details of these standards differ, they are all founded on basic circuit principles that dictate some common characteristics. The next example takes a brief look at some of the high-speed serial I/O standards.

Example 7-4 High-speed serial I/O standards

The best-known serial I/O standard is Ethernet, the well-known local area network. The original 10 Mbit/sec Ethernet standard has developed into a series of standards: the original Ethernet standard is now known as 802.3 and is available from the IEEE at *standards.ieee.org*; Gigabit Ethernet; and 10 Gigabit Ethernet (*www.10gea.org*).

Serial interfaces are also used to connect to disk drives. SCSI (Small Computer Systems Interface) was an early example of this category of serial interface (*www.t10.org*). More recently, Fibre Channel (*www.fibrechannel.org*) and Infiniband (*www.infinibandta.org*) have been developed for advanced disk drive interfaces.

As it turns out, Gigabit Ethernet uses the physical interface of Fibre Channel while using the frame structure of Ethernet.

High-speed interfaces generally require specialized circuitry to drive the line. Many high-speed standards use optical interconnections, which means that a laser must be driven.

High-speed interfaces also make use of advanced modulation schemes. Traditional digital logic uses very simple encodings for data. However, the higher noise levels, asynchronous communication, and desire for speed in serial I/O systems demand using fancier schemes to encode the data as it is sent over the bus. Encodings can be used to ensure that 0-1 and 1-0 transitions happen at some minimum intervals to allow a clock to be extracted from the data; they also improve error correction. An example is the 8B/10B scheme used in Fibre Channel and Gigabit Ethernet. During transmission, each 8 bits of data are encoded into 10 bits using a standard table; at the receiver, 10 bits are translated back into the original 8 bits of data.

Serial I/O subsystems on platform FPGAs typically do not directly support higher-level networking tasks like filling the packet with data. Higher-level networking tasks can generally be performed in parallel, lowering the clock rate and making it feasible to perform them with the

FPGA array. The serial I/O subsystem takes in a set of bits, serializes them, performs low-level transformations like 8B/10B encoding and error code generation/checking, and drives the transmitter circuit.

7.3.3 Memories

memory blocks vs LE-based RAM

The large memory blocks in platform FPGAs complement the smaller memory blocks found in FPGA fabrics. Because SRAM-based FPGAs use memory to hold the configurations of lookup tables, it is easy to allow them to be used as RAMs. However, the lookup table memories are very small, typically 16 bits.

Adding large amounts of memories requires building large memory arrays. The addressing logic around the edge of the memory core can take up substantial room; the peripheral logic takes up a larger fraction of the total SRAM area for smaller memories. Large memory blocks allow the FPGA to hold more on-chip memory than would be possible using lookup tables alone.

memory blocks vs. off-chip memory

On-chip memory is clearly faster than off-chip memory. On-chip memory blocks also provide higher bandwidth. However, on-chip memories do not provide as much memory as off-chip bulk memory. On-chip memory is SRAM, which uses a larger, 6-transistor core cell than the one-transistor cell used for DRAM. Bulk DRAM is typically both denser (more bits per chip) and cheaper than the on-chip memory in FPGAs.

Some applications may in fact be able to run in the memory available on-chip. However, some applications easily outrun the available FPGA memory. Consider, for example, a single video frame with 352 x 240 pixels. Even if only a black-and-white version of this frame is stored using 8 bits per pixel, this still requires 675,840 bits. If color information is required or multiple frames need to be stored, then even more storage is required. The architecture is also likely to need additional bits for storage of intermediate results.

memory hierarchy in FPGA systems

We can build a memory hierarchy in FPGAs similar to the memory hierarchy in CPUs as shown in Figure 7-18. Closest to the logic elements are the LE memory blocks. Like the registers in CPUs, they have a very small capacity but run in a single cycle. The large memory blocks play a role akin to on-chip cache in a CPU. Larger quantities of memory are supplied by off-chip DRAM, as in CPUs. These memories are not caches but they do display a range of capacity/performance trade-offs.

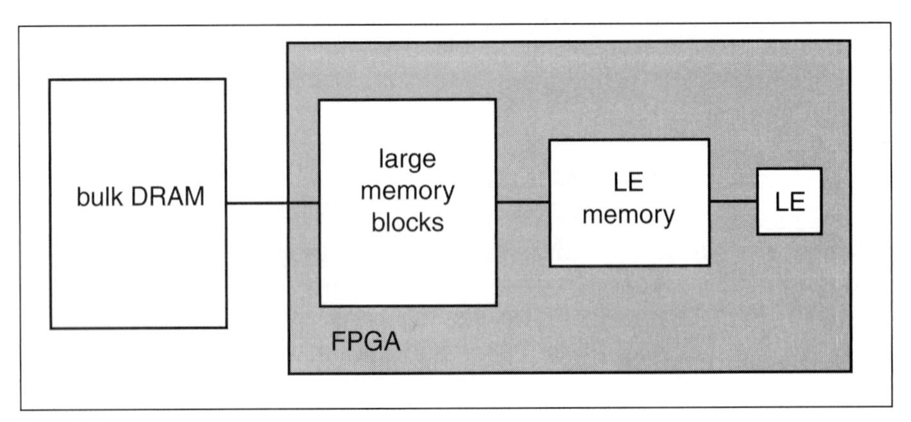

Figure 7-18 The memory hierarchy in platform FPGAs.

7.3.4 CPUs and Embedded Multipliers

CPUs vs programmable logic

CPUs are a powerful complement to FPGA fabrics on platform FPGAs. CPUs, like FPGA fabrics, are programmable, but CPUs and FPGAs are good at very different types of operations. FPGA fabrics can handle a wide variety of data widths and can perform specialized operations on that data. CPUs can be easily programmed, can execute large and complex programs, and can take advantage of pipelining and other design optimizations.

Of course, we do not need to rely on CPUs built into the FPGAs. We can put a CPU into any FPGA using a piece of soft or hard IP. Many Verilog or VHDL models for CPUs are available on the Internet. One advantage of using an IP CPU is that we can modify the CPU's architecture to add instructions or modify the cache organization. However, an IP CPU requires a lot of logic elements and we need a fairly large FPGA to have room left over for non-CPU logic or memory.

hardware/software partitioning

Given an application, how do we decide what parts of the application should go into the CPU as software and what parts should go into the FPGA fabric? This problem is known as **hardware/software partitioning**. As shown in Figure 7-19, we are trying to fit the application into a pre-existing architecture consisting of a CPU and an FPGA fabric connected by a bus. We refer to the logic on the FPGA side as an **accelerator**; in contrast, a co-processor would be dispatched by the execution unit of the CPU.

Figure 7-19 CPU/FPGA architecture for hardware/software partitioning.

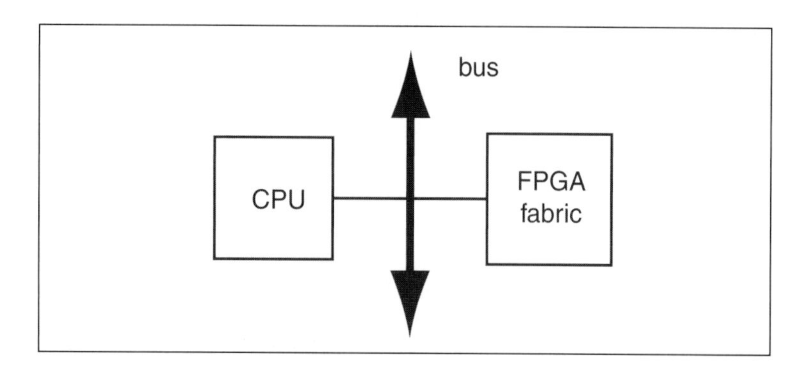

We must partition the application into two pieces, the CPU side of the bus and the FPGA side of the bus. There are many different ways to do this and a careful analysis is required to find a partitioning that results in a higher-performance system. There are several ways to approach this problem [DeM01], depending in part on the type of application you are trying to implement on the platform architecture.

loop analysis

One way to approach the problem is to look at the critical parts of a program and try to decide what parts (if any) should be moved to the FPGA. Loop-intensive programs are good candidates since they perform a large number of operations using a small number of operators. We can build a small piece of the hardware in the FPGA that can perform these operations and use it many times as the loop executes over and over.

However, there are costs to take into account. First, we have to determine how many clock cycles the FPGA logic will take to perform one iteration of the loop and be sure that it is in fact faster than the software implementation. Second, we have to determine the size of the FPGA logic. Third, we have to take into account the data communication required. We have to pump data into the FPGA side, either from main memory or from the CPU itself; we also have to pull data out of the FPGA and send it either to the CPU or to main memory. We can write the net performance gain for the loop as [Ern93]

$$t_s = n[(t_{SW} - t_{HW}) - (t_i + t_o)] \qquad \textbf{(EQ 7-6)}$$

where n is the number of iterations executed by the loop, t_{HW} and t_{SW} are the execution times of the FPGA and CPU implementations, respectively, and t_i and t_o are the times required to send data from the CPU to the FPGA and back from the FPGA to the CPU. If the total speedup t_s is negative, then the FPGA actually slows down the system. If the speedup is positive, then we can compare the speedup time with the amount of FPGA logic used to determine if this is a good use of the available FPGA logic.

If we decide to use the FPGA logic as an accelerator, we must have some method for the accelerator and the CPU to synchronize. It is possible to have the CPU perform a fixed number of instructions while waiting for the FPGA logic to finish. However, it is safer to use special registers to handshake between the software on the CPU and the accelerator in the FPGA fabric.

The next example illustrates the analysis of a loop for hardware/software partitioning.

Example 7-5 Speeding up a loop

Consider this loop:

```
for (i=0; i<N; i++)
    for (j=0; j<M; j++)
        x[i] = a[i] * b[i][j];
```

A simplistic analysis would conclude that each iteration of the loop performs one multiplication. However, the arrays must be indexed. The *x[i]* and *a[i]* accesses are straightforward, but the *b[i][j]* access requires the software to compute $i \times N + j$. However, if we know the values of *M* and *N* when we design the hardware, we can use a counter to more simply perform this address calculation.

At each iteration, we must:

- fetch *a[i]* and *b[i][[j]*;
- compute *x[i]*;
- store *x[i]*.

A hardware accelerator for the body of the inner loop looks something like this:

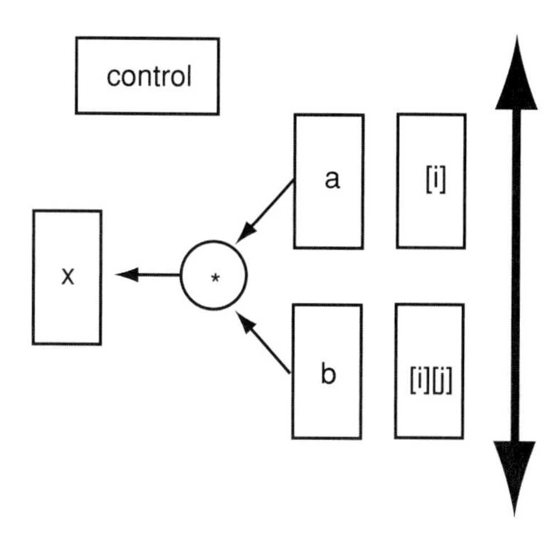

There are registers for the a, b, and x values. It also contains registers for the two loop indexes used. We can use counters to update the values of these two index registers since the arrays are accessed in very regular patterns. A controller sequences the fetches, multiply, and store.

Let us assume that we can access main memory in one clock cycle. The hardware implementation of the loop body takes four clock cycles: one to fetch *a[i]*, one to fetch *b[i][j]*, one to compute *x[i]*, and one to store *x[i]*. For consistency with the software measurements we will count the array load and store times in the execution time $t_{HW} = 4$, $t_i = 0, t_i = 0$.

The software implementation has to perform the same operations, but the index calculations take longer: *b[i][j]* takes two clock cycles to compute; we must also use two extra clock cycles to update i and j. (This estimate is optimistic and ignores a little of the loop overhead; a more accurate estimate requires analysis of the assembly language code.) In this case, $t_{HW} = 7$.

The total speedup for this system is

$$N \cdot M[(7-4)-(0+0)].$$

This is a considerable speedup, even for this simple loop body. (A loop body that performed more operations on the data could provide additional parallelism and opportunities for hardware speedup.) The major challenge in this design will be providing enough memory bandwidth to keep the unit supplied with new operands and saving the results.

From our design experience in Section 4.6.5 we know the cost of various multiplier designs. The counters for the array indexes are small and the control is simple. Based on this information we can decide whether to add this accelerator to the system.

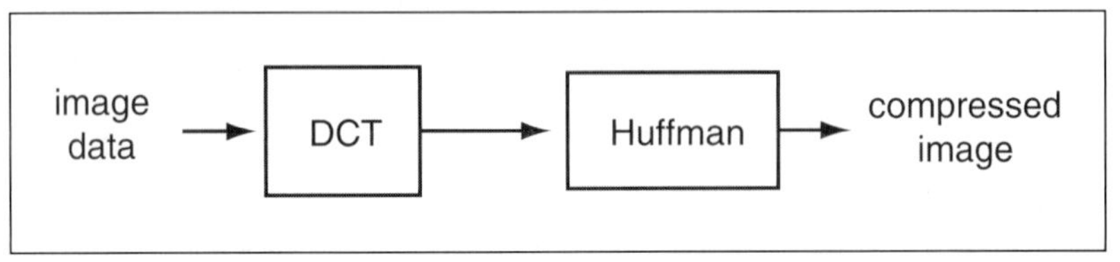

Figure 7-20 A simple example of task-level parallelism.

task-level parallelism

Loops provide low-level parallelism. Some applications provide parallelism at higher levels of abstraction in the form of **task-level parallelism**. Consider the simplified image compression application shown in Figure 7-20. This system compresses images in two stages: DCT and then Huffman. The two stages can be operated in a pipeline fashion if we have separate computational units, either a CPU or a piece of logic configured into the FPGA fabric, to perform these operations.

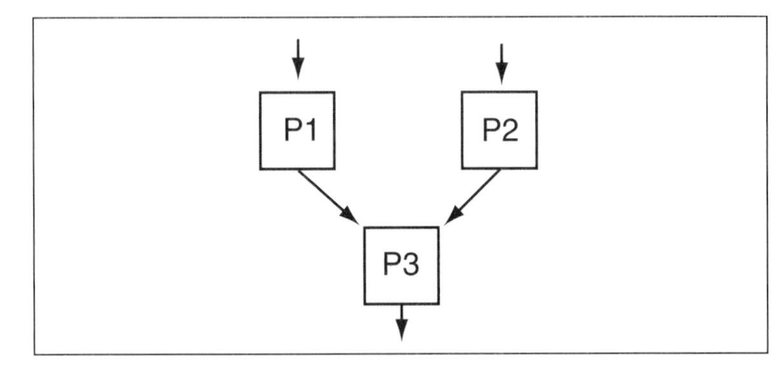

Figure 7-21 Another example of task-level parallelism.

In general, task-level parallelism is not limited to pipelining. Systems with arbitrary communication can exhibit parallelism. Consider the block diagram shown in Figure 7-20. *P3* accepts data from both *P1* and *P2*. However, *P1* and *P2* can operate in parallel to feed data to *P3*. If we have enough hardware to perform *P1* and *P2* simultaneously, we can speed up the system.

When we take into account task-level parallelism, we must analyze the potential speedup for each block in the system, much as we did for loop-level parallelism. However, we must also analyze the system-level performance carefully to determine how much of that raw speedup will contribute to the overall system speedup. In the example of Figure 7-20, the total system speedup is not the sum of the speedups for *P1* and *P2*; it is the *minimum* of the two speedups. If *P3* must wait for both *P1* and *P2* to finish blocks of data before it can continue, then it does not help if one block is considerably faster than the other. System-level performance analysis may show that speeding up one stage with special-purpose hardware is not worthwhile; it may also show that one special-purpose block can be made simpler and smaller in order to balance its execution time with that of another limiting block in the system.

embedded multipliers

Both the Altera and Xilinx platform FPGAs support dedicated multipliers. These multiplier arrays are less flexible than CPUs but offer large amounts of parallelism to applications that can make use of these multipliers. High-speed digital filtering is one example of an application that requires many multiplications.

One way to recognize opportunities to use these multipliers is by analyzing the application's task-level parallelism. A task that is naturally dominated by multiplications and multiply-accumulate operations can be mapped into the multiplier array. Another way to find opportunities to use the multiplier array is by **loop unrolling**, as illustrated in the next example.

Example 7-6 Loop unrolling

Consider this simple loop:

```
for (i=0; i<N; i++)
    x[i] = a[i] * b[i];
```

If we know the value of N, we can transform this loop into an equivalent form known as an unrolled loop. For simplicity, let's assume that $N = 4$; then the unrolled loop is

```
x[0] = a[0] * b[0];
x[1] = a[1] * b[1];
x[2] = a[2] * b[2];
x[3] = a[3] * b[3];
```

We can now see that each statement is independent—the x[1] computation does not, for example, depend on x[0]. We therefore have a great deal of freedom to perform these operations in arbitrary order.

Even if there are dependencies between loop iterations, such as the code

```
for (i=1; i<N; i++)
    x[i] = a[i] * x[i-1];
```

loop unrolling still exposes some parallelism. We can take advantage of that parallelism in the hardware.

7.4 Multi-FPGA Systems

Digital systems are typically built with multiple FPGAs for one of two reasons. First, some systems are simply too large to fit into a single FPGA. Second, the multi-FPGA system may be a prototype for a custom design. Multi-FPGA systems are sufficiently common that methods have been developed to map designs into multiple FPGAs. The large design and prototype cases have somewhat different constraints: large designs may be concerned about the efficient use of the FPGAs while prototypers may be as interested in design compilation time as they are in efficiency.

A multi-FPGA system may be designed especially for an individual application. In this case, the types of FPGAs and the interconnection between them can be optimized for that application. However, quite a few machines have been built for prototyping and other purposes that are designed to handle a broad range of designs. **Emulators** and **dynamically reconfigurable systems** are examples of FPGA-based machines that are designed to accommodate many different designs. In this case, your design must be mapped into an existing structure with particular FPGAs and pre-designed interconnections between them. General-purpose interconnection networks and multi-FPGA partitioning algorithms are important tools in designing FPGA-based host systems.

In this section we will look at the factors that must be taken into account in a multi-FPGA design, interconnection networks, and algorithms for partitioning applications into multiple FPGAs.

7.4.1 Constraints on Multi-FPGA Systems

pinout

The first constraint on multi-FPGA systems is pinout. Each FPGA has a limited number of pins, which puts a strict limit on the number of connections into or out of the system. The design requires a certain number of inputs and outputs, but beyond that pins must be used to make connections between the FPGAs. Pins are even scarcer than on-chip wires, so pinout limitations can severely hurt the design.

Rent's Rule can help us predict our pinout requirements if we know the style of design that we intend to use and have some data on that type of design.

If you don't have enough pins, you cannot route the signals in your design. It may be possible to make the design routable by changing the way that logic is partitioned into FPGAs. We saw in Section 4.8.1 how partitioning can change the number of wires going between partitions; we will discuss algorithms specifically designed for multi-FPGA partitioning in Section 7.4.3. Repartitioning does not require a change in the design's logic. If partitioning does not work, then the logic design itself must be modified to use multiplexing in some form—busses, control signals, etc.—to reduce the number of wires needed between the FPGAs.

Pin multiplexing is one solution to pinout problems. Babb et al. introduced **virtual wires** [Bab97] to increase the number of signals that could be passed between FPGAs. Virtual wires are time-multiplexed signals on pins. Virtual wires, which were originally proposed for emulation, increase the available signals between chips at the cost of performance.

performance

Pin performance can also be a limiting factor in multi-FPGA systems. I/O pins do not run as fast as on-chip logic so a wire between FPGAs cannot carry as much information as a wire in the FPGA. If the design is mapped into a multi-FPGA system directly, then the speed of the logic in the FPGAs is limited by the chip-to-chip speed. A design that is created with multi-FPGA implementation in mind can organize communication to minimize the effects of long chip-to-chip delays.

7.4.2 Interconnecting Multiple FPGAs

interconnection circuits

A multi-FPGA system requires connections from chip-to-chip. In a small system, those interconnections may be made directly on a board. In larger systems that require more than one board, a backplane or other set of wires may be used to connect one board to another. Some connections between FPGAs may go across these global connections.

There are several ways to organize chip-to-chip connections. We can rely on fixed connections (known as **traces**) on the printed circuit board. This is adequate for a single design but generally doesn't work for boards designed to host many different designs. We could also use the FPGAs on the board to route connections—we could send a signal into one pin, through the on-chip wiring, and out another pin without using any of the on-chip logic. We could also dedicate some of the chips on the board to routing connections without implementing logic. Some of those dedicated routing chips may, in fact, be FPGAs that are programmed to provide interconnections only. Specialized routing chips may also be used on the board to provide connections between FPGAs.

network topologies

If we want to build a board or set of boards that can host multiple designs, then general-purpose interconnection networks are critical. The elements to be connected by the network are often called **processing elements** (**PEs**). A great deal of effort has gone into studying the properties of interconnection networks and this work can be used to design an interconnection network that fits the class of designs that we want to implement. Pins are sufficiently scarce on FPGAs that we must carefully design the interconnect network that goes across chip boundaries.

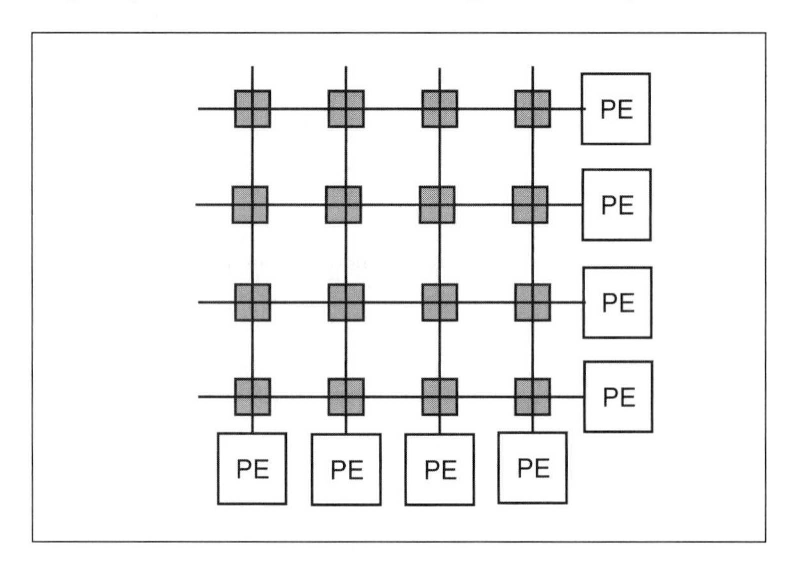

Figure 7-22 A crossbar network.

One type of network that can be used for chip-to-chip connections is the **crossbar**, shown in Figure 7-22. The crossbar provides full connectivity—any combination of connections between inputs and outputs can be made by the crossbar. The crossbar can be thought of (and is often built as) a grid of wires with a switch at each intersection between horizontal and vertical wires. Horizontal wires connect to the crossbar outputs while vertical wires connect to the inputs. To make a connection between an input and output, you simply turn on the switch at the intersection between your input's vertical wire and your output's horizontal wire. The size of the crossbar is proportional to the square of the number of nodes connected (inputs x outputs), but because FPGAs have a limited number of pins, it may be feasible to use a crossbar to connect chips. An FPGA can be programmed to implement a crossbar or special-purpose crossbar chips can be used.

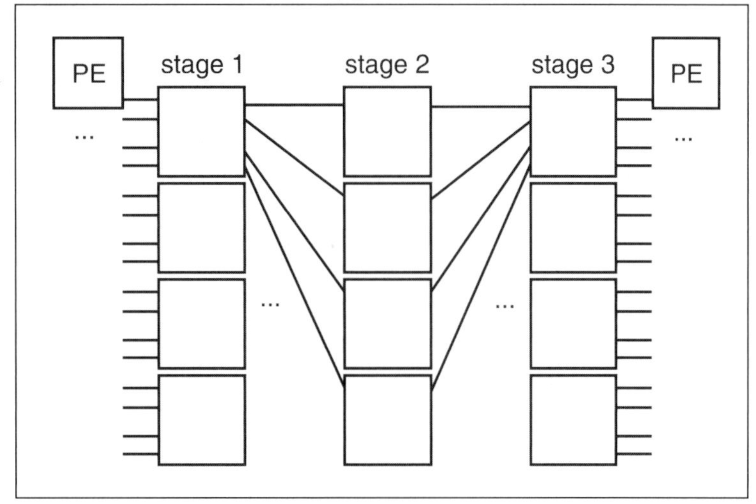

Figure 7-23 A Clos network.

Another well-known network is the **Clos network**, shown in Figure 7-23. This simple example shows a network with 16 inputs and 16 outputs; other sizes are possible. The network itself consists of three stages of switches, each of which is, in this case, a 4 x 4 crossbar. Each crossbar in the first stage is connected to four of the Clos network inputs on one side and to the four stage two crossbars in the second stage. Each of the stage two crossbars is connected to each of the stage three crossbars, which in turn are connected to the Clos network outputs. This network is very flexible but not quite as rich as the crossbar. Like the crossbar, for appropriate parameters it provides full connectivity for two-point connections: every combination of connection between one input and one output can be routed without interference. However, multi-point connections are not guaranteed to be conflict-free. If, for example, one input needs to be connected to several outputs, that connection may conflict with another multi-point connection. The Clos network is smaller than a crossbar network with the same number of inputs and outputs.

A set of processing elements can also be connected in a tree, as shown in Figure 7-24. Processing elements are at the leaves of the tree and switches are at the non-leaf intersections of branches. The processing elements communicate by passing a message first up the tree and then back down to the destination. The **fat tree** [Lei85] is a tree in which the communication capacity of a branch increases with height in the tree.

Figure 7-24 A tree network.

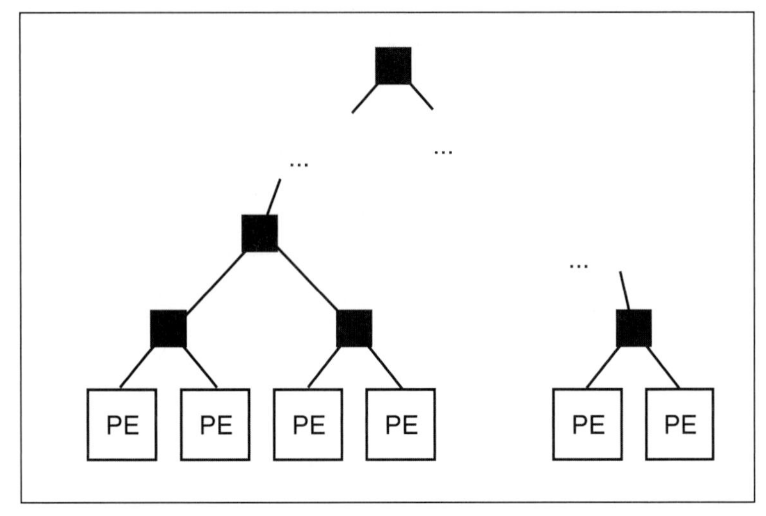

The fat tree has some useful theoretical properties and can be used to simulate any other network topology.

7.4.3 Multi-FPGA Partitioning

The basic partitioning techniques we discussed for logic in Section 4.8.1 are applicable to multi-FPGA systems as well. However, several methods that specifically address the challenges of multi-FPGA systems—notably the small number of pins available to connect the large blocks of logic within the FPGAs—have been developed to handle multi-FPGA partitioning.

hypergraph models

Partitioning algorithms sometimes use the terminology of **hypergraphs**. A hypergraph is a generalization of a graph. While an edge in a graph connects exactly two nodes, a hypergraph's edge can connect more than two edges. For example, the hypergraph edge *{p1, p2, p3, p4}* connects the hypergraph nodes *p1*, *p2*, *p3*, and *p4*. Hypergraphs are harder to draw. But they do eliminate the problem of how to represent a net in the graph. When using conventional graphs to model nets, we must break the net into a set of pairwise connections. A hypergraph edge, in contrast, directly models the edge.

Multi-FPGA partitioning generally assumes that the logic has been mapped into logic elements. While some improvements might be possible by performing logic optimization during partitioning, the design space for that problem is very large and the possible improvements fairly small.

partitioning metrics

Remember that the basic metric for partitioning of netlists is the net cut—the number of nets that cross the partition boundary. That metric is even more valid for multi-FPGA partitioning than it was for placement. Each FPGA has a given number of pins; that number determines the number of signals that can go into or out of the chip. Net cut is a direct measure of whether a given partitioning satisfies that requirement.

k-way partitioning

Multi-FPGA partitioning is different from placement in two important ways. First, placement does not need to put a strict upper bound on the number of nets going between two partitions, while multi-FPGA partitions does. Second, placement is generally performed by recursively dividing the each partition into two subpartitions. Multi-FPGA partitioning, on the other hand, is generating a set of k partitions that correspond to FPGAs in the multi-FPGA system. This is known as the ***k*-way partitioning** problem.

There are several ways to perform k-way partitioning. One can directly divide the gates into k different sets. Alternatively, one can extract a single block of k gates from the logic, then repeat this procedure until all the logic has been divided into k-sized blocks.

The Fiduccya-Matthesys algorithm [Fid82] is the basis for many partitioning algorithms. Their algorithm is itself based on the Kernighan-Lin algorithm [Ker70] that we discussed in Section 4.8.1. Fiduccia-Mathesys measures cell gain as the change in the number of nets crossing the partition boundary when a cell is moved. They found an efficient algorithm for updating the gains of all the cells, which in turn allowed more partitions to be tried and larger designs to be placed.

Chou *et al.* [Cho95] developed an improvement on Fiduccya-Matthesys partitioning for multi-FPGA systems; they were particularly interested in emulators. They model the problem as set-covering—this allows some logic to be contained in more than one FPGA. (In practice, that logic would be replicated in each FPGA.) Each set must have no more logic than can fit into the FPGA. They also require that each set obey the FPGA's pinout restrictions—no FPGA can have more nets going into it or coming out of it than there are available pins. They call such a set of logic a **feasible FPGA**. Their goal is to minimize the number of sets required to cover all the logic in the user's design.

The quality of a cut can be measured by the ratio of the number of nets crossing the cut and the product of the number of nodes in the two partitions [Wei91]. The local ratio-cut metric relies on local clustering but tries to retain some global information to help improve the search results. The cluster size is a parameter given to the algorithm. They extract a subcircuit C' from the circuit by growing it from a randomly selected seed. They then search through the neighbors of C'—nodes outside of C' that are directly connected to nodes in C'—to grow the cluster. After one cluster has been built, all that logic is removed from consideration and a new cluster is built from the remaining logic. This process continues until all the logic has been placed in a cluster.

Their set covering algorithm was inspired by Espresso. They start with a cover for the circuit. They identify the essential candidate clusters that must be included in order to completely cover the logic. A reduce operation then eliminates enough unessential clusters that removing any more would destroy the cover. They then try to expand each cluster as much as possible. They repeat this process until they reach a plateau.

Cong and Lim [Con98] improved the Fiduccya-Matthesys algorithm to be able to handle larger logic designs. Their algorithm pre-processes an initial partition to organize blocks into pairs. They then consider moves that involve two blocks that are paired. They choose pairs that produced a maximum or minimum cutsize reduction during previous partitioning runs.

Karypis and Kumar [Kar99] developed a hypergraph partitioning algorithm. In the first phase of their algorithm, they build smaller hypergraphs by merging nodes together. They use several schemes to select nodes for merging: they look for pairs of vertices that are shared in many hyperedges; they find a maximal independent set of hyperedges; and they find a set of node such that each node is highly connected to at least one other node in the set. They finish coarsening when they have reduced the grape to a small, pre-chosen number of nodes. They then partition the reduced hypergraph. Finally, they uncoarsen the partitioned set. As sets are expanded, new degrees of freedom can be exploited to improve the partitioning result. They use a simplified version of Fiduccya-Matthesys to help improve the results of each phase of uncoarsening.

7.5 Novel Architectures

In this section we will consider two topics. First, we will look at machines built from FPGAs that take advantage of the characteristics of SRAM-based FPGAs. We will then look at some different architectures that have been proposed for FPGAs themselves.

7.5.1 Machines Built From FPGAs

The Quickturn emulator [Sam92] was an early machine built from multiple FPGAs. It was designed as a tool for verifying integrated circuit designs. Software mapped the IC design into the network of FPGAs in the emulator. The emulator also included other facilities, such as logic to apply stimuli to the machine and a built-in logic analyzer.

The Splash machine [Gok91] was designed to be a reconfigurable scientific computing engine. The machine was organized as a linear array, with each stage in the array comprising one FPGA and one memory. The Splash board had 32 Xilinx 3090 FPGAs, each with 320 CLBs, and 32 memory chips. Splash was programmed using a hardware description language. It was used for DNA sequence comparison, among other applications.

The DECPeRLe-1 [Vul96] was also designed for scientific computing. Its inventors referred to it as a **programmable active memory** because of the reconfigurability offered by SRAM-based FPGAs. Its FPGAs were organized in a two-dimensional matrix; the original machine had a 4 x 4 matrix of Xilinx 3090 FPGAs. SRAM was attached to the edges of the matrix. It was programmed using a C++-based hardware description language. It was used for cryptography, DNA sequencing, finite-element computations, neural network simulation, and video compression, among other tasks.

7.5.2 Alternative FPGA Fabrics

The FPGA architectures we have studied are for the most part very fine grained—their building blocks are about as complex as a logic gate. Several groups have experimented with FPGAs that built from larger computational units. This approach can both make more efficient use of the available silicon and make it possible to reconfigure the fabric on the fly. In this section we will look at some examples of novel architectures.

PipeRench [Cop00] is a **pipeline reconfiguration** machine. A PipeRench implementation is built from a pipeline of reconfigurable stages. Each stage is comparable in complexity and capability to a pipeline stage in a CPU or a signal processing system. However, each stage and the connections between stages can be reconfigured.

If the application to be run on a PipeRench pipeline has the same number of stages as the chip's pipeline, then the application runs much as it would on a non-reconfigurable pipeline. However, PipeRench can use a

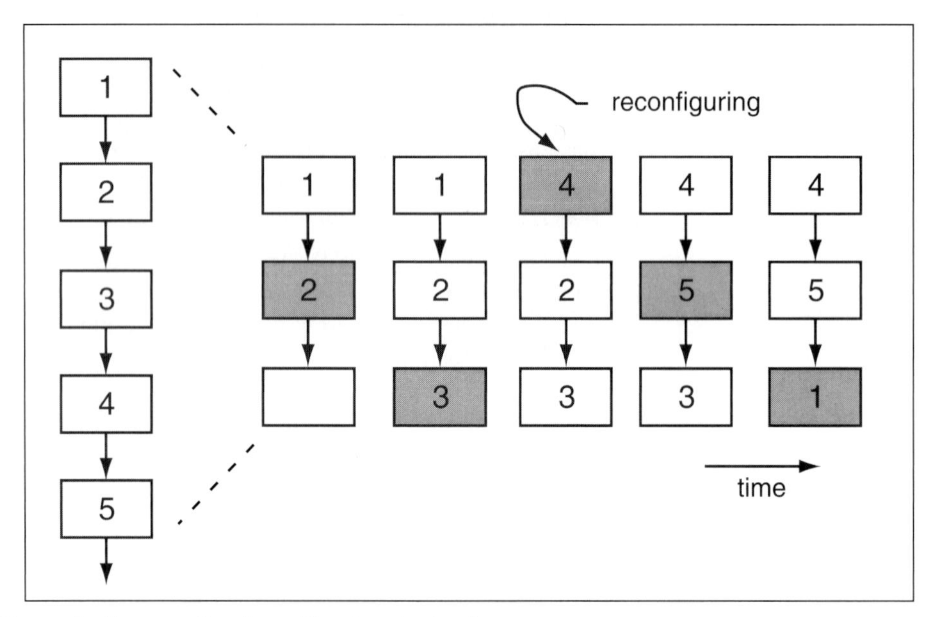

Figure 7-25 Dynamically reconfiguring a PipeRench pipeline.

virtual pipeline to execute applications that have more pipeline stages than are available on-chip. Because each pipeline stage can be independently reconfigured, we can reconfigure the stages during execution to complete the pipeline. As shown in Figure 7-25, stages can be reconfigured ahead of the data to complete the pipeline. The stage being reconfigured is not available for useful work, so the pipeline has one fewer useful stages than physical stages.

The RaPiD architecture [Cro99] is also designed with coarser-grained computational elements but it takes a very different approach to microarchitecture. As shown in Figure 7-26, a RaPiD machine contains a large number of function units connected by reconfigurable interconnect. The reconfigurable interconnection network can be used to build a linear pipeline from the function units, placing whatever function units you want in your desired order. The pipeline control is a mixture of CPU-style instructions and FPGA-style reconfiguration. Units that are likely to need to be flexible, such as the multiplexers that connect the function units to the busses, are designed as **soft control** that can be reconfigured on every cycle. Other units are designed with **hard control** that can be changed only in configuration mode. A configurable instruction decoder translates instructions into the control bits used in the data path and interconnect.

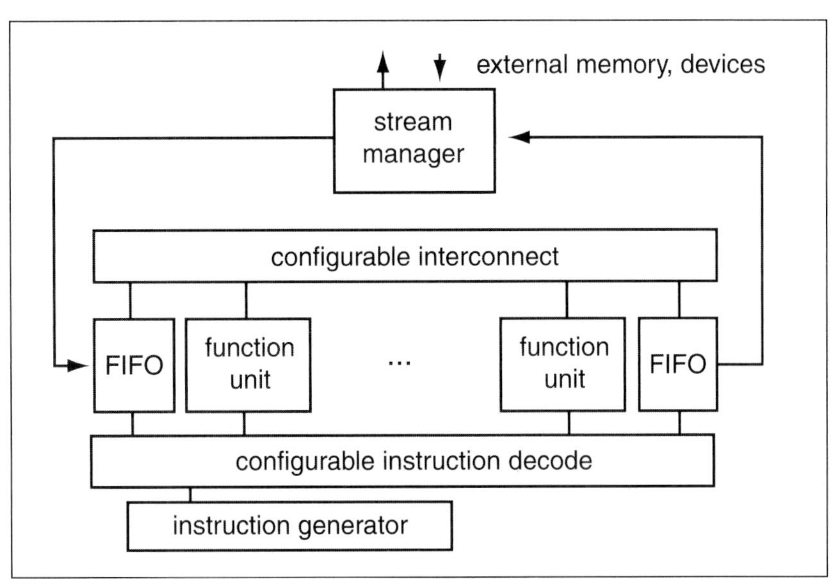

Figure 7-26 The RaPiD architecture.

7.6 Summary

Complex digital systems generally contain more than just arrays of logic. They also contain a mixture of different types of computational elements and special-purpose interconnect like busses. In this section we have looked at complex components for large-scale systems, such as platform FPGAs and interconnection networks; we have also looked at design techniques, like bus interfacing and multi-FPGA partitioning.

7.7 Problems

Q7-1. Write inequalities for the timing constraints of Figure 7-8 that show how t_1 and t_2 are determined from t_s, the flip-flop setup time, and t_h, the flip-flop hold time.

Q7-2. Draw a timing diagram similar to the diagram of Figure 7-8 that shows the timing requirements on a latch-based bus receiver. Explain the origin of each of the timing requirements.

Q7-3. Show the timing constraints that might be required between the end of one four-cycle handshake to the beginning of the next.

Q7-4. Draw a state transition graph for a simple bus that performs read operations but does not allow wait states—all transfers take one cycle. The bus uses a clock signal. Define all other signals required on the bus.

Q7-5. Draw a state transition graph for the PCI read operation shown in Example 7-1.

Q7-6. Draw a timing diagram for a PCI bus transfer that moves two words; the first word incurs one wait cycle while the second word incurs two wait cycles.

Appendices

A Glossary

Thanks to John Redford and Derek Beatty for many colorful terms.

ALU Arithmetic logic unit, which can perform several different arithmetic and logic operations as determined by control signals. (See Section 4.6.4.)

ALAP As-late-as-possible, a schedule that performs operations at the last possible time. (See Section 6.2.2.)

ASAP As-soon-as-possible, a schedule that performs operations at the earliest possible time. (See Section 6.2.2.)

ASIC Application-specific integrated circuit. (See Section 1.3.3.)

ATPG See *automatic test pattern generation.*

allocation The assignment of operations to function units. (See Section 6.2.2.)

aggressor net In crosstalk, the net that generates the noise. (See Section 2.5.6.)

antifuse An electrically programmable connection that is manufactured as broken and is then unbroken by electrically burning the antifuse. (See Section 3.4.)

antifuse-based FPGA
An FPGA whose logic elements and interconnect are programmed with antifuses. (See Section 3.4.)

architecture-driven voltage scaling
A technique for reducing power consumption in which the power supply voltage is reduced and logic operating in parallel is increased to make up for the performance deficiency. (See Section 6.2.3.)

array multiplier
A multiplier built from a two-dimensional array of adders and additional logic. (See Section 4.6.5.)

arrival time The time at which a signal transition arrives at a given point in a logic network. (See Section 4.6.5.)

automatic test pattern generation
Use of a program to generate a set of manufacturing tests.

BIST See built-in self-test.

Baugh-Wooley multiplier
A multiplication algorithm for two's-complement signed numbers. (See Section 4.6.5.)

behavioral synthesis See *high-level synthesis.*

binding In high-level synthesis, synonym for *allocation.*

body effect Variation of threshold voltage with source/drain voltage. (See Section 2.4.2.)

Booth encoding A technique for reducing the number of stages in array multipliers. (See Section 4.6.5.)

buffer An amplifier inserted in a wiring network to improve performance. (See Chapter 3.)

bus A common connection. (See Section 7.2.)

CLB See *logic element.*

CPU Central processing unit. (See Section 7.3.)

carry-lookahead adder
An adder that evaluates propagate and generate signals in a carry-lookahead network which directly computes the carry out of a group of bits. (See Section 4.6.3.)

carry-select adder
An adder that first generates alternate results for different possible carry-ins, then selects the proper result based on the actually carry-in. (See Section 4.6.3.)

carry-skip adder
An adder that recognizes certain conditions for which the carry into a group of bits may be propagated directly to the next group of bits. (See Section 4.6.3.)

chaining Performing two data operations, such as two additions, in the same clock cycle. (See Section 6.2.2.)

channel A rectangular routing region. (See Section 3.3.3.)

clock A signal used to load data into a memory element.

clock distribution
The problem of distributing a clock signal to all points within a chip with acceptable delay, skew, and signal integrity. (See Section 3.3.3.)

clocking discipline
A set of rules that, when followed, ensure that a sequential system will operate correctly across a broad range of clock frequencies. (See Section 5.4.2.)

Clos network An interconnection network built from multiple smaller crossbars. (See Section 7.4.2.)

combinational logic block See *logic element*.

control dependency
A set of control decisions, one of which depends on the other. (See Section 6.2.2.)

controller A state machine designed primarily to generate control signals. (See Section 6.2.2.)

crossbar A network that provides arbitrary connections between inputs and outputs. (See Section 7.4.2.)

crosstalk Noise generated by one line interfering with another. (See Section 2.5.6.)

DRAM Dynamic random-access memory. A three-transistor cell was an early form; the one-transistor cell is universal in commodity DRAM and increasingly used in logic chips. See *embedded RAM*.

data dependency A relationship between two data computations in which the result of one is needed to compute the other. (See Section 6.2.2.)

data path A unit designed primarily for data-oriented operations. (See Section 6.2.2.)

data path-controller architecture A sequential machine built from a data path plus a controller that responds to the data path's outputs and provides the data path's control inputs. (See Section 6.2.2.)

delay In logic gate design, particularly measured between 50% points in the waveform. (See Section 2.4.2 for gate delay and Section 4.4.1 for delay in combinational networks.)

departure time The time at which a signal transition leaves a given point in a logic network. (See Section 4.4.1.)

design flow A series of steps used to design a chip. (See Section 6.3.)

design methodology Generally similar to a design flow, though this is perhaps a more general term. (See Section 6.3.)

distributed control
A controller built from several communicating machines. (See Section 6.2.2.)

dog and pony show
A presentation to management.

don't-care A value that could be 0 or 1. See also *input don't-care* and *output don't-care*. (See Section 1.2.1.)

drain One of the transistor terminals connected to the channel. (See Section 2.3.)

driver error A bug caused by the designer, not by the tools. (See Example 4-12).

effective capacitance
A capacitance value chosen to estimate the gate delay induced by a wiring load. (See Section 2.5.2.)

Elmore delay A wiring delay model for RC transmission lines. (See Section 2.5.4.)

embedded CPU A CPU used in a larger system design.

embedded RAM Memory fabricated on the same die as logic components.

FPGA Field-programmable gate array. (See Section 1.1.)

fabric A regular computational structure, such as the interconnection of logic elements and programmable wires in an FPGA. (See Section 3.1.)

fanin All the gates which drive a given input of a logic gate.

fanout All the gates driven by a given gate.

flash Electrically in-circuit reprogrammable read-only memory.

flash-based FPGA
An FPGA that is configured using flash memory. (See Section 3.4.)

flip-flop A type of memory element not normally transparent during clocking. (See Section 5.4.1.)

full adder An adder that generates both a sum and a carry. (See Section 4.6.3.)

gate The transistor terminal that controls the source-drain current. (See Section 2.3.)

global routing Determining the paths of wires through channels or other routing areas without determining the exact layout of those wires; compare to *detailed routing*. (See Section 4.8.2.)

half adder An adder that puts out only a sum. (See Section 4.6.3.)

hardware/software co-design
The simultaneous design of an embedded CPU system and the software that will execute on it. (See Section 7.3.4.)

high-level synthesis
CAD techniques for allocation, scheduling, and related tasks.

hit by a truck The canonical means of losing a key technical person at a critical point in a project.

hold time The interval for which a memory element data input must remain stable after the clock transition. (See Section 5.4.1.)

input don't-care Shorthand for 0 and 1 inputs having the same output. (See Section 1.2.1.)

LE See *logic element*.

latch A type of memory element that is transparent when the clock is active. (See Section 5.4.1.)

Lee/Moore router A common algorithm for area routing. (See Section 4.8.2.)

linear region The region of transistor operation in which the drain current is a strong function of the source/drain voltage. (See Section 2.3.)

logic block See *logic element.*

logic element In an FPGA, a programmable circuit used to implement logic functions. (See Section 3.3.2 and Section 3.4.3.)

logic synthesis The automatic design of a logic network implementation. (See Section 4.7.3.)

memory element A generic term for any storage element: flip-flop, latch, etc. (See Section 5.4.1.)

metal migration A failure mode of metal wires caused by excessive current relative to the size of the wire. (See Section 2.5.1.)

multiplexer A combinational logic unit that selects one out of n inputs based on a control signal.

n-type diffusion An n-doped region. (See Section 2.3.)

no-op A no-operation instruction. A useless person.

observability don't-care A don't-care internal to a Boolean network; used in logic synthesis. (See Section 4.7.4.)

one-hot code A unary code used for state assignment or other codes in which each symbol is represented by a single true bit.

one-transistor DRAM
A dynamic RAM circuit that uses one capacitor to store the value and one transistor to access the value. Also called *one-T DRAM*.

output don't-care
An incompletely specified function; for these inputs the implementer may choose 0 or 1 output. (See Section 1.2.1.)

π model A model for the load on a gate that uses two capacitors bridged by a resistor. (See Section 2.5.4.)

p-type diffusion A p-doped region. (See Section 2.3.)

PCB Printed circuit board.

PLL See *phase-locked loop*.

package Any carrier for an integrated circuit, typically made of ceramic or plastic.

pad A large metal region used to make off-chip connections. (See Section 2.7.2.)

pass transistor A single transistor (usually n-type) used for switch logic. (See Section 2.4.6.)

phase A clock signal that has a specified relationship to other clock phases. (See Section 5.4.)

phase-locked loop
A circuit that is often used to generate an internal clock from a slower external clock source.

pin The connection between a package and a board.

pipelining A logic design technique that adds ranks of memory elements to reduce clock cycle time at the cost of added latency. (See Section 6.2.4.)

placement The physical arrangement of elements. (See Section 4.8.1.)

plate capacitance
A capacitance between two parallel plates. The capacitance mechanism for transistor gates and metal capacitance. (See Section 2.4 for transistors and Section 2.5 for wires.)

platform FPGA A chip with an FPGA fabrics and other system-level components. (See Section 7.3.)

polysilicon Material used for transistor gates and wires. (See Section 2.2.)

power-down mode
An operating mode of a digital system in which large sections are turned off.

primary input An input to the complete system, as opposed to an input to a logic gate in the system.

primary output An output of the complete system, as opposed to an output of a logic gate in the system.

propagation time
The time required for a signal to travel through combinational logic. (See Section 4.4.)

pulldown Any transistor used to pull a gate output toward V_{SS}. (See Section 2.4.)

pulldown network
The network of transistors in a logic gate responsible for pulling the gate output toward V_{SS}. (See Section 2.4.)

pullup Any transistor used to pull a gate output toward V_{DD}. (See Section 2.4.)

pullup network
The network of transistors in a logic gate responsible for pulling the gate output toward V_{DD}. (See Section 2.4.)

RAM Random-access memory. May be dynamic or static. (See Section 2.6.)

ROM Read-only memory.

real estate Chip area.

redundant In combinational logic, an expression that is not minimal. (See Section 1.2.1.)

refresh Restoring the dynamically-stored value in a memory. (See Section 2.6.)

register Typically a memory element used in a data path, but may be used synonymously with memory element.

retiming Moving memory elements through combinational logic to change the clock period. (See Section 5.5.4.)

routing The physical design of wiring. (See Section 4.8.2.)

SCR See *silicon-controlled rectifier.*

SoC See *system-on-chip.*

SRAM Static read-only memory. (See Section 2.6.2.)

SRAM-based FPGA

An FPGA whose logic elements and interconnect are programmed by static RAM. (See Section 3.3.)

satisfiability don't-care

A don't-care internal to a Boolean network; used in logic synthesis. (See Section 4.7.4.)

saturation region

The region of transistor operation that is roughly independent of the source/drain voltage. (See Section 2.3.)

scheduling The assignment of operations to clock cycles. (See Section 6.2.2.)

sense amplifier A differential amplifier used to sense the state of bit lines in memories. (See Section 2.6.2.)

setup time The time by which a memory element's data input must arrive for it to be properly stored by the memory element. (See Section 5.4.)

shifter A logic unit designed for shift (and perhaps rotate) operations.

short circuit power
The power consumed by a logic gate or network when both pullup and pulldown transistors are on.

sidewall capacitance
Junction capacitance from the side of a diffusion region to the substrate. (See Section 2.5.2.)

silicide An improved gate material.

silicon-controlled rectifier
In VLSI circuits, a parasitic device that can cause the chip to latch up. (See Section .)

sign-off The approval of a design for manufacturing (or possibly some intermediate point in the design).

skin effect The result of electromagnetic fields in low-resistance conductors that causes current to be carried primarily along the conductor's skin. (See Section 2.5.2.)

solder bump A technique for making connections to a chip across its entire surface, not just at the periphery.

source One of the transistor terminals connected to the gate. (See Section 2.3.)

spin A workaholic's term for a *turn*.

state The current values of the memory elements. (See Section 5.2.)

state assignment
The selection of binary codes for symbolic states. (See Section 5.3.3.)

state transition graph A specification of a sequential machine, equivalent to a *state transition table*. (See Section 5.2.)

state transition table
A specification of a sequential machine, equivalent to a *state transition graph*. (See Section 5.2.)

static logic Logic that does not rely on dynamically-stored charge.

suit A manager. See *no-op*.

switchbox A rectangular routing region with pins on all four sides.

synthesis subset A subset of a hardware description language that can be synthesized into hardware. (See Section 4.3.)

system-on-chip A complete electronic system on a single chip.

tapered wire A wire whose width varies along its width, usually to reduce the wire delay. (See Section 2.5.4.)

threshold voltage
The gate voltage at which a transistor's drain current is deemed to be significant. (See Section 2.3.)

transmission gate
A pair of n-type and p-type transistors connected in parallel and used to build switch logic. (See Section 2.4.6.)

transition time The time it takes a gate to rise or fall, often measured from 10% to 90% for rise time and visa versa for fall time. (See Section 2.4.2.)

turn One iteration of the complete design cycle.

underdamped An RLC circuit that oscillates.

unknown voltage
A voltage that represents neither logic 0 nor logic 1. (See Section 2.4.2.)

VHDL A hardware description language. (See Section 4.3.3.)

vector Inputs applied to a chip.

Verilog A hardware description language. (See Section 4.3.2.)

via A hole in the chip's insulating layer that allows connections between different layers of interconnect. (See Section 2.2.)

victim net In crosstalk, the net that receives the noise. (See Section 2.5.6.)

voltage scaling
Any one of several techniques for reducing the power supply voltage of a chip to lower its power consumption.

Wallace tree A design for high-speed multiplication. (See Section 4.6.5.)

win the lottery To get a much higher salary from a competitor.

xter, xstr Synonyms for transistor.

B Hardware Description Languages

B.1 Introduction

This section briefly reviews the Verilog and VHDL hardware description languages. These are both complex languages and this section is not intended to be a complete guide by any means. Hopefully, these sections can help remind you of some basic syntactic elements of the languages. Section 4.3 and Section 5.3.4 talk about HDL design and synthesis.

B.2 Verilog

The IEEE standard defines Verilog. Books by Thomas and Moorby [Tho98], Smith and Franzon [Smi00] , and Ciletti [Cil03] are useful guides to the language.

B.2.1 Syntactic Elements

Verilog has two forms of comments:

```
/* this is a
  multiline comment */
// this is a comment
```

Verilog defines the value set [0 1 x z] of signal values. The value x is the unknown value while z is a high-impedance.

B.2.2 Data Types and Declarations

The type *wire* is used to carry signal values. If the wire is not driven it is assigned the default value *z*.

A hardware register is of type *reg*. A register is assigned the default value *x*.

An integer can be written in a variety of bases; the general form for an integer is

```
size'base number
```

The timescale statement can be used to specify the units of time in printouts, etc:

```
'timescale 10 ns / 1 ns
```

The first number is the units used and the second number is the least-significant digit.

A wire or reg may be declared as an array:

```
wire [ expr1 : expr2 ] wire_name ;
reg [ expr1 : expr2 ] reg_name ;
```

A parameter declaration defines a constant in a module:

```
parameter param_name = value ;
```

A preprocessor directive can be used to define constants that can be used in a variety of ways:

```
'define const_name value
```

B.2.3 Operators

Boolean logical perators include:

```
&& (and) || (or) ~ (not)
```

Verilog provides bitwise Boolean operators that can be applied to wire arrays:

```
& (and) ~& (nand) ^ (xor) ~^ (xnor) | (or) ~| (nor)
```

If these operators are used as binary operators, then they perform bitwise operations. If they are used as unary operators, then they combine the bits in the wire array using the operator, such as ANDing together all the bits in a wire array.

Shift operators include

<< (left shift) >> (right shift)

Relational operators include

< (less than) <= (less than or equal to) >= (greater than or equal to) > (greater than) == (equal) != (not equal)

Arithmetic operators include

+ - * (multiply) / (divide) % (modulus)

Synthesis of multiply, divide, and modulus require access to hardware modules for these operators. They are, of course, large blocks of logic.

Curly braces can be used to concatenate signals:

```
{a,b}
```

forms a vector from *a* and *b*.

B.2.4 Statements

An assignment statement has the form

assign *net_name* = *expression* ;

The concatenation operator can be used to put together signals into a bundle, for example

```
assign {asig, bsig} = w1 & w2;
```

Blocking assignments are performed in order:

```
v1 = val1;
v2 = val2;
```

Non-blocking assignments are performed concurrently:

```
sig1 <= a;
sig2 <= b;
```

A statement block is a set of statements in between *begin* and *end*.

The *always* block repeats a block of code until the simulation terminates:

always @(*event_expression*)
    statement_block ;

The event controlling the *always* may be one of several types: a level type triggers the block whenever a named set of signals changes; an edge type, such as *posedge sig* or *negedge sig* looks for an edge in a particular direction.

The *if* statement has the form

```
if (expression) block
{ elsif (expression) block }
[ else block ] ;
```

The *case* statement has the form

```
case (expression)
    { value : block; }
    [ default: block;]
endcase
```

The *case* statement has two important variants: *casez* treats *z* or *?* values as don't-cares; *casex* treates *z*, *x*, or *?* values as don't-cares.

The *for* loop has the form

```
for (initial_index; terminal_index; step) block;
```

The *for* statement can be synthesized if it is used to iterate in space over an array of signals, using an integer for the index.

B.2.5 Modules and Program Units

A module is the basic unit of hardware specification. A module description has the form

```
module module_name( port_list );
parameter_list
port_declarations
wire wire_declarations
reg reg_declarations
submodule_instantiations
body
endmodule
```

A port may be declared to be *in*, *out*, or *inout*:

```
module foo(a, b, c, d)
input a;
output b, c;
inout d;
```

```
endmodule;
```

Submodule instantiations include functions and tasks. A function is a single-output, executes in zero time, and cannot contain timing control statements. A *function* has the form

```
function [range] function_name;
    parameters
    input input_declarations
    reg reg_declarations
    body
endfunction
```

A task is more general, though its outputs must be registered:

```
task task_name;
    parameters
    input input_declarations
    output output_declarations
    reg reg_declarations
    body
endtask
```

B.2.6 Simulation Control

The $monitor statement prints a formatted string every time one of the signals in its list changes. The $monitor statement is similar to the C printf statement.:

```
$monitor(format_string,signal,...);
```

The formatting string is enclosed by quotes (" and "). Formatting directives in the monitor statement include %d (decimal), %b (binary), %x (hex), and %o (octal). A newline is denoted by *\n* and a tab by *\t*.

The pound sign can be used to advance the simulation clock:

```
#10
```

This statement advances the simulation clock by 10 time units.

The initial block defines a set of code that is executed once at the start of simulation:

```
initial begin
end
```

The $stop command suspends simulation. The $finish command terminates the simulation run. Both are terminated by a semicolon.

B.3 VHDL

The IEEE standard defines VHDL [IEE93]. Bhasker's book [Bha95] is a useful introduction to the language.

B.3.1 Syntactic Elements

A comment in VHDL looks like this:

```
-- This is a comment until the end of the line.
```

VHDL is case-insensitive and generally provides free-form syntax.

A library is used in a module with this declaration:

library *library_name* [, *library_name_list*];

B.3.2 Data Types and Declarations

VHDL allows the declaration of enumeration types, for example:

```
type enum_1 is (a, w, xxx);
```

The language defines several enumeration types: character, bit (with values '0' and '1'), boolean (with values true and false), severity_level, file_open_kind, and file_open_status.

VHDL also allows the declaration of integer subranges:

```
type subrange1 is range 1 to 32;
```

An array declaration may make use of any base type:

```
type array1 is array (0 to 15) of bit;
```

A record in VHDL is similar to the structures or records of other modern programming languages:

```
type rec1 is
    field1 : integer;
    field2 : bit;
    field3 : array (0 to 31) of bit;
```

A constant declaration looks like this:

constant *const_name* := *value* ;

A variable declaration has the form:

variable *variable_name* : *type_name* ;

A signal declaration has a similar form:

signal *signal_name* : *type_name* ;

B.3.3 Operators

Logical operators include:

and or nand nor xor xnor not

Relational operators include:

= /= <= < > >=

The /= operator is the not equals operator.

Shift operators include:

sll srl sla sra rol ror

Addition operators include:

+ - &

The & operator is the concatenation operator.

Multiplication operators include:

* / mod rem

Other operators include:

abs **

The ** operator is the exponentiation operator.

B.3.4 Sequential Statements

A signal assignment looks like this:

signal <= *expression* [after *delay_value*];

The *wait* statement has several forms:

```
wait on sensitivity_list;
wait until boolean_expression;
wait for time_expression;
```

The *wait on* statement waits for an event on one of the signals on the sensitivity list. The *wait until* statement waits until the expression becomes true. The *wait for* statement waits for the specified amount of time.

The *if* statement has the form

```
if boolean_expression then
    sequential_statements
{elsif boolean_expression then
    sequential_statements}
[else
    sequential_statements]
end if;
```

The *case* statement has the form

```
case expression is
    when choices => sequential_statements
    [ when others => sequential_statements ]
end case;
```

The *for* statement has the form

```
for identifier in range loop
    sequential_statements
end loop;
```

The *while* statement has the form

```
while boolean_expression loop
    sequential_statements
end loop;
```

The general *loop* statement has the form

```
label: loop
    sequential_statements
exit when boolean_expression;
end loop label;
```

The *assertion* statement has the form

```
assert boolean_expression
    [ report string_expression ]
    [ severity expression ];
```

If the assertion's condition fails, the run time system puts out a warning message.

B.3.5 Structural Statements

A declaration of a component instance looks like this:

instance_name: *type_name* port map (*pin1*, *pin2*);

The *instance_name* is the name of this instantiation of the component while *type_name* is the name of the type of component to be instantiated. The list of pins shows how signals are to be connected to the instance's pins.

B.3.6 Design Units

VHDL defines five types of design units:

- Entity declaration.
- Architecture body.
- Configuration declaration.
- Package declaration.
- Package body.

An entity declaration is a form of type declaration for a hardware unit. It defines the name of the entity and its ports. An entity declaration looks like this:

```
entity entity_name is
     port (a, b : in bit; c : inout bit; d, e : out bit);
end entity_name;
```

The port list following the *port* keyword gives all the ports for the entity. *in*, *out*, and *inout* are directions for the ports. The name *bit* is a type of a signal; other types of signals are also possible.

An architecture body describes the internal organization of an entity and looks like this:

```
architecture arch_name of entity_name is
     { component_list }
begin
     { structural_statements | sequential_statements }
```

```
end arch_name;
```

The *arch_name* parameter is the name of this architecture; an entity may have several different architectures defined for it. If the architecture uses structural statements to connect components, the components needed are declared like this:

```
component component_name
    port ( port_list );
end component;
```

A configuration declaration declares which architecture to use for an entity and to bind components:

```
configuration config_name of entity_name is
    for arch_name
        for comp1:type1
            use entity lib1.entity1(arch);
        end for;
    end for;
```

A variety of statements can be used in the configuration declaration to determine the binding of components.

A package is a language unit that facilitates code reuse. A package declaration looks like this:

```
package package_name is
    type_declaration;
    component_declaration;
    constant_declaration;
    function_declaration;
end package_name;
```

A package body fills in the information behind the package declarations:

```
package_body package_name is
    package_contents;
end package_name;
```

B.3.7 Processes

Processes are used to model behavior. A typical process looks like this:

```
process (a, b) is
    begin
```

```
x <= a or b;
wait for 2 ns;
y <= not b;
end process;
```

The signal list following the *process* keyword is the sensitivity list of signals to be observed by the process. The process is activated when any signal on the sensitivity list changes. The process body may include any sequential statement.

References

[Act02] Actel, "Axcelerator Family FPGAs Detailed Specifications," version 2.0, 2002. Available at http://www.actel.com.

[Act02b] Actel, "ProASIC500K Family," version 3.0, February 2002. Available at http://www.actel.com.

[Alt02] Altera, "APEX-II Programmable Logic Device Family," version 2.0, May 2002. Available at http://www.altera.com.

[Alt03] Altera, "Stratix Device Handbook, volume 1," April 2003. Available at http://www.altera.com.

[Bab97] Jonathan Babb, Russell Tessier, Matthew Dahl, Silvina Zimi Hanono, David M. Hoki, and Anant Agarwal, "Logic emulation with virtual wires," *IEEE Transactions on CAD/ICAS*, 16(6), June 1997, pp. 609-626.

[Bak90] H. B. Bakoglu, *Circuits, Interconnections, and Packaging for VLSI*, Addison-Wesley, 1990.

[Bau73] Charles R. Baugh and Bruce A. Wooley, "A two's complement parallel array multiplication algorithm," *IEEE Transactions on Computers*, C-22(12), December, 1973, pp. 1045-1047.

[Bet98] Vaughn Betz and Jonathan Rose, "How much logic should go into an FPGA logic block?," *IEEE Design & Test of Computers*, January-March 1998, pp. 10-15.

[Bet98b] Vaughn Betz and Jonathan Rose, "Effect of prefabricated routing track distribution on FPGA area-efficiency," *IEEE Transactions on VLSI Systems*, 6(3), September 1998, pp. 445-456.

[Bet99] Vaughn Betz and Jonathan Rose, "Circuit design, transistor sizing and wire layout of FPGA interconnect," in *Proceedings, IEEE 1999 Custom Integrated Circuits Conference*, IEEE, 1999, pp. 171-174.

[Bha95] J. Bhasker, *A VHDL Primer*, revised edition, Englewood Cliffs NJ: Prentice Hall, 1995.

[Boe93] K. D. Boese, A. B. Kahng, B. A. McCoy, and G. Robins, "Fidelity and near-optimality of Elmore-based routing constructions," in *Proceedings, ICCD '93*, IEEE Comptuer Society Press, 1993, pp. 81-84.

[Boo51] Andrew D. Booth, "A signed binary multiplication technique," *Quart. Journal of Mech. and Appl. Math.*, Vol. IV, Pt. 2, 1951, pp. 236-240.

[Bor95] G. Borriello, Ebeling, C.; Hauck, S.A.; Burns, S., "The Triptych FPGA architecture," *IEEE Transactions on VLSI Systems*, 3(4) , Dec 1995, pp. 491-501.

[Bra84] R. K. Brayton, C. McMullen, G. D. Hachtel, and A. Sangiovanni-Vincentelli, *Logic Minimization Algorithms for VLSI Synthesis*, Kluwer Academic Publishers, Norwell, MA, 1984.

[Bra90] R. K. Brayton, G. D. Hachtel, and A. L. Sangiovanni-Vincentelli, "Multilevel logic synthesis," *Proceedings of the IEEE*, 78(2), February, 1990, pp. 264-300.

[Bre77] Melvin A. Breuer, "A class of min-cut placement algorithms," *Proceedings, 14th Design Automation Conference*, ACM/IEEE, 1977, pp. 284-290.

[Bro92] S. Brown, R. Francis, J. Rose, and Z. Vranesic, *Field-Programmable Gate Arrays*, Boston: Kluwer Academic Publishers, 1992.

[Bro96] Stephen Brown, Muhammad Kellah, and Zvonko Vranesic, "Minimizing FPGA interconnect delays," *IEEE Design & Test of Computers*, Winter 1996, pp. 16-23.

[Cal96] Thomas K. Callaway and Earl E. Schwartzlander, Jr., "Low Power Arithmetic Components," Chapter 7 in Jan M. Rabaey and Massoud Pedram, eds., *Low Power Design Methodologies*, Kluwer Academic Publishers, Norwell MA, 1996.

[Cha72] T. J. Chaney and C. E. Molnar, "Anomalous behavior of synchronizer and arbiter circuits," *IEEE Transactions on Computers*, C-22(4), April 1973, pp. 421-422.

[Cha92] Anantha P. Chandrakasan, Samuel Sheng, and Robert W. Brodersen, "Low-power CMOS digital design," *IEEE Journal of Solid-State Circuits*, 27(4), April, 1992, pp. 473-484.

[Cha02] Vikas Chandra and Herman Schmit, "Simultaneous optimization of driving buffer and routing switch sizes in an FPGA using an iso-area approach," in *Proceedings, IEEE Computer Society International Symposium on VLSI*, IEEE Computer Society Press, 2002.

[Che92] Kuang-Chien Chen, Jason Cong, Yuzheng Ding, Andrew B. Kahng, and Peter Trajmar, "DAG-Map: graph-based FPGA technology mapping for delay optimization," *IEEE Design & Test of Computers*, September 1992, pp. 7-20.

[Che00] Chung-Kuan Cheng, John Lillis, Shen Lin, and Norman Chang, *Interconnect Analysis and Synthesis*, New York: Wiley Interscience, 2000.

[Cho95] Nan-Chi Chou, Lung-Tien Liu, Chung-Kuan Cheng, Wei-Jin Dai, and Rodney Lindelhof, "Local ratio cut and set covering partitioning for huge logic emulation systems," *IEEE Transactions on CAD/ICAS*, 14(9), September 1995, pp. 1085-1092.

[Cho99] Paul Chow, Soon Ong Seo, Jonathan Rose, Kevin Chung, Gerard Paez-Monzon, and Immanuel Rahardja, "The design of a SRAM-based field-programmable gate array—part II: circuit design and layout," *IEEE Transactions on VLSI Systems*, 7(3), September 1999, pp. 321-330.

[Cil03] Michael D. Siletti, *Advanced Digital Design with the Verilog HDL*, Prentice Hall, 2003.

[Con94] Jason Cong and Yuzheng Ding, "FlowMap: An optimal technology mapping algorithm for delay optimization in lookup-table based FPGA designs," *IEEE Transactions on CAD/ICAS*, 13(1), January 1994, pp. 1-12.

[Con98] Jason Cong and Sung Kyu Lim, "Multiway partitioning with pairwise movement," in *Proceedings, ICCAD '98*, IEEE Computer Society Press, 1998, pp. 512-516.

[Cop00] Seth Copen Goldstein, Herman Schmit, Mihai Budiu, Srihari Cadambi, Matt Moe, and R. Reed Taylor, "PipeRench: A reconfigurable architecture and compiler," *IEEE Computer*, April 2000, pp. 70-77.

[Cro99] Darren C. Cronquist, Chris Fisher, MIguel Figueroa, Paul Franklin, and Carl Ebeling, "Architecture design of reconfigurable pipelined datapaths," in *Advanced Research in VLSI, 1999, 20th Anniversary Conference on*, MIT Press, pp. 23-40.

[De01] Vivek De, Yibin Ye, Ali Keshavarzi, Siva Narendra, James Kao, Dinesh Somasekhar, Raj Nair, Shekhar Borkar, "Techniques for leakage power reduction," Chapter 3 in Anantha Chanddrakasan, William J. Bowhill, and Frank Fox, eds., *Design of High-Performance Microprocessor Circuits*, New York: IEEE Press, 2000.

[DeM01] Giovanni De Micheli, Rolf Ernst, and Wayne Wolf, eds., *Readings in Hardware/Software Co-Design*, Morgan Kaufman, 2001.

[Den85] Peter Denyer and David Renshaw, *VLSI Signal Processing: A Bit-Serial Approach*, Addison-Wesley, 1985.

[Dev88] Srinivas Devadas, Hi-Keung Ma, A. Richard Newton, and A. Sangiovanni-Vincentelli, "MUSTANG: State assignment of finite state machines targeting multilevel logic implementations," *IEEE Transactions on CAD/ICAS*, CAD-7(12), December, 1988, pp. 1290-1300.

[Die78] Donald L. Dietmeyer, *Logic Design of Digital Systems*, second edition, Allyn and Bacon, 1978.

[Dun85] Alfred E. Dunlop and Brian W. Kernighan, "A procedure for placement of standard-cell VLSI circuits," *IEEE Transactions on CAD/ICAS*, CAD-4(1), January, 1985, pp. 92-98.

[Ebe95] C. Ebeling, L. McMurchie, S.A. Hauck, and S. Burns, "Placement and routing tools for the Triptych FPGA," *IEEE Transactions on VLSI Systems*, 3(4) , Dec 1995, pp. 473 –482.

[ElG88] Abbas El Gamal, Khaled A. El-Ayat, Jonathan W. Greene, and Ta-Pen R. Guo, "Universal logic module comprising multiplexers," U. S. Patent 4,910,417, March 20, 1990.

[ElG90] Abbas El Gamal, Khaled A. El-Ayat, and Amr Moshen, "User programmable integrated circuit interconnect architecture and test method," U. S. Patent 4,758,745, July 19, 1988.

[Elm48] W. C. Elmore, "The transient response of damped linear networks with particular regard to wideband amplifiers," *Journal of Applied Physics*, 19, January, 1948, pp. 55-63.

[Ern93] Rolf Ernst, Joerg Henkel, and Thomas Benner, "Hardware-software cosynthesis for microcontrollers," IEEE Design and Test of Computers, 10(4), December 1993, pp. 64-75.

[Fag76] M. E. Fagan, "Design and code inspections to reduce errors in program development," *IBM Systems Journal*, 15(3), 1976,m pp. 219-248.

[Fid82] C. M. Fiduccia and R. M. Mathesys, "A linear-time heuristic for improving network partitions," in *Proceedings, 19th Design Automation Conference*, IEEE Computer Society Press, 1982, pp. 175-181.

[Fis90] John P. Fishburn, "Clock skew optimization," *IEEE Transactions on Computers*, 39(7), July, 1990, pp. 945-951.

[Fis95] J. P. Fishburn and C. A. Schevon, "Shaping a distributed-RC line to minimize Elmore delay," *IEEE Transactions on CAS-I*, 42, December, 1995, pp. 1020-1022.

[Fra90] Robert J. Francis, Jonathan Rose, and Kevin Chung, "Chortle: a technology mapping program for lookup table-based field programmable gate arrays," in *Proceedings 27th Design Automation Conference*, IEEE, 1990, pp. 613-619.

[Fra92] J. Frankle, "Iterative and adaptive slack allocation for performance-driven layout and FPGA routing," in *Proceedings, 29th Design Automation Conference*, IEEE Computer Society Press, 1992, pp. 536-542.

[Fre89] Ross H. Freeman, "Configurable electrical circuit having configurable logic elements and configurable interconnect," U. S. Patent 4,870, 302, September 26, 1989.

[Gla85] Lancer A. Glasser and Daniel W. Dobberpuhl, *The Design and Analysis of VLSI Circuits*, Addison-Wesley, 1985.

[Gok91] Maya Gokhale, William Holmes, Andrew Kosper, Sara Lucas, Ronald Minnich, Douglas Sweely, and Daniel Lopresti, "Building and using a highly parallel programmable logic array," *IEEE Computer*, January 1991, pp. 81-89.

[Hac96] Gary D. Hachtel and Fabio Somenzi, *Logic Synthesis and Verification Algorithms*, Kluwer Academic Publishers, 1996.

[Hil89] Dwight Hill , Don Shugard , John Fishburn, and Kurt Keutzer, *Algorithms and Techniques for VLSI Layout Synthesis*, Kluwer Academic Publishers, Norwell, MA, 1989.

[Hod83] David A. Hodges and Horace G. Jackson, *Analysis and Design of Digital Integrated Circuits*, McGraw-Hill, 1983.

[IEE93] IEEE, *IEEE Standard VHDL Language Reference Manual*, Std 1076-1993, New York: IEEE, 1993.

[Int94] Intel Corporation, *Microprocessors: Vol. III*, 1994.

[Kar99] George Karypis and Vipin Kumar, "Multilevel k-way hypergraph partitioning," in *Proceedings, Design Automation Conference '99*, ACM Press, 1999, pp. 343-348.

[Ker70] B. W. Kernighan and S. Lin, "An efficient heuristic procedure for partitioning graphs," *Bell System Technical Journal*, 49(2), 1970, pp. 291-308.

[Kor93] Israel Koren, *Computer Arithmetic Algorithms*, Prentice Hall, New York, 1993.

[Kur96] T. Kuroda, T. Fujita, S. Mita, T. Nagamatsu, S. Yoshioka, K. Suzuki, F. Sano, M. Norishima, M. Murota, M. Kato, M. Kinugawa, M. Kakumu, and T. Sakurai, "A 0.9V, 150 MHz, 10-mW, 4mm^2, 2-D discrete cosine transform core processor with variable threshold-voltage (VT) scheme," *IEEE Journal of Solid-State Circuits*, 31(11), November 1996, pp. 1770-1779.

[Lei83] Charles E. Leiserson, Flavio M. Rose, and James B. Saxe, "Optimizing synchronous circuitry by retiming," *Proceedings, Third Caltech Conference on VLSI* , Randal Bryant, ed., Computer Science Press, Rockville, MD, 1983, pp. 87-116.

[Lei85] C. E. Leiserson, "Fat trees: Universal networks for hardware-efficient supercomputing," *IEEE Transactions on Computers*, C-34(10), October 1985, pp. 892-901.

[Lig88] Michael Lightner and Wayne Wolf, "Experiments in logic optimization," *Proceedings, ICCAD-88*, ACM/IEEE, 1988, pp. 286-289.

[McW80] T. M. McWilliams, *Verification of Timing Constraints on Large Digital Systems*, Ph.D. Thesis, Stanford University, May , 1980.

[Mea80] Carver Mead and Lynn Conway, *Introduction to VLSI Systems*, Addison-Wesley, 1980.

[Min95] Mindshare, Inc., *PCI System Architecture*, third edition, Reading MA: Addison-Wesley, Inc., 1995.

[Mon93]Jose' Montiero, Srinivas Devadas, and Abhijit Ghosh, "Retiming sequential circuits for low power," in *Proceedings, ICCAD-93*, IEEE Computer Society Press, 1993, pp. 398-402

[Mur90] Rajeev Murgai, Yoshihito Nishizaki, Narenda Shenoy, Robert K. Brayton, and Alberto Sangiovanni-Vincentelli, "Logic synthesis for programmable gate arrays," in *Proceedings 27th Design Automation Conference*, IEEE, 1990, pp. 620-625.

[Mur93] Shyam P. Murarka, *Metallization: Theory and Practice for VLSI and ULSI*, Butterworth-Heinemann, 1993.

[Mut95] S. Mutoh, T. Douseki, Y. Matsuya, T. Aoki, S. Shigematsu, and J. Yamada, "1-V power supply high-speed digital circuitry with multithreshold CMOS," *IEEE Journal of Solid-State Circuits*, 30(8), August 1995, pp. 147-854.

[Nai87] Ravi Nair, "A simple yet effective technique for global wiring," *IEEE Transactios on CAD/ICAS,* CAD-6(3), March 1987, pp. 165-172.

[Noi82] David Noice, Rob Mathews, and John Newkirk, "A Clocking Discipline for Two-Phase Digital Systems," *Proceedings, International Conference on Circuits and Computers*," IEEE Computer Society, 1982, pp. 108-111.

[Ram65] Simon Ramo, John R. Whinnery, and Theodore van Duzer, *Fields and Waves in Communication Electronics*, New York: John Wiley and Sons, 1965.

[Ost84] John K. Osterhout, Gorton T. Hamachi, Robert N. Mayo, Walter S. Scott, and George S. Taylor, "Magic: A VLSI Layout System," *Proceedings, 21st Design Automation Conference*, ACM/IEEE, 1984, pp. 152-159.

[Reg70] William M. Regitz and Joel A. Karp, "Three-transistor-cell 1024-bit 500-ns MOS RAM," *IEEE Journal of Solid-State Circuits*, SC-5(5), October, 1970, pp. 181-186.

[Roy00] Kaushik Roy and Sharat C. Prasad, *Low-Power CMOS VLSI Circuit Design*, New York:Wiley Interscience, 2000.

[Sak92] K. A. Sakallah, T. N. Mudge, and O. A. Olukotun, “Analysis and design of latch-controlled synchronous digital circuits,” *IEEE Transactions on CAD/ICAS*, 11(3), March, 1992, pp. 322-333

[Sak93] Takayasu Sakurai, "Closed-form expressions for interconnect delay, coupling, and crosstalk in VLSI's," *IEEE Transactions on Electron Devices*, 40(1), January 1993, pp. 118-124.

[Sam92] Stephen P. Sample, Michael R. d’Amour, and Thomas S. Payne, “Apparatus for emulation of electronic hardware system,” U. S. Patent 5, 109, 353, April 28, 1992.

[San93] Alberto Sangiovanni-Vincentelli, Abbas El Gamal, and Jonathan Rose, “Synthesis methods for field programmable gate arrays,” *Proceedings of the IEEE*, 81(7), July 1993, pp. 1057-1083.

[San99] Y. Sankar and J. Rose, "Trading Quality for Compile Time: Ultra-Fast Placement for FPGAs," in *FPGA `99, ACM Symp. on FPGAs*, ACM Press, 1999, pp. 157-166.

[Ser89] Donald P. Seraphim, Ronald Lasky, and Che-Yu Li, *Principles of Electronic Packaging*, McGraw-Hill, 1989.

[Sha38] Claude E. Shannon, "A symbolic analysis of relay and switching circuits," *Transactions AIEE*, 57, 1938, pp. 713-723.

[Sin88] Kanwar Jit Singh, Albert R. Wang, Robert K. Brayton, and Alberto Sangiovanni-Vincentelli, "Timing optimization of combinational logic," *Proceedings, ICCAD-88* , IEEE Computer Society Press, Los Alamitos, CA, 1988, pp. 282-285.

[Smi00] David R. Smith and Paul D. Franzon, *Verilog Styles for Synthesis of Digital Systems*, Upper Saddle River NJ: Prentice Hall, 2000.

[Sun84] Hideo Sunami, Tokuo Kure, Norikazu Hashimoto, Kiyoo Itoh, Toru Toyabe, and Shojiro Asai, "A corrugated capacitor cell (CCC)," *IEEE Transactions on Electron Devices*, ED-31(6), June, 1984, pp. 746-753.

[Sut99] Ivan Sutherland, Bob Sproull, and David Harris, *Logical Effort: Designing Fast CMOS Circuits*, Morgan Kaufman, 1999.

[Swa98] J. Swartz, V. Betz, and J. Rose, “A Fast Routability-Driven Router for FPGAs,” in *FPGA `98, ACM Symp. on FPGAs,* ACM Press, 1998, pp. 140-149.

[Tak85] Yoshihiro Takemae, Taiji Ema, Masao Nakano, Fumio Baba, Takashi Yabu, Kiyoshi Miyakasa, and Kazunari Shirai, "A 1 Mb DRAM with 3-dimensional stacked capacitor cells," in *Digest of Technical Papers, 1985 IEEE International Solid-State Circuits Conference*, IEEE, 1985, pp. 250-251.

[Tho98] Donald E. Thomas and Philip R. Moorby, The Verilog Hardware Description Language, Fourth Edition, Boston: Kluwer, 1998.

[Tri94] Steve Trimberger, "SRAM Programmable FPGAs," Chapter 2 in Stephen M. Trimberger, ed., *Field-Programmable Gate Array Technology*, Boston: Kluwer, 1994.

[Tri98] Stephen M. Trimberger, "Configurable electronic device which is compatible with a configuration bitstream of a prior generation electronic device," U. S. Patent 5,773,993, June 30, 1998.

[Vin98] James E. Vinson and Juin J. Liou, "Electrostatic discharge in semiconductor devices: an overview," *Proceedings of the IEEE*, 86(2), February 1998, pp. 399-418.

[Vul96] Jean E. Vullemin, Patrice Bertin, Didier Roncin, Mark Shand, Herve H. Touati, and Philippe Boucard, "Programmable active memories: reconfigurable systems come of age," *IEEE Transactions on VLSI Systems*, 4(1), March 1996, pp. 56-69.

[Wal64] C. S. Wallace, "A suggestion for a fast multiplier," *IEEE Transactions on Electronic Computers*, EC-13(1), February, 1964, pp. 14-17.

[Wei91] Yen-Chuen Wei and Chung-Kuan Cheng, "Ratio cut partitioning for hierarchical designs," *IEEE Transactios on CAD/ICAS,* CAD-10(7), July 1991, pp. 911-921.

[Xil01] Xilinx, "Spartan-II 2.5V FPGA Family: Functional Description," DS001-2 (v2.1), March 5, 2002. Available at http://www.xilinx.com.

[Xil02] Xilinx, "Virtex-II Pro Platform FPGAs: Functional Description," DS083-2 (v2.0), June 13, 2002. Available at http://www.xiilnx.com.

[Xin98] Shanzeng Xing and William W. H. Yu, "FPGA adders: Performance evaluation and optimal design," *IEEE Design & Test of Computers*, January-March 1998, pp. 24-29.

Index

Q

R

T

XILINX END USER LICENSE

PLEASE READ THIS DOCUMENT CAREFULLY BEFORE USING THE SOFTWARE. BY USING THE SOFTWARE, YOU ARE AGREEING TO BE BOUND BY THE TERMS OF THIS LICENSE. IF YOU DO NOT AGREE TO THE TERMS OF THIS LICENSE, PROMPTLY RETURN THE SOFTWARE TO THE PLACE WHERE YOU OBTAINED IT AND YOUR MONEY WILL BE REFUNDED. IF YOU HAVE OBTAINED THIS SOFTWARE AS AN UPDATE TO SOFTWARE FOR WHICH YOU HAVE PREVIOUSLY OBTAINED A LICENSE, THE TERMS OF THAT PRIOR LICENSE WILL CONTINUE TO CONTROL YOUR USE OF THE SOFTWARE. IF YOU ARE A QUALIFIED UNIVERSITY USER, YOU MAY OBTAIN AN EXTENSION OF THIS LICENSE BY REGISTERING WITH THE XILINX UNIVERSITY PROGRAM. LIMITED WARRANTY AND DISCLAIMER XILINX WARRANTS THAT, FOR A PERIOD OF NINETY (90) DAYS FROM THE DATE OF DELIVERY TO YOU OF THE SOFTWARE AS EVIDENCED BY A COPY OF YOUR RECEIPT, THE MEDIA ON WHICH THE SOFTWARE IS FURNISHED WILL, UNDER NORMAL USE, BE FREE FROM DEFECTS IN MATERIAL AND WORKMANSHIP. SUBJECT TO APPLICABLE LAWS:(1) XILINX'S AND ITS LICENSORS' ENTIRE LIABILITY TO YOU AND YOUR EXCLUSIVE REMEDY UNDER THIS WARRANTY WILL BE FOR XILINX, AT ITS OPTION, AFTER RETURN OF THE DEFECTIVE SOFTWARE MEDIA, TO EITHER REPLACE SUCH MEDIA OR TO REFUND THE PURCHASE PRICE PAID THEREFOR AND TERMINATE THIS LICENSE;(2) EXCEPT FOR THE ABOVE EXPRESS LIMITED WARRANTY, THE SOFTWARE IS PROVIDED TO YOU "AS IS"; (3) XILINX AND ITS LICENSORS MAKE AND YOU RECEIVE NO OTHER WARRANTIES OR CONDITIONS, EXPRESS, IMPLIED, STATUTORY OR OTHERWISE, AND XILINX SPECIFICALLY DISCLAIMS ANY IMPLIED WARRANTIES OF MERCHANTABILITY, NONINFRINGEMENT, OR FITNESS FOR A PARTICULAR PURPOSE. XILINX DOES NOT WARRANT THAT THE FUNCTIONS CONTAINED IN THE SOFTWARE WILL MEET YOUR REQUIREMENTS, OR THAT THE OPERATION OF THE SOFTWARE WILL BE UNINTERRUPTED OR ERROR FREE, OR THAT DEFECTS IN THE SOFTWARE WILL BE CORRECTED. FURTHERMORE, XILINX DOES NOT WARRANT OR MAKE ANY REPRESENTATIONS REGARDING USE OR THE RESULTS OF THE USE OF THE SOFTWARE IN TERMS OF CORRECTNESS, ACCURACY, RELIABILITY OR OTHERWISE. LIMITATION OF LIABILITY SUBJECT TO APPLICABLE LAWS:(1) IN NO EVENT WILL XILINX OR ITS LICENSORS BE LIABLE FOR ANY LOSS OF DATA, LOST PROFITS, COST OF PROCUREMENT OF SUBSTITUTE GOODS OR SERVICES, OR FOR ANY SPECIAL, INCIDENTAL, CONSEQUENTIAL OR INDIRECT DAMAGES ARISING FROM THE USE OR OPERATION OF THE SOFTWARE OR ACCOMPANYING DOCUMENTATION, HOWEVER CAUSED AND ON ANY THEORY OF LIABILITY; (2) THIS LIMITATION WILL APPLY EVEN IF XILINX HAS BEEN ADVISED OF THE POSSIBILITY OF SUCH DAMAGE; (3) THIS LIMITATION SHALL APPLY NOTWITHSTANDING THE FAILURE OF THE ESSENTIAL PURPOSE OF ANY LIMITED REMEDIES HEREIN. THE LIMITATIONS OF REMEDIES AND DAMAGES IN THIS SOFTWARE LICENSE SHALL NOT APPLY TO PERSONAL INJURY (INCLUDING DEATH) TO ANY PERSON CAUSED BY XILINX'S NEGLIGENCE AND ARE SUBJECT TO THE PROVISIONS SET OUT BELOW UNDER THE HEADING "GOVERNING LAW".

1. License. XILINX, Inc. ("XILINX") hereby grants you a nonexclusive license to use the application, demonstration, and system software included on this disk, diskette, tape or CD ROM, and related documentation (the "Software") solely for your use in developing designs for XILINX Programmable Logic devices. No right is granted hereunder to use the Software, or its products, to develop designs for non-XILINX devices. You own the media on which the Software is recorded, but XILINX and its licensors retain title to the Software and to any patents, copyrights, trade secrets and other intellectual property rights therein. 2. Registration and Transfer. Each licensed user must register with Xilinx, and the Software may be used solely by such licensed user, provided that any licensed user may install a copy of the Software on multiple computers, none of which will be used simultaneously by such user. You may transfer the Software, including any backup copy of the Software you may have made, the related documentation, and a copy of this License to another user only with Xilinx's consent (which shall not be unreasonably withheld), and provided (i) the subsequent user reads and agrees to accept the terms and conditions of this License and registers with Xilinx, and (ii) you retain no copies of the Software yourself. 3. Restrictions. The Software contains copyrighted material, trade secrets, and other proprietary information. In order to protect them you may not decompile, reverse engineer, disassemble, or otherwise reduce the Software to a human-perceivable form. You agree not for any purpose to transmit the Software or display the Software's object code on any computer screen or to make any hard copy memory dumps of the Software's object code. You may not modify or prepare derivative works of the Software in whole or in part. You may not publish or disclose the results of any benchmarking of the Software, or use such results for your own competing software development activities, without the prior written permission of Xilinx. You may not make any copies of the Software, except to the extent necessary to be used on separate non-simultaneous computers as permitted herein, and one copy of the Software in machine-readable form for backup purposes only. You must reproduce on each copy of the Software the copyright and any other proprietary legends that were on the original copy of the Software. 4. Term; Termination. Subject to the terms of Section 11 below, this License is effective for one (1) year from your registration of the Software. The activation date shall be deemed to be the date you agree to be bound by the terms of this License. If you wish to extend this License, you must contact XILINX or its authorized sales and/or distribution representatives to determine the additional fees, terms and conditions which may be applicable to such an extended term. You may terminate this License at any time by destroying the Software and all copies thereof. This License will terminate immediately without notice from XILINX if you fail to comply with any provision of this License. Subject to the terms of Section 11 below, upon termination you must destroy the Software and all copies thereof, provided that you may retain one copy of the Software together with the right to use such copy for modifying and debugging any designs created during the license term. 5. Governmental Use. The Software is commercial computer software developed exclusively at Xilinx's expense. Accordingly, pursuant to the Federal Acquisition Regulations (FAR) Section 12.212 and Defense FAR Supplement Section 227.2702, use, duplication and disclosure of the Software by or for the Government is subject to the restrictions set forth in this License. Manufacturer is XILINX, INC. , 2100 Logic Drive, San Jose, California 95124. 6. Export Restriction. You agree that you will not export or re-export the Software, reference images or accompanying documentation in any form without the appropriate government licenses. Your failure to comply with this provision is a material breach of this License. 7. Third Party Beneficiary. You understand that portions of the Software and related documentation have been licensed to XILINX from third parties and that such third parties are intended third party beneficiaries of the provisions of this License. 8. Interoperability. If you acquired the Software in the European Union (EU), even if you believe you require information related to the interoperability of the Software with other programs, you shall not decompile or disassemble the Software to obtain such information, and you agree to request such information from Xilinx at the address listed above. Upon receiving such a request, Xilinx shall determine whether you require such information for a legitimate purpose and, if so, Xilinx will provide such information to you within a reasonable time and on reasonable conditions. 9. Governing Law. This License shall be governed by the laws of the State of California, without reference to conflict of laws principles, provided that if the Software is acquired in the EU, this License shall be governed by the laws of the Republic of Ireland. The local language version of this License shall apply to Software acquired in the EU. Irish law provides that certain conditions and warranties may be implied in contracts for the sale of goods and in contracts for the supply of services. Such conditions and warranties are hereby excluded, to the extent such exclusion, in the context of this transaction, is lawful under Irish law. Conversely, such conditions and warranties, insofar as they may not be lawfully excluded, shall apply. Accordingly nothing in this License shall prejudice any rights that you may enjoy by virtue of Sections 12, 13, 14 or 15 of the Irish Sale of Goods Act 1893 (as amended). Nothing in this Agreement will be interpreted or construed so as to limit or exclude the rights or obligations of either party (if any) which it is unlawful to limit or exclude under the relevant national laws and, where applicable, the laws of any Member State of the EU which implement relevant European Communities Council Directives. Nothing in this Agreement will be interpreted or construed so as to limit or exclude the rights or obligations of either party (if any) which it is unlawful to limit or exclude under the relevant national laws and, where applicable, the laws of any Member State of the EU which implement relevant European Communities Council Directives. 10. General. If for any reason a court of competent jurisdiction finds any provision of this License, or portion thereof, to be unenforceable, that provision of the License shall be enforced to the maximum extent permissible so as to effect the intent of the parties, and the remainder of this License shall continue in full force and effect. This License constitutes the entire agreement between the parties with respect to the use of this Software and related documentation, and supersedes all prior or contemporaneous understandings or agreements, written or oral, regarding such subject matter. 11. Legacy Devices. Notwithstanding termination of this License pursuant to Section 4 above at the end of its term, you shall continue to have the right to use the Software for the sole purpose of creating and modifying designs for Xilinx's 4k/Spartan/SpartanXL device families. This extended right shall be void and of no force or effect if this License terminates due to your default

ISE Student Edition Version 4.2i

Product ID **IIB960699999**

Your Registration ID Number is: 9990-6970-9169

FOR EDUCATIONAL PURPOSES ONLY
The ISE Student Edition Software is intended for student use (Home or Dorm), one copy per student, it is not for use in the school lab. The standard ISE software should be used in the school lab. University staff members may obtain an ISE standard package through the Xilinx University Program by visiting http://www.university.xilinx.com

TECHNICAL SUPPORT
No live technical support (telephone or e-mail) is offered for the Student Edition package. For technical support, students should refer to the Xilinx University Resource Center website at xup.msu.edu.

Available Resources

DEVELOPMENT BOARDS

Xilinx distributors and third party partners offer an array of state-of-the-art boards which help you design and interface to specific system and application level functions.

LOW COST BOARDS

• XESS Corp,	www.xess.com
• Digilent Inc,	www.digilentinc.com
• Burch Electronic Designs,	www.burched.com.au

SPECIALTY BOARDS

• Celoxica,	www.celoxica.com
• Nallatech,	www.nallatech.com
• Virtual Computer Corp,	www.vcc.com

** This is a partial list of available board resources*

TECHNICAL VENUES (RESOURCES, FORUMS & USER GROUPS)

• XUP Resource Center	www.xup.msu.edu
• Comp.arch.fpga Users Group	http://groups.google.com/
• XESS Users Forum	www.xess.com/list_reg.html
• FPGA CPU News	www.fpgacpu.org
• Xilinx Technical Answers Database (Q&A)	www.xilinx.com/support/support.htm
• FREE Software Tutorials	www.xilinx.com/support/techsup/tutorials/index.htm
• Technical Lectures	www.xilinx.com/support/education-home.htm
• Xilinx Tech Tips	www.xilinx.com/xlnx/xil_tt_home.jsp
• Xilinx Tech Xclusives	www.xilinx.com/support/techxclusives/techX-home.htm

Your Registration ID Number is: 9990-6970-9169

Xilinx Student Edition Software version 4.2i Installation Notes

To ensure the optimum use and operation of your new design tools, install Student Edition v4.2i on the recommended hardware with sufficient memory (RAM and hard disk swap space). If you experience problems with the installation, operation, or verification of your installation, see http://xup.msu.edu for FAQs and links to more support.

Supported Operating Systems

The Student Edition v4.2i software supports the following operating systems and versions.

Platform and System Requirements
IBM PC or Compatible
Windows: 98 SE , NT 4.0 (With Service Pack 4-6a), ME, 2000
The memory requirements for both RAM and hard disk space will vary depending on your target device family and size as well as the unique characteristics of your design.

Equipment and Permissions

The following table provides information about related equipment, permissions, and network connections.

Item	Requirement
Directory Permissions	Write permissions must exist for all directories containing design files to be edited. **Note** To view and/or edit the example projects included in the $XILINX\ISEexamples directory, the directory must be writable. Move the projects to a writable directory before working on them in the ISE Project Navigator. Open the project in ISE. To move the project to a writable location, select **Project → Save As**.
Hardware Component	Xilinx recommends that you have an IBM-compatible Pentium class machine.
Monitor	Color VGA operating in the following modes. Minimum Resolution: 640 x 480 Minimum Recommended: 1024 x 768
Mouse	You should have a 2-button (Microsoft Windows compatible) or 3-button (Microsoft compatible) mouse.
CD Drive	You should have an ISO9660 compliant drive on your system.
Ports	You should have two ports (one for a mouse and one parallel port for the parallel download cable, if needed).
Network Compatibility	The Xilinx Installation program supports TCP/IP networks. If you are using a Windows NT operating system, then the TCP/IP protocol needs to be installed first. For more information, see the solution record at the following location: http://www.support.xilinx.com/techdocs/2510.htm

System Memory Requirements

RAM - 128 MB to 256 MB (dependent on device)

Virtual Memory - 128 MB to 256 MB (dependent on device)

***Note**: Due to the size and complexity of the Virtex and Virtex-II devices, Xilinx recommends that these designs be compiled using a high-performance computer. 128 MB of RAM as well as 128 MB of virtual memory is required to compile XC9500 designs. For Spartan II and Virtex devices up to 300K gates, Xilinx recommends 256 MB of RAM.*

Virtual memory size requirements also vary with the design and constraint set size. By default, Windows 95/98 manages its swap filesize automatically, but for Windows NT, you may need to increase it. Typically, your Windows NT swap file size should be twice as large as your system RAM amount.

It is important to note that slower systems or systems with less than the recommended RAM and/or swap space may exhibit longer runtimes.

Required disk space, 2 GB recommended

Running Setup

1. To start the installation, insert the CD #2 into the CD-ROM drive.
2. Run the installation program. The installer should automatically start when the CD is inserted. If it does not, from Windows select **Start_Run**. Type **D:\setup.exe** in the Open field of the Run window and click OK . (If your CD-ROM drive is not the D drive, substitute the appropriate drive designation.)
3. The Welcome window will open, prompting you to register either via the Website, E-Mail, or FAX. **Registration is NOT required for this product**. Click Next to continue.
4. Accept the software license agreement by clicking the white box next to I accept the terms of this software license. so that a check mark appears. Click Next to continue.
5. In the Registration window, enter your Registration ID: **9990-6970-9169**. Click Next to continue.
6. Either choose the default directory (recommended) or select another directory for the installation. Click Next to continue. (Note that if another version of Xilinx software is installed in the indicated directory, you will be prompted to uninstall the previous version.)
7. Perform a **Typical** installation. Click Next to continue.
8. Click on the white box to place a check mark next to each device family that you wish to install. Note that each device family requires a certain amount of memory on your system. Click Next to continue. If you have selected to install the Virtex-2 Pro device, you need to accept the Virtex-2 Pro License Agreement. Click the white box next to I accept the terms of this software license so that a check mark appears. Click OK .
9. When the Update Environment window opens, leave the default settings and click Next to continue.
10. On the following screen, click on Install to install the software.
11. After the Design Environment is installed, you will be prompted to insert the Xilinx Docs CD. If you wish to install the documentation, insert CD #1 and click OK , otherwise click Cancel . When complete, remove the CD #1.
12. If you have installed the Documentation, you will be prompted to insert the Xilinx ISE Design Environment Tools CD into the CD drive when installation of the Documentation is complete. Insert CD #1 back into the CD drive and click OK .
13. The MultiLINX cable installer will start. If you use this cable for development, click Yes to install the drivers, otherwise click No .

You may need to reboot your PC to allow the new/modified environment variables and path statement to take effect before you can run the design implementation tools. The Install program will inform you if you need to reboot.

Service Packs

Download the latest service packs for XSE 4.2i at http://university.xilinx.com/univ/xse42.html. The downloads will include a service pack for the XSE tools and an IP update for Core Generator. To install the service pack for the XSE tools, double-click on the executable file and run the installer to completion. To install the IP Update, extract the contents of the ZIP file using **Winzip**

7.0 SR-1 or later and extract the contents of the file to the ISE installation directory. For more information about installing the service packs, as well as the installation process, visit the online tutorial at http://xup.msu.edu/license/v42i/index.htm.

Modelsim

Modelsim is **not included** with the Student Edition. Modelsim can be downloaded at http://www.xilinx.com/ise/mxe2. For information about installing and licensing Modelsim, visit the online tutorial at http://xup.msu.edu/license/v42i/index.htm.

Licensing

*********IMPORTANT**********

If you register and receive a license file via email, **DO NOT INSTALL IT**. Student Edition does not require a license file to run the software.

Technical Support

There are **no updates** and **no live technical support** for this product. For technical support, abide by the following process: (1) First contact your Professor, and then (2) refer to the Xilinx University Resource Center at http://xup.msu.edu.